£86.55
ML

INDUCTIVE MINING PROSPECTING
Part I: Theory

SERIES

Methods in Geochemistry and Geophysics

METHODS IN GEOCHEMISTRY AND GEOPHYSICS, 20A

INDUCTIVE MINING PROSPECTING Part I: Theory

ALEXANDER A. KAUFMAN and GEORGE V. KELLER

Department of Geophysics,
Colorado School of Mines,
Golden,
Colorado 80401,
U.S.A.

ELSEVIER

Amsterdam — Oxford — New York — Tokyo, 1985

ELSEVIER SCIENCE PUBLISHERS B.V.
Molenwerf 1
P.O. Box 211, 1000 AE Amsterdam, The Netherlands

Distributors for the United States and Canada:

ELSEVIER SCIENCE PUBLISHING COMPANY INC.
52, Vanderbilt Avenue
New York, NY 10017

Library of Congress Cataloging in Publication Data

Kaufman, Alexander A., 1931-
  Inductive mining prospecting.

  (Methods in geochemistry and geophysics ; 20A-   )
  Bibliography: v.1, p.
  Includes index.
  Contents: pt. 1. Theory.
  1. Electric prospecting. 2. Magnetic prospecting.
I. Keller, George Vernon, 1927-   . II. Title.
III. Series.
TN269.K338 1985   622'.154   84-28730
ISBN 0-444-42271-4 (U.S. : v. 1)

ISBN 0-444-42271-4 (Vol. 20A)
ISBN 0-444-41690-0 (Series)

Printed in The Netherlands

CONTENTS

LIST OF SYMBOLS

a	radius
a, b	semi-radii of ellipsoid or spheroid
A	magnetic vector potential, defined by $H = \text{curl } A$
B	magnetic induction vector, $B = \mu H$
B_{2k}	Bernoulli's numbers
D	dielectric displacement vector $D = \epsilon E$
e	charge
E	vector electric field, volts/meter
E_n	electric field component normal to surface
E_0	a primary electric field
$\mathscr{E}$	electromotive force, volts
h	a linear dimension, meters
h	skin depth, $(2/\sigma\mu\omega)^{1/2}$
h	normalized magnetic field strength
F	force
H_t	total magnetic field vector
H_0	primary magnetic field vector
H_s	secondary magnetic field vector
$I_\nu(ikR), K_\nu(ikR)$	modified Bessel functions of the first and second kind of argument ikR
In	in-phase part of
I	current, amperes
J, j	current density vector, ampere/meter^2
j_i	component of current density
$J_\nu(), Y_\nu()$	Bessel functions of the first and second kind of argument ()
$k^2 = -i\omega\mu(\sigma + i\omega\epsilon)$	square of a complex wave number
$K(k), E(k)$	complete elliptical integrals of the first and second kind of argument k
L	vector distance
M	magnetic moment, ampere-meter^2
n	unit vector normal to a surface
L	inductance, Henries
$P_1{}^{(1)}(\cos\theta)$	Legendre polynomial
Q, P	points at which a field is observed
$p = (\omega\mu\sigma a^2/2)^{1/2}$	scaled radius
r	radius of a current-carrying loop
R	resistance, ohms
$R = (r^2 + z^2)^{1/2}$	radius in spherical coordinates
Res	residue of
Q	quadrature part of
S	conductance, mhos
s	surface area

t	time, seconds
t_r	ramp time
T	period, seconds
U	scalar potential, volts
Z	impedance, ohms
Z	dummy representing a linear combination of Bessel functions
$\alpha = 1/\sigma\mu a^2$	
δ	volume charge density, coulombs/meter^3
Δ	difference
σ	electrical conductivity, mhos/meter
ϵ	normalized electric field
ϵ_0	dielectric permittivity of free space
$\Phi(\ )$	probability integral function
Φ, ϕ	magnetic flux
Φ	scalar potential of the magnetic type
λ	linear charge density, coulombs/meter
μ	magnetic permeability, henrys/meter
μ_0	magnetic permeability of free space, $4\pi \times 10^{-7}\,\text{H/m}$
ρ	resistivity, ohm-meters
$\psi_1(\ ), \xi_1(\ )$	Legendre polynomials
Σ	surface charge density, coulombs/meter^2
Π	magnetic vector potential defined by $H = \text{curl } \Pi$
$\tau_0 = \rho\epsilon_0$	time constant for a medium
$\tau = 2\pi(2t/\sigma\mu)^{1/2}$	scaled variable used in time domain
$\tau_0 = \sigma\mu a^2/\pi$	time constant for a spherical conductor
$\tau_0 = \sigma\mu b^2/q_1$	generalized time constant for a body of arbitrary shape, where b is a linear dimension and q_1 is a shape correction factor
$\tau_0 = \mu Sa/7.70$	time constant for a spheriod with $S = 2ab$, a and b being the principal radii of the spheroid
ω	frequency, radians/second
ω	solid angle

Chapter 1

INTRODUCTION

For well over half a century, it has been recognized that certain types of ore bodies can be located by electromagnetic induction occurring in them. In order for induction methods to work, the ore body which is being sought must have a relatively high conductivity in comparison to the host rock in which it occurs. Not all useful minerals have this characteristic; the principal ore minerals which form ore bodies that can be discovered with inductive methods are the sulfides, both the base-metal sulfides and the uneconomic sulfides. Certain other valuable minerals also have the property of good electrical conductivity, including the oxides of iron, the native metals, and carbon in the form of graphite. Induction methods might be used for a wide variety of applications in which an underground structure is known to have a different electrical conductivity than the surrounding rock, but primarily, we view the electromagnetic induction methods as being best suited for use in the exploration for relatively massive base-metal sulfide deposits.

In discussing the inductive methods of prospecting, we can divide the subject area into four principal categories, namely, the theory, the interpretation of field surveys, the description of equipment and procedures used in acquiring and processing field data, and the consideration of case histories.

This particular monograph will cover some of the essential features of the theory for induction methods, while the rest of the topic will be covered in a separate monograph, which the authors plan to prepare in the very near future.

In view of the very limited scope of this monograph, it is important at this point to answer the question as to what is the main objective in presenting the theoretical development here. Our point of view, generally, is that the main reason for examining the theory on which induction prospecting methods are based will be its use in an insightful analysis of the relationships that exist between the fields that are measured during a survey on the one hand, and the properties of the geoelectric section which is being studied on the other hand. If well handled, this examination of theory should deliver an understanding of highly important aspects of the application of induction methods including:

(1) The generality of the principles on which the methods are based, such as the laws of Coulomb, Ampère, Biot, Savart, Faraday, and Ohm, and the principle of conservation of charge.

(2) The capability of knowing the primary electromagnetic fields contributed by various types of sources, as for example, by circular or rectangular loops through which current flows, magnetic and electric dipoles, and so on.

(3) A deep and thorough understanding of the behavior of the electromagnetic induction in both the frequency and time domains caused by currents in various types of confined conductors surrounded by an insulating host medium. These fields will be the most useful source of information about the characteristics of an ore body.

(4) A complete understanding of the behavior of the field at low, intermediate, and high frequencies in the frequency domain, or during the early, intermediate and late stages of transient response in the time domain. This understanding covers the relationships which must exist between the fields observed within these ranges of frequencies and times, and the parameters characterizing in conductive bodies, such as their dimensions, conductivity, precise location, and orientation with respect to the source of the primary field. Such an understanding will permit us to specify an optimal range of frequencies or times over which the tightest relationship between the parameters of the conductor and the behavior of the field will be observed. When this can be done, one can hope to obtain a maximum resolution with electromagnetic methods both in the simple models which we can treat mathematically and more complicated geoelectric sequences when geologic noise is present.

(5) An understanding of frequency and transient responses caused by currents induced in the medium surrounding an ore body. An understanding of the contribution by these fields is a matter of great practical interest, inasmuch as these fields, as for example would be caused by currents flowing in an overburden, in a uniform half-space, in a two-layer medium, or in other more complex geometries, represent geologic noise (with all due apologies to geologists) which inhibits the purpose of a survey in finding an ore body and which ultimately will limit the maximum depth of investigation which can be achieved. For this reason, perhaps the most important purpose in studying the theory on which inductive methods are based is that of being able to formulate the essential differences in behavior in the frequency- and time-domain responses caused by current flowing in a confined conductor such as an ore body, and currents flowing in a surrounding medium of infinite extent.

(6) A quantitative description of the relationship between the location of a conductive body, its orientation, and its geometric form on one hand, and the characteristic features of the profiles for the various components of the electromagnetic field over that body on the other.

(7) An understanding of the influence of electrical charges that arise at interfaces and the relationship between the galvanic part of the field caused by these charges and the geoelectric properties of the medium, as well as the relationship between vortex and galvanic parts of the fields over various frequency ranges or times in the time domain.

(8) An awareness of the effect of the method of excitation on the behavior of the primary field, whether it be inductive or galvanic, and the characteristic zones for behavior of the electromagnetic field, whether it be the near zone, the intermediate zone, or the wave zone, where measurements are carried out, on the resolution and depth of investigation that can be achieved.

(9) The effect of the geometry of a transmitter-receiver array, as for example the shape and dimensions of the transmitter on the depth of investigation.

(10) An understanding of the role played by the distance between the transmitter and receiver, as well as by which component of the field is measured on the depth of investigation in cases where geological noise cannot be ignored.

Investigation of all of these factors, as well as many others of similar character, represents the use of the theory for induction methods in electrical prospecting. Information derived from such studies permits us to develop interpretative procedures for various types of induction methods, to choose the proper parameters in the design of equipment, and in many cases, to predict with a reasonable reliability the best system as well as the probable chance to success for a given set of field conditions before carrying out surveys.

In order to develop the theory for induction methods it is necessary first to know the electromagnetic field, either in the frequency or the time domain, for a conducting medium characterized by various geoelectric parameters. Usually such models of the earth include such things as a confined conductor (that is, an ore body) or a system of confined conductors embedded within a layered medium (usually consisting of an overburden and a uniform half-space or a two-layer medium). In other words, we should have available to us from theory various representations that contain information about the electromagnetic field in a nonuniform conducting medium of various types when the primary field is generated by various types of sources. This most general problem can be solved by at least two powerful methods, namely by numerical modelling, or by physical modelling. Numerical modelling involves the solution of a boundary-value problem based on Maxwell's equations, while physical modelling consists of an imitation of actual conditions and arrays along with the equipment in the laboratory. If the scaling laws are applied correctly, and if the accuracy of calculations and measurements are sufficiently high, both approaches will provide equivalent results, since they in principle solve the same boundary-value problem and deliver the same information about the behavior of the electromagnetic field.

The solution of general boundary-value problems in electrodynamics is a subject that more properly belongs in the field of applied mathematics, inasmuch as it involves the use of methods in mathematical physics. It includes such approaches as the method of separation of variables in wave equations, various applications of the integral equation technique, the method of finite differences, various asymptotic approaches, such as the method of successive approximations or various approximate methods of calculation based on understanding the behavior of fields over certain restricted ranges in frequency or in time.

On the other hand, physical modelling belongs in experimental physics and can be carried out using a number of approaches such as scale modelling with highly conductive materials, as for example, metals, or scale modelling with solutions having various conductivities, providing that the interfaces between zones with different fluids do not distort the field. Membranes which have high longitudinal resistance and low transversal resistance can be constructed to assist in modelling using an electrolytic approach. For relatively simple models which are characterized by axial symmetry, the use of finite ring integrators can be very simple and very useful.

The two approaches, that of numerical and that of physical modelling, are the only recognized approaches which permit us to obtain the information that we desire about the behavior of electromagnetic fields in a nonuniform conducting medium in the most general case, and this information will be basic in developing a theory of inductive methods for electrical prospecting. However, in this monograph we will not describe methods of solving boundary-value problems based on either numerical or physical models, but instead, restrict our attention to those problems which have been solved either using the method of separation of variables or the method of integral equations. It might be argued that studying the electromagnetic field behavior about a few relatively simple mathematical models will not tell us much about the more general problem of electromagnetic prospecting, but it is our viewpoint that the principles one needs to understand the more complicated problems can be developed one by one by first handling the simpler problems. However, before leaving the subject, we must stress that a monograph devoted to various methods of numerical and physical modelling with details describing each of the available approaches and the comparison of them with a thoughtful analysis of the advantages and limitations of each method along with numerical examples would be extremely useful to geophysicists involved in developing electromagnetic methods in applied geophysics.

The analytical solutions to boundary-value problems that we will treat in this monograph have existed for a very long period of time, having appeared in the literature even before there was any concern with the use of electromagnetic methods in prospecting. Boundary-value problems such as an electromagnetic field in a horizontally stratified medium, in the presence of a sphere embedded in a uniform medium, or the field in the presence of a right circular cylinder embedded in a uniform conducting medium have been in principle so. i' for every type of excitation of the primary field long, long ago. The o solutions have been obtained in explicit form and expressed in terms of e own and tabulated functions, such as Bessel's functions and the Legendre polynomials, as well as elementary functions. For this reason, the analysis of these particular problems has always been an initial step in developing the theory of induction methods, and can be used to develop extremely valuable insight into the behavior of electromagnetic fields as they are used in electrical prospecting. The most important steps in the solution of these problems, such as the method of separation of variables, the introduction of various types of vector potential of the electric and magnetic types, the formulation of the correct boundary conditions, the presentation of the primary fields in terms of eigen functions, have all been described in detail in numerous monographs covering various aspects of electrodynamics. In particular, the necessary information concerning the solution of these boundary value problems is covered in two well known texts on mathematical physics, that by Stratton (1941) and that by Smythe (1950).

Even though the fundamental solutions to these simple boundary value problems have been known since early in this century, a great deal of effort has been devoted in recent time to developing numerical means for calculating these fields, to obtain the responses in the frequency and time domains, to derive asymptotic expressions, and to prepare numerous tables and in some cases, as for example for a horizontally layered medium, collections of reference curves describing the frequency- and time-domain responses. Literally hundreds of references might be cited, and in view of the great diversity of references available in literature, it would be difficult to cite everyone who has contributed to this development over the past half century. However, we feel compelled to mention a few outstanding developments in the literature such as:

(1) Papers by S. S. Stefanescu, describing the quasi-stationary field of an electric dipole situated on the surface on a uniform half-space, published in 1935.

(2) A paper by S. M. Sheinmann concerning transient fields in the earth, published in 1947.

(3) Numerous publications by A. N. Tikhonov and colleagues describing frequency- and time-domain responses in a horizontally stratified medium, published in the two decades between 1950 and 1970.

(4) The extensive publications by J. R. Wait devoted to an investigation of fields caused by currents flowing in spherical and cylindrical conductors, as well as in horizontally layered media.

(5) A single excellent paper by H. W. March on the field caused by a magnetic dipole in the presence of a conducting sphere published in Geophysics in 1953.

These particular studies in addition to the others which might be cited, have had an especially strong influence on the commercial development of induction methods of prospecting. We have relied heavily on these particular sources in developing the present monograph.

The rapid growth in the size and availability of computing machines has permitted researchers to solve more and more complicated boundary-value problems of electrodynamics in both the frequency and time domain. During the last decade, with the availability of large-capacity computers, such numerical methods as the method of integral equations, the method of finite difference, and the method of finite elements have lead to the calculation of fields about conductors with more complicated shapes surrounded by a conducting medium with one or more interfaces. Among many groups of researchers working in this particular area, we need to mention at least three in view of the importance of their work on the development of the theory of induction methods:

(1) The University of Utah group (see references by Hohmann).

(2) Moscow State University researchers (see references listed under Dmitriev).

(3) The Institute of Geology and Geophysics at Novosibersk (see references listed under Tabarovskiy).

The University of Utah group has made significant progress in the application of the integral equation method based on the volumetric current elements in a conductor, and their technique is being applied successfully for computation of fields measured with a variety of experimental methods. The groups at the University of Moscow and the Novosibersk Institute of Geology and Geophysics have been using the integral equation approach based on tangential components of the electric and magnetic fields on the surface of the conductor. Numerous sets of calculations carried out by these groups have been published in several collections of frequency- and time-domain responses for various models of the earth and which are relatively easily available to the geophysical community. In the monograph (Tabarovskiy, 1975) "Application of the integral equation method in problems of geoelectric exploration", the approach has been described in detail. In addition, Tabarovskiy and his coworkers have used a finite-difference method to solve several two- and three-dimensional problems which are of practical interest in transient methods of mineral prospecting.

While we have mentioned only three groups, there are in fact many more researchers around the world devoting time to this study. However, a detailed analysis of the state of the art in the application of numerical methods as applied to problems for induction prospecting is beyond our selected scope for this present monograph.

In addition to numerical modelling, physical modelling has been carried out at a number of institutions around the world. It is difficult to find any university with a geophysics department or research laboratory where there are no facilities for carrying out physical modelling. Many commercial companies are also involved in physical modelling, because in some respects it is much more practical to solve exploration problems on short notice than is numerical modelling. We recognize that there are several groups which have contributed significantly to the development of induction prospecting methods by applying physical modelling, including the University of Toronto (G. West), U.S. Geological Survey (F. Frischknecht), Leningrad University (G. Molochnov), Leningrad Mining Institute (V. Zakharov), the Leningrad Institute of Techniques and Pospecting (A. Velikin), the Boliden Mining Company, the Commonwealth Scientific and Industrial Research Organization, Australia (K. McCracken), and the National Geophysical Research Institute, Hyderabad, India.

All of these efforts to develop the theoretical background for electromagnetics have brought us to our present state of the art. With this in mind, let us now describe the content of this particular monograph. In the chapter which follows this introduction, we will describe the basic laws governing the behavior of electromagnetic fields as they are used in the inductive methods of electrical prospecting. These include Coulomb's law, the Biot-Savart and Ampère's laws, Faraday's law and the principle of conservation of charge. In discussing these basic laws, we will pay particular attention to their physical meaning, their range of application, their limitations, and we will give examples showing the application of these laws.

The examples we will use have been chosen in such a way that in spite of their simplicity they help in developing an insight into some of the principal characteristics of the frequency- and time-domain responses of electromagnetic fields contributed by currents in even more complicated situations.

We will pay particular attention to those mechanisms that give rise to the electromagnetic field, that is the currents and electric charges created. Quite often, Coulomb's law is used only for static field behavior in a non-conducting medium. This may have created the impression that when direct or alternating fields are considered in a conducting medium there can be no electric charges, and as a consequence, the Coulomb law will not play any role in governing the field. However, in order to understand the behavior of the galvanic part of the field, which gives rise to a phenomenon sometimes called the "channeling effect" by geophysicists, it is necessary to assume a distribution of electrical charges within a non-uniform conducting medium. It should be noted that in the geophysical literature, there are publications where the existence of electric charges is treated.

Also in Chapter 2, special attention is paid to the behavior of the electromagnetic field when one can neglect the existence of displacement currents, and as a consequence, can assume that the velocity of propagation of the electromagnetic field is unlimited. In this approximation, the field will appear to arrive instantaneously at all points within the range of observation. In many of the publications that have appeared in North America, this field is named the "quasi-static field". This term should be understood as meaning the field in a non-conducting medium. As a matter of fact, while neglecting displacement currents, the electromagnetic field — regardless of the frequency or time at which it is measured — coincides at every instant with the static field, if in both cases the current at the source is the same. For this reason, one can say that the term "quasi-static" field is proper provided that a non-conducting medium is being considered. However, if the source is placed near or inside a conducting medium, as a consequence of electromagnetic induction, the behavior of the field even when displacement currents are neglected is not the same as that of the static field. Therefore, in most cases that are of interest in inductive electrical prospecting, the use of this term can result in a misunderstanding. In the European literature, to avoid this misunderstanding, the term "quasi-stationary" field is often used. We will use this term in this monograph.

Establishing the relationship between the principal physical laws of electrodynamics and Maxwell's equations will be one of the main developments in Chapter 2. Making use of Maxwell's equations and the principle of charge conservation, we will discuss various forms of these equations, taking into account the behavior of particular components of the field at interfaces between media with different resistivity.

Chapter 3 is devoted to an analysis of the frequency- and time-domain responses caused by currents induced within a confined conductor having various shapes (that of a sphere, spheroid, right circular cylinder, elliptical cylinder, or plate) embedded in an insulating host rock, for several types of excitation of the primary field. This analysis is based on previously published solutions by the method of separation of variables and the various calculations which are done using the method of integral equations.

We are concerned primarily in this chapter with the relationships between the characteristics of the conductive body (such as its conductivity, dimensions, shape, and location) and the behavior of various components of the electromagnetic field (that is, the quadrature and inphase components in the frequency domain, and the transient response in the time domain), measured over different ranges of the spectrum or different parts of the time domain. The late stage of transient response and the low-frequency part of the spectrum are considered in detail because of their particular applicability in exploration. Common features as well as some differences in their behaviors are discussed. The high-frequency part of the spectrum and the early stage of the transient response are considered in somewhat less detail. The total frequency and transient responses of the magnetic field and electromotive force are dsecribed in several curve sets.

A simple but approximate method of computing the field at low frequencies and during late stages of the time-domain response is developed. In essence, the third chapter is devoted only to fields caused by currents induced in a single conductive body, meant to simulate an ore body, situated in free space. From these considerations, it is a simple matter to arrive at the conclusion that field survey done in the frequency and time domains will provide exactly equivalent information, if one can neglect geologic noise or ambient noise in either method in comparison with the size of the signal which is being measured.

In Chapter 4, we will examine the resolution and depth of investigation obtainable in the frequency and time domains when one type of geological noise is present. The type of geological noise considered is that contributed by a confined conductor that is of no interest in the exploration project. In spite of its simplicity, this special case permits us to recognize some essential and important differences between the frequency domain methods now being used which are based on detection of quadrature and inphase components of the field, and the transient method in terms of the depth of investigation and the sensitivity of the fields to the parameters characterizing the conductor. In addition, in this chapter, we formulate conditions which will lead to the maximum possible rejection of geological noise. This chapter will demonstrate the types of parameters that must be measured in the frequency domain in order to achieve the same depth of investigation and resolution that can already be obtained using the late stage of the transient response in the time domain. We also consider situations in which neither the frequency nor transient method can detect an ore body under any circumstances.

Chapter 5 will include a description of the effect of eddy currents generated in a full-space, in a uniform half-space, in an overburden and in a two-layer medium in measuring the quadrature and inphase components in the frequency domain and the transient response in the time domain. The purpose of this chapter, first of all, is to provide a determination of the optimum range of frequencies or time as such types of geological noise are most highly suppressed, and therefore for which the maximum depth of investigation can be achieved. Following this, we present a comparison of resolution capabilities as well as of depth of investigation for measurements made of various components of the field in both the time and frequency domains. The principal results derived in this chapter are based on calculations of fields using the integral equation method.

Some understanding of the behavior of eddy currents in a confined conductor and in the host medium has permitted us to develop in this chapter an approximate method for computing the field. In turn, as a practical result is has helped us establish conditions when one can consider the field in the frequency or time domain to be a simple sum:

$$H_t = H_0 + H_s$$

where H_0 is the magnetic field due only to currents in the host medium without the conductor being present and H_s is the field due only to currents in the conductor when it is situated in free space. This consideration is of great importance in interpretation.

This chapter also describes some methods of evaluating the maximum depth of investigation for various conductors in various types of geologic noise. These considerations permit us to determine the optimum array, configuration, range of time or frequency, and other factors for which we can detect an ore body, and also demonstrate the maximum depth of investigation for a given set of geologic conditions. In other words, it demonstrates

10

the limits beyond which the application of frequency or transient methods will be useless for all practical purposes.

In Chapter 6, in contrast to the preceding chapters, frequency- and time-domain responses are studied when both sources of the field are present, namely electric charges and induced currents. The electric charges are responsible for the galvanic part of the field. Inasmuch as this part of the field is related with the properties of the conductor and of the host medium in different ways than the vortex part of the field, we have paid attention mainly to evaluating the role of electric charges under various conditions at different ranges of frequencies and time. This section of the theory for induction methods is not yet as well developed as we would wish, and we can expect that with progress in numerical and physical modelling, much more information will be obtained in the future. Understanding the effect of the galvanic part of the field is extremely important to proper interpretation and almost all of the information available to us concerning the role of electrical charges is described in this chapter.

We are fully aware that this monograph does not cover all aspects of the theory of various induction prospecting methods. Some of them will be described in the next volume, as for example, electromagnetic profiling and the VLF methods. However, some elements of theory will still be beyond the scope of our consideration.

In the bibliography which accompanies this volume, we have collected references to monographs and papers which describe the results of developing numerical and physical modelling and various aspects of the theory of inductive mineral prospecting.

In conclusion to our introduction, we would like to acknowledge an expression of gratitude for the invaluable assistance of Dr. Mark Goldman, who has done many of the numerical calculations presented here, and who has worked with us for the past six months at Golden, and to Dr. Frank Frischknecht of the U. S. Geological Survey, who has been kind enough to read and comment on portions of the text. We are both indebted to the Colorado School of Mines for providing the opportunity to prepare this volume, and to the following companies who have provided financial support: Mining Geophysical Surveys, Saskatchawan Mining Development Corporation, Geo-Physi-Con Company, Ltd., and Zonge Engineering and Research. We wish particularly to express thanks to Dorothy Nogues, who typed the manuscript, and to George S. Keller, who drafted the illustrations.

th

Chapter 2

BASIC ELECTROMAGNETIC LAWS AND MAXWELL'S EQUATIONS

INTRODUCTION

This chapter describes the principal laws of electromagnetism which are important in inductive methods. Although these laws are treated in numerous excellent text books, the examples and models given are usually not very appropriate for understanding the behavior of the fields in a non-uniform conductive earth. The purpose of this chapter is to present the basic laws of electromagnetism from a point of view which will facilitate application of the theory to geophysical problems.

First we will consider the laws of Coulomb, Biot-Savart and Faraday, emphasizing their experimental origins and areas in which they can be applied. Then the relationship between these laws and Maxwell's equations will be described to further explore the physical meaning of the laws and especially to describe the sources of electric and magnetic fields. Special attention will be paid to the form of equations which describes the quasi-stationary or quasi-static field since, except at very high frequencies or early times, this special form provides an accurate description of the fields measured in inductive mineral prospecting. Finally, we will consider the formulation of the Helmholtz equations and magnetic and electric vector potentials, which are useful in solving boundary value problems in conducting media.

2.1. COULOMB'S LAW

As a starting point, we will assume that the reader accepts the concept that an electric charge is a source of an electric field. As a consequence, the distribution of electric charges is a primary factor in controlling the field. In describing electric fields, we will make use of such functional descriptions of charge as volume, surface and linear densities of charge. The volume density of charge, δ, is defined by the equation:

$$\delta = \lim_{dV \to 0} \frac{de}{dV} \qquad (2.1)$$

where de is the charge in an elementary volume dV.

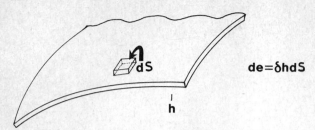

Fig. 2.1. Definition of an element of charge within a thin layer.

It is clear that as the element of volume dV is decreased, the charge in the elementary volume, de, will also decrease. In the limit, if the ratio of total charge to volume remains nearly constant, we arrive at a charge density which is non-zero.

The volume density of charge is the most general way in which to describe a charge distribution, but for particular cases, we may also wish to define such functions as a surface or a linear density of charge. Suppose that charges are distributed through a very thin layer, so that the volume density, δ, is invariant in any direction perpendicular to the surface of the thin layer (see Fig. 2.1). The elementary volume charge can then be written as:

$$de = \delta h dS$$

where h is the thickness of the thin layer and dS is an elementary area of its surface. By letting the thickness h tend to zero while the charge density, δ, increases without limit in such a way that product δh remains constant, we obtain a definition for an elementary surface charge density:

$$de = \Sigma dS \tag{2.2}$$

where Σ is the surface density of charge.

Similarly, when charges are distributed in a rod-like volume of small diameter as shown in Fig. 2.2, and we are only concerned with the field at distances which are far greater than the diameter of the rod, it is often convenient to define a linear elementary charge, de, and a linear density, λ, as follows:

$$de = \lambda dl \tag{2.3}$$

In doing so, we replace the volume within the rod by a line that carries the same amount of charge.

Occasionally it is also convenient to define a point charge, e, by assuming that the whole charge density under consideration is concentrated within an infinitesimal distance about a single point in the medium.

Elementary volume, surface, and linear charges have a common feature in that they are situated within volumes that have dimensions that are much less than the distance from the charge to a point at which the field is being observed. They differ from each other in unit dimensions. With the proper

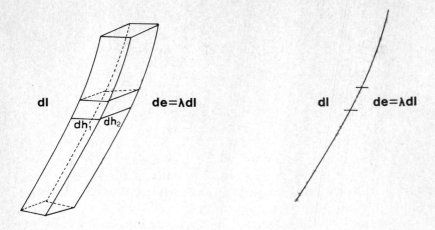

Fig. 2.2. Definition of linear charge density.

description of volume charge density, the volume charge always remains finite, while for elementary surface and line charges, the volume density must be assumed to increase without limit within the volume which is charged. Actually, in accord with eq. 2.2:

$$\delta = \Sigma/h, \quad \text{as } h \to 0$$

Inasmuch as Σ is finite, the volume density of the surface charge becomes infinite as the function $1/h$ becomes infinite.

For an elementary linear charge, we have:

$$\delta = \lambda/\mathrm{d}h_1\mathrm{d}h_2$$

Here $\mathrm{d}h_1$ and $\mathrm{d}h_2$ are linear dimensions of a cross-section (Fig. 2.2). As $\mathrm{d}h_1$ and $\mathrm{d}h_2$ tend to zero, the volume density of the linear charge tends to increase without limit more rapidly than was the case for a surface charge.

The dimensions for charge densities are also different for the different types of geometry being considered. The proper unit for volume charge density is Coulombs per cubic meter. For surface and linear charge densities, the units become Coulombs per square meter, and Coulombs per meter. These differences in units must be looked after carefully in problems in which these approximations are used.

As one might expect, these various degrees of concentration of charge into linear or sheet-like volumes result in different behavior for the electric field about these charges. A point charge is the distribution characterized by the maximum concentration of charges in a small volume, with the volume density of charge going to infinity as $1/h^3$ (here, h is taken to be the linear dimension of an elementary volume about the point where the charge is concentrated).

Now let us discuss the main subject of this section, that is, Coulomb's law. Experimental investigations carried out by Coulomb and other researchers in the 19th century showed that the force acting between an elementary electric charge situated at the point q and another elementary charge situated at a point a is described by an extremely simple expression:

$$F = \frac{1}{4\pi\epsilon_0} \frac{de(q)de(a)}{L_{qa}^3} L_{qa} \tag{2.4}$$

where L_{qa} is the vector:

$$L_{qa} = L_{qa} L_{qa}^0$$

with L_{qa} being the distance between the points q and a, while L_{qa}^0 is a unit vector directed along the line connecting points q and a. Also, ϵ_0 is a constant known as the dielectric permeability or electrical permittivity of free space. In the practical system of units, this constant is:

$$\epsilon_0 = \frac{1}{36\pi} 10^{-9} \frac{F}{m}$$

Also,

$$\frac{1}{4\pi\epsilon_0} = 9 \times 10^9 \frac{m}{F}$$

Equation 2.4 can be rewritten as:

$$F = \frac{1}{4\pi\epsilon_0} \frac{de(q)de(a)}{L_{qa}^2} L_{qa}^0 \tag{2.5}$$

The electric force of interaction between two elementary charges is directly proportional to the product of the charge strengths, inversely proportional to the square of the distance between them, and is in the same direction as the unit vector L_{qa}^0, when the charges are of the same sign, or is directed in the opposite direction, when the product is negative (see Fig. 2.3).

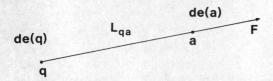

Fig. 2.3. Definition of the sign of the force defined by Coulomb's law.

This simple expression is valid, of course, only so long as the distances between charges are far greater than the dimensions of the volume within which the charges are situated. In order to define the electrical force of interaction between charges when one or both are contained in volumes possessing a dimension comparable to the distance between the charges, one must make use of the principle of superposition. According to this principle, each charge exerts a force on every other charge, the size of the force being independent of the presence of additional charges. Using this principle, an arbitrary volume distribution of charges can be represented as a sum of elementary volumes. For example, the force between an elementary charge at point a, $de(a)$ and a distributed charge in a volume V as shown in Fig. 2.4 can be written as:

$$F(a) = \frac{de(a)}{4\pi\epsilon_0} \int_V \frac{\delta(q)dV}{L_{qa}^3} L_{qa} \tag{2.6}$$

where q indicates the position of any point within the volume V. The total electric force $F(a)$ is the vector sum of all the individual forces contributed by the individual elementary charges.

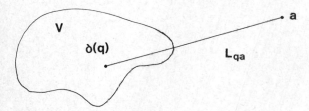

Fig. 2.4. Distribution of charge in a volume.

Extending this approach to a more general case in which all types of charges are present (volume, surface, linear, and point charges) and again applying the principle of superposition, we obtain the following expression for the electrical force of interaction between an elementary charge $de(a)$ and a completely arbitrary distribution of charges:

$$F(a) = \frac{de(a)}{4\pi\epsilon_0} \left[\int_V \frac{\delta(q)dV}{L_{qa}^3} L_{qa} + \int_S \frac{\Sigma(q)dS}{L_{qa}^3} L_{qa} + \int_L \frac{\lambda(q)dl}{L_{qa}^3} L_{qa} \right.$$

$$\left. + \sum_{i=1}^{M} \frac{e_i(q)}{L_{qa}^3} L_{qa} \right] \tag{2.7}$$

where δdV, ΣdS, λdl, and e_i are the symbols representing elementary volume, surface, linear and point charges respectively.

At this point, we will define the strength of the electric field, $E(a)$ as being the ratio between the force of electrical interaction, F, and the size of the elementary charge (considered to be a test charge) $de(a)$ at the point a:

$$E(a) = \frac{F}{de(a)} \tag{2.8}$$

For convenience, the strength of the electric field is usually referred to merely by the term "electric field". It does not have the same dimensions as force, but in the practical system of units it has the dimensions of volts per meter.

The electric field, E, can be thought of as the electric force acting on a test charge, de, inserted into a region of interest. If the electric field is known, it is a simple manner using eq. 2.8 to calculate the force of interaction, F. As follows from eq. 2.7 the expression for the electric field can be written as:

$$E(a) = \frac{1}{4\pi\epsilon_0} \left[\int_V \frac{\delta dV}{L_{qa}^3} L_{qa} + \int_S \frac{\Sigma dS}{L_{qa}^3} L_{qa} + \int_L \frac{\lambda dl}{L_{qa}^3} L_{qa} + \sum_{i=1}^{M} \frac{e_i(q)}{L_{qa}^3} L_{qa} \right] \quad (2.9)$$

If the distribution of charges is given, the function E depends only on the coordinates at which the test point is located. Because it depends only on position, the function is termed a "field".

When the electric field does not vary with time, it depends only on the charge density, and calculation of the field E using eq. 2.9 presents no fundamental difficulties. Considering only the portion of the field contributed by charges, a change in electric field as a function of time indicates that at some place in space, there has been an instantaneous change in charge density. In order to have a complete description of field behavior, it is necessary to investigate a second source of the electric field, a source which acts when a time varying magnetic field is present, but before considering this, let us further investigate the nature of the electric field caused by charges only.

First of all, let us consider several examples of fields caused by specific distribution of electric charge.

Normal component of the electric field caused by a planar charge distribution

Suppose that there is a surface charge distribution on a plane surface as shown as Fig. 2.5. Introduce a vector:

$$dS = dSn$$

where n is the unit vector directed away from the under side of the plane (1) toward the top side of the plane on which the charge is distributed (2). We need only consider the normal component of the field, that is, the component which is perpendicular to the surface. In accord with Coulomb's law as expressed in eq. 2.4, every elementary charge, $\Sigma(q)dS$ creates a field described by the equation:

$$dE(a) = \frac{1}{4\pi\epsilon_0} \frac{\Sigma(q)dS}{L_{qa}^3} L_{qa} \quad (2.10)$$

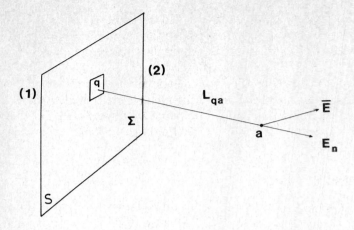

Fig. 2.5. Distribution of charge on a plane surface.

Therefore, the normal component of this field is:

$$dE_n = dE\cos(L_{qa}\boldsymbol{n}) = \frac{1}{4\pi\epsilon_0}\frac{\Sigma(q)dS}{L_{qa}^2}\cos(L_{qa}\boldsymbol{n}) \qquad (2.11)$$

$$= \frac{1}{4\pi\epsilon_0}\frac{\Sigma(q)dS\,L_{qa}}{L_{qa}^3}\cos(L_{qa}\boldsymbol{n})$$

Here $(L_{qa}\boldsymbol{n})$ is angle between directions L_{qa}^0 and $\boldsymbol{n}$. It is clear that the product $dS\,L_{qa}\cos(L_{qa}\boldsymbol{n})$ can be written as a scalar product as follows:

$$dS\,L_{qa}\cos(L_{qa}\boldsymbol{n}) = (dS\cdot L_{qa}) = -(dS\cdot L_{aq}) \qquad (2.12)$$

because $L_{qa} = -L_{aq}$. Thus, the normal component of the electric field can be written as:

$$dE_n(a) = -\frac{1}{4\pi\epsilon_0}\frac{dS\cdot L_{aq}\Sigma(q)}{L_{aq}^3}$$

because $L_{aq} = L_{qa}$.
As can readily be seen, the function $d\omega$, when:

$$d\omega_{aq} = \frac{dS\cdot L_{aq}}{L_{aq}^3} \qquad (2.13)$$

represents a solid angle subtending the element dS from the point a.

In a similar fashion, the solid angle subtended by the entire surface S as viewed from the point a is:

$$\omega_a = \int_S \frac{dS\cdot L_{aq}}{L_{aq}^3} \qquad (2.14)$$

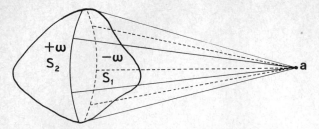

Fig. 2.6. Representation of a closed surface by two open surfaces.

This expression allows us to find the solid angle when the surface is of arbitrary shape. For example, with an observation point inside a closed surface, the solid angle is 4π. If the observation point is situated outside of the closed surface, the solid angle subtended by the surface is zero. This can be derived from the fact that the closed surface could also be represented as two open surfaces, as shown in Fig. 2.6, which are viewed from any external point with the same solid angle being subtended by each of the two surfaces, but with the two solid angles being of equal size and opposite sign. In doing so, we must remember that the sign for solid angle is defined by the angle between the direction of the vector L and the vector dS.

Returning again to the calculation of the normal component, E_n (Fig. 2.5), we can write it as:

$$E_n = \frac{-1}{4\pi\epsilon_0} \int_S \Sigma \, d\omega_{aq} \tag{2.15}$$

In particular, if the charge is distributed uniformly on the surface ($\Sigma = $ constant), we have:

$$E_n = - \frac{1}{4\pi\epsilon_0} \omega_a \Sigma \tag{2.16}$$

where ω_a is the solid angle subtended by the surface S when viewed from the point a. It is obvious (see Fig. 2.7) that the solid angle ω_a is either positive or negative depending on whether the front side or the back side of the surface is viewed.

With increasing distance from the surface S, the solid angle decreases, and correspondingly, the normal component of the field becomes smaller. In the opposite case, when the point a is considered to approach the plane surface S, the solid angle increases, and in the limit becomes equal to -2π and $+2\pi$, when the observation point a is located either on the front face (2) or the reverse face (1) of the surface, respectively. Thus, we have the following expression for the normal component of the electric field on either side of the surface:

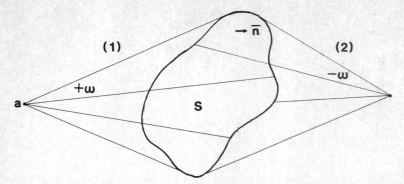

Fig. 2.7. Illustrating the fact that the angles subtended by a surface can be either positive or negative depending on the view point.

$$E_n^{(2)} = \Sigma/2\epsilon_0$$
$$E_n^{(1)} = -\Sigma/2\epsilon_0$$

(2.17)

These two expressions indicate that the normal component of the electric field is discontinuous across the surface S. Let us examine this behavior of the normal component in some detail. The normal component of the electric field can be written as the sum of two terms:

$$E_n = E_n^p + E_n^{s-p}$$

(2.18)

where E_n^p is the part of the normal component caused by the elementary charge $\Sigma(p)dS$ located in the immediate vicinity of the point p, and E_n^{s-p} is the part of the normal component contributed by all of the other surface charge. It is clear that:

$$E_n^{s-p}(a) = -\frac{1}{4\pi\epsilon_0} \Sigma \int_{s-p} d\omega = -\frac{1}{4\pi\epsilon_0} \Sigma \, \omega^{s-p}(a)$$

where $\omega^{s-p}(a)$ is the solid angle subtended by the plane surface S minus the element of surface $dS(p)$ as viewed from the point a. Letting the point a approach the elementary area $dS(p)$, the solid angle subtended by the rest of the surface tends to zero, and the normal component is defined only by the charge located on the elementary surface $dS(p)$:

$$E_n^{s-p} \to 0, \quad \text{as } a \to p$$

During this same process, the solid angle subtended by the surface element $dS(p)$, no matter how small that area is, when viewed from an infinitesimally small distance from the point p, tends to subtend solid angles of $\pm 2\pi$:

$$\omega^p \to \pm 2\pi, \quad \text{as } a \to p$$

Therefore, the normal component of the field on either side of the surface is determined only by the elementary charge located in the immediate vicinity of the point p:

$$E_n^{(2)}(p) = \frac{1}{2\epsilon_0} \Sigma(p)$$

$$E_n^{(1)}(p) = -\frac{1}{2\epsilon_0} \Sigma(p)$$

(2.19)

The difference in sign of the fields on either side of the surface reflects the fundamental fact that the electric field vector shows the direction along which an elementary positive charge will move under the force of the field. Therefore, the discontinuity in the normal field as a test point passes through the surface is caused only by the elementary charge located near the observation point. For example, if there is a hole in the surface, the normal component on either side of the surface is E_n^{s-p}, and therefore, the field is continuous along a line passing through the hole.

We can generalize these results to the case in which the surface carrying the charge is not planar. Making use of the same approach based on the principle of superposition and the definition of solid angles, we arrive at the following expressions for the normal components on either side of a surface:

$$E_n^{(2)}(p) = \frac{\Sigma(p)}{2\epsilon_0} + E_n^{(2)(s-p)}$$

$$E_n^{(1)}(p) = -\frac{\Sigma(p)}{2\epsilon_0} + E_n^{(1)(s-p)}$$

(2.20)

In contrast to the previous case, the normal component $E_n^{s-p}(p)$ caused by charges located on the surface but outside the element $dS(p)$ is not necessarily zero at the point p (see Fig. 2.8). However, we can readily recognize a very important feature of this part of the field. Inasmuch as these charges are located at some distance from the point p, their contribution to the field is a continuous function when the observation point, a, passes through the element $dS(p)$, and therefore:

$$E_n^{(1)(s-p)}(p) = E_n^{(2)(s-p)} = E_n^{(s-p)}$$

(2.21)

Correspondingly, eq. 2.20 can be written as:

$$E_n^{(2)}(p) = \frac{\Sigma(p)}{2\epsilon_0} + E_n^{s-p}$$

$$E_n^{(1)}(p) = -\frac{\Sigma(p)}{2\epsilon_0} + E_n^{s-p}$$

(2.22)

This means that the discontinuity in the normal component, as before, is:

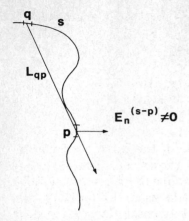

Fig. 2.8. Illustrating the fact that the normal component of electric field caused by charges located on the surface but outside the element dS is not necessarily zero at a point p.

$$E_n^{(2)} - E_n^{(1)} = \Sigma(p)/\epsilon_0 \qquad (2.23)$$

and it is caused only by charges located within the elementary surface area, $dS(p)$.

It should be stressed that eq. 2.23 is a fundamental equation describing electromagnetic field behavior, with the equation being valid for any rate of change of the field with time. In essence, one might say even though we risk getting ahead of ourselves that eq. 2.23 is a surface analogy of the third Maxwell equation.

Effect of a conductor situated within an electric field

We will now consider a second example illustrating electrostatic induction. Suppose that a conductive body of arbitrary shape is situated within the region of influence of an electric field $E_\bullet$, as shown in Fig. 2.9. Under the action of the field, the positive and negative charges residing inside the conductor move in opposite direction. As a consequence of this movement, electric charges develop on both sides of the conductor. In doing so, these

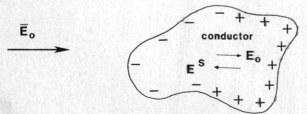

Fig. 2.9. Under the action of an applied field, positive and negative charges residing inside a closed conductor move in opposite directions.

charges create a secondary electric field, which is directed in opposition to the primary field inside the conductor. The induced surface charges distribute themselves in such a way that the total electric field inside a conductor will disappear, that is:

$$E_i \equiv 0 \tag{2.24}$$

where E_i indicates the electric field strength within the conductor. This process is termed "electrostatic induction". At this point it is important to make two comments:

(1) In our description of this phenomenon, we have given a very approximate description of the process in which only the electrostatic field is considered to be present. In fact, the process of accumulation of surface charges involves other phenomena in electromagnetic field behavior, including in particular the change of the magnetic field with time, which plays an important role that will be examined later.

(2) The phenomenon of electrostatic induction is observed in any conductive body, regardless of its electrical resistivity. For example, the conductive body could be composed of metal, or of an electrolytic solution, minerals, or rocks. It is fundamental that any charge source of a constant primary electric field must be situated outside of a conductor. We will see later that the magnitude of the resistivity plays a role in determining the time which is required for the electric field inside the conductor to disappear, but it does not change the final result of the electrostatic induction, that is, that the internal electric field will go to zero ($E_i \equiv 0$).

It should be obvious that the secondary electric field contributed by the surface charges which developed can be defined from the equation:

$$E^s(a) = \frac{1}{4\pi\epsilon_0} \int_S \frac{\Sigma(q)\mathrm{d}S}{L_{qa}^3} L_{qa} \tag{2.25}$$

where $\Sigma(q)$ is a surface density of charges. Correspondingly, condition 2.24 can be rewritten as:

$$E_0 + \frac{1}{4\pi\epsilon_0} \int_S \frac{\Sigma(q)\mathrm{d}S}{L_{qa}^3} L_{qa} = 0 \tag{2.26}$$

where E_0 is the primary field contributed by external sources. For instance, if a single point charge, e, is situated outside the conductor, in point q, its electric field at any point inside the conductor is:

$$E_0(a) = \frac{1}{4\pi\epsilon_0} \frac{e}{L_{qa}^3} L_{qa} \tag{2.27}$$

where "a" is the point at which E is observed.

The electric field caused by a given system of charges does not depend on the electrical properties of the medium. If the field changes, this means that new charge develops.

In our case, positive and negative charges arise on the surface of a conductor. At the same time, the total charge of a neutral conductor remains zero:

$$\sum e^s \equiv 0 \tag{2.28}$$

When the density of the surface charge is known, the electric field outside of the conductor can be calculated from eq. 2.25. However, the distribution of charges caused by electrostatic induction is not known beforehand. For this reason, in order to make use of eq. 2.25, we must determine the function $\Sigma(q)$. Below, we will describe an integral equation approach for determining by numerical means the surface charge density $\Sigma(q)$. From eq. 2.23 we have:

$$E_n^e(p) - E_n^i(p) = \Sigma(p)/\epsilon_0 \tag{2.29}$$

where E_n^e and E_n^i are the normal components of the electric field from the external and internal sides of the surface of the conductor in the vicinity of the point p. However, because of electrostatic induction, the electric field inside the conductor is zero, and therefore $E_n^i = 0$. As a consequence, eq. 2.29 is simplified:

$$E_n^e(p) = \Sigma(p)/\epsilon_0 \tag{2.30}$$

Using the principle of superposition, let us write E_n^e as the sum of three terms:

$$E_n^e = E_n^0 + E_n^p + E_n^{s-p} \tag{2.31}$$

where E_n^0 is the normal component of the primary field at the point p, E_n^p is the normal component of the field caused by the elementary surface charge $\Sigma(p)dS$ situated in the immediate vicinity of the point p, and E_n^{s-p} is the normal component contributed by the rest of the surface charge. In accord with eq. 2.17:

$$E_n^p = \Sigma(p)/2\epsilon_0$$

where n is a unit normal vector directed outwards from the conductor surface. Inasmuch as:

$$E^{s-p} = \frac{1}{4\pi\epsilon_0} \int_S \frac{\Sigma(q)}{L_{qp}^3} L_{qp} \, dS, \quad \text{if } q \neq p$$

we have:

$$E_n^{s-p} = \frac{1}{4\pi\epsilon_0} \int_S \frac{\Sigma(q)(L_{qp} \cdot n_p)}{L_{qp}^3} \, dS, \quad \text{if } q \neq p \tag{2.32}$$

where n_p is the unit normal vector at the point p and $L_{qp} \cdot n_p$ is the scalar product of the vectors L_{qp} and n_p:

$L_{qp} \cos(L_{qp} n_p)$

Collecting all the terms in eq. 2.31, and taking eq. 2.30 into account, we obtain:

$$\frac{\Sigma(p)}{\epsilon_0} = E_n^0 + \frac{\Sigma(p)}{2\epsilon_0} + \frac{1}{4\pi\epsilon_0} \int_S \Sigma(q) K(q,p) dS$$

or

$$\Sigma(p) = 2\epsilon_0 E_n^0(p) + \frac{1}{2\pi} \int_S \Sigma(q) K(q,p) dS, \quad \text{if } q \neq p \tag{2.33}$$

where:

$$K(q,p) = \frac{(L_{qp} \cdot n_p)}{L_{qp}^3} = \frac{\cos(L_{qp} n_p)}{L_{qp}^2}$$

and

$$2\epsilon_0 E_n^0(p)$$

are known functions.

Equation 2.33 is an integral equation of the Fredholm type of the second kind with respect to an unknown charge density at any point p on the conductor surface. In practice, one can conceptually replace the surface of the conductor with a system of small cells within each of which the charge density is practically constant. In doing so, the integral equation 2.33 can be rewritten approximately as:

$$\Sigma(p) \approx 2\epsilon_0 E_n^0(p) + \frac{1}{2\pi} \sum_{n=1}^N \Sigma(q) K(q,p) \Delta S, \quad \text{if } q \neq p \tag{2.34}$$

Having written this equation for every cell, we obtain a system of N linear equations with N unknown terms. The solution is determinate. When the charge density is known, for example, using eq. 2.25, the secondary field is calculated in terms of a surface integral.

As an application of this behavior of the electrostatic induction phenomenon, let us consider an example of its use in electrical prospecting with direct current. Suppose that we consider a model of a conducting medium which consists of a sequence of layers each characterized by its own resistivity (see Fig. 2.10). Above the surface of the Earth, electrical charges due to atmospheric processes which do not vary with time, are present. In accord with Coulomb's law, they create the same field in the conductive medium as would have been observed if that volume were free space. But, because of electrostatic induction, induced charges appear on the Earth's surface which exactly compensate the primary field inside the conductive medium. Because

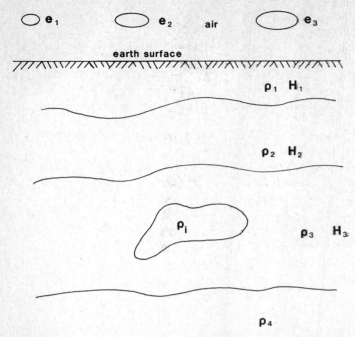

Fig. 2.10. Model of a conducting medium that consists of a sequence of layers each characterized by its own resistivity.

of surface charges that accumulate on the interface between the upper half space and the conductive medium, the charges situated above the Earth's surface do not have any effect on the electric field within the Earth, and therefore, the field within the medium can only be caused by sources existing within the medium.

At this point, we can describe some general problems involved in electric field behavior caused by charges. It has already been mentioned that when the charge distribution is unknown, we cannot make use of Coulomb's law to calculate field behavior. Unfortunately, in most cases of interest in electrical prospecting, the distribution of charges is unknown, and Coulomb's law is of no use in determining the field. This is why we must consider some general features of electric field behavior caused by charges.

Proceeding from Coulomb's law, we can obtain two fundamental equations for this field. First of all, let us introduce the concept of electric flux as being the surface integral of the scalar product of the electric field E and the vector dS, Fig. 2.11:

$$N = \int_S E \cdot dS \qquad (2.35)$$

where

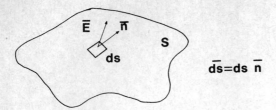

$$\overline{ds} = ds\ \overline{n}$$

Fig. 2.11. Representation of electric flux as being the surface integral of the scalar product of the electric field and the surface.

$$E \cdot dS = EdS \cos(EdS)$$

Assume that an elementary charge de situated at point q is the sole source of an electric field. In accord with Coulomb's law, the flux of the electric field through an arbitrary surface is:

$$N = \frac{de}{4\pi\epsilon_0} \int_S \frac{L_{qp} \cdot dS}{L_{qp}^3} \qquad (2.36)$$

where p is an arbitrary point on the surface S. Inasmuch as this integral represents the solid angle ω subtended by the surface S as seen from the point q, we can write:

$$\int_S E \cdot dS = \frac{de}{4\pi\epsilon_0} \omega_q$$

In the particular case in which the surface S is closed, and the charge de is situated inside the surface, the solid angle ω_q is 4π and we have:

$$\oint_S E \cdot dS = de/\epsilon_0 \qquad (2.37)$$

It should be clear that when the charge de is located outside the surface S, the flux of the electric field caused by the charge is zero.

Equation 2.37 has been obtained for the case of an elementary charge. Using superposition, we derive the following equation for an arbitrary distribution of charges:

$$\oint_S E \cdot dS = \frac{1}{\epsilon_0} \left[\int_V \delta dV + \int_S \Sigma dS + \int_L \lambda dl + \Sigma e_i \right] \qquad (2.38)$$

where δ, Σ, and λ are volume, surface, and linear charge densities, along with the point charges e_i, all of which are situated inside the surface S. The flux caused by charges outside of the surface is zero.

The following comments should be made about eq. 2.38:

(1) A change in position of the charge within the volume V surrounded by

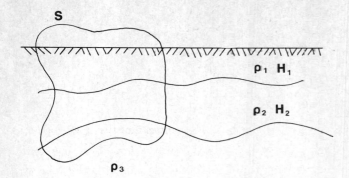

Fig. 2.12. The surface S referred to in the text can be quite arbitrary in shape and position, such that it can intersect portions of the various media characterized by different electrical properties.

the surface S alters the value of the field E at the surface, but the flux does not change because it is a function only of the total charge within the surface.

(2) The surface S can have quite an arbitrary shape and position. In particular, it can intersect portions of the media characterized by varying electrical properties, as shown in Fig. 2.12. Assuming an actual distribution of charge described by the volume density δ, we have:

$$\oint_S E \cdot dS = \frac{1}{\epsilon_0} e_V \qquad (2.39)$$

where e_V is the volume charge within the volume V:

$$e_V = \int_V \delta \, dV$$

Usually, the total charge e_V is the sum of two types of charges, one being free charges, e_0, which are free to move, and the other being polarization charges, in which the displacement is measured in micro dimensions. The displacement of charges will not routinely be considered here, and therefore, we will normally mean that the charge e_V is the free charge.

Equation 2.39, which was developed directly from Coulomb's law, is in fact the third of Maxwell's equations valid both for constant and time-varying electromagnetic fields. Omitting the subscript "V" on the parameter e, we have:

$$\oint_S E \cdot dS = e/\epsilon_0 \qquad (2.40)$$

This equation shows the relationship between field values observed on various points of the surface S and can be interpreted from two points of view. If the charge e is known, eq. 2.40 can be considered to be an integral equation in an unknown variable, the field intensity E. In contrast, when the electric field is known, use of the flux allows one to determine the sources of the field. If we wish to find the relationship between the flux and the sources within an elementary volume, we can make use of Gauss's theorem:

$$\oint_S E \cdot dS = \int_V \operatorname{div} E \, dV \tag{2.41}$$

where S is a closed surface surrounding the volume V. Applying this equation to an elementary volume where the function div E is nearly constant, we have:

$$\oint_S E \cdot dS \approx \operatorname{div} E \, dV$$

or in the limit:

$$\operatorname{div} E = \frac{1}{dV} \oint_S E \cdot dS \tag{2.42}$$

Thus, the divergence of the electric field characterizes the flux of the vector E through a surface surrounding the elementary volume. In accord with eqs. 2.40 and 2.42, we have:

$$\operatorname{div} E = \delta/\epsilon_0 \tag{2.43}$$

Divergence of the electric field along with the flux through an arbitrary closed surface characterizes the distribution of charges. However, eq. 2.43 describes the volume density of charges in the vicinity of any point; that is, it has a differential character, different than that of eq. 2.40. Equation 2.43, like eq. 2.40, is valid for electromagnetic fields regardless of the rate of change of the field with time. It is Maxwell's third equation expressed in differential form.

We must stress that there is a fundamental difference between the integral and differential forms of Maxwell's third equation. While the integral form (eq. 2.40) can be applied everywhere, it is necessary to be careful in the use of the differential form (eq. 2.43). This caution must be exercised because the function div E does not exist at certain points, lines, or surfaces. As must be recognized, div E is expressed in terms of spatial derivatives of the field components. For example, in the cartesian coordinate system, it is:

$$\operatorname{div} E = \frac{\partial E_x}{\partial x} + \frac{\partial E_y}{\partial y} + \frac{\partial E_z}{\partial z}$$

At points where one of the derivatives is not properly behaved, eq. 2.43 cannot be applied. In other words, it does not permit us to describe the nature of the sources at such locations. A very important example from electrical prospecting where this equation cannot be applied is provided by any model in which electrical charges are distributed at interfaces representing a step-wise change in electrical properties. As was shown in the first example in this section, the normal component of the electric field is a discontinuous function of the spatial variables in passing through a surface charge, and respectively the normal derivative $\partial E_n / \partial n$ does not exist at such points. Therefore, in order to characterize sources on an interface, one must use Maxwell's third equation in the integral form (eq. 2.40). Applying it to an elementary cylindrical surface enclosing a small piece of the interface as shown in Fig. 2.13, we obtain a well-known relationship:

$$E_n^{(2)} - E_n^{(1)} = \Sigma / \epsilon_0 \tag{2.44}$$

where Σ is the surface charge density, $E_n^{(2)}$ and $E_n^{(1)}$ are the normal components of the field on either side of the surface, and n is a unit normal directed from the reverse side to the front side of the surface. Equation 2.44 can be interpreted as being a surface analogy of eq. 2.43, and it, too, is a form of Maxwell's third equation. Comparing eqs. 2.44 and 2.23 we see that they coincide exactly. This follows directly from the fact that the discontinuity in the normal component, E_n, is due to the presence of surface charges. In particular, if the surface charge is absent at some point, the normal component of the field is found to be continuous.

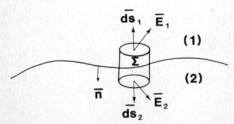

Fig. 2.13. Definition of an elementary cylindrical surface that encloses a small piece of an interface between two regions with different resistivity.

By starting with Coulomb's law, we have obtained three useful forms of Maxwell's third equation:

$$\oint E \cdot dS = e / \epsilon_0$$

$$\operatorname{div} E = \delta / \epsilon_0 \tag{2.45}$$

$$E_n^{(2)} - E_n^{(1)} = \Sigma / \epsilon_0$$

Each of them characterizes the distribution of charge and one can say that they are the same tool of analysis written in three ways.

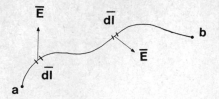

Fig. 2.14. An arbitrary path along which the dot product of electric field and direction is integrated.

Another highly useful concept which illustrates some of the fundamental characteristics of the electric field can be introduced as follows:

$$\int_a^b E \cdot dl \qquad (2.46)$$

The integral is the voltage between points a and b measured along some arbitrary path L (see Fig. 2.14) caused by an electric field. The product $E \cdot dl$ can be written as:

$$E \cdot dl = E \, dl \cos(E \, dl) = E \, dl \cos\alpha$$

where α is the angle between the electric field vector and the tangent to the path L at every point. From the principal point of view, the product $E \cdot dl$ is an element of work performed by the electric field in transporting a unit positive charge along the elementary displacement. This product has the dimensions of work, per unit charge, and in the practical system of units has the dimensions of volts. Therefore, the integral in eq. 2.46 represents the work or voltage done in carrying a charge between two points, a and b. In the general case, for a given function E, this work integral depends on the particular path of integration L which is chosen, and on the upper and lower limits, that is, on the terminal points a and b of the path L. Starting from Coulomb's law, it can be demonstrated that the voltage of the electric field caused only by static charges is independent of the path of integration.

Assume that the source for the field is a single elementary charge, de. In accord with Coulomb's law, the electric field is:

$$E(p) = \frac{1}{4\pi\epsilon_0} \frac{de(q)}{L_{qp}^3} L_{qp} \qquad (2.47)$$

where L_{qp} is the vector directed from the point q to the point p. If both terminal points a and b are situated on the same radius vector, L_{qp} and the path of integration is along this radius (see Fig. 2.15), the voltage between these points is very easily calculated:

$$V = \int_a^b E \cdot dl = \frac{de(q)}{4\pi\epsilon_0} \int_a^b \frac{dl \cdot L_{qp}}{L_{qp}^3} = \frac{de(q)}{4\pi\epsilon_0} \int_a^b \frac{dL}{L_{qp}^2}$$

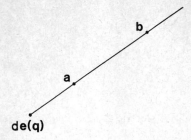

Fig. 2.15. Example of a case in which the points a and b are situated on a common radius vector from the position of the electric charge.

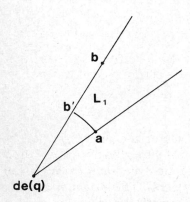

Fig. 2.16. Example of a case in which the points a and b are situated on different radius vectors.

inasmuch as:

$$d\boldsymbol{l} \cdot \boldsymbol{L}_{qp} = dl \, L_{qp} \cos 0 = L_{qp} dL$$

Carrying out the integration as indicated, we obtain:

$$V = \frac{de(q)}{4\pi\epsilon_0} \left(\frac{1}{L_{qa}} - \frac{1}{L_{qb}} \right) \tag{2.48}$$

Now assume that the points a and b are situated on two different radius vectors, L_{qa} and L_{qb}, as shown in Fig. 2.16. Let us choose the path L_1. The path of integration L_1 consists of two parts. The first part is a simple arc, ab', and the second element of the path is along the radius vector L_{qb}. In this case, the voltage can be written as:

$$V = \frac{1}{4\pi\epsilon_0} de(q) \left[\int_{\text{arc } ab'} \frac{d\boldsymbol{l} \cdot \boldsymbol{L}_{qp}}{L_{qp}^3} + \int_{b'}^{b} \frac{d\boldsymbol{l} \cdot \boldsymbol{L}_{qp}}{L_{qp}^3} \right]$$

The integral along the arc ab' is clearly zero since the scalar product is:

Fig. 2.17. An arbitrary path that can be represented as the sum of radius vectors and arcs.

$$d\boldsymbol{l} \cdot L_{\mathrm{qp}} = 0$$

Thus, the voltage between points a and b is again equal to:

$$V = \frac{de(q)}{4\pi\epsilon_0}\left(\frac{1}{L_{\mathrm{qa}}} - \frac{1}{L_{\mathrm{qb}}}\right) \tag{2.49}$$

If instead of the path L_1, we consider a more arbitrary path L_2, it should be clear that this path could be represented as the sums of arcs and elements of radius vectors as shown in Fig. 2.17. All integral contributions along simple arcs are zero, while the sum of integrals along all radius vectors is:

$$V = \frac{de(q)}{4\pi\epsilon_0}\left(\frac{1}{L_{\mathrm{qa}}} - \frac{1}{L_{\mathrm{qb}}}\right)$$

that is, they are equal to the voltage along the path L_1.

We have established the second fundamental characteristic of an electric field, namely, that the voltage between two points does not depend on the particular path along which integration is carried out, but is determined by the terminal points only. This fact can be written formally as:

$$\int_{\substack{a\\L_1}}^{b} \boldsymbol{E} \cdot d\boldsymbol{l} = \int_{\substack{a\\L_2}}^{b} \boldsymbol{E} \cdot d\boldsymbol{l} = \ldots = \int_{\substack{a\\L_n}}^{b} \boldsymbol{E} \cdot d\boldsymbol{l} \tag{2.50}$$

Making use of the principle of superposition, this result can be generalized to a field caused by any arbitrary distribution of charges. It must be stressed again that this result is valid only for electric fields caused by constant electric charges, and that it cannot be applied to time-varying fields.

The independence of a voltage on the path of integration can be written in

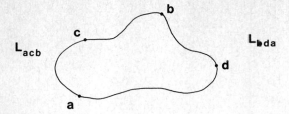

Fig. 2.18. A path of integration along a closed contour L which can be broken down into the summations of other open contours.

another form. Consider a closed contour L as shown in Fig. 2.18 as consisting of two other contours, L_{acb} and L_{bda}.
   In accord with eq. 2.50 we have:

$$\int_{acb} E \cdot dl = \int_{adb} E \cdot dl \qquad (2.51)$$

In these integrals, the element dl is directed from a to b. Changing the direction of integration in the integral on the right-hand side of eq. 2.51, we can write:

$$\int_{adb} E \cdot dl = - \int_{bda} E \cdot dl, \quad \text{i.e.}$$

$$\int_{acb} E \cdot dl = - \int_{bda} E \cdot dl \quad \text{or}$$

$$\int_{acb} E \cdot dl + \int_{bda} E \cdot dl = 0 \quad \text{or}$$

$$\oint E \cdot dl = 0 \qquad (2.52)$$

Thus, the voltage along an arbitrary closed path is zero. Sometimes the quantity:

$$\oint E \cdot dl$$

is called "the circulation" of the electric field or the electromotive force (EMF). The path L can have an arbitrary shape, and it can intersect media characterized by various physical properties (see Fig. 2.19). In particular, it can be completely contained within a conducting medium. Because of the fact that the electromotive force caused by electric charges is zero, the Coulomb charge force, E^c, cannot alone cause an electric current. This is the

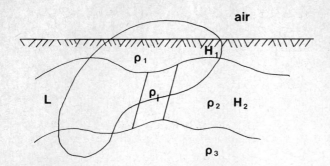

Fig. 2.19. The path L can have an arbitrary shape and it can intersect media characterized by different physical properties.

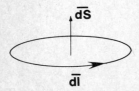

Fig. 2.20. Use of the right-hand rule to define the directions of dl and dS.

reason that non-Coulomb forces must be considered in order to understand the creation of current flow. Equation 2.52 is the first equation for the electric field caused by charges, but written in integral form, in which it relates the values of the field at various points in the medium. To obtain eq. 2.52 in differential form, we will make use of Stoke's theorem, according to which, for any vector A, when the first spatial derivatives exist, the following relationship holds:

$$\int_L A \cdot \mathrm{d}l = \int_S (\mathrm{curl}\, A) \cdot \mathrm{d}S \tag{2.53}$$

In this expression, the orientations for dl and dS are considered to be related through the right-hand rule as indicated in Fig. 2.20.

The function curl A is a vector expressed in terms of the spatial derivatives of the components of A. As an example, in cartesian coordinates, the curl A is:

$$\mathrm{curl}\, A = \left(\frac{\partial A_z}{\partial y} - \frac{\partial A_y}{\partial z} \right) i + \left(\frac{\partial A_x}{\partial z} - \frac{\partial A_z}{\partial x} \right) j + \left(\frac{\partial A_y}{\partial x} - \frac{\partial A_x}{\partial y} \right) k$$

Other relationships can be written in other orthogonal systems of coordinates.

One can demonstrate that curl A characterizes the maximum change of voltage in the vicinity of a source point in the direction perpendicular to the field.

In accord with eqs. 2.52 and 2.53, we obtain a differential form of the first equation for the electric field:

$$\text{curl } E = 0 \tag{2.54}$$

This equation, as well as eq. 2.52, reflects the fact that the voltage along a closed path must be zero. It is appropriate to emphasize that both follow directly from Coulomb's law.

As has been previously mentioned, Stoke's theorem (eq. 2.53) is valid only when the first spatial derivatives exist. Thus, this last equation cannot be used at points where one of the field components is a discontinuous function of position. In order to obtain a differential form of eq. 2.52 valid at such points, we will apply this equation along an elementary path as shown in Fig. 2.21. Considering that the elements dl' and dl'' are separated by the distance dh which tends to zero, we obtain:

$$(E \cdot dl'') + (E \cdot dl') + (E \cdot dh) = 0$$

$$E_t^{(2)} dl'' - E_t^{(1)} dl' = 0$$

and finally:

$$E_t^{(2)} - E_t^{(1)} = 0 \tag{2.55}$$

The tangential components of the field are continuous as the path passes through a surface charge.

We have now derived three forms of the first equation based on Coulomb's law:

$$\oint E \cdot dl = 0, \quad \text{curl } E = 0, \quad E_t^{(2)} = E_t^{(1)} \tag{2.56}$$

Each of them expresses the same fact, that is, that the electromotive force caused by electric charges is zero, or in other words, the voltage between two arbitrary points does not depend on the path of integration.

We must make an important comment about eq. 2.56. The first two relationships are not valid when the field is time-varying, since the second

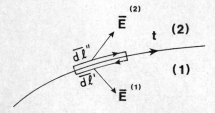

Fig. 2.21. Application of Stoke's theorem along an elementary path.

source for an electric field, that is the change of magnetic field intensity with time, is not taken into account. On the other hand, the surface analog for these equations is valid for any electromagnetic field. This reflects the fact that in the development of this particular form of the equation, it was assumed that the area surrounded by the integration path was zero and therefore the flux of the magnetic field through this area vanished.

Let us note one further thing. Although the equations $\oint E \cdot dl = 0$, and curl $E = 0$ are not valid for time-varying electromagnetic fields, this does not mean that Coulomb's law is inapplicable. In the further analysis here, we will have a chance to demonstrate that in many cases, time-varying charges create electric fields which are very nearly described by Coulomb's law.

Returning to the first field equation, let us consider one more important feature of the electric field caused only by charges. In fact, according to eq. 2.54, the field can be written as:

$$E = - \operatorname{grad} U \tag{2.57}$$

inasmuch as

$$\operatorname{curl\ grad} U \equiv 0$$

is an identity relationship.

The scalar function U is called the "potential" for the electric field. In accord with eq. 2.57, the electric field E coincides with the direction of maximum decrease in potential, and any projection of the field can be expressed in terms of the potential as follows:

$$E_1 = - \partial U / \partial l \tag{2.58}$$

Equations 2.57 and 2.58 are useful in determining the field when the potential is known.

At this point, one will write an expression for voltage using the potential. For this purpose, let us write down an obvious equality for the differentiation of the potential:

$$dU = \frac{\partial U}{\partial l} dl = dl \cdot \operatorname{grad} U = -E \cdot dl \tag{2.59}$$

Here $dl = dl\, I_0, I_0$ is unit vector.
Integrating the last of these terms along any path between two arbitrary points, and considering that voltage does not depend on the choice of path, we obtain:

$$\int_a^b E \cdot dl = - \int_a^b dU = U(a) - U(b) \tag{2.60}$$

The voltage of the electric field between two points can be written as the difference of potential between these points.

Now we can use eq. 2.60 to define the potential caused by a distribution of charges. From eq. 2.60 we have:

$$U(a) = U(b) + \int_a^b E \cdot dl \tag{2.61}$$

It is obvious that we can assume that at great distance from the charge, as the distance increases without limit, the potential will vanish. Then, letting b equal infinity in eq. 2.61, and assuming that the potential at this distance is zero, we have:

$$U(a) = \int_a^\infty E \cdot dl \tag{2.62}$$

Assume that the source for an electric field is a single elementary charge, de, situated at the point q. By using eqs. 2.47 and 2.62, we obtain:

$$U(a) = \frac{1}{4\pi\epsilon_0} \frac{de}{L_{qa}} \tag{2.63}$$

Making use of superposition for an arbitrary distribution of volume, surface, linear and point charges, we arrive at the following expression for the potential:

$$U(a) = \frac{1}{4\pi\epsilon_0} \left[\int_V \frac{\delta dV}{L_{qa}} + \int_S \frac{\Sigma dS}{L_{qa}} + \int_L \frac{\lambda dl}{L_{qa}} + \sum \frac{e_i}{L_{qa}} \right] \tag{2.64}$$

Comparing this last expression with eq. 2.9, we see that the potential, U, is related with charges in a much simpler way than is the electric field. This simplicity is an important reason for making use of the potential. Having defined U as we have, the electric field is very easily found by applying eq. 2.58.

In conclusion, we will derive a series of equations reflecting the behavior of the potential. Substituting eq. 2.57 into eq. 2.43, we obtain:

div grad $U = -\delta/\epsilon_0$

or

$$\nabla^2 U = \Delta U = -\delta/\epsilon_0 \tag{2.65}$$

where the operator ∇^2 or Δ is the Laplacian operator. As an example, in cartesian coordinates eq. 2.65 is:

$$\frac{\partial^2 U}{\partial x^2} + \frac{\partial^2 U}{\partial y^2} + \frac{\partial^2 U}{\partial z^2} = -\frac{\delta}{\epsilon_0} \tag{2.66}$$

that is, the simple sum of the second derivatives of the potential with respect to each of the spatial variables is directly proportional to the volume density of charge taken with the opposite sign. Equation 2.65 is most commonly called Poisson's equation for the potential and describes the behavior of the potential at points where the volume density of charge is nonzero. In areas where charge is not present, it simplifies and becomes Laplace's equation for the potential:

$$\frac{\partial^2 U}{\partial x^2} + \frac{\partial^2 U}{\partial y^2} + \frac{\partial^2 U}{\partial z^2} = 0 \tag{2.67}$$

At this point, we can demonstrate one remarkable feature of the behavior of potential at points where there is no charge. We replace the derivatives in eq. 2.67 by finite differences. For sufficiently small values of Δx, we have:

$$\frac{\partial U}{\partial x} = \frac{1}{\Delta x}\left[U\left(x + \frac{\Delta x}{2}, y, z\right) - U\left(x - \frac{\Delta x}{2}, y, z\right)\right] \tag{2.68}$$

$$\frac{\partial^2 U}{\partial x^2} = \frac{\partial}{\partial x}\frac{\partial U}{\partial x} = \frac{1}{\Delta x}\left[\frac{\partial U}{\partial x}\left(x + \frac{\Delta x}{2}, y, z\right) - \frac{\partial U}{\partial x}\left(x - \frac{\Delta x}{2}, y, z\right)\right] \tag{2.69}$$

where

$$\frac{\partial U}{\partial x}\left(x + \frac{\Delta x}{2}, y, z\right) = \frac{1}{\Delta x}[U(x + \Delta x, y, z) - U(x, y, z)]$$

$$\frac{\partial U}{\partial x}\left(x - \frac{\Delta x}{2}, y, z\right) = \frac{1}{\Delta x}[U(x, y, z) - U(x - \Delta x, y, z)] \tag{2.70}$$

Using the notation:

$$U(x, y, z) = U(a); \quad U(x - \Delta x, y, z) = U_x^{(\prime)}; \quad U(x + \Delta x, y, z) = U_x^{(\prime\prime)}$$

we can write:

$$\frac{\partial^2 U}{\partial x^2} = \frac{1}{(\Delta x)^2}[U_x^{(\prime\prime)} + U_x^{(\prime)} - 2U(a)] \tag{2.71}$$

By analogy, we can write similar expressions for the other second derivatives:

$$\frac{\partial^2 U}{\partial y^2} = \frac{1}{(\Delta y)^2}[U_y^{(\prime\prime)} + U_y^{(\prime)} - 2U(a)]$$

and $\tag{2.72}$

$$\frac{\partial^2 U}{\partial z^2} = \frac{1}{(\Delta z)^2}[U_z^{(\prime\prime)} + U_z^{(\prime)} - 2U(a)]$$

where

$$U_y^{(")} = U(x, y + \Delta y, z), \quad U_y^{(')} = U(x, y - \Delta y, z)$$
$$U_z^{(")} = U(x, y, z + \Delta z), \quad U_z^{(')} = U(x, y, z - \Delta z)$$

Adding eqs. 2.71 and 2.72, in letting the first differences, Δx, Δy, and Δz be h, we obtain:

$$\nabla^2 U = \frac{1}{h^2} [U_x^{(')} + U_y^{(')} + U_z^{(')} + U_x^{(")} + U_y^{(")} + U_z^{(")} - 6U(a)]$$

$$= \frac{6}{h^2} [U_{av} - U(a)] \tag{2.73}$$

where

$$U_{av} = \frac{U_x^{(')} + U_x^{(")} + U_y^{(')} + U_y^{(")} + U_z^{(')} + U_z^{(")}}{6}$$

From this, we see that Laplace's equation can be written in the form:

$$U_{av} = U(a) \tag{2.74}$$

The value of the potential U at a specific point (where there is no charge) is the average of values of the potential at six other points situated at a short distance h from the point along three mutually perpendicular lines intersecting at this point (see Fig. 2.22). It is common convention that functions satisfying Laplace's equation or eq. 2.74 are termed "harmonic" functions. Correspondingly, the potential for the electric field caused by charge out of them is a harmonic function characterized by numerous useful and interesting features which follow from eq. 2.74.

Fig. 2.22. The value of potential U at a specific point (a) can be taken as being the average value of the potential of six other points situated a short distance away along three mutually perpendicular lines.

We can derive a general solution for Poisson's equation when the source of the field, and therefore of the potential, consists of volume charges only. In accord with eq. 2.64, the potential U caused by such charges is the volume integral:

$$U(a) = \frac{1}{4\pi\epsilon_0} \int_V \frac{\delta(q)dV}{L_{qa}} \tag{2.75}$$

On the other hand, Poisson's equation (eq. 2.65) describes the potential everywhere, whether charge is present or not. Therefore, the right-hand side of eq. 2.75 satisfies this equation and it is a solution. It is obvious that Poisson's equation along with Laplace's equation describes the potential when the second derivative of the potential or the first derivatives of the electric field exist. Unfortunately there are many cases when this condition is not met, and when eqs. 2.66–2.67 cannot be used. Among these, the most important case is that of a surface distribution of charge. As has been shown, near a surface carrying charge the tangential component of the electric field is continuous, while the normal component is discontinuous. Therefore, the derivative with respect to the normal does not exist. For this reason, we will define another equation describing the potential behavior near surface charges.

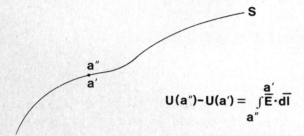

Fig. 2.23. Application of a boundary condition on potential observed on either side of a surface S.

In accord with eq. 2.62 the potential on either side of a surface (see Fig. 2.23) is:

$$U(a') = \int_{a'}^{\infty} E \cdot \mathrm{d}l \quad \text{and} \quad U(a'') = \int_{a''}^{\infty} E \cdot \mathrm{d}l$$

or
$$U(a'') - U(a') = \int_{a''}^{a'} E \cdot \mathrm{d}l \qquad (2.76)$$

Inasmuch as the field on both sides of the surface is finite, but the distance between the points a' and a'' is vanishingly small, the difference in potential between the two sides tends to zero. Therefore, the potential of the electric field on any surface carrying charge with a density Σ is continuous:

$$U_1 = U_2 \qquad (2.77)$$

This condition can be considered to be the surface analogy of Poisson's equation.

So far we have considered mostly electric fields caused by specified charges in free space. We have also investigated the field of charges that accumulate on the surface of a conductor which along with other source charges create a static field. Now we will show that Coulomb's law manifests itself when there is a direct current flowing in a conducting medium. In doing so, we will make use of Ohm's law which relates current density to electric field intensity:

$$j = \sigma E \tag{2.78}$$

where j is current density, which is a vector quantity with a magnitude equal to the amount of charge passing through a unit area oriented perpendicularly to the flux of charges during a unit time interval. The direction of this vector, j, is coincident with the normal to the area. It is clear that a total current I is related to the current density as:

$$I = \int_S j \cdot dS \tag{2.79}$$

When the current I is measured in amperes, the current density has the unit of amperes per square meter.

In Ohm's law, the coefficient of proportionality, indicated by the symbol σ, and which is determined experimentally, is defined as being the conductivity of the medium. The units for conductivity are siemens/meter or mho/meter. Often in geophysical practice, the reciprocal of conductivity, $\rho = 1/\sigma$ is used and is called the resistivity. The units for resistivity are ohm-meters.

In Ohm's law, the electric field, E, can be written as consisting of parts:

$$E = E^c + E^{\text{other}} \tag{2.80}$$

where E^c is the electric field contributed by electric charges, and is governed by Coulomb's law, and E^{other} is the electric field caused by sources other than charges (electric fields of electro-chemical origin, caused by diffusion of ions in rocks, or induction electric fields caused by variation of magnetic fields with time). In unusual cases, such physical phenomena may give rise to electric fields, as in the piezo-electric phenomenon, the magneto-electric phenomenon, the thermo-electric phenomenon, and etc. Assuming that the field does not depend on time, and that the observation point is located well away from other sources, we can take the total electric field in Ohm's law as being the Coulomb field:

$$E = E^c \tag{2.81}$$

In addition, we can make use of the principle of conservation of charge, reflecting the fact that the direct current density flux through any closed surface is zero:

$$\oint_S j \cdot dS = 0 \tag{2.82}$$

This equation is amenable to a direct interpretation. The integral on the left-hand side is the amount of charge passing through a closed surface per unit time, including those charges which leave the volume as well as those charges which enter the volume. If the total of the two is not zero, we would observe an accumulation of charge during any interval of time and correspondingly the field would not be constant. This is the reason that eq. 2.82 is valid for direct currents.

Applying Gauss's theorem, we obtain the principle of charge conservation for direct currents in differential form:

$$\text{div} \, j = 0 \tag{2.83}$$

In contrast to the case for eq. 2.82, this equation is applied at points where the first spatial derivatives of components of current density exist. However, there are places such as interfaces between media with differing conductivity where the tangential component of the current density is a discontinuous function. According to eq. 2.55 we have (see Fig. 2.24):

$$\sigma_i E_{t,i} \neq \sigma_{i+1} E_{t,i+1}$$

The derivative of the tangential component of the vector j is not defined at such interfaces. To obtain a surface analog of eq. 2.83, we will repeat the algebra carried out in deriving eq. 2.44. In doing so, we obtain:

$$j_n^{(1)} = j_n^{(2)} \tag{2.84}$$

That is, the normal component of current density is continuous at interfaces.

We again have obtained three forms of the equation describing conservation of electrical charge:

$$\int_S j \cdot dS = 0, \quad \text{div} \, j = 0. \quad j_n^{(2)} = j_n^{(1)} \tag{2.85}$$

It is remarkable that the equation listed above is valid for time-varying electromagnetic fields, so long as the time variation has a quasi-stationary character.

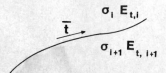

Fig. 2.24. Example of a location on an interface between two media with different conductivity where the tangential component of current density is discontinuous.

At this point we are prepared to demonstrate that in a conducting medium, the current field j is accompanied by the appearance of electric charges, these being the sole source of an electric field at places where $E^{other} = 0$. First of all, let us assume that the conductivity of the medium is a continuous function of position and that discontinuous interfaces are absent. In accord with eqs. 2.78 and 2.83:

$$\text{div } j = \text{div } \sigma E = 0$$

Making use of the conventional rules for taking derivatives of the product of a scalar by a vector, we obtain:

$$\text{div } \sigma E = \sigma \text{ div } E + E \cdot \text{grad } \sigma = 0$$

and hence:

$$\text{div } E = -\frac{E \cdot \text{grad } \sigma}{\sigma} \tag{2.86}$$

Making use of eq. 2.43, we have finally:

$$\delta = -\epsilon_0 \frac{E \cdot \text{grad } \sigma}{\sigma} \tag{2.87}$$

Thus, a volume distribution of charge appears in a conductive medium when the medium is non-uniform and when the electric field is not oriented perpendicular to the direction of maximum change in conductivity. It is clear that at points where the medium is uniform, there are no charges and therefore div E is zero. In accord with eq. 2.87, motion of charge with zero net charge density at each point in the medium can be accompanied by the formation of fixed charges. They in fact are the source of an electric field which governs the behavior of the current density field j.

Now assume that an interface between media characterized by different values of conductivity is present. We can show that surface charges arise. This is done by taking the third form of eq. 2.85, that is, the continuity of the normal component of current density:

$$j_n^{(1)} = j_n^{(2)}$$

or $\tag{2.88}$

$$\sigma_1 E_n^{(1)} = \sigma_2 E_n^{(2)}$$

The normal component of the electric field is discontinuous at the interface as was shown earlier. This discontinuity in the electric field is caused by an electric charge on the surface with a density Σ which generates a normal component of the field having opposite signs on either side of the surface. Making use of eq. 2.19, we can write eq. 2.88 as follows:

44

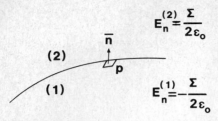

Fig. 2.25. Definition of the boundary conditions as applied in eq. 2.89.

$$\sigma_1 \left(-\frac{\Sigma}{2\epsilon_0} + E_n^{s-p} + E_n^0 \right) = \sigma_2 \left(\frac{\Sigma}{2\epsilon_0} + E_n^{s-p} + E_n^0 \right) \tag{2.89}$$

where $\pm \Sigma/2\epsilon_0$ is the normal component of the field caused by surface charges situated near point p (see Fig. 2.25), E_n^{s-p} is the normal component of the field caused by the rest of the surface charges, and E_n^0 is the normal component of the field caused by charges located outside the surface. It should be noted that the components E_n^{s-p} and E_n^0 are continuous at the point p. Solving eq. 2.89 we have:

$$\Sigma = 2\epsilon_0 \frac{\sigma_1 - \sigma_2}{\sigma_1 + \sigma_2} (E_n^0 + E_n^{s-p}) \tag{2.90}$$

or

$$\Sigma = 2\epsilon_0 \frac{\rho_2 - \rho_1}{\rho_2 + \rho_1} (E_n^0 + E_n^{s-p}) \tag{2.91}$$

A surface charge arises at every point on the interface where the normal component of the field contributed by the charges located outside this point is not zero.

We might say that the surface and volume charges which develop within a conductive medium play a vital role in forming both the electric field and the current field. Without the appearance of these charges, the normal component of the current density could not be a continuous function of the spatial variables at an interface and we would observe an accumulation of charge. Correspondingly, no constant current could occur. It is clear that volume and surface charges that develop in a conducting medium create an electric field which obeys Coulomb's law. However, the actual use of Coulomb's law in calculating the electric field in a conducting medium is usually impracticable inasmuch as the manner of distribution of these charges is unknown.

2.2. BIOT-SAVART LAW

In the preceding section it was shown that electric charges create an electric field which behaves in a manner described by Coulomb's law. The next step in our consideration of the behavior of electromagnetic fields will be the consideration of the magnetic field associated with constant electrical current. Experimentally it has been shown that the magnetic field generated by a direct (constant) current can be described with the equation:

$$dH(a) = \frac{I}{4\pi} \frac{dl \times L_{pa}}{L_{pa}^3}$$

(2.92)

which is generally known as the Biot-Savart law. In this expression, I is the total current flowing in a linear element dl located at a point p as shown in Fig. 2.26 and L_{pa} is the distance from point p to point a.

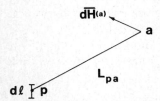

Fig. 2.26. Geometry relative to eq. 2.92.

Considering the definition of vector product, eq. 2.92 can be written as:

$$dH(a) = \frac{Idl}{4\pi} \frac{s_0}{L_{pa}^2} \sin(dl\, L_{pa})$$

(2.93)

where s_0 is a unit vector perpendicular to the plane in which dl and L_{pa} are located. The magnetic field caused by an elementary linear current is directly proportional to that current and to the sine of the angle between the vectors dl and L_{pa}. It can be seen that there is no component of the magnetic field along the direction of current flow, dl, because of the presence of the multiplier $\sin(dl \cdot L_{pa})$.

By integrating eq. 2.93 along a path L, we obtain an expression for the magnetic field caused by a filamentary current in a closed loop:

$$H(a) = \frac{I}{4\pi} \int_L \frac{dl \times L_{pa}}{L_{pa}^3}$$

(2.94)

Let us write the expression for current I as a product:

$$I = j \cdot dS$$

where j is the current density and dS is the cross-sectional area of an elementary tube. Then one can write:

$$I(dl \times L_{pa}) = (j \cdot dS)(dl \times L_{pa}) = (dl \cdot dS)(j \times L_{pa}) = (j \times L_{pa})dV$$

inasmuch as:

$$d\boldsymbol{l} \cdot d\boldsymbol{S} = dV$$

Therefore, the magnetic field caused by currents distributed through a volume of conducting material is written as:

$$H(a) = \frac{1}{4\pi} \int_V \frac{\boldsymbol{j} \times \boldsymbol{L}_{\mathrm{pa}}}{L_{\mathrm{pa}}^3} \, dV \qquad (2.95)$$

In the same way that we considered the distribution of charges on the surface rather than in the volume, let us now assume that we have only surface currents, flowing through relatively thin conductive zones as shown in Fig. 2.27. In this case, we replace the product $\boldsymbol{j}dV$ by a surface element $\boldsymbol{i}dS$ representing the surface currents ($\boldsymbol{i}$ is the surface current density), and we obtain the following equation for the resulting magnetic field:

$$H(a) = \frac{1}{4\pi} \int_S \frac{\boldsymbol{i} \times \boldsymbol{L}_{\mathrm{pa}}}{L_{\mathrm{pa}}^3} \, dS \qquad (2.96)$$

Applying the principle of superposition, we obtain an expression for the combined effects of magnetic fields caused by linear, volume, and surface currents:

$$H(a) = \frac{1}{4\pi} \int_V \frac{\boldsymbol{j} \times \boldsymbol{L}_{\mathrm{pa}}}{L_{\mathrm{pa}}^3} \, dV + \frac{1}{4\pi} \int_S \frac{\boldsymbol{i} \times \boldsymbol{L}_{\mathrm{pa}}}{L_{\mathrm{pa}}^3} \, dS + \frac{1}{4\pi} \sum I_i \oint_{L_i} \frac{d\boldsymbol{l} \times \boldsymbol{L}_{\mathrm{pa}}}{L_{\mathrm{pa}}^3} \qquad (2.97)$$

In accord with this expression, we can state that the magnetic field is specified completely by distribution of currents analogously to the way in which the distribution of electric charges defines the constant electric field.

We should remember that the experimental investigations of the magnetic field behavior were carried out using closed loops of current, and therefore, eq. 2.92 which deals with current flowing only between end points is actually an assumption, but fortunately it appears to provide the correct results.

We should note again that eq. 2.97 was developed from experiments in which the direct currents were used. However, as will be shown later, the equation is valid also for quasi-stationary fields, the type which is of most interest in inductive mineral prospecting. In the practical system of units, the magnetic field intensity is measured in amperes per meter. By comparing eqs. 1.9 and 2.97 we can see that the calculation of a magnetic field will usually be a more complicated procedure than the determination of the electric field due to charges because of the presence of the vector product in the calculations. In order to simplify such calculations, as well as to derive some useful relationships for the magnetic field, we can introduce an auxiliary function called the vector potential for the magnetic field due to constant currents. With this in mind, we will show that the magnetic field

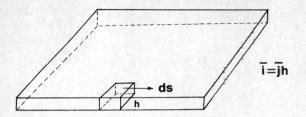

Fig. 2.27. Illustration of surface currents flowing through relatively thin conductive zones.

can be represented as being the curl of some vector function of space. In particular, the following identities will be used:

$$\frac{L_{pa}}{L_{pa}^3} = \text{grad}_p \frac{1}{L_{pa}} = -\text{grad}_a \frac{1}{L_{pa}} \qquad (2.98)$$

or, in operational notation:

$$\frac{L_{pa}}{L_{pa}^3} = \nabla_p \frac{1}{L_{pa}} = -\nabla_a \frac{1}{L_{pa}} \qquad (2.99)$$

Inasmuch as the function L_{pa} can vary as the points p and a are changed, one can consider the gradient of this function when either the point a or the point p is fixed. As an example, in cartesian coordinates we have:

$$\text{grad}_p \frac{1}{L_{pa}} = \frac{\partial}{\partial x_p} \frac{1}{L_{pa}} i + \frac{\partial}{\partial y_p} \frac{1}{L_{pa}} j + \frac{\partial}{\partial z_p} \frac{1}{L_{pa}} k$$

where i, j, and k are unit vectors directed along the x, y, and z axes, and:

$$L_{pa} = [(x_a - x_p)^2 + (y_a - y_p)^2 + (z_a - z_p)^2]^{1/2}$$

Carrying out the differentiation, we obtain eq. 2.98. Substituting 2.98 into 2.95, we have:

$$H(a) = \frac{1}{4\pi} \int_V \left(j \times \nabla_p \frac{1}{L_{pa}} \right) dV = \frac{1}{4\pi} \int_V \left(\nabla_a \frac{1}{L_{pa}} \times j \right) dV \qquad (2.100)$$

because the vector product changes sign when the relative position of the two vectors is changed.

Now we will make use of the following identity:

$$\nabla_a \times \frac{j}{L_{pa}} = \nabla_a \frac{1}{L_{pa}} \times j + \frac{\nabla_a j}{L_{pa}} \qquad (2.101)$$

which can be obtained by use of the vector identity:

$$\nabla \times (uv) = u \nabla \times v + \nabla u \times v$$

Applying eq. 2.101, we can write eq. 2.100 as:

$$H(a) = \frac{1}{4\pi} \int_V \left(\nabla_a \times \frac{j}{L_{pa}} \right) dV - \frac{1}{4\pi} \int_V \frac{(\nabla_a \times j)}{L_{pa}} dV \tag{2.102}$$

The current density j is a function of the location of the point p and it does not depend on the location of the observation point a. Therefore, the integrand of the second integral is zero and:

$$H(a) = \frac{1}{4\pi} \int_V \mathrm{curl}_a \frac{j(p)}{L_{pa}} dV \tag{2.103}$$

Because the integration and differentiation indicated in eq. 2.103 are carried out with respect to the two mutually independent points, p and a, we can interchange the order of operations and obtain:

$$H(a) = \mathrm{curl}_a \frac{1}{4\pi} \int_V \frac{j(p)}{L_{pa}} dV = \mathrm{curl}_a A \tag{2.104}$$

where

$$A = \frac{1}{4\pi} \int_V \frac{j(p)}{L_{pa}} dV \tag{2.105}$$

Thus, the magnetic field H caused by current flow can be expressed in terms of the vector potential A defined in eq. 2.105. The field A is more simply related to the distribution of current than is the magnetic field.

We can derive expressions for the vector potential A directly from eq. 2.105, for either surface or linear current flow. Making use of the obvious relationships:

$$jdV = idS \quad \text{or} \quad jdV = Idl$$

we have:

$$A = \frac{1}{4\pi} \int_S \frac{idS}{L_{pa}} \quad \text{and} \quad A = \frac{I}{4\pi} \int_L \frac{dl}{L_{pa}} \tag{2.106}$$

Applying the principle of superposition, we obtain the following expression for the vector potential caused by volume, surface and linear current:

$$A = \frac{1}{4\pi} \int_V \frac{jdV}{L_{pa}} + \frac{1}{4\pi} \int_S \frac{idS}{L_{pa}} + \frac{1}{4\pi} \sum_{i=1}^{N} I_i \oint_{L_i} \frac{dl}{L_{pa}} \tag{2.107}$$

The components of the vector potential can be derived directly from this last expression. For example, in cartesian coordinates, they would be:

$$A_x = \frac{1}{4\pi} \int_V \frac{j_x dV}{L_{pa}} + \frac{1}{4\pi} \int_S \frac{i_x dS}{L_{pa}} + \frac{1}{4\pi} \sum_{i=1}^{N} I_i \int_{L_i} \frac{dx}{L_{pa}}$$

$$A_y = \frac{1}{4\pi} \int_V \frac{j_y dV}{L_{pa}} + \frac{1}{4\pi} \int_S \frac{i_y dS}{L_{pa}} + \frac{1}{4\pi} \sum_{i=1}^{N} I_i \int_{L_i} \frac{dy}{L_{pa}} \tag{2.108}$$

$$A_z = \frac{1}{4\pi} \int_V \frac{j_z dV}{L_{pa}} + \frac{1}{4\pi} \int_S \frac{i_z dS}{L_{pa}} + \frac{1}{4\pi} \sum_{i=1}^{N} I_i \int_{L_i} \frac{dz}{L_{pa}} \tag{2.108}$$

Similar expressions can be written for the vector potential components in other systems of coordinates.

It can be seen from eq. 2.108 that if current flows along a single straight line path, the vector potential has but a single component parallel to this line. Also, when currents are situated in a single plane, the vector potential A is parallel to this plane at every point in space. Later, we will consider several examples illustrating the behavior of vector potential A, but at this point, we will derive several useful relationships that characterize both the vector potential function and the magnetic field. First, we will determine the divergence of the vector potential, A. In accord with eq. 2.105:

$$\mathrm{div_a}\, A(a) = \mathrm{div_a}\, \frac{1}{4\pi} \int_V \frac{j(p)dV}{L_{pa}}$$

Because the differentiation and integration in this expression are performed at various points in space, we can change the order of operation so that we have:

$$\mathrm{div_a}\, A(a) = \frac{1}{4\pi} \int_V \mathrm{div_a}\, \frac{j(p)}{L_{pa}} dV \tag{2.109}$$

The volume over which the integration is carried includes all the currents that are present, and therefore, the volume can be enclosed by a surface outside of which there are no currents. The normal component of current density on this surface S must be zero:

$$j_n = 0, \quad \mathrm{on}\ S \tag{2.110}$$

The integrand in eq. 2.18 can be written as:

$$\nabla_a \frac{j}{L_{pa}} = \frac{\nabla_a j}{L_{pa}} + j \cdot \nabla_a \frac{1}{L_{pa}} = j \cdot \nabla_a \frac{1}{L_{pa}},$$

because $\mathrm{div_a}\, j(p) = 0$. This last expression can also be written as:

$$j \cdot \nabla_a \frac{1}{L_{pa}} = -j \cdot \nabla_p \frac{1}{L_{pa}} = -\nabla_p \cdot \frac{j}{L_{pa}} + \frac{\nabla_p \cdot j}{L_{pa}} = -\mathrm{div_p} \frac{j}{L_{pa}} + \frac{1}{L_{pa}} \mathrm{div_p}\, j$$

Inasmuch as total charge must be conserved (2.83), it follows that:

$$\mathrm{div_p}\, j = 0$$

and therefore:

$$j \cdot \nabla_a \frac{1}{L_{pa}} = - \text{div}_p \frac{j}{L_{pa}} \tag{2.111}$$

Thus, eq. 2.109 can be written as:

$$\text{div} \, A = -\frac{1}{4\pi} \int\limits_V \text{div}_p \frac{j}{L_{pa}} \, dV \tag{2.112}$$

Both the integration and the differentiation operations carried out on the right-hand side of eq. 2.112 are performed with respect to the same point p, so that one can apply Gauss's theorem with the result:

$$\text{div} \, A = -\frac{1}{4\pi} \int\limits_V \text{div}_p \frac{j}{L_{pa}} \, dV = -\frac{1}{4\pi} \oint\limits_S \frac{j \cdot dS}{L_{pa}} = -\frac{1}{4\pi} \oint\limits_S \frac{j_n dS}{L_{pa}}$$

Considering that the normal component of current density j_n is zero on the surface S surrounding the currents, we obtain:

$$\text{div} \, A = 0 \tag{2.113}$$

That is, the flux lines for the vector potential field A are closed.

In following chapters when we consider electromagnetic fields, several types of vector potentials will be introduced, and in most cases, the divergence for these fields will not be zero.

In the previous paragraph, it was shown that the potential for the electric field, U, satisfies Poisson's equation:

$$\nabla^2 U = - \delta/\epsilon_0$$

which has the solution of the form:

$$U = \frac{1}{4\pi\epsilon_0} \int\limits_V \frac{\delta dV}{L_{pa}} \tag{2.114}$$

According to eq. 2.108, the components of vector potential expressed in cartesian coordinates will also satisfy Poisson's equation:

$$\nabla^2 A_x = -j_x; \quad \nabla^2 A_y = -j_y; \quad \nabla^2 A_z = -j_z \tag{2.115}$$

Multiplying each of these equations by the corresponding unit vector, i, j, or k, and adding the three equations together, we obtain an equation for the total vector potential A:

$$\nabla^2 A = -j \tag{2.116}$$

The forms of eqs. 2.113 and 2.116 reflect the basic features of vector potential A, and permit one to derive equations for the behavior of the magnetic field.

Now, making use of eq. 2.104, we discover that the divergence of the magnetic field is also zero. In fact, we have the following identity:

$$\text{div } H = \text{div curl } A \tag{2.117}$$

As is well known in vector algebra, the right-hand side of eq. 2.117 is identically zero. Therefore:

$$\text{div } H = 0 \tag{2.118}$$

This can be interpreted physically as indicating that magnetic charges do not exist and that magnetic flux lines are closed.

Applying Gauss's theorem, we obtain the integral form of this equation:

$$\oint_S H \cdot dS = 0 \tag{2.119}$$

That is, the total flux of the magnetic field through an enclosed surface is zero.

Making use of eq. 2.119 in calculating the magnetic flux through an elementary cylindrical surface as shown in Fig. 2.28, we have:

$$H_n^{(2)} = H_n^{(1)} \tag{2.120}$$

As indicated by eq. 2.120, the normal component of the magnetic field is always a continuous function of the spatial variables at an interface of a nonmagnetic medium. This behavior is in contrast to that of the normal component of electric field, specified in eq. 2.88. It is obvious that surface magnetic charge cannot exist.

Thus, we have obtained three forms of the first equation describing the magnetic field caused by constant currents:

$$\text{div } H = 0, \quad \oint_S H \cdot dS = 0, \quad H_n^{(2)} - H_n^{(1)} = 0 \tag{2.121}$$

Each of them expresses the same fact, that is, that magnetic charges do not exist. Equations 2.121 have been derived by algebraic manipulation of the Biot-Savart law for direct current behavior, but in actuality, they are still

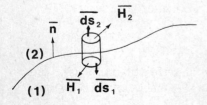

Fig. 2.28. Calculation of the magnetic flux through an elementary cylindrical surface.

valid for alternating electromagnetic fields, and are in effect, the fourth of the Maxwell equations.

At this point, we will develop a second equation for the magnetic field. Making use of the identity:

$$\text{curl curl } M = \text{grad div } M - \nabla^2 M$$

in eq. 2.104, we have:

$$\text{curl } H = \text{curl curl } A = \text{grad div } A - \nabla^2 A$$

Considering that:

$$\text{div } A = 0$$

we have:

$$\text{curl } H = -\nabla^2 A = j$$

Thus, a second equation for the magnetic field is:

$$\text{curl } H = j \tag{2.122}$$

Physically, this equation expresses the fact that currents are the source for magnetic fields. Making use of Stoke's theorem, we can rewrite this equation in a second form, which is Ampere's law:

$$\int_L H \cdot dl = \oint_S \text{curl } H \cdot dS = \int_S j \cdot dS = I$$

or $\tag{2.123}$

$$\oint_L H \cdot dl = I$$

where I is the current flowing through the surface S bounded by the path L (see Fig. 2.29). It should be clear that the mutual orientation of the vectors dl and dS is not arbitrary, but must be taken in accord with the right-hand rule. Thus, the circulation of the magnetic field is defined by the amount of current I piercing the surface surrounded by the contour L as shown in Fig. 2.29, and it does not depend on currents which are located outside the perimeter of this area.

It should be obvious though from the fact that the circulation is zero, it does not necessarily follow that the magnetic field is also zero at every point along L. It is appropriate at this point to emphasize that the path L can cut through media with differing electrical properties. For example, applying eq. 2.113 along a parameter L enclosing an interface between two media (see Fig. 2.30), we obtain:

$$\oint H \cdot dl = H_t^{(2)} dl - H_t^{(1)} dl + j dl\, dh$$

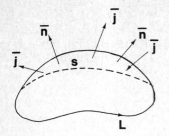

Fig. 2.29. Definition of the surface S bounded by the path L used in eq. 2.123.

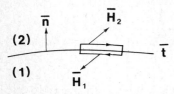

Fig. 2.30. Path for the application of the boundary condition in eq. 2.113.

Letting dh tend to zero, we have:

$$H_t^{(2)} - H_t^{(1)} = 0 \qquad (2.124)$$

We see that the tangential components of magnetic field intensity are a continuous function of position. At this point, we have derived three forms of the second equation for the magnetic field caused by direct currents, showing that the circulation of the magnetic field is defined by the current flux through any surface bounded by a path of integration. These forms are:

$$\oint_l H \cdot dl = I, \quad \mathrm{curl}\, H = j, \quad H_t^{(2)} - H_t^{(1)} = 0 \qquad (2.125)$$

It is interesting to note that the last of these equations is valid also for any alternating current field, and it is usually taken to be a boundary condition for the magnetic field. On occasion, it is convenient to assume that there is a surface current density, i, at an interface. Then, repeating the operations carried out above, we find that the tangential component of the magnetic field is discontinuous at such an interface:

$$H_t^{(2)} - H_t^{(1)} = i_l \qquad (2.126)$$

where t and l represent two mutually perpendicular directions tangent to the interface. Referring to eqs. 2.55 and 2.124, we can see that the continuity of tangential components of the electric and magnetic fields follows directly from Coulomb's and Biot-Savart's laws, respectively.

Although the first two equations in eq. 2.125:

$$\oint \boldsymbol{H} \cdot \mathrm{d}\boldsymbol{l} = I \quad \text{and} \quad \operatorname{curl} \boldsymbol{H} = \boldsymbol{j}$$

were derived from expressions for magnetic fields caused by direct-current flow, they remain valid for quasi-stationary electromagnetic fields such as are used in minerals prospecting. Thus, the Biot-Savart and Ampere's laws apply for quasi-stationary magnetic field behavior.

At this point, it may be fruitful to illustrate the use of these equations in terms of several examples.

Example 1: The magnetic field of a straight wire line

Consider a current I flowing through a vertical line as shown in Fig. 2.31. We will define the magnetic field at an arbitrary point, a, in a cylindrical system of coordinates, r, ϕ, z, with the origin situated on the current-carrying line.

Starting with the Biot-Savart law, we can say that the magnetic field has axial symmetry and is represented by a single component H_ϕ. From the principle of superposition, we can say that the total field is the sum of a number of fields contributed by current elements, $I\mathrm{d}z$. Then we have:

$$H_\phi = \frac{I}{4\pi} \int_{z_1}^{z_2} \frac{\mathrm{d}z \times L_{\mathrm{pa}}}{L_{\mathrm{pa}}^3} \tag{2.127}$$

where

$$L_{\mathrm{pa}} = (r^2 + z^2)^{1/2}$$

It should be clear that absolute value of $\mathrm{d}z \times L_{\mathrm{pa}}$ is:

$$|\mathrm{d}z \times L_{\mathrm{pa}}| = \mathrm{d}z \, L_{\mathrm{pa}} \sin(\mathrm{d}z \, L_{\mathrm{pa}}) = \mathrm{d}z \, L_{\mathrm{pa}} \sin\beta = \mathrm{d}z \, L_{\mathrm{pa}} \cos\alpha$$

Thus:

$$H_\phi = \frac{I}{4\pi} \int_{z_1}^{z_2} \frac{\mathrm{d}z}{L_{\mathrm{pa}}^2} \cos\alpha \tag{2.128}$$

Inasmuch as:

$$z = r \tan\alpha$$

we have:

$$\mathrm{d}z = r \sec^2\alpha \, \mathrm{d}\alpha \quad \text{and} \quad L_{\mathrm{pa}}^2 = r^2(1 + \tan^2\alpha) = r^2 \sec^2\alpha$$

Substituting these expressions into eq. 2.128, we obtain:

$$H_\phi(a) = \frac{I}{4\pi r} \int_{\alpha_1}^{\alpha_2} \cos\alpha \, \mathrm{d}\alpha = \frac{I}{4\pi r}(\sin\alpha_2 - \sin\alpha_1)$$

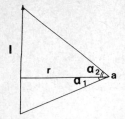

Fig. 2.31. Current flowing through a vertical line.

The final expression for the magnetic field caused by current flowing along a straight line has the form:

$$H_\phi(a) = \frac{I}{4\pi r}(\sin\alpha_2 - \sin\alpha_1) \qquad (2.129)$$

where α_1 and α_2 are the angles subtended by the radii from point a to the ends of the line. It is readily seen that Ampere's law cannot be used here because the current flow is not closed.

Next, suppose that the current carrying line is infinitely long so that the two angles α_2 and α_1 have the values $\pi/2$ and $-\pi/2$, respectively. Then:

$$H_\phi(a) = I/2\pi r \qquad (2.130)$$

In this case, one might think that the current is closed at infinity, and Ampere's law will instantly give the correct result. Considering the circuit to be closed and in view of the axial symmetry, we have:

$$\oint H \cdot dl = H_\phi \oint dl = 2\pi r H_\phi = I$$

and hence

$$H_\phi = I/2\pi r$$

In the case of a long line which is only semi-infinite, $\alpha_1 = 0$ and $\alpha_2 = \pi/2$, one cannot apply Ampere's law, but using eq. 2.129, we obtain a field which is half that for the case of an infinitely long current-carrying wire; that is:

$$H_\phi = I/4\pi r \qquad (2.131)$$

Now, suppose that $\alpha_2 = \alpha$ and $\alpha_1 = -\alpha$. Then, in accord with eq. 2.129, we have:

$$H_\phi(r) = \frac{I}{2\pi r}\sin\alpha = \frac{I}{2\pi r}\frac{l}{(r^2 + l^2)^{1/2}} \qquad (2.132)$$

where $2l$ is the length of the current-carrying line. If l is significantly greater than the distance, r, the right-hand side of eq. 2.132 can be expanded in a power series in terms of the term $(r/l)^2$. Then we obtain:

$$H_\phi = \frac{I}{2\pi r} \frac{1}{(1 + r^2/l^2)^{1/2}} \approx \frac{I}{2\pi r} \left[1 - \frac{1}{2} \left(\frac{r}{l} \right)^2 + \frac{3}{8} \left(\frac{r}{l} \right)^4 - \cdots \right]$$

We see that if the length of the current line, $2l$, is four to five times larger than the separation r, the resulting field is practically the same as that from an infinitely long current-carrying wire.

Example 2: Magnetic field of a current flowing in a rectangular loop

The equation giving the magnetic field from a current-carrying line with finite length (eq. 2.129) can be used to calculate the magnetic field caused by current flowing in a rectangular loop, such as that shown in Fig. 2.32. The calculation is particularly simple for the case in which the observation point is situated in the same plane as the loop. The field from the loop is the sum of four contributions to the field for the four sides of the loop:

$$H = \sum_{i=1}^{4} H^{(i)} \tag{2.133}$$

where

$$H^{(1)} = \frac{I}{4\pi r_1} (\sin \alpha_2 - \sin \alpha_1) \varphi_{01}$$

$$H^{(2)} = \frac{I}{4\pi r_2} (\sin \alpha_4 - \sin \alpha_3) \varphi_{02}$$

$$\tag{2.134}$$

$$H^{(3)} = \frac{I}{4\pi r_3} (\sin \alpha_6 - \sin \alpha_5) \varphi_{03}$$

$$H^{(4)} = \frac{I}{4\pi r_4} (\sin \alpha_8 - \sin \alpha_7) \varphi_{04}$$

Here φ_{0i} is a unit vector normal to the plane defined by r and the i-th side of the loop.

When the observation point is situated in the same plane as the rectangular loop, all the terms in eq. 2.134 are directed perpendicularly to this plane, and we obtain a simple scalar sum:

$$H_n = \sum_{i=1}^{4} H_n^{(i)}$$

It should be noted that when the observation point lies outside the plane of the current loop, it is more convenient to make use of the vector potential A, and then differentiate the vector potential to calculate the magnetic field intensity.

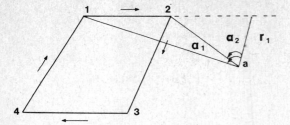

Fig. 2.32. Example of current flowing in a rectangular loop.

Example 3: Vector potential A and magnetic field H of the current flowing in a circular loop

Assume that the observation point is situated on the axis of a loop with a radius a, as shown in Fig. 2.33a. Then, in accord with eq. 2.106:

$$A = \frac{I}{4\pi} \oint_L \frac{dl}{L_{pq}}$$

Inasmuch as the distance L_{pq} is the same for all points on the loop, we have:

$$A = \frac{I}{4\pi L_{pq}} \oint dl$$

By definition, the sum of the elementary vectors, dl, along any closed path is zero. Therefore, the vector potential A on the z-axis of a circular current-carrying loop vanishes.

Now, we can calculate the magnetic field on the z-axis. From Fig. 2.33a, it can be seen that with a cylindrical system coordinate system, r, ϕ, z, each current element $I\,dl$ creates two field components, dH_z and dH_r. However, it is always possible to find two current elements, $I\,dl$, which contribute the same horizontal component or points at a z-axis, but of opposite sign. Therefore, the magnetic field has only a vertical component along the axis. As can be seen from Fig. 2.33a, we have:

$$dH_z = \frac{I}{4\pi} \frac{dl}{L^2} \frac{a}{L} = \frac{Ia\,dl}{4\pi L^3}$$

since

$$|dl \times L| = L\,dl \quad \text{and} \quad L = (a^2 + z^2)^{1/2}$$

Having integrated along a closed path, we finally obtain:

$$H_z = \frac{Ia2\pi a}{4\pi(a^2 + z^2)^{3/2}} = \frac{Ia^2}{2(a^2 + z^2)^{3/2}} = \frac{M}{2\pi(a^2 + z^2)^{3/2}} \tag{2.135}$$

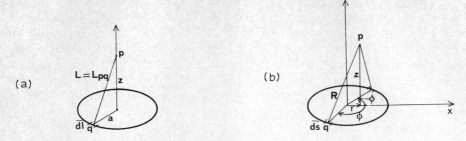

Fig. 2.33. a. Geometry for an observation point situated on the axis of a loop with radius a. b. Geometry for an observation point located arbitrarily with respect to the axis z.

where

$$M = I\pi a^2 = IS$$

with S being the area enclosed by the loop.

When the distance z is much greater than the radius of the loop, a, we obtain an expression appropriate for the magnetic field of a magnetic dipole situated on the z-axis. Neglecting a in comparison with z, we have:

$$H_z = M/2\pi z^3$$

We see that a relatively small current-carrying loop with a radius a creates the same magnetic field as a physical magnetic dipole having the moment $M = \pi a^2 I$, oriented along the z-axis.

It can be seen from eq. 2.135 that when the distance z is at least three times the radius a, the treatment of the loop as a magnetic dipole situated at the center of the loop contributes an error of no more than 5%. In the preceding section, this behavior has been proven only for points located on the z-axis, but in fact it remains valid for any arbitrary position of the observation point if the distance of the point from the loop is considerably greater than the radius of the loop.

Making use of eq. 2.135, let us explore the influence of loop radius, a, on the magnetic field beneath the loop along the z-axis. This will be useful in choosing the optimum radius for the current-carrying loop used in inductive mining prospecting.

As we see from eq. 2.135, for constant current and for small values of the ratio a/z, the field increases in proportion to a^2, as the radius increases, because the current loop behaves as though it were magnetic dipole. In the case in which a/z is much larger than unit, the magnetic field decreases in inverse proportion to a. Therefore, at some critical distance z, there is an optimum radius for the loop providing the maximum magnetic field at this point along the z-axis.

So far, we have considered the vector potential and the magnetic field only along the z-axis. We can now investigate a more general case. We will calculate the vector potential at any point p (see Fig. 2.33b). In view of the symmetry the vector potential A does not depend on the coordinate ϕ. Therefore, for simplicity, we can choose the point p in the x-z plane where $\phi = 0$. As can be seen from Fig. 2.33b, every pair of current-carrying elements, Idl, equally distant from p, and having coordinates ϕ and $-\phi$, create a vector potential dA directed perpendicularly to the x-z plane. Inasmuch as the whole loop can be represented as the sum of such pairs, we conclude that the vector potential A caused by the current-carrying loop has only the component, A_ϕ. Therefore, from eq. 2.106, it follows that:

$$A_\phi = \frac{I}{4\pi} \oint_L \frac{dl_\phi}{R} = \frac{I}{2\pi} \int_0^\pi \frac{a \cos\phi d\phi}{(a^2 + r^2 + z^2 - 2ar \cos\phi)^{1/2}} \tag{2.136}$$

where dl_ϕ is the component of dl in the ϕ direction and:

$$dl_\phi = a \cos\phi d\phi, \quad R = (a^2 + r^2 + z^2 - 2ar \cos\phi)^{1/2}$$

If the distance from the center of the current-carrying loop to the observation point is considerably greater than the loop radius, then:

$$R_0 = (r^2 + z^2)^{1/2} \gg a$$

and eq. 2.136 can be simplified so that we have:

$$A_\phi \approx \frac{Ia}{2\pi} \int_0^\pi \frac{\cos\phi d\phi}{(R_0^2 - 2ar \cos\phi)^{1/2}} \approx \frac{Ia}{2\pi R_0} \int_0^\pi \frac{\cos\phi d\phi}{[1 - (2ar/R_0^2) \cos\phi]^{1/2}} \approx$$

$$\approx \frac{Ia}{2\pi R_0} \int_0^\pi \left(1 + \frac{ar}{R_0^2} \cos\phi\right) \cos\phi d\phi =$$

$$= \frac{Ia}{2\pi R_0} \int_0^\pi \cos\phi d\phi + \frac{Ia^2 r}{2\pi R_0^3} \int_0^\pi \cos^2\phi d\phi$$

The first integral in this last equation is zero, so that we obtain:

$$A_\phi = \frac{Ia^2 r}{4R_0^3}$$

or

$$A_\phi = \frac{M}{4\pi R_0^2} \sin\theta \tag{2.137}$$

where M is the dipole moment given by $\pi a^2 I$ and $\sin\theta = r/R_0$.

Thus, at distances considerably greater than the loop radius, the vector potential and the corresponding components of the magnetic field intensity are the same as those for a physically real magnetic dipole with the moment M directed along the z-axis.

Returning to the general case and letting $\phi = \pi + 2\alpha$, we have:

$$d\phi = 2d\alpha \quad \text{and} \quad \cos\phi = 2\sin^2\alpha - 1$$

and, therefore:

$$A_\phi = \frac{aI}{\pi} \int_0^{\pi/2} \frac{(2\sin^2\alpha - 1)\,d\alpha}{[(a+r)^2 + z^2 - 4ar\sin^2\alpha]^{1/2}}$$

Introducing a new parameter:

$$k^2 = \frac{4ar}{[(a+r)^2 + z^2]} \tag{2.138}$$

and carrying out some fairly simple algebraic operations, we obtain:

$$A_\phi = \frac{kI}{2\pi}\left(\frac{a}{r}\right)^{1/2}\left[\left(\frac{2}{k^2} - 1\right)\int_0^{\pi/2}\frac{d\alpha}{(1 - k^2\sin^2\alpha)^{1/2}}\right.$$

$$\left. - \frac{2}{k^2}\int_0^{\pi/2}(1 - k^2\sin^2\alpha)^{1/2}\,d\alpha\right] = \tag{2.139}$$

$$= \frac{I}{\pi k}\left(\frac{a}{r}\right)^{1/2}[(1 - \tfrac{1}{2}k^2)K - E]$$

where K and E are the complete elliptical integrals of the first and second kind:

$$K(k) = \int_0^{\pi/2}\frac{d\alpha}{(1 - k^2\sin^2\alpha)^{1/2}}$$

$$E(k) = \int_0^{\pi/2}(1 - k^2\sin^2\alpha)^{1/2}\,d\alpha \tag{2.140}$$

Values for the elliptical integrals $K(k)$ and $E(k)$ are listed in Table 2.I.

Using the relationship between vector potential and the magnetic field as given in eq. 2.104 in cylindrical coordinates, we have:

$$H_r = -\frac{\partial A_\phi}{\partial z}, \quad H_z = \frac{1}{r}\frac{\partial}{\partial r}(rA_\phi), \quad H_\phi = 0$$

The derivatives for the elliptical integrals, as given in mathematical handbooks, are:

$$\frac{\partial K}{\partial k} = \frac{E}{k(1 - k^2)} - \frac{K}{k}, \quad \frac{\partial E}{\partial k} = \frac{E}{k} - \frac{K}{k}$$

TABLE 2.I

Values for elliptical integrals

k^2	$E(k^2)$	$K(k^2)$	k^2	$E(k^2)$	$K(k^2)$
0.00	1.57079 6327	1.57079 6326	0.25	1.46746 2209	1.68575 0354
0.01	1.56686 1942	1.57474 5561	0.26	1.46308 5873	1.69120 8199
0.02	1.56291 2645	1.57873 9912	0.27	1.45868 8155	1.69674 8620
0.03	1.55894 8244	1.58278 0342	0.28	1.45426 8698	1.70237 3977
0.04	1.55496 8546	1.58686 7847	0.29	1.44982 7128	1.70808 6731
0.05	1.55097 3352	1.59100 3453	0.30	1.44536 3064	1.71388 9448
0.06	1.54696 2456	1.59518 8221	0.31	1.44087 6115	1.71978 4808
0.07	1.54293 5653	1.59942 3244	0.32	1.43636 5871	1.72577 5609
0.08	1.53889 2730	1.60370 9654	0.33	1.43183 1919	1.73186 4778
0.09	1.53483 3465	1.60804 8619	0.34	1.42727 3821	1.73805 5373
0.10	1.53075 7637	1.61244 1348	0.35	1.42269 1133	1.74435 9597
0.11	1.52666 5017	1.61688 9090	0.36	1.41808 3394	1.75075 3802
0.12	1.52255 5369	1.62139 3137	0.37	1.41345 0127	1.75726 8504
0.13	1.51842 8454	1.62595 4829	0.38	1.40879 0839	1.76389 8388
0.14	1.51428 4027	1.63057 5548	0.39	1.40410 5019	1.77064 7323
0.15	1.51012 1831	1.63525 6732	0.40	1.39939 2139	1.77751 9371
0.16	1.50594 1612	1.63999 9865	0.41	1.39465 1652	1.78451 8804
0.17	1.50174 3101	1.64480 6490	0.42	1.38988 2992	1.79165 0116
0.18	1.49752 6026	1.64967 8205	0.43	1.38508 5568	1.79891 8039
0.19	1.49329 0109	1.65461 6667	0.44	1.38025 8774	1.80632 7559
0.20	1.48903 5058	1.65962 3598	0.45	1.37540 1972	1.81388 3936
0.21	1.48476 0581	1.66470 0785	0.46	1.37051 4505	1.82159 2726
0.22	1.48046 6375	1.66985 0086	0.47	1.36559 5691	1.82945 9798
0.23	1.47615 2126	1.67507 3429	0.48	1.36064 4814	1.83749 1363
0.24	1.47181 7514	1.68037 2822	0.49	1.35566 1135	1.84569 3998
			0.50	1.35064 3881	1.85407 4677

and

$$\frac{\partial k}{\partial z} = -\frac{zk^3}{4ar}, \quad \frac{\partial k}{\partial r} = \frac{k}{2r} - \frac{k^3}{4r} - \frac{k^3}{4a}$$

Therefore, after differentiation we have:

$$H_r = \frac{I}{2\pi} \frac{z}{r[(a+r)^2 + z^2]^{1/2}} \left[-K + \frac{a^2 + r^2 + z^2}{(a-r)^2 + z^2} E \right]$$

$$\tag{2.141}$$

$$H_z = \frac{I}{2\pi} \frac{1}{[(a+r)^2 + z^2]^{1/2}} \left[K + \frac{a^2 - r^2 - z^2}{(a-r)^2 + z^2} E \right]$$

Thus, in general the magnetic field caused by current flowing in a circular loop can be expressed in terms of elliptical integrals, but only along the axis of the loop or at great distances can the field be described in terms of elementary functions.

Let us note that the parameter k as indicated in eq. 2.138 varies over the range $0 \leqslant k \leqslant 1$. By carrying out operations similar to the derivations described above, it is a direct matter to show that the vector potential A as well as the magnetic field, H, caused by a closed current flowing within a plane along a path of arbitrary shape, are equivalent correspondingly to those, caused by a magnetic dipole with a moment $M = IS$ where S is the area enclosed by the current flow path, provided that the distance to the observation point from the current is much greater than the dimensions of the current flow pattern.

At this point, we have examined the magnetic fields contributed by direct current flowing in an infinitely long wire, and in rectangular and circular loops. However, the expressions obtained for these fields are quite useful in calculating quasi-stationary magnetic fields for these same sources. In electrical prospecting, these expressions yield for what is termed the primary magnetic field.

Example 4: The magnetic field caused by current flowing in the wire grounded at the surface of a horizontally layered medium (Fig. 2.34)

Using the principle of superposition, this pattern of current flow can be represented as the sum of three types of excitation, namely:

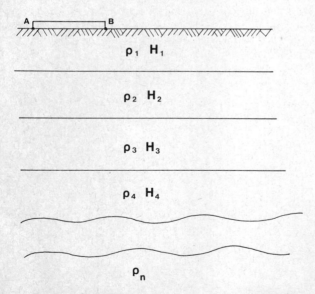

Fig. 2.34. Excitation of a magnetic field caused by current flowing in a grounded wire on the surface of a horizontally layered medium.

(1) A current flowing through a vertical infinitely long wire grounded at a point A.

(2) A current of an infinitely long wire grounded at a point B.

(3) A current in a nongrounded line, $\infty\,BA\,\infty$, situated in the vertical plane.

Inasmuch as the currents along the neighboring vertical lines have opposite directions, these three systems of currents are equivalent to a current-carrying wire grounded at the points A and B, as shown in Fig. 2.35.

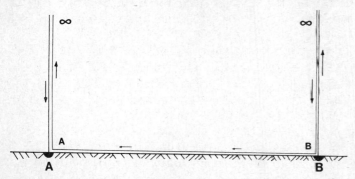

Fig. 2.35. Consideration of the total circuit when current flows into the ground at the ends of the grounded wire.

First of all, consider the magnetic field contributed by a vertical current-carrying wire grounded at the point A. Above the earth's surface, this current is linear, but beneath the surface, the current flows in a volume. In view of the axial symmetry, the calculation of this part of the magnetic field can be carried out easily by making use of Ampere's law (eq. 2.123):

$$\oint H \cdot dl = I$$

Consider any horizontal path L as being a circle with a radius r located above or on the earth's surface so that we obtain:

$$H_{\phi_1}^{(1)} 2\pi r_{Aa} = I$$

or

$$H_{\phi_1}^{(1)} = I/2\pi r_{Aa} \tag{2.142}$$

This expression is the same as that for an infinitely long current-carrying line (see eq. 2.130). According to eq. 2.131, the current in a half infinitely long line situated above the earth's surface creates on the surface a magnetic field which is:

$$H_{\phi_1} = I/4\pi r_{Aa}$$

Therefore, the currents in the conducting medium flowing from the electrode A generate a magnetic field which has only an tangential component which is:

$$H^A_{\phi_1} = \frac{I}{2\pi r_{Aa}} - \frac{I}{4\pi r_{Aa}} = \frac{I}{4\pi r_{Aa}} \tag{2.143}$$

It is interesting to note that these currents which are distributed through a volume create exactly the same field as the half infinitely long current-carrying wire directed vertically downwards with a current I. Also, it is important that the magnetic field on the earth's surface contributed by currents flowing from the electrode A into a horizontally stratified medium does not depend on the sequence of conductivities in a stratified earth. This follows directly from the property of axial symmetry, and is valid at every point in the upper half space including the surface of the earth.

Similarly the magnetic field for currents flowing from the conductive medium at the electrode B is:

$$H^B_{\phi_2} = \frac{I}{4\pi r_{Ba}} \tag{2.144}$$

Thus, the currents flowing in the stratified medium contribute only a horizontal component of the magnetic field, and in accord with eqs. 2.143 and 2.144, at the earth's surface, we have the following expression for the tangential component, H_t:

$$H_t = \frac{I}{4\pi r_{Aa}}\, \boldsymbol{\varphi}_{01} + \frac{I}{4\pi r_{Ba}}\, \boldsymbol{\varphi}_{02} \tag{2.145}$$

where r_{Aa} and r_{Ba} are the distances from the electrodes A and B to an observation point, a, respectively and $\boldsymbol{\varphi}_{01}$ and $\boldsymbol{\varphi}_{02}$ are unit vectors as defined in Fig. 2.36.

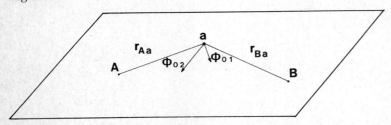

Fig. 2.36. Definition of vectors $\boldsymbol{\varphi}_{01}$ and $\boldsymbol{\varphi}_{02}$ as used in eq. 2.145.

Now we can consider the third and last element of the system, that is, the current flowing through the wire connecting the electrodes and situated on the earth's surface. According to the Biot-Savart law (eq. 2.92), this current causes only a vertical component for the magnetic field at points on the surface:

$$H_z = \frac{I}{4\pi}\, z_0 \int\limits_B^A \frac{dl}{L^2_{pa}} \sin\,(dl\, L_{pa}) \tag{2.146}$$

where z_0 is a unit vector perpendicular to the earth's surface, dl is an element of the current-carrying line, and L_{pa} is the distance from any element dl to the observation point. Equations 2.145 and 2.146 completely describe the magnetic field for direct current flow when the field is observed on the earth's surface, and the current-carrying line is grounded on a laterally homogeneous stratified earth. Let us notice, that this elegant method was suggested by Stefanescu (1935).

Suppose that the magnetic field is observed at distances considerably greater than the separation between the electrodes, A and B. In this case, the current flow path from A to B which originally was assumed to follow an arbitrary path, can be replaced by a straight line. In this case, the wire AB is considered to be an electric dipole (see Fig. 2.37).

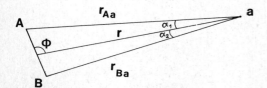

Fig. 2.37. Basis for trigonometric relationships listed in eq. 2.147.

In order to derive the appropriate expressions for the magnetic field from eqs. 2.145 and 2.146, we will use the following notation: r is the distance from the middle of the source dipole to the observation point; and ϕ is the angle between the line BA and the radius r.

As may be seen from Fig. 2.37, the following relationships exist:

$$r_{Aa} = \left[r^2 + \left(\frac{AB}{2} \right)^2 - ABr \cos \phi \right]^{1/2}$$

$$r_{Ba} = \left[r^2 + \left(\frac{AB}{2} \right)^2 + ABr \cos \phi \right]^{1/2} \qquad (2.147)$$

$$\frac{AB}{2r_{Aa}} = \frac{\sin \alpha_1}{\sin \phi}, \quad \frac{AB}{2r_{Ba}} = \frac{\sin \alpha_2}{\sin \phi}$$

Considering that $r \gg AB$ we can arrive at the following approximations:

$$r_{Aa} \approx r - \frac{AB}{2} \cos \phi, \quad r_{Ba} \approx r + \frac{AB}{2} \cos \phi$$

and

$$\sin \alpha_1 = \frac{AB \sin \phi}{2 \left(r - \frac{AB}{2} \cos \phi \right)}, \quad \sin \alpha_2 = \frac{AB \sin \phi}{2 \left(r + \frac{AB}{2} \cos \phi \right)}$$

or

$$\sin \alpha_1 \approx \sin \alpha_2 \approx \sin \alpha = \frac{AB}{2r} \sin \phi$$

It follows from Fig. 2.37 that:

$$H_r = -\frac{I}{4\pi r} \sin \alpha - \frac{I}{4\pi r} \sin \alpha = -\frac{2I}{4\pi r} \frac{AB}{2r} \sin \phi = -\frac{IAB}{4\pi r^2} \sin \phi$$

$$H_\phi = \frac{I}{4\pi r_{Aa}} \cos \alpha - \frac{I}{4\pi r_{Ba}} \cos \alpha = \frac{I}{4\pi} \left(\frac{1}{r_{Aa}} - \frac{1}{r_{Ba}} \right)$$

inasmuch as $\cos \alpha = 1$, so that finally we have:

$$H_\phi = \frac{IAB}{4\pi r^2} \cos \phi$$

From eq. 2.93, the vertical component of the magnetic field for the dipole is:

$$H_z = \frac{IAB}{4\pi r^2} \sin \phi$$

We see that the magnetic field of an electric dipole on the Earth's surface when the earth is stratified can be described by the equations:

$$H_r = -\frac{IAB}{4\pi r^2} \sin \phi, \quad H_\phi = \frac{IAB}{4\pi r^2} \cos \phi, \quad H_z = \frac{IAB}{4\pi r^2} \sin \phi \qquad (2.148)$$

We see that it depends on the current I, the distance AB, and the location of the observation point, but on no other parameters.

The analysis of the magnetic field caused by direct currents, as carried out for this example, vividly illustrates that measurements of the magnetic field on the earth's surface contain no information about the subsurface electrical structure in a horizontally stratified medium. However, one must recognize that this conclusion does not hold when the magnetic field is observed beneath the earth's surface. Also, measurements of the magnetic field on the earth's surface can be useful in detecting non-horizontal structures which are frequently of particular interest in minerals prospecting.

Again making use of the principle of superposition, the magnetic field contributed by current in a line with finite length and arbitrary shape grounded on the Earth's surface can be represented as being the sum of field contributions described by the eqs. 2.148.

In investigating alternating magnetic fields contributed by grounded wire sources, it will later be shown that the equations derived for the magnetic field (eqs. 2.97, 2.129, 2.130, 2.134, and 2.148) play a very important role when induction in the field is not particularly significant.

In concluding this section, it is appropriate to make the following comments:

(1) Direct current acts as a source for constant magnetic field according to the Biot-Savart law.

(2) Starting with the Biot-Savart law and making use of the principle of conservation of charge, we are able to write two equations describing the constant magnetic field, each of which can be represented in three forms:

$$\oint_L H \cdot dl = I, \quad \text{curl } H = j, \quad H_t^{(2)} - H_t^{(1)} = 0 \tag{A}$$

$$\oint_S H \cdot dS = 0, \quad \text{div } H = 0, \quad H_n^{(2)} - H_n^{(1)} = 0 \tag{B}$$

The equation set B reflects the fact that magnetic charges do not exist. This set is valid for alternating electromagnetic fields, and is the third of Maxwell's equations. Equation set A is valid for constant fields, and there will be different effects to consider for alternating fields. When the electromagnetic field is time varying, there will be other sources for the magnetic field in addition to the conduction currents, j. However, for the so-called quasi-stationary range of behavior, the influence of the additional magnetic field sources can be negligible and it will be convenient to use quasi-stationary electromagnetic field behavior for many problems in practical electrical prospecting.

2.3. THE POSTULATE OF CONSERVATION OF CHARGE, AND THE DISTRIBUTION OF CHARGES IN A CONDUCTING MEDIUM

This section will show that for certain conditions electric charges can exist in a conducting medium. In order to explore this problem, we will make use of the postulate of conservation of charge for time-varying fields:

$$\oint_S j \cdot dS = -\frac{\partial e}{\partial t} \tag{2.149}$$

where j is the current density at any point on an arbitrary closed surface S as shown in Fig. 2.38, and e is the charge distributed within the volume V surrounded by the surface S, and $\partial e / \partial t$ is the derivative of the charge with respect to time. The scalar product:

$$j \cdot dS = j_n \, dS$$

represents the amount of charge crossing an element of surface dS during a time period of one second. Similarly, the integral:

$$\oint_S j \cdot dS = \oint_S j_n \, dS$$

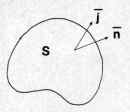

Fig. 2.38. Current density at an arbitrary point on an arbitrary closed surface S.

indicates the electric charge flux through the surface S during a time period of one second also. In general, at some points on the surface S, the vector j can be directed outwards, while at other points, it can be directed inwards. The current density flux given by eq. 2.149 is usually the algebraic sum of positive and negative fluxes through the surface S. For example, if the flux:

$$\oint_S j \cdot dS$$

is positive, the physical meaning is that during any time interval, a certain amount of charge leaves the volume V, and the derivative $\partial e/\partial t$ is negative, that is, the volume charge, e, decreases. In the opposite case, when the total flux of j is negative, the derivative $\partial e/\partial t$ is positive and the amount of charge contained in the volume V increases with time. Moreover, one can imagine a case when the positive and negative fluxes through a closed surface are equal, and the total flux will be zero. Then the derivative $\partial e/\partial t$ vanishes, and the amount of charge contained within the volume does not change with time.

We will now write eq. 2.149 in various forms which can be used in further applications. Applying Gauss's theorem:

$$\oint_S A \cdot dS = \int_V \operatorname{div} A \, dV$$

at points in the medium where the divergence for the vector j exists, we have:

$$\oint_S j \cdot dS = \int_V \operatorname{div} j \, dV = -\frac{\partial e}{\partial t} = -\frac{\partial}{\partial t} \int_V \delta \, dV = -\int_V \frac{\partial \delta}{\partial t} \, dV$$

where δ is the charge density. Thus:

$$\int_V \left(\operatorname{div} j + \frac{\partial \delta}{\partial t} \right) dV = 0$$

or

$$\operatorname{div} j = -\frac{\partial \delta}{\partial t}$$

(2.150)

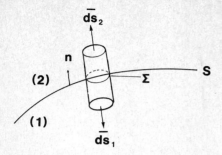

Fig. 2.39. Presence of a charge density Σ on a surface S.

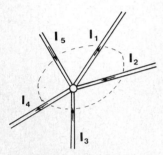

Fig. 2.40. A system of linear or almost linear currents.

This last equation is the differential form of eq. 2.149, and is valid at points where current density is a continuous function of the spatial variables. It has the same physical meaning as does eq. 2.149, but unlike that equation, it describes the relationship between currents and charges in the immediate neighborhood of a single point.

Now assume that some surface has a surface charge with a density Σ as shown in Fig. 2.39. Defining the flux of current density through an elementary cylindrical surface as shown in this figure, we obtain a surface analog to eqs. 2.150:

$$j_n^{(2)} - j_n^{(1)} = -\frac{\partial \Sigma}{\partial t} \tag{2.151}$$

Thus, the difference between normal components of current density on either side of a surface is equal to the rate of change of the surface charge density, taken with a negative sign. In eq. 2.151, the normal vector n is directed from side (1) to side (2).

Finally, assume that we have a system of linear or almost linear currents as shown in Fig. 2.40. Making use of eq. 2.149, we have:

$$\oint_S \boldsymbol{j} \cdot \mathrm{d}\boldsymbol{S} = \sum_{i=1}^{N} \oint \boldsymbol{j} \cdot \mathrm{d}\boldsymbol{S} = \sum_{i=1}^{N} I_i = -\frac{\partial e}{\partial t}, \text{ i.e.}$$

$$\sum_{i=1}^{N} I_i = -\frac{\partial e}{\partial t} \tag{2.152}$$

where I_i is the current passing through the ith current tube and taken with the appropriate sign, and e is the charge at the point a where all the tubes intersect. We now have four forms for the postulate of conservation of charge:

$$\oint_S \boldsymbol{j} \cdot \mathrm{d}\boldsymbol{S} = -\frac{\partial e}{\partial t}, \quad \operatorname{div} \boldsymbol{j} = -\frac{\partial \delta}{\partial t}$$

$$j_n^{(2)} - j_n^{(1)} = -\frac{\partial \Sigma}{\partial t}, \quad \sum_{i=1}^{N} I_i = -\frac{\partial e}{\partial t} \tag{2.153}$$

It should be emphasized that the first of these equations is the most general, being applicable everywhere. The second equation expressed in terms of derivatives of $\boldsymbol{j}$ can be used only at points where the current density, $\boldsymbol{j}$, is a continuous function of the spatial variables. The third equation describes the behavior of the normal component of current density at interfaces which carry a surface charge, and the fourth expression is appropriate for a system of linear currents. The equations listed in eq. 2.153 are extremely useful in determining at which places in a conductive medium electric charges will accumulate, and will provide a means for determining the charge density.

It is convenient to start our investigation of this problem with a simple case such as a conductive medium in which the electromagnetic field does not depend on time, and therefore all derivatives with respect to time will be zero. Correspondingly, we will repeat some results obtained in the first paragraph. Then, eqs. 2.153 take the form:

$$\oint_S \boldsymbol{j} \cdot \mathrm{d}\boldsymbol{S} = 0 \qquad \operatorname{div} \boldsymbol{j} = 0$$

$$\text{if} \quad \frac{\partial e}{\partial t} = \frac{\partial \delta}{\partial t} = \frac{\partial \Sigma}{\partial t} = 0 \tag{2.154}$$

$$j_n^{(2)} - j_n^{(1)} = 0 \quad \sum I_i = 0$$

Thus, for a constant field, the flux of current density through a closed surface is always zero; that is, the amount of charge arriving in a volume V is exactly equal to the amount of charge leaving this volume. In the case of an interface, the normal component of current density will be a continuous function. It is important to note that the last equation in the group 2.154 is Kirchoff's law for currents.

In order to determine the distribution of volume charges, we can use equations derived previously:

$$\operatorname{div} j = 0, \quad \operatorname{div} E = \delta/\epsilon_0 \tag{2.155}$$

along with Ohm's law:

$$j = \sigma E$$

Here, E is the electric force caused by the existence of charges.

In accord with eq. 2.155 we have:

$$\operatorname{div} j = \operatorname{div} \sigma E = E \operatorname{grad} \sigma + \sigma \operatorname{div} E = 0$$

Whence

$$\operatorname{div} E = -\frac{E \cdot \operatorname{grad} \sigma}{\sigma}$$

Comparing this result with the second equation in set 2.155, we have:

$$\delta = -\epsilon_0 E \cdot \frac{\operatorname{grad} \sigma}{\sigma} = -\epsilon_0 \frac{\nabla \sigma}{\sigma} \cdot E \tag{2.156}$$

Thus, when the current flows through a conducting medium, electric charges arise at places where the medium is nonuniform, providing that the electric field has a component in the direction of grad σ. The sign of the volume charge depends on the mutual orientations of the electric field E and the gradient σ. Electric charges will not appear at points where the medium is uniform in properties, and so:

$$\operatorname{div} E = 0 \tag{2.157}$$

We will now derive expressions for the surface charge starting from the two equations:

$$j_n^{(2)} - j_n^{(1)} = 0, \quad E_n^{(2)} - E_n^{(1)} = \Sigma/\epsilon_0 \tag{2.158}$$

where Σ is the surface charge density. Let us write the first equation in 2.158 in the form:

$$\sigma_2 E_n^{(2)} - \sigma_1 E_n^{(1)} = \tfrac{1}{2}[(\sigma_2 + \sigma_1)(E_n^{(2)} - E_n^{(1)}) + (\sigma_2 - \sigma_1)(E_n^{(2)} + E_n^{(1)})] = 0$$

Making use of the second equation of 2.158, we have:

$$(\sigma_2 + \sigma_1)\frac{\Sigma}{2\epsilon_0} + (\sigma_2 - \sigma_1)E_n^{av} = 0$$

where

$$E_n^{av} = (E_n^{(1)} + E_n^{(2)})/2$$

is the average magnitude of the normal component of the electric field at a point on the interface. Thus, we have the following expression for the surface charge density:

$$\Sigma = -2\epsilon_0 \frac{\sigma_2 - \sigma_1}{\sigma_2 + \sigma_1} E_n^{av} = -2\epsilon_0 K_{12} E_n^{av} \tag{2.159}$$

where

$$K_{12} = (\sigma_2 - \sigma_1)/(\sigma_2 + \sigma_1) \tag{2.160}$$

or

$$\Sigma = 2\epsilon_0 \frac{\rho_2 - \rho_1}{\rho_2 + \rho_1} E_n^{av} \tag{2.161}$$

The quantities ρ_1 and ρ_2 are the resistivities for the two media.

The normal component of the field on either side of the interface can be written as:

$$E_n^{(1)}(p) = E_n^0(p) + E_n^{s-p}(p) - \Sigma(p)/2\epsilon_0$$
$$E_n^{(2)}(p) = E_n^0(p) + E_n^{s-p}(p) + \Sigma(p)/2\epsilon_0 \tag{2.162}$$

where

$$E_n^0(p) + E_n^{s-p}(p)$$

is the normal component of the field at a point p, contributed by all charges except the charge at this point. It follows from eq. 2.162 that:

$$E_n^{av} = E_n^0 + E_n^{s-p} \tag{2.163}$$

where the normal is directed from the reverse side (1) to the top side (2) of the interface.

We see that the charge density which arises at the interface is directly proportional to the normal component of the field, E_n^{av}, with the constant of proportionality being represented by the symbol K_{12}. As has been shown, the coefficient of proportionality, K_{12} can vary within the range:

$$-1 \leqslant K_{12} \leqslant +1 \tag{2.164}$$

We should note that due to the presence of the surface electric charge, the normal component of current density is a continuous function of the spatial variable, while the normal component of the electric field has a discontinuity caused by the presence of the surface charge.

Let us now consider a general case in which the electromagnetic field varies with time. In determining the charge density, we can make use of eqs. 2.150 and 2.43:

$$\text{div } j = -\partial\delta/\partial t, \quad \text{div } E = \delta/\epsilon_0 \tag{2.165}$$

From these we see:

$$\operatorname{div} \boldsymbol{j} = \operatorname{div} \sigma \boldsymbol{E} = \sigma \operatorname{div} \boldsymbol{E} + \boldsymbol{E} \operatorname{grad} \sigma = -\frac{\partial \delta}{\partial t}$$

or

$$\frac{\sigma}{\epsilon_0} \delta + \boldsymbol{E} \operatorname{grad} \sigma = -\frac{\partial \delta}{\partial t}$$

Finally, we obtain the following differential equation for the volumetric charge density:

$$\frac{\partial \delta}{\partial t} + \frac{1}{\rho \epsilon_0} \delta = -\boldsymbol{E} \operatorname{grad} \sigma \tag{2.166}$$

where ρ is resistivity and ϵ_0 is the dielectric permittivity of free space. Now assume that the nonpolarizable medium in the neighborhood of a point is uniform, or that the electric field is perpendicular to the gradient of conductivity. In either case:

$$\boldsymbol{E} \operatorname{grad} \sigma = 0$$

and eq. 2.166 takes a simpler form:

$$\frac{\partial \delta}{\partial t} + \frac{1}{\rho \epsilon_0} \delta = 0 \tag{2.167}$$

This particular equation has a well known solution:

$$\delta = \delta_0 e^{-t/\rho \epsilon_0}$$

or

$$\delta = \delta_0 e^{-t/\tau_0} \tag{2.168}$$

where δ_0 is the charge at the initial instant of time and:

$$\tau_0 = \rho \epsilon_0 \tag{2.169}$$

The quantity τ_0 is the time constant, and in a conducting medium its value is usually very small. For example, if $\rho = 100$ ohm-meters:

$$\tau_0 = 100 \frac{1}{36\pi} 10^{-9} = \frac{10^{-7}}{36\pi} \approx 10^{-9} \text{ s}$$

In accord with eq. 2.168, a charge placed within a conducting medium will disappear very quickly. If we are concerned only with charge densities which exist at times greater than τ_0, we can assume that for all practical purposes they are absent. In addition, it is appropriate to point out that with the excitation provided by the kinds of electromagnetic fields used in minerals prospecting, there is no initial volumetric charge in the conducting medium, i.e.:

$$\delta_0 = 0$$

Therefore, we can conclude that at points where the medium is uniform or which meet the condition:

$$E \cdot \nabla \sigma = 0$$

there are no electric charges, and so:

$$\text{div } E = 0 \tag{2.170}$$

A much different situation exists when the medium is nonuniform and:

$$E \cdot \nabla \sigma \neq 0$$

In this case, the right-hand side of eq. 2.166 does not vanish and we have a first order nonhomogeneous differential equation of the form:

$$\frac{dy}{dt} + \frac{1}{\tau_0} y = f(t) \tag{2.171}$$

where

$$y = \delta(t), \quad f(t) = -E \cdot \text{grad } \sigma = -E \cdot \nabla \sigma$$

The general solution of eq. 2.171 is known to be of the form:

$$y = y_0 e^{-t/\tau_0} + e^{-t/\tau_0} \int_0^t e^{t/\tau_0} f(t) dt \tag{2.172}$$

where y_0 is the value of the function y at the instant $t = 0$.
In accord with eq. 2.166, we have:

$$\delta = \delta_0 e^{-t/\tau_0} - e^{-t/\tau_0} \int_0^t e^{t/\tau_0} (E \cdot \text{grad } \sigma) dt \tag{2.173}$$

This last equation can also be written as:

$$\delta = \delta_0 e^{-t/\tau_0} - e^{-t/\tau_0} \int_0^t e^{t/\tau_0} E(t) dt (e_0 \cdot \nabla \sigma) \tag{2.174}$$

where $E(t)$ is the magnitude of the electric field, and e_0 is a unit vector indicating the direction of this field:

$$E = E e_0$$

In general we can recognize two types of charges which behave quite differently as functions of time. The second term in eq. 2.174 describes the behavior of the second type of charges which can exist in nonuniform regions in a medium with the aid of electric forces, while the first kind represents charges in any regions in the medium.

As may be seen from eqs. 2.169 and 2.174, the time rate of change of the first kind of charges is independent of the uniformity or nonuniformity of the medium, and depends only on the time constant, τ_0.

In contrast to the behavior of the first kind of charges, the second type of charges occurs only as a consequence of the existence of an electromagnetic field.

Let us rewrite eq. 2.174 as:

$$\delta = \delta_1 + \delta_2 \tag{2.175}$$

where

$$\delta_1 = \delta_0 e^{-t/\tau_0}, \quad \delta_2 = -e_0 \cdot \nabla \sigma e^{-t/\tau_0} \int_0^t e^{t/\tau_0} E(t)\,dt \tag{2.176}$$

Inasmuch as measurements are usually made at times significantly greater than τ_0, and moreover, in most cases, $\delta_0 = 0$, we will consider only the second types of charge, δ_2. According to eq. 2.176, a volumetric charge density will arise in the neighborhood of any point in a nonuniform medium, provided the primary field is not normal to the direction of grad σ. Assuming that the condition:

$$t \gg \tau_0 \tag{2.177}$$

holds, we will expand the right-hand side of the expression for δ_2 in a power series in terms of small values of the parameter τ_0. To do so consider the integral:

$$\int_0^t e^{t/\tau_0} E(t)\,dt \tag{2.178}$$

Integrating this integral by parts, we obtain:

$$\int_0^t e^{t/\tau_0} E(t)\,dt = \tau_0 \left\{ E(t) e^{t/\tau_0} \big|_0^t - \int_0^t E'(t) e^{t/\tau_0}\,dt \right\}$$

$$= \tau_0 \left\{ E(t) e^{t/\tau_0} \big|_0^t - \tau_0 \left[E'(t) e^{t/\tau_0} \big|_0^t - \int_0^t E''(t) e^{t/\tau_0}\,dt \right] \right\}$$

$$= \tau_0 E(t) e^{t/\tau_0} - \tau_0 E(0) - \tau_0^2 E'(t) e^{t/\tau_0} + \tau_0^2 E'(0) + \tau_0^2 \int_0^t E''(t) e^{t/\tau_0}\,dt$$

The volume density δ_2 can be written as:

$$\delta_2(t) = -(e_0 \cdot \nabla \sigma) \left\{ \tau_0 E(t) - \tau_0^2 E'(t) + \tau_0^2 e^{-t/\tau_0} \int_0^t E''(t) e^{t/\tau_0} dt \right\} \qquad (2.179)$$

Continuing this process, it is possible to obtain higher order terms of the series. Considering that the time constant, τ_0, is normally extremely small, and that the condition in eq. 2.177 usually applies, we can discard all of these terms but the first one, and we have:

$$\delta_2(t) = -(e_0 \cdot \nabla \sigma) \tau_0 E(t) \qquad (2.180)$$

In this case, the charge density changes synchronously with changes in the electric field; that is, it is determined by the instantaneous values of the electric field at the same point. Such relationships between volume charge density and electric field strength are essential to the definition of quasi-stationary field behavior, which as has already been mentioned, is of importance in electrical prospecting. One can conclude that alternating charges develop in the quasi-stationary case in the same manner that they develop for a constant field. Unfortunately, this type of charge has not been sufficiently considered in the published literature concerning electromagnetic methods up to this time and in view of this fact, one of the two basic sources for the field turns out practically to be beyond the realm of consideration at this time. In order to illustrate these results, consider two examples:

Example 1
Assume that an electric field varies exponentially with time:

$$E = E_0 e^{-t/\tau} e_0 \qquad (2.181)$$

with τ being the parameter characterizing the rate at which the field changes. Then eq. 2.176 becomes:

$$\delta_2(t) = -e^{-t/\tau_0}(e_0 \cdot \nabla \sigma) \int_0^t e^{[(1/\tau_0)-(1/\tau)]t} dt E_0$$

and after carrying out the indicated integration:

$$\delta_2(t) = -\frac{\tau_0 e^{-t/\tau} E_0}{1 - \tau_0/\tau} [1 - e^{-t[(1/\tau_0)-(1/\tau)]}] (e_0 \cdot \nabla \sigma)$$

Further assuming that $\tau_0 \ll \tau$ and $t \gg \tau_0$, that is, the rate of field decay is considerably slower than τ_0 and the time at which measurements are made is considerably greater than the time constant, we have:

$$\delta_2(t) = -\tau_0 e^{-t/\tau}(e_0 \cdot \nabla \sigma) E_0 \qquad (2.182)$$

The volumetric charge density δ_2 decays exponentially at exactly the same rate as the electric field. For example, when the time constant of the field, τ, is one second, the function δ_2 decreases at the same rate, with the time constant of one second, regardless of the conductivity of the medium.

Example 2

Now assume that an electromagnetic field varies as $\sin \omega t$:

$$E = E_0 \sin \omega t \, e_0$$

Substituting this expression into eq. 2.176 we have:

$$\delta_2(t) = -e^{-t/\tau_0} E_0(e_0 \cdot \nabla \sigma) \int_0^t e^{t/\tau_0} \sin \omega t \, dt$$

The integral is well known:

$$\int e^{t/\tau_0} \sin \omega t \, dt = \frac{e^{t/\tau_0}}{1/\tau_0^2 + \omega^2} \left(\frac{1}{\tau_0} \sin \omega t - \omega \cos \omega t \right)$$

Whence:

$$\int_0^t e^{t/\tau_0} \sin \omega t \, dt = \frac{1}{1/\tau_0^2 + \omega^2} \left[\omega + e^{t/\tau_0} \left(\frac{1}{\tau_0} \sin \omega t - \omega \cos \omega t \right) \right]$$

Therefore we have:

$$\delta_2(t) = -\frac{E_0}{1/\tau_0^2 + \omega^2} \left[\omega e^{-t/\tau_0} + \left(\frac{1}{\tau_0} \sin \omega t - \omega \cos \omega t \right) \right] (e_0 \cdot \nabla \sigma)$$

In the quasi-stationary approximation, when the period of oscillation $T(\omega = 2\pi/T)$ is much greater than the relaxation time, τ_0, and t is also much greater than τ_0, we have:

$$\delta_2(t) = -(\tau_0 \sin \omega t - \omega \tau_0^2 \cos \omega t) E_0(e_0 \cdot \nabla \sigma)$$

Neglecting the second term and assuming that the field is not zero, that is, that ωt is not equal to $2\pi n$, we finally obtain the expression for the volumetric charge under quasi-stationary harmonic conditions:

$$\delta_2(t) = -\tau_0 E_0 \sin \omega t \, e_0 \cdot \nabla \sigma = -\tau_0 E(t) \cdot \mathrm{grad}\, \sigma \qquad (2.183)$$

So far we have investigated only the volume density of charge. Next let us consider alternating surface charges. Applying the equations:

$$j_n^{(2)} - j_n^{(1)} = -\frac{\partial \Sigma}{\partial t}, \quad E_n^{(2)} - E_n^{(1)} = \frac{\Sigma}{\epsilon_0}$$

we have:

$$\sigma_2 E_n^{(2)} - \sigma_1 E_n^{(1)} = \tfrac{1}{2}[(\sigma_2 + \sigma_1)(E_n^{(2)} - E_n^{(1)}) + (\sigma_2 - \sigma_1)(E_n^{(2)} + E_n^{(1)})]$$

$$= -\frac{\partial \Sigma}{\partial t}$$

or

$$\sigma^{av}\frac{\Sigma}{\epsilon_0} + (\sigma_2 - \sigma_1)E_n^{av} = -\frac{\partial \Sigma}{\partial t}$$

where $\sigma^{av} = (\sigma_1 + \sigma_2)/2$.

Thus, the equation for surface charge density is a differential equation of first order similar to that for volume charge density:

$$\frac{\partial \Sigma}{\partial t} + \frac{1}{\tau_{0s}}\Sigma = (\sigma_1 - \sigma_2)E_n^{av} \tag{2.184}$$

where $\tau_{0s} = \epsilon_0/\sigma^{av}$ is the relaxation time for the surface charge.

In accord with eq. 2.172 the solution to eq. 2.184 can be written as:

$$\Sigma = \Sigma_0 e^{-t/\tau_{0s}} + e^{-t/\tau_{0s}}(\sigma_1 - \sigma_2)\int_0^t E_n^{av}(t)e^{t/\tau_{0s}}dt \tag{2.185}$$

or

$$\Sigma = \Sigma_1 + \Sigma_2$$

where

$$\Sigma_1 = \Sigma_0 e^{-t/\tau_{0s}}, \quad \Sigma_2 = (\sigma_1 - \sigma_2)e^{-t/\tau_{0s}}\int_0^t E_n^{av}(t)e^{t/\tau_{0s}}dt \tag{2.186}$$

Thus, there are two types of surface charge that occur in this case. The first type, Σ_1, corresponds to the case in which some charge with a density Σ_0 is placed on the interface. In accord with eq. 2.186, it decays exponentially in time with the time constant τ_{0s}, that is, its behavior is controlled by the conductivity and dielectric constant in the medium, and it is independent of the existence or nonexistence of an electromagnetic field. Inasmuch as the relaxation time τ_{0s} is usually very small and measurements are usually made at much greater times than τ_{0s}, again we will not consider this type of charge. Moreover, with the exception of charges on the surface of electrodes at the ends of current-carrying lines, the charges are not placed at interfaces in electrical prospecting practice. Therefore, let us concentrate our attention on the charge of the second type Σ_2.

As in the case for volume charge density, the surface charge of the second kind arises only as the consequence of the existence of an electromagnetic field. Considering that the time constant τ_{0s} is usually very short and that measurements are made at much longer times or that the period of oscillations will be much greater than τ_{0s}, it will be appropriate to represent the right-hand side of the equation for Σ_2 (eq. 2.186) as a series in the parameter τ_{0s}. Carrying out this expansion in analogy to eq. 2.178 in discarding all terms except the first, we have:

$$\Sigma_2(t) = \tau_{0s}(\sigma_1 - \sigma_2)E_n^{av}(t) \tag{2.187}$$

Taking into account expression for τ_{0s}, it can readily be seen that at a given time eq. 2.187 is exactly the same as 2.159, which describes the surface density of charges when the field is independent of time. The surface density of charge in this case (eq. 2.187) is controlled by the instantaneous values of the field and the condition is that for quasi-stationary field behavior.

It is clear that the expression for the surface charges (eq. 2.187) follows directly from the differential eq. 2.184 if we can neglect the derivative $\partial\Sigma/\partial t$ in comparison with the term Σ/τ_{0s}. This is equivalent to the condition:

$$\frac{\partial\Sigma}{\partial t} \ll \frac{1}{\tau_{0s}}\Sigma \tag{2.188}$$

In accord with eq. 2.159, the function $E_n^{av}(p)$ is the electric field contributed by all sources except a charge situated in the immediate neighborhood of the point p. For this reason, the right-hand side of eq. 2.184 can be interpreted as the flux of the current density through a closed surface with unit cross-sectional area (see Fig. 2.39) caused by external sources only. The term Σ/τ_{0s} can be written as:

$$\frac{1}{\tau_{0s}}\Sigma(p) = \frac{(\sigma_1 + \sigma_2)}{2\epsilon_0}\Sigma(p) = \sigma_1\frac{\Sigma(p)}{2\epsilon_0} + \sigma_2\frac{\Sigma(p)}{2\epsilon_0} \tag{2.189}$$

As was shown previously, the term $\Sigma(p)/2\epsilon_0$ indicates the magnitude of the normal component of the electric field caused by the surface charge near point p. Therefore, the term Σ/τ_{0s} describes the current density flux through the closed surface shown in Fig. 2.39 and caused by only the charge $\Sigma(p)dS$ located inside. Thus, in accord with eq. 2.184, the flux of current density caused by external sources is compensated by two fluxes, namely: (1) the change in surface density with time, $\partial\Sigma(p)/\partial t$; and (2) the flux caused by the electric field from the charge $\Sigma(p)$, that is, $\Sigma(p)/\tau_{0s}$.

Obviously, in a quasi-stationary approximation when condition 2.188 applies, only the flux caused by the electric fields arising from charge distributions cancels the flux from external sources. In this case, using eq. 2.189, eq. 2.184 can be rewritten as:

$$\sigma_1(E_n^{(1)}(p) + E_n^{av}) = \sigma_2(E_n^{(2)}(p) + E_n^{av})$$

where

$$E_n^{(1)}(p) = -\frac{\Sigma}{2\epsilon_0}, \quad E_n^{(2)}(p) = \frac{\Sigma}{2\epsilon_0}$$

or

$$j_n^{(2)} = j_n^{(1)} \tag{2.190}$$

In the quasi-stationary approximation, the normal component of current density, as well as in the direct current field, is a continuous function of position. It is obvious that this result follows directly from eq. 2.151 by neglecting the right-hand side, $\partial \Sigma / \partial t$. However, our consideration allows one to understand that condition:

$$\frac{\partial \Sigma}{\partial t} \to 0 \quad \text{as} \quad \frac{\tau_{0s}}{t} \to 0$$

does not necessarily mean the absence of surface charges, and in the quasi-stationary approximation, the derivate $\partial \Sigma / \partial t$ need only be small with respect to the flux, Σ / τ_{0s}.

The postulate of charge conservation has permitted us to investigate in detail the distribution of charge in a conducting medium. It also serves to introduce the concept of displacement currents which play a vital role in propagating electromagnetic fields. We will start from two equations, one of which is:

$$\text{curl } H = j$$

which is derived for the magnetic field of direct currents, and the other being the postulate of charge conservation, written in the differential form as:

$$\text{div } j = -\partial \delta / \partial t$$

It can readily be seen that these two equations are contradictory inasmuch as:

$$\text{div curl } H = \text{div } j = 0$$

To solve this problem, we must add an additional term on the right-hand side of eq. 2.122, so that we obtain:

$$\text{curl } H = j + X \tag{2.191}$$

where X is an undetermined function at this time.

We can choose this function to satisfy eq. 2.150. Performing the divergence operation on both sides of eq. 2.191, we have:

$$0 = \text{div } j + \text{div } X$$

or

$$\text{div } X = \partial \delta / \partial t$$

As is known:

$$\text{div } D = \delta$$

where

$$D = \epsilon E$$

is the electric displacement vector, and $\epsilon = \epsilon_r \epsilon_0$; while δ is a free charge:

$$\operatorname{div} X = \frac{\partial}{\partial t} \operatorname{div} D = \operatorname{div} \frac{\partial D}{\partial t} = \operatorname{div} \dot{D}$$

One possible solution to the problem is the vector:

$$X = \frac{\partial D}{\partial t} = \dot{D}$$

Substituting $X = \dot{D}$ on the right-hand side of eq. 2.191, we obtain the second of Maxwell's equations:

$$\operatorname{curl} H = j + \frac{\partial D}{\partial t} \tag{2.192}$$

Numerous experiments have shown the appropriateness of selecting the vector X in this form. It is termed a displacement current. As follows from Maxwell's second equation (eq. 2.192), there are two sources for the magnetic field, namely: (1) conduction currents; and (2) displacement currents. Applying Stoke's theorem, we can obtain Maxwell's second equation in an integral form:

$$\oint_L H \cdot dl = \int_S (j + \dot{D}) \cdot dS \tag{2.193}$$

Using an approach as was previously described, it can readily be seen that the inclusion of displacement currents makes no change in the surface analog for this equation derived for the magnetic field of direct currents, that is, the tangential components of the alternating magnetic field remain continuous functions at an interface:

$$H_t^{(2)} - H_t^{(1)} = 0 \tag{2.194}$$

Thus, in the general case, we have three forms for Maxwell's second equation:

$$\oint_L H \cdot dl = \int_S (j + \dot{D}) \cdot dS$$

$$\operatorname{curl} H = j + \partial D / \partial t \tag{2.195}$$

$$H_t^{(2)} - H_t^{(1)} = 0$$

2.4. FARADAY'S LAW AND THE FIRST MAXWELL EQUATION

It was observed by early investigators of electric and magnetic fields that when the magnetic induction vector B through a surface S bounded by a contour $L[S]$ as shown in Fig. 2.41 changes with time, an electromagnetic force will be observed along the contour in the amount:

$$\mathscr{E} = -\frac{\partial \Phi}{\partial t} \tag{2.196}$$

where $\mathscr{E}$ is electromotive force, Φ is the magnetic flux through the surface bounded by $L[S]$, equal to:

$$\int_S B \cdot dS$$

here

$$B = \mu H$$

and $\partial \Phi / \partial t$ is the derivative of the flux, Φ, with respect to time. The contour L can be of any form and it need not be located solely within a conducting medium. To the contrary, the contour L can be an arbitrary path crossing media with various properties, including media which are insulating. Moreover, there is no relationship between the orientation of this contour and the current flow lines.

As is well known, the electromotive force can also be represented as:

$$\mathscr{E} = \oint_L E \cdot dl \tag{2.197}$$

where E is the electric field intensity defined at each point along the contour L. Equation 2.196 can therefore be rewritten as:

$$\oint_L E \cdot dl = -\frac{\partial \Phi}{\partial t} \tag{2.198}$$

This expression can be interpreted in a natural sense as follows: a change of the magnetic flux, Φ, with time gives rise to an electric field. This phenomenon was first observed and reported by Faraday and has been called electromagnetic induction. The relationship between electric field intensity and the rate of change of magnetic flux as described by eq. 2.198 is one of the most fundamental relationships in physics.

By convention, the electric field which is due to electromagnetic induction is called the inductive field, E^{ind}, emphasizing the origin of this particular field component. One can rewrite eq. 2.198 in the form:

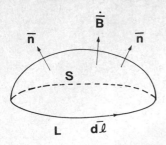

Fig. 2.41. The time rate of change of magnetic induction through a surface S bounded by a contour L causes an electromagnetic force to be observed along that contour.

$$\oint_L E^{\text{ind}} \cdot dl = -\frac{\partial \Phi}{\partial t} \qquad (2.199)$$

It is a basic fact that a change in magnetic flux with time gives rise to a specific electromotive force. However, to determine the electrical field, additional information must be provided. Up to this point, we have considered only one source of electric field, namely, electric charges. In addition to charges, a change in magnetic flux with time provides a second mechanism for the development of an electric field. This fact is the fundamental basis of electromagnetic induction. The electric field, as well as the magnetic field generally can be attributed to two forces. Of course, one can readily think of particular cases in which one of these forces of the electrical field does not exist, as for example:

(1) A constant field in which derivatives with respect to time are zero, and an electric field can arise only from the presence of electrical charges.

(2) An alternating electromagnetic field in which the electric field intensity has only a tangential component at interfaces between media with different values of conductivity. In this case, the electric field arises only from the variation of magnetic field with time.

However, both sources for an electric field play an essential role in the general case, and this must be understood in order to solve significant interpretation problems such as arise with a variety of methods for electrical prospecting. In this respect, suppose that an electric field arises from both types of sources, namely: (a) electrical charges which alter with time, but at every given moment they create the field E^c, described by Coulomb's law (the quasi-stationary field); and (b) a change of the flux of the magnetic field with time, $\partial \Phi / \partial t$. So, the total electrical field, E, can be presented as a sum:

$$E = E^c + E^{\text{ind}} \qquad (2.200)$$

whence

$$E^{\text{ind}} = E - E^c \qquad (2.201)$$

Combining eqs. 2.199 and 2.201, we have:

$$\oint_L \boldsymbol{E} \cdot \mathrm{d}\boldsymbol{l} - \oint_L \boldsymbol{E}^{\mathrm{c}} \cdot \mathrm{d}\boldsymbol{l} = -\frac{\partial \Phi}{\partial t}$$

As was shown in the first paragraph, the circulation of Coulomb's electrical field is equal to zero and therefore:

$$\oint_L \boldsymbol{E}^{\mathrm{ind}} \cdot \mathrm{d}\boldsymbol{l} = \oint_L \boldsymbol{E} \cdot \mathrm{d}\boldsymbol{l} = -\frac{\partial \Phi}{\partial t}$$

This result sometimes leads to a misunderstanding of the role played by electrical charges in forming the electromagnetic field. Actually this consideration merely shows that the electromagnetic force due to the Coulomb electric field is zero. But this conclusion cannot be extended to the electric field itself. The Coulomb electric field has an influence on the distribution of currents in the conducting medium which in turn create an alternating magnetic field. Therefore, in general, the inductive electric field, $\boldsymbol{E}^{\mathrm{ind}}$, does depend on the charges.

Now, we will write Faraday's law in various forms. First, using the definition of magnetic flux:

$$\Phi = \int_S \boldsymbol{B} \cdot \mathrm{d}\boldsymbol{S}$$

we have:

$$\oint_L \boldsymbol{E} \cdot \mathrm{d}\boldsymbol{l} = -\frac{\partial}{\partial t} \int_S \boldsymbol{B} \cdot \mathrm{d}\boldsymbol{S}$$

We will not consider any electromagnetic force caused by moving the integration paths, and therefore, the last equation can be written:

$$\oint_L \boldsymbol{E} \cdot \mathrm{d}\boldsymbol{l} = -\int_S \frac{\partial \boldsymbol{B}}{\partial t} \mathrm{d}\boldsymbol{S} = -\int_S \dot{\boldsymbol{B}} \cdot \mathrm{d}\boldsymbol{S} \qquad (2.202)$$

where $\dot{\boldsymbol{B}} = \partial \boldsymbol{B}/\partial t$.

This equation is an exact formulation of Faraday's law and is also considered to be the first of Maxwell's equations when they are written in an integral form. In this equation the vector quantity, $\mathrm{d}\boldsymbol{l}$, indicates the direction in which integration is carried along the contour L, while the vector $\mathrm{d}\boldsymbol{S}$ represents the direction normal, n, to the surface ($\mathrm{d}\boldsymbol{S} = \mathrm{d}S \cdot \boldsymbol{n}$). It should be clear that there is a specific relationship between the vectors $\mathrm{d}\boldsymbol{l}$ and $\mathrm{d}\boldsymbol{S}$, otherwise in changing the direction of one of the vectors $\mathrm{d}\boldsymbol{l}$ or $\mathrm{d}\boldsymbol{S}$, the same rate of change of flux density, $\partial \Phi/\partial t$, would create an electromagnetic force, having the same magnitude, but opposite sign.

In order to retain the physical meaning of Faraday's law, the vectors $\mathrm{d}\boldsymbol{l}$ and $\mathrm{d}\boldsymbol{S}$ in eq. 2.202 are chosen according to a right-hand rule; that is, an observer facing in the direction of the vector dS sees that the path along the contour L is traversed in a counterclockwise sense. It is only when this is true that eq. 2.202 correctly describes the electromagnetic induction phenomenon.

Next, by making use of Stoke's theorem, we can obtain a differential form for the first of Maxwell's equations:

$$\oint_L E \cdot dl = \int_S \text{curl } E \cdot dS = -\int_S \frac{\partial B}{\partial t} \cdot dS$$

Whence

$$\text{curl } E = -\partial B/\partial t \tag{2.203}$$

where the functions E and B are considered in the near proximity of a same point.

Both eqs. 2.202 and 2.203 and their several other forms describe the same physical law, but the differential form of Maxwell's first equation (eq. 2.203) is applied only at points, and only in the case in which all the components of the electric field are continuous functions of the spatial variables.

Considering that in many problems we must examine electromagnetic fields in media with discontinuous changes in properties (interfaces) it is desirable to derive a surface analogy of the first Maxwell equation. It is clear that eq. 2.203 cannot be applied for points on an interface between two media with different conductivity values, since the normal component of the electric field is a discontinuous function of the spatial variables. For this reason, we will proceed further with eq. 2.202, evaluating it along the path shown in Fig. 2.42 so that we have:

$$E_t^{(2)} - E_t^{(1)} = 0$$

where t is an arbitrary direction, tangential to the interface. In its general form, this equation is:

$$n \times (E_2 - E_1) = 0 \tag{2.204}$$

The right-hand side of this equation is zero, inasmuch as the flux of the magnetic field goes to zero when the area enclosed by the path vanishes. In accord with this last equation, one can say that the tangential components of the alternating electric field are continuous functions of position, as well as would be the case in which the field is caused only by electrical charges.

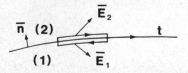

Fig. 2.42. Evaluation of Faraday's law along a path t.

Thus, we have obtained three different forms of the first Maxwell equation:

$$\oint_L E \cdot dl = -\int_S \dot{B} \cdot dS$$

$$\operatorname{curl} E = -\partial B/\partial t \qquad\qquad (2.205)$$

$$n \times (E_2 - E_1) = 0$$

It should be emphasized that each of these equations describes the electromagnetic induction phenomenon. We will now examine a few examples that demonstrate some of the features of electromagnetic induction, when it is assumed that the electromagnetic field changes relatively slow such that the displacement currents can be neglected.

Example I

Suppose that a magnetic field arises as the consequence of an alternating current flowing in an infinitely long cylindrical solenoid as shown in Fig. 2.43. It is well known that the magnitude of the field is non-zero and uniform inside the solenoid, and that at an external location, the magnetic field is absent. Also, inasmuch as both the vectors B and $\partial B/\partial t$ are directed along the z-axis, an induction (vortex) electric field is observed in horizontal planes. Moreover, due to the axial symmetry, the vector lines for the electric field are circles with a common center located on the solenoid axis. The electric field caused by the variation of the magnetic field inside the solenoid has only the component E_ϕ, which is a function of the radius, r. Making use of eq. 2.202 along any circle with the radius r, located in a horizontal plane, we have:

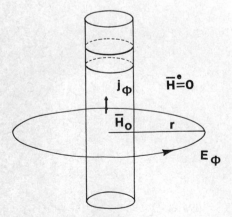

Fig. 2.43. Excitation of a magnetic field by an alternating current flowing in an infinitely long cylindrical solenoid.

$$\oint E \cdot dl = E_\phi 2\pi r = -\frac{\partial \Phi}{\partial t}$$

or

$$E_\phi = -\frac{1}{2\pi r}\frac{\partial \Phi}{\partial t} \qquad (2.206)$$

where $\partial\Phi/\partial t$ is the rate of change of the magnetic flux within the area bounded by the circle with radius r. Suppose that the magnetic field varies with time as follows:

$$H = H_0 f(t)$$

Then, in accord with eq. 2.206, the vortex field inside the solenoid (r less than a) is:

$$E_\phi^{(i)} = -\frac{\pi r^2}{2\pi r} B_0 f'(t) = -\frac{B_0}{2} r f'(t), \quad r \leqslant a \qquad (2.207)$$

That is, the electric field inside the solenoid increases linearly with the radius, r.

For all horizontal circles with a radius r exceeding the radius a of the solenoid, the flux, Φ, and the corresponding derivative $\partial\Phi/\partial t$, remain the same at a given instant in time:

$$\Phi = B_0 \pi a^2 f(t)$$

$$\frac{\partial \Phi}{\partial t} = B_0 \pi a^2 f'(t)$$

Therefore, the voltage $\oint E \cdot dl$ along any of these circles does not change with increasing radius and in accord with eq. 2.206 we have:

$$E_\phi^e = -\frac{B_0}{2\pi r}\pi a^2 f'(t) = -\frac{B_0 a^2}{2r} f'(t), \quad r \geqslant a \qquad (2.208)$$

The vortex electric field outside the solenoid decreases linearly with increasing radius, r.

This example is one of practical interest, inasmuch as it clearly demonstrates a case in which a vortex electric field is nonzero at points where the magnetic field is absent.

In the next several examples, we will consider vortex electrical fields caused by the change of magnetic fields of the type of source which are often used in inductive mining prospecting.

88

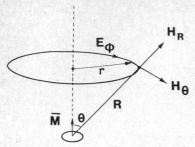

Fig. 2.44. A magnetic dipole with a moment $M(t)$ directed along the z-axis and situated at the origin of a spherical coordinate system.

Example II. The vortex electric field of a magnetic dipole in free space

In this example, we will consider a magnetic dipole with the moment $M(t)$ directed along the z-axis and situated at the origin of a spherical coordinate system (see Fig. 2.44). We reject displacement currents by considering that in this approximation the magnetic field is defined by the instantaneous magnitude of the dipole moment; therefore one can make use of calculations based on a static magnetic field. In accord with eq. 2.137, we have the following expressions for the alternating magnetic field caused by a magnetic dipole in free space:

$$H_R(t) = \frac{2M(t)}{4\pi R^3} \cos\theta, \quad H_\theta(t) = \frac{M(t)}{4\pi R^3} \sin\theta, \quad H_\phi = 0 \qquad (2.209)$$

Inasmuch as the vector representing this magnetic field lies in longitudinal planes of the spherical coordinate system and as a consequence of the axial symmetry, the vortex electric field arising as a result of the change of this magnetic field with time has only a single component, E_ϕ. Therefore, the vector lines representing this field are circles centered on the z-axis.

Making use of eq. 2.202, we have:

$$E_\phi = -\frac{1}{2\pi r} \dot{\Phi} \qquad (2.210)$$

where Φ is the flux penetrating the area bounded by a circle with the radius r (see Fig. 2.44). Inasmuch as the vector normal to this area is parallel to the z-axis, we have the following expressions for the flux Φ:

$$\Phi = \int_S \boldsymbol{B} \cdot \mathrm{d}\boldsymbol{S} = \int_S B_z \, \mathrm{d}S = \int_0^r B_z \, 2\pi r \, \mathrm{d}r \qquad (2.211)$$

where B_z is the vertical component of the magnetic field.

As may be seen from Fig. 2.44:

$$B_z = B_R \cos\theta - B_\theta \sin\theta$$

and considering eq. 2.209, we obtain:

$$B_z = \frac{\mu M}{4\pi R^3} (3 \cos^2 \theta - 1) \tag{2.212}$$

Substituting the result from eq. 2.212 into eq. 2.211 and integrating, we obtain:

$$\dot{\Phi} = \frac{\partial \Phi}{\partial t} = \frac{1}{2} \mu \frac{\dot{M}(t)}{R^3} r^2 \tag{2.213}$$

where $R = (r^2 + z^2)^{1/2}$.

Therefore, we can write the expression for the inductive electric field as:

$$E_\phi = -\mu \frac{\dot{M}(t)}{4\pi R^2} \sin \theta \tag{2.214}$$

It should not be unexpected that the electric field is zero on the z-axis ($\theta = 0$) since the flux through a surface bounded by a circle with a radius r tends to zero when this radius decreases. With increasing radius of the electric lines, there is always some radius r for which the magnetic vector lines begin to intersect the surface S surrounded by the circle twice. For this reason, with a further increase of the radius of the circle, the magnetic flux and the corresponding vortex electric field begin gradually to decrease.

Thus, neglecting displacement currents, the electromagnetic field of an alternating magnetic dipole situated in a free space is described as follows:

$$H_R = \frac{2M(t)}{4\pi R^3} \cos \theta, \quad H_\theta = \frac{M(t)}{4\pi R^3} \sin \theta, \quad E_\phi = -\mu \frac{\dot{M}(t)}{4\pi R^2} \sin \theta \tag{2.215}$$

It is an essential feature of the behavior of this field that along with the magnetic field at every point in free space, there is also an electric field. One might suspect that if the medium has a non-zero conductivity, this electric force will give rise to current flow.

The field described by eq. 2.215 has as its cause only the current in the magnetic dipole, and for this reason by convention it has been called the "primary field". Several examples of primary fields will be considered in this section.

Now let us describe briefly the electromagnetic field of a magnetic dipole when its moment varies with time in a relatively simple way.

(A) Suppose that the current in the dipole changes sinusoidally, that is:

$$M = M_0 \sin \omega t \tag{2.216}$$

where M_0 is the magnitude of the moment and $\omega = 2\pi f = 2\pi/T$ is the radial frequency, with T being the period of oscillation. In accord with eqs. 2.215 and 2.216, we have for the magnetic field:

$$H_R = \frac{2M_0}{4\pi R^3} \cos\theta \sin\omega t, \quad H_\theta = \frac{M_0}{4\pi R^3} \sin\theta \sin\omega t \qquad (2.217)$$

The expression for the electrical field can be written as:

$$E_\phi = \frac{\mu\omega M_0}{4\pi R^2} \sin\left(\omega t - \frac{\pi}{2}\right) \sin\theta$$

and therefore one can say that the primary electric field exhibits a phase shift of $-90°$ with respect to the current flowing in the dipole source or to the primary magnetic field.

(B) Now let us consider a dipole moment which varies with time as (see Fig. 2.45):

$$M = \begin{cases} M_0 & t \leqslant 0 \\ M_0 - at & 0 \leqslant t \leqslant t_r \\ 0 & t \geqslant t_r \end{cases} \qquad (2.218)$$

where $a = M_0/t_r$.

This relationship describes a primary magnetic field which is constant for times less than zero, and which then decreases linearly over the interval $0 \leqslant t \leqslant t_r$, and which is exactly equal to zero when $t \geqslant t_r$. A primary vortex electric field will exist only within the time interval over which the magnetic field is varying, $0 < t < t_r$ and, in view of the linear dependence of time, the electric field is constant over this interval. Thus we have:

$$H_R = \begin{cases} \dfrac{2M_0}{4\pi R^3} \cos\theta & t \leqslant 0 \\[2ex] \dfrac{2M(t)}{4\pi R^3} \cos\theta & 0 \leqslant t \leqslant t_r \\[2ex] 0 & t \geqslant t_r \end{cases}$$

$$H_\theta = \begin{cases} \dfrac{M_0}{4\pi R^3} \sin\theta & t \leqslant 0 \\[2ex] \dfrac{M(t)}{4\pi R^3} \sin\theta & 0 \leqslant t \leqslant t_r \\[2ex] 0 & t \geqslant t_r \end{cases} \qquad (2.219)$$

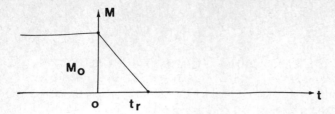

Fig. 2.45. Variation of magnetic moment with time as defined by the expression in eq. 2.218.

and

$$E_\phi = \begin{cases} 0 & t < 0 \\ \dfrac{\mu a}{4\pi R^2} \sin\theta = \dfrac{\mu M_0}{4\pi R^2 t_{\mathrm{r}}} \sin\theta & 0 < t < t_{\mathrm{r}} \\ 0 & t > t_{\mathrm{r}} \end{cases}$$

The curves given in Fig. 2.46 illustrate the behavior of the moment as well as of the electric and magnetic fields as functions of time.

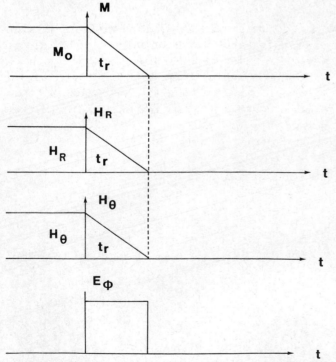

Fig. 2.46. Variation with time of the electromagnetic field components when a source moment varies as shown in Fig. 2.45.

Example III. The inductive electric field caused by the magnetic field from a current flowing in a circular loop

In this case, we assume that the source of an electromagnetic field is an alternating current $I(t)$ flowing in a circular loop with a radius a as is shown in Fig. 2.47. Expressions for the magnetic field caused by a constant current flowing in the loop were derived previously in paragraph 2, where it was shown that they can be represented in terms of an expression containing elliptical integrals. Inasmuch as displacement currents are ignored, the magnetic field is defined as follows by the instantaneous value for the current flowing in the loop:

$$B_r(t) = \frac{\mu I(t)}{2\pi} \frac{z}{r\left[(a+r)^2 + z^2\right]^{1/2}} \left[-K + \frac{a^2 + r^2 + z^2}{(a-r)^2 + z^2} E\right]$$

$$B_z(t) = \frac{\mu I(t)}{2\pi} \frac{1}{\left[(a+r)^2 + z^2\right]^{1/2}} \left[K + \frac{a^2 - r^2 - z^2}{(a-r)^2 + z^2} E\right]$$

(2.220)

and

$$B_\phi = 0$$

Considering the axial symmetry, as well as the fact that the magnetic field is located in longitudinal planes, the electric field caused by the time variation of the magnetic field has only a single component, $E_\phi(r, z)$, or that is, in a cylindrical coordinate system, r, ϕ, z:

$$E = (0, E_\phi, 0)$$

To describe the vortex electric field, one can in principle make use of Maxwell's equation in its integral form:

$$\oint_L E \cdot dl = -\int_S \dot{B} \cdot dS$$

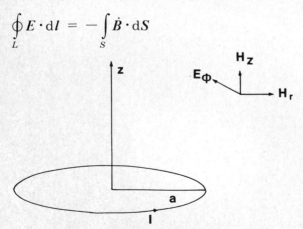

Fig. 2.47. Electromagnetic field components caused by an alternating current flowing in a circular loop.

However, this approach leads to integration of elliptical functions, which is a cumbersome approach. Fortunately, there is a much more efficient way to derive an expression for the electric field. Considering that the vector potential A, $(H = \operatorname{curl} A)$, as well as the magnetic field, is defined by the instantaneous value of current flowing in the source loop, one can write:

$$A_\phi(t) = \frac{I(t)}{4\pi} \int_L \frac{dl_\phi}{R}$$

or

$$A_\phi(t) = \frac{I(t)}{\pi k} \left(\frac{a}{r}\right)^{1/2} [(1 - \tfrac{1}{2}k^2)K - E]$$

From the first Maxwell equation:

$$\operatorname{curl} E = -\dot{B}$$

it follows that:

$$\operatorname{curl} E = -\mu \frac{\partial}{\partial t} \operatorname{curl} A = -\mu \operatorname{curl} \frac{\partial A}{\partial t}$$

or

$$\operatorname{curl} \left(E + \mu \frac{\partial A}{\partial t}\right) = 0$$

From this, we have:

$$E = -\mu \frac{\partial A}{\partial t} - \operatorname{grad} U \tag{2.221}$$

where U is a scalar potential. In accord with this last equation, we have the following expression for the component E_ϕ:

$$E_\phi = -\mu \frac{\partial A_\phi}{\partial t} - \frac{1}{r} \frac{\partial U}{\partial \phi}$$

However, in view of the axial symmetry, the second term vanishes so that we have:

$$E_\phi = -\mu \frac{\partial A_\phi}{\partial t} \tag{2.222}$$

It should be noted at this point that the vector potential A which we have used here is identical at any instant with the vector potential caused by a constant current flow. Equation 2.222 permits us to write down the expression for E_ϕ:

94

$$E_\phi = -\mu \frac{\mathrm{d}I(t)}{\mathrm{d}t} \frac{1}{\pi k} \left(\frac{a}{r}\right)^{1/2} [(1 - \tfrac{1}{2}k^2)K - E] \tag{2.223}$$

In conclusion, it should be emphasized that the vector lines describing the electric field are circles lying in horizontal planes with centers situated on the z-axis. It is an easy matter to show that the expression for the electric field given in eq. 2.223 is practically identical with that for an electric field caused by a magnetic dipole (see eq. 2.215) when the distance measured from an observation point to the center of the source loop is significantly greater than the radius of that source loop.

Example IV. The vortex electric field caused by an infinitely long current filament

Let us suppose that an alternating current $I(t)$ flows along a straight and infinitely long wire, along the z-axis as shown in Fig. 2.48. Considering a quasi-stationary approximation, the magnetic field from this current in free space is (eq. 2.130):

$$H_\phi(t) = \frac{I(t)}{2\pi r} \tag{2.224}$$

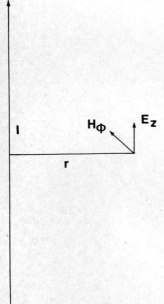

Fig. 2.48. Electromagnetic field components caused by an alternating current flowing in an infinitely long wire.

Applying the equation:

$$H = \operatorname{curl} A$$

we find that the vector potential has but a single component, A_z, which is:

$$A_z = -\frac{I(t)}{2\pi}\ln r \tag{2.225}$$

here a constant is chosen to be zero.

From this, we recognize that the vortex electric field and the current density have the same components, that is:

$$E_z = \frac{\mu}{2\pi}\frac{dI(t)}{dt}\ln r \tag{2.226}$$

We see that the primary vortex electric field has a logarithmic singularity at the current source. The assumption of an infinitely long current-carrying wire also leads to the existence of a singularity in the electric field with increasing r. This behavior will be investigated in more detail below, but here consider the magnetic field caused by the two currents $I(t)$ and $-I(t)$ as shown in Fig. 2.49. If the magnetic field is measured at points situated along a profile between two current-carrying wires, then the individual magnetic field contributed by each current has the same direction so that for the total field we obtain:

$$H(t) = \frac{I(t)}{2\pi}\left(\frac{1}{r_1}+\frac{1}{r_2}\right) \tag{2.227}$$

Here r, and r_2 are distances from lines $I(t)$ and $-I(t)$, respectively. However, the vortex electric field generated by these current filaments has opposite directions at any observation point and so we have:

$$E_z = E_{1z} + E_{2z} = \frac{\mu}{2\pi}\frac{dI(t)}{dt}\ln\frac{r_1}{r_2} \tag{2.228}$$

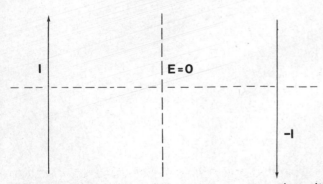

Fig. 2.49. Configuration of two current-carrying wires with currents I and $-I$.

In this case, the total electric field does not exhibit a singularity, when the distance from the current-carrying wire increases without limit. Moreover, it is interesting to note that the electric field disappears at points in a plane for which $r_1 = r_2$.

Example V. The vortex electric field caused by a uniform magnetic field

Often in order to facilitate an analysis of field behavior it is assumed that the primary magnetic field is uniform; that is, that it does not depend on the coordinates of an observation point. It is obvious that a uniform magnetic field cannot exist throughout space, or the source would need to have an infinitely large expenditure of energy. One can, however, consider that at distances relatively far away from any current source, the magnetic field becomes nearly uniform over a volume in space whose dimensions are much less than the distance from the source. Thus, we can consider special cases when the magnetic field can be approximated over certain ranges by a uniform magnetic field.

(A) Consider that the source for a magnetic field is a current flowing in a circular loop with a radius a, as shown in Fig. 2.50. The magnetic field near the z-axis can be considered as being almost uniform. The electric field can be derived in at least two different ways. The first way is one based on an analysis of an asymptote of eq. 2.223 as the distance r tends to zero. To obtain the equation for the electric field near the z-axis, the elliptical integrals must be expanded as series with respect to the small parameter, r, which designates the distance between the z-axis and the observation point.

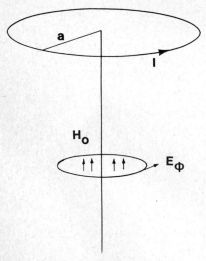

Fig. 2.50. The magnetic field is generated by current flowing in a circular loop with radius a, while the electric field is measured around a co-axial ring.

The second way we might consider is a much simpler approach. Making use of axial symmetry in applying eq. 2.199 to an arbitrary circle lying in a horizontal plane and situated within the range where the magnetic field can be considered to be uniform, we obtain:

$$2\pi r E_\phi = -\mu \pi r^2 \frac{\partial H_z}{\partial t}$$

or

$$E_\phi = -\mu \frac{r}{2} \frac{\partial H_z}{\partial t}$$

(2.229)

where H_z is the intensity for the uniform magnetic field along the z-axis. As has previously been shown, for points on the z-axis we have:

$$H_z = \frac{Ia^2}{2(a^2 + z^2)^{3/2}}$$

and taking eq. 2.229 into account, we obtain:

$$E_\phi = -\frac{\mu a^2 r}{4(a^2 + z^2)^{3/2}} \frac{\partial I(t)}{\partial t}$$

(2.230)

We see that the asimuthal component of the electric field is zero on the z-axis and increases linearly with r within the range where the magnetic field can be considered to be uniform. This example demonstrates the case in which the magnetic field is uniform, but the vortex electric field changes in magnitude and in sign on either side of the plane $\phi = \pi/2$.

(B) If we consider the electromagnetic field at a great distance from a current-carrying loop, one can use equations derived for a magnetic dipole source (eq. 2.215). It should be clear that we can always choose a relatively small volume of space where both the magnetic and vortex electric fields are essentially uniform.

(C) Assume that a magnetic field is generated by a linearly current filament, as shown in Fig. 2.48. It is clear that at relatively large distances from a current filament, there will be some small volume over which both the magnetic and electric fields can be considered to be almost uniform, and in accord with eqs. 2.224 and 2.226, we have:

$$H_\phi = \frac{I(t)}{2\pi r}, \quad E_z = \frac{\mu}{2\pi} \frac{dI(t)}{dt} \ln r$$

(2.231)

where r is the distance from the current filament to some median point within the volume under consideration.

In another case, if we examine the field at points lying between two current-carrying lines having opposite directions, the magnetic field midway between the two lines is almost uniform, while the electric field, E_z is zero at points on the mid-plane and has opposite signs on either side of this plane, with a magnitude that increases linearly with distance from the median plane.

Each of these cases has illustrated that essentially the same uniform magnetic field can be accompanied by electrical fields of various forms, the direction of which is determined by the geometry of the current flow. In other words, a uniform magnetic field does not uniquely define a single vortex electrical field, and so, in order to determine it, one has to have additional information on the geometry of current flow.

This feature of the behavior of electric and magnetic fields is of considerable practical interest in understanding the secondary electromagnetic field; caused by the presence of a inhomogeneity within a conductive medium.

Up to this point, we have considered examples of quasi-stationary electromagnetic fields caused by a given distribution of currents, in cases in which displacement currents can be neglected. These alternating currents, in contrast to direct current, are sources of both magnetic and vortex electric fields. One can argue the reality of these electric fields only to the same extent as the reality of the magnetic fields. Thus, in a nonconductive medium, the electromagnetic induction phenomenon manifests itself by the appearance of a vortex electric field.

Next we will investigate some features of the behavior of induction currents which appear due to vortex electric fields. This problem will be examined in detail in following chapters, so that here we will consider only the very simplest case; that is, inductive or eddy currents that arise within a conducting ring.

Example VI. Induction currents in a thin conducting ring situated within an alternating field (Fig. 2.51)

The induction of currents in a conducting ring can be described as follows. With a change of primary magnetic field with time, a primary vortex electric field arises. For simplicity, we will assume that this electric field has but a single component, E_ϕ^0, which is tangential to the surface of the ring. This field is the cause of induction currents in the ring. In turn, these induction currents generate a secondary electromagnetic field, and it should be obvious that the density of the induced currents in the ring is defined by both the primary and secondary vortex electric fields. In accord with Ohm's law, we can write that:

$$j_\phi = \sigma(E_\phi^0 + E_\phi^s) \tag{2.232}$$

where j_ϕ is current density, σ is the conductivity of the ring, E_ϕ^0 is the primary electric field, and E_ϕ^s is the secondary electric field.

We will use Faraday's law (eq. 2.196) to find the current in the ring:

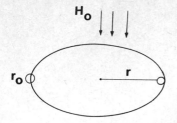

Fig. 2.51. Induction of current in a thin conducting ring located within an alternating field.

$$\mathscr{E} = -\frac{\partial \Phi}{\partial t} \tag{2.233}$$

The flux, Φ, enclosed in an area bounded by the ring can be written as the sum:

$$\Phi = \Phi_0 + \Phi_s \tag{2.234}$$

where Φ_0 is the flux in the primary magnetic field caused by external source and Φ_s is the flux in the magnetic field caused by the induction currents in the ring. Thus, eq. 2.233 can be rewritten as:

$$\mathscr{E} = -\frac{\partial \Phi_0}{\partial t} - \frac{\partial \Phi_s}{\partial t} \tag{2.235}$$

In this equation, only the term $\partial \Phi_0 / \partial t$ is known, while the electromotive force, $\mathscr{E}$, and the rate of change of the secondary magnetic flux, $\partial \Phi_s / \partial t$, are unknown. Our objective is to determine the current, I, flowing in the ring, and so, we will attempt to express both unknowns in terms of this parameter. First of all, making use of Ohm's law in integral form, we have:

$$\mathscr{E} = RI \tag{2.236}$$

where R is the resistance of the ring, given by:

$$R = \rho \frac{l}{S} \tag{2.237}$$

Here ρ is the resistivity (inverse of conductivity), l is the circumference of the ring, and S is its cross-sectional area. It should be clear that the magnetic flux Φ_s, caused by the current flow in the ring is directly proportional to I, and can be written as:

$$\Phi_s = LI \tag{2.238}$$

where L is a coefficient of proportionality, known as the inductance of the ring. According to eq. 2.238, one could say that the inductance of the ring is the ratio of the magnetic flux through the ring and the current:

$$L = \Phi_s / I$$

The inductance is determined by geometrical parameters of the ring. In general, determination of the inductance involves solution of some very complicated problems. But in some special cases, including very thin circular rings, this task is relatively easy and the following expression for the total inductance of such a ring in a free space where $\mu = \mu_0$ can be derived:

$$L = r\mu_0 \left(\ln \frac{8r}{r_0} - 1.75 \right) \tag{2.239}$$

Inductance is measured in henrys per meter in the MKS system of units.

If, instead of one ring, we have n turns, the inductance increases as the square of the number of turns:

$$L = r\mu_0 n^2 \left(\ln \frac{8r}{r_0} - 1.75 \right) \tag{2.240}$$

The simple form of the conductive volume, a thin uniform circular ring, and the assumption that the current density is uniform over the cross-section of the ring, has permitted us to find simple analytical expressions for the coefficients of proportionality between electromotive force and secondary magnetic flux on the one hand, and the current in the ring on the other hand. Substituting eqs. 2.235 and 2.237 into eq. 2.238, we arrive at a differential equation from which the current I can be determined:

$$L \frac{dI}{dt} + RI = -\frac{\partial \Phi_0}{\partial t}$$

or

$$\frac{dI}{dt} + \frac{1}{\tau_0} I = f(t) \tag{2.241}$$

where

$$\tau_0 = L/R$$

and

$$f(t) = -\frac{1}{L} \frac{\partial \Phi_0}{\partial t}$$

is known. Making use of results obtained in paragraph 3 (see eq. 2.172) we have the following solution for the induction current:

$$I = I_0 e^{-t/\tau_0} - e^{-t/\tau_0} \frac{1}{L} \int_0^t e^{t/\tau_0} \frac{\partial \Phi_0}{\partial t} dt \tag{2.242}$$

Now we can consider the behavior of the induction currents for two special cases.

(A) Assume that the primary magnetic field varies with time as shown in Fig. 2.46, so that we have the following expressions for $\partial\Phi_0/\partial t$:

$$\frac{\partial\Phi_0}{\partial t} = \begin{cases} 0 & t < 0 \\ -\dfrac{\Phi_0}{t_r} & 0 \leqslant t \leqslant t_r \\ 0 & t \geqslant t_r \end{cases} \tag{2.243}$$

where t_r is called the "ramp time". During the time interval, over which the primary magnetic flux, Φ_0, does not change with time, $(t < 0)$, there are no induced currents in the ring, that is:

$$I(t) = 0 \quad \text{if } t < 0$$

During the ramp time, the primary flux, Φ_0, changes with time, and the induced current is defined by the rate of change of the primary magnetic flux, as well as by the characteristics of the ring (R and L). After the primary flux has disappeared, the behavior of the induced current is controlled by only the parameters R and L. In this case, eq. 2.241 is simplified, and we have:

$$\frac{\mathrm{d}I}{\mathrm{d}t} + \frac{I}{\tau_0} = 0 \tag{2.244}$$

A solution to this equation is:

$$I(t) = C\,e^{-t/\tau_0}, \quad \text{if } t_r \geqslant t \tag{2.245}$$

The parameter, τ_0, is commonly called the time constant for the ring, inasmuch as it represents the rate at which the current decays in the absence of external sources.

In order to find the constant C, we will investigate the behavior of induced current flow during the ramp time. In accord with eqs. 2.242 and 2.243, we have:

$$I(t) = I_0\,e^{-t/\tau_0} + e^{-t/\tau_0}\,\frac{\Phi_0}{t_r L}\int_0^t e^{t/\tau_0}\,\mathrm{d}t$$

$$= I_0\,e^{-t/\tau_0} + \frac{\tau_0}{t_r}\frac{\Phi_0}{L}(1 - e^{-t/\tau_0}), \quad 0 \leqslant t \leqslant t_r \tag{2.246}$$

In eq. 2.246, the parameter I_0 represents the amplitude of the current at the instant $t = 0$. Inasmuch as at this instant, the eddy current is absent ($I_0 = 0$) we have:

$$I(t) = \frac{\tau_0}{t_r} \frac{\Phi_0}{L} (1 - e^{-t/\tau_0}), \quad 0 \leqslant t < t_r \tag{2.247}$$

The constant C is readily found from eq. 2.245 and 2.246. In fact, we have:

$$I(t_r) = C e^{-t_r/\tau_0} = \frac{\tau_0}{t_r} \frac{\Phi_0}{L} (1 - e^{-t_r/\tau_0})$$

Thus:

$$C = \frac{\tau_0}{t_r} \frac{\Phi_0}{L} (e^{t_r/\tau_0} - 1) \tag{2.248}$$

Therefore, we obtain the following expressions for the current induced in the ring:

$$I(t) = \begin{cases} 0 & t \leqslant 0 \\[2ex] \dfrac{\tau_0}{t_r} \dfrac{\Phi_0}{L} (1 - e^{-t/\tau_0}) & 0 \leqslant t \leqslant t_r \\[2ex] \dfrac{\tau_0}{t_r} \dfrac{\Phi_0}{L} (e^{t_r/\tau_0} - 1) e^{-t/\tau_0} & t \geqslant t_r \end{cases} \tag{2.249}$$

In accord with those last equations, during the ramp time, the current induced within the ring increases gradually, reaching a maximum at the moment $t = t_r$, and then begins to decrease exponentially.

Suppose that the ramp time, t_r, is much less than the time constant, τ_0. Then, expanding the exponential terms in eq. 2.249 as a power series and discarding all terms but those of first and second order, we have:

$$I(t) = \begin{cases} 0 & t < 0 \\[2ex] \dfrac{t}{t_r} \dfrac{\Phi_0}{L} & 0 \leqslant t \leqslant t_r \\[2ex] \dfrac{\Phi_0}{L} e^{-t/\tau_0} & t \geqslant t_r \quad \text{if } t_r \ll \tau_0 \end{cases} \tag{2.250}$$

In another special case, in which $t_r \geqslant \tau_0$, at the beginning, the induced current increases linearly and then slowly approaches the maximum value for $t = t_r$ equal to:

$$\frac{\tau_0}{t_r} \frac{\Phi_0}{L} \ll \frac{\Phi_0}{L}$$

after which the induced current decays exponentially. Curves representing the behavior of such induced currents are shown in Fig. 2.52.

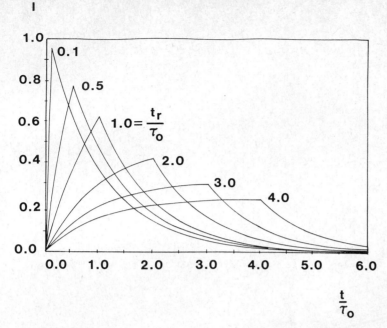

Fig. 2.52. Curves showing the behavior of induced currents.

We will investigate induced currents in the ring when the primary current and the primary magnetic flux change as a step function (see Fig. 2.53). It might already be obvious that the behavior of the induced current in this case is described by the last expression in eq. 2.250 in the limit as t_r approaches zero, that is:

$$I(t) = \frac{\Phi_0}{L}\, e^{-t/\tau_0}, \quad t > 0 \tag{2.251}$$

Thus, the initial value of the induced current does not depend on the resistance, R, but is determined by the primary flux, Φ_0, and the inductance of the ring, L.

Because in practice there always is a non-zero ramp time, the initial value of the current, Φ_0/L should be interpreted as being its value at the instant $t = t_r$, provided that t_r is much less than τ_0.

It is interesting to obtain the same result directly from eq. 2.241. Integrating both parts of eq. 2.241, we obtain:

$$R \int_0^{t_r} I\, \mathrm{d}t + L \int_0^{t_r} \frac{\partial I}{\partial t}\, \mathrm{d}t = -\int_0^{t_r} \frac{\partial \Phi_0}{\partial t}\, \mathrm{d}t$$

whence

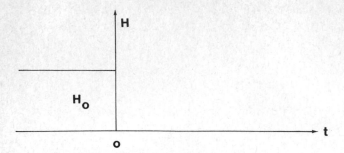

Fig. 2.53. The variation of magnetic field with time for an ideal step function.

$$R \int_0^{t_r} I \, dt + L[I(t_r) - I(0)] = \Phi_0(0) - \Phi_0(t_r) \tag{2.252}$$

Inasmuch as we have the following condition at the initial instant:

$$\Phi_0(0) = \Phi_0, \quad I(0) = 0$$

and at the instant $t = t_r$, the primary flux disappears so eq. 2.252 can be rewritten as:

$$R \int_0^{t_r} I \, dt + LI(t_r) = \Phi_0 \tag{2.253}$$

The integrand $I dt$ indicates the total quantity of charge passing through the ring during the time dt. It is obvious that with decreasing ramp time, the quantity of charge tends to zero, and in the limit, when the primary magnetic flux changes as a step function, we have:

$$LI(0) = \Phi_0, \quad \text{as } t_r \to 0$$

or, the initial current is:

$$I(0) = \Phi_0/L \tag{2.254}$$

This is exactly the same as obtained with eq. 2.251.

The analysis carried out earlier, shows that errors caused by discarding the integral become smaller as the ratio t_r/τ_0 becomes smaller, that is, with increasing inductance, L, or with decreasing resistance, R, eq. 2.254 becomes all the more precise.

Considering that for t greater than zero, the current satisfies a homogeneous differential equation, we again obtain:

$$I(t) = \frac{\Phi_0}{L} e^{-t/\tau_0} \tag{2.255}$$

Thus, at the initial instant, the current which is induced in the ring does not depend on the conductivity, and is defined by the primary flux and the geometric parameters describing the ring.

The equality:

$$LI(0) = \Phi_0 \tag{2.256}$$

is an essential feature of electromagnetic induction. In fact, the left-hand side of eq. 2.256 defines the magnetic flux of induced current included in the area bounded by the ring at the instant $t = 0$, when the primary flux disappears. Thus, an induced current arises in the ring of such magnitude $[I(0)]$ that at the first instant, its magnetic flux, $LI(0)$ is exactly and precisely equal to the primary flux, Φ_0. Later, this result will be generalized to include more complicated models of conducting material.

(B) Suppose that the primary magnetic field varies sinusoidally:

$$H_0 \sin \omega t \tag{2.257}$$

where H_0 is the amplitude of the field, f is the frequency, $\omega = 2\pi f$, and $f = 1/T$, with T being the period of the oscillation. In contrast to previous cases, here we are examining a field which has already been established, inasmuch as it is assumed that the sinusoidal process began a very long time before the present, and is repeating itself periodically. In order to find the induced current in the ring, we will make use of eq. 2.242. Since the primary flux can be written as $\Phi_0 \sin \omega t$, we have:

$$I(t) = I_0\, e^{-t/\tau_0} - \frac{\omega \Phi_0}{L}\, e^{-t/\tau_0} \int_0^t e^{t/\tau_0} \cos \omega t\, dt \tag{2.258}$$

Because:

$$\int e^{ax} \cos bx\, dx = \frac{e^{ax}}{a^2 + b^2}(a \cos bx + b \sin bx)$$

we obtain:

$$e^{-t/\tau_0} \int_0^t e^{t/\tau_0} \cos \omega t\, dt = \frac{1}{\dfrac{1}{\tau_0^2} + \omega^2}\left(\frac{1}{\tau_0}\cos \omega t + \omega \sin \omega t\right)$$

$$-\frac{1}{\tau_0}\frac{e^{-t/\tau_0}}{\dfrac{1}{\tau_0^2} + \omega^2}, \quad \text{where } \tau_0 = L/R$$

Thus, the induced current in the ring is:

$$I(t) = I_0 e^{-t/\tau_0} - \frac{\omega\Phi_0 R}{R^2 + \omega^2 L^2} \cos \omega t - \frac{\omega^2 \Phi_0 L}{R^2 + \omega^2 L^2} \sin \omega t$$

$$+ \frac{\omega\Phi_0 R}{R^2 + \omega^2 L^2} e^{-t/\tau_0}$$

Inasmuch as we are interested in the induced current for an established sinusoidal process, that is for t much greater than τ_0, we have:

$$I(t) = - \frac{\omega\Phi_0}{R^2 + \omega^2 L^2} (R \cos \omega t + \omega L \sin \omega t) \tag{2.259}$$

Let us introduce the following notations:

$$a = - \frac{\omega\Phi_0 R}{R^2 + \omega^2 L^2} \quad \text{and} \quad b = - \frac{\omega^2 \Phi_0 L}{R^2 + \omega^2 L^2} \tag{2.260}$$

Correspondingly, we have:

$$I = a \cos \omega t + b \sin \omega t \tag{2.261}$$

that is, the induced current can be represented as the sum of two separate oscillations. One oscillation is of the form $b \sin \omega t$ which changes synchronously with the primary magnetic field, and which is usually called the "inphase" component of the current; that is:

In $I = b \sin \omega t$

The other, $a \cos \omega t$, represents a term shifted in phase by 90° with the current in the primary source, and is called the "quadrature" component of the current:

$QI = a \cos \omega t$

Equation 2.261 suggests that it is desirable to treat the induced current in the ring as being the sum of an inphase and a quadrature component, the intensity of which are given by eq. 2.260. One can write the parameters a and b as:

$$a = A \sin \phi, \qquad b = A \cos \phi \tag{2.262}$$

so that we obtain:

$$I = A (\sin \phi \cos \omega t + \cos \phi \sin \omega t) = A \sin (\omega t + \phi) \tag{2.263}$$

Therefore, the induced current and the primary field through the loop are both sinusoidal functions, each having the same frequency ω, and characterized by two parameters A and ϕ.

The parameter A is the amplitude of the secondary current, with the oscillation reaching its maximum value A each time the argument $\omega t + \phi$ is equal to an odd multiple of $\pi/2$.

The presence of the phase, ϕ, indicates that the two oscillations, one being the primary field and the other the induced current, do not change in precise synchronism with each other, but rather, there is a phase shift between them, as illustrated in Fig. 2.54.

In accord with eqs. 2.260 and 2.262, we have:

$$A = (a^2 + b^2)^{1/2} = \frac{\omega\Phi_0}{(R^2 + \omega^2 L^2)^{1/2}} \qquad (2.264)$$

and

$$\tan\phi = a/b \quad \text{or} \quad \phi = \tan^{-1}(R/\omega L)$$

Curves for the quadrature and inphase components of these frequency responses, as well as curves for amplitude and phase are shown in Fig. 2.55.

In spite of the apparent simplicity of the very thin circular ring, the frequency responses of the induced current contain some general features which are inherent in much more complicated cases as will be demonstrated in later chapters.

To further examine the response of the ring, first consider the low-frequency part of the spectrum. Assuming that ωL is less than R, we can expand eq. 2.260 as a series:

$$a = -\frac{\omega\Phi_0}{R}\left\{1 + \left(\frac{\omega L}{R}\right)^2\right\}^{-1} = -\frac{\omega\Phi_0}{R}\left\{1 - \frac{\omega^2 L^2}{R^2} + \frac{\omega^4 L^4}{R^4} - \frac{\omega^6 L^6}{R^6} + \cdots\right\}$$

or

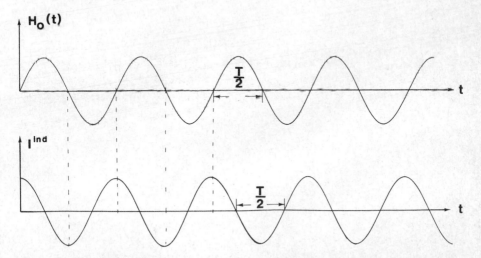

Fig. 2.54. Sinusoidal variation of a primary field and of induced current showing the presence of a phase shift between the two (case: $\omega \to 0$).

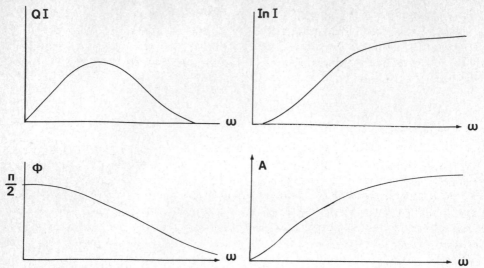

Fig. 2.55. Typical curves for quadrature and inphase components of frequency response, and for amplitude and phase.

$$a = -\frac{\Phi_0}{R}\omega + \frac{L^2}{R^3}\Phi_0\omega^3 - \frac{\Phi_0 L^4}{R^5}\omega^5 + \frac{\Phi_0 L^6}{R^7}\omega^7 - \ldots$$

and $\hspace{9cm}$ (2.265)

$$b = -\frac{\Phi_0 L}{R^2}\omega^2 \left\{1 + \left(\frac{\omega L}{R}\right)^2\right\}^{-1} = -\frac{\Phi_0 L\omega^2}{R^2} + \frac{\Phi_0 L^3}{R^4}\omega^4$$

$$-\frac{\Phi_0 L^5}{R^6}\omega^6 + \ldots$$

At the low-frequency part of the spectrum, the quadrature and inphase components of the induced currents can be represented as series containing either odd or even powers of ω. As will be shown later, this feature remains valid for induced currents arising in any confined conductor, as well as in some other special cases.

It is clear that both series converge only when the condition:

$$\omega L/R < 1 \hspace{8cm} (2.266)$$

is met. In other words, the radius of convergence for these series is defined from the equation:

$$\omega = R/L \hspace{8.5cm} (2.267)$$

From the general behavior of power series approximation, it is very well known that the radius of convergence for such series is the distance from the origin to the first singularity of the function being represented, located in the complex plane for the argument. In our case, the origin is at the point $\omega = 0$, with ω being considered as a complex variable. In order to find this singularity, we must investigate the denominator in eq. 2.260. It becomes zero when:

$$R^2 = -\omega^2 L^2 \quad \text{or} \quad \omega^2 = -\frac{R^2}{L^2}, \quad \text{i.e.} \quad \omega = \pm i\frac{R}{L}$$

where $i = \sqrt{-1}$. Thus, in this case, there are two singular points, located on the imaginary axis as shown in Fig. 2.56. It is interesting to note that the radius of convergence for the series describing the low-frequency part of the spectrum coincides with the power of the exponent for the transient field given in eq. 2.251. This fact will be considered in detail in subsequent chapters, and reflects one of the more important aspects of the relationship between the low-frequency part of the spectrum and the late stage of the transient coupling.

Next, suppose that the frequency is sufficiently low that we need only consider the first term in the series expansion in eq. 2.265:

$$a = -\frac{\omega\Phi_0}{R} \quad \text{and} \quad b = -\frac{\Phi_0 L}{R^2}\omega^2 \tag{2.268}$$

or

$$QI = -\frac{\Phi_0\omega}{R}\cos\omega t \quad \text{and} \quad \text{In}\,I = -\frac{\Phi_0 L}{R^2}\omega^2\sin\omega t$$

From this, it is apparent that at low frequencies, the quadrature component of the induced current is dominant and is directly proportional to the conductivity of the ring and to the frequency, while it does not depend on the inductance. This behavior can readily be explained as follows: If we neglect the flux caused by induced currents, the total flux is the same as the primary flux, $\Phi_0\sin\omega t$. As it changes with time, we have:

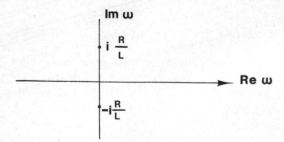

Fig. 2.56. The positions of singular points on the imaginary axis for the behavior of the spectrum.

$$\frac{\partial \Phi}{\partial t} = \frac{\partial \Phi_0}{\partial t} = \omega \Phi_0 \cos \omega t$$

and therefore, in accord with Ohm's law, we have:

$$QI = -\frac{\omega \Phi_0}{R} \cos \omega t$$

Thus, at low frequencies, the quadrature component is directly proportional to the primary field, frequency and conductivity. It is important to state that this behavior will not change even when more complicated conductors are being considered.

In contrast to the situation with the quadrature component, the inphase component arises due to the secondary magnetic flux. In the approximation under consideration, the magnetic flux caused by the quadrature component of the current is:

$$\Phi_1 = L\,QI = -\frac{\omega \Phi_0 L}{R} \cos \omega t$$

and its rate of change with time is:

$$\frac{\partial \Phi_1}{\partial t} = \frac{\omega^2 \Phi_0}{R} L \sin \omega t$$

Correspondingly, for the inphase component of the current caused by this magnetic flux, we have:

$$\mathrm{In}I = -\frac{\omega^2 \Phi_0 L}{R^2} \sin \omega t$$

which is identical with the first term of the series in eq. 2.265. Let us note that by applying the same approach we can obtain the subsequent terms in these series.

Next, we will investigate the high-frequency portion of the spectrum for the induced currents for the case with ωL much greater than R. In accord with eq. 2.260:

$$a \to 0, \quad \text{and} \quad b \to -\Phi_0/L \tag{2.269}$$

In this portion of the spectrum, the inphase component dominates, and it approaches a constant value which is determined by the magnetic flux in the primary field and by the geometric parameters of the ring. Comparing eqs. 2.254 and 2.269, one can say that the magnitudes of the induced current at the early stage of transient coupling and at the high-frequency part of the spectrum coincide. This result is not accidental, and can be explained by the next example.

Here it is appropriate to make the following comment. The way in which both components approach their asymptotic values at high frequencies differs from that when solid confined conductors are considered. This is related to the fact that the cross-section of the ring has been assumed to be infinitesimal. However, the general features of the behavior of the frequency responses shown in Fig. 2.55 are similar to those for induced currents in confined conductors. Inasmuch as, in accord with the Biot-Savart law, the quadrature and inphase components of the secondary magnetic field are generated by the corresponding components of the induced currents, it can readily be understood that the frequency responses for the corresponding components of the magnetic field and the induced currents are similar.

Example VII. Behavior of the electromagnetic field at the early stage and at high frequencies in a conductive medium

Suppose that we have an arbitrarily oriented system of n conducting rings. The equation for the induced current in the kth ring can be written as:

$$R_k I_k = -\frac{\partial \Phi_k}{\partial t} = -\frac{\partial \Phi_{0k}}{\partial t} - \frac{\partial \Phi_{sk}}{\partial t} \tag{2.270}$$

where R_k and I_k are the resistance and current in the kth ring, and Φ_{0k} and Φ_{sk} are magnetic fluxes of the primary and secondary fields, respectively. It is obvious that the magnetic flux Φ_{sk} can be written as:

$$\Phi_{sk} = M_{1k} I_1 + M_{2k} I_2 + \ldots + L_k I_k + \ldots + M_{nk} I_n$$

where L_k is the inductance of the kth ring, M_{ik} are mutual inductance values, that is, the magnetic flux through the area of the kth ring, caused by the current I_i in the ith ring, as defined from the equation:

$$\Phi_{ik} = M_{ik} I_i$$

Correspondingly, eq. 2.270 can be rewritten as:

$$R_k I_k + L_k \frac{\partial I_k}{\partial t} + \sum_{n=1}^{N} M_{nk} \frac{\partial I_n}{\partial t} = -\frac{\partial \Phi_{0k}}{\partial t}, \quad n \neq k \tag{2.271}$$

Now assume that the primary flux Φ_0 caused by external sources starts to change from a value of Φ_0 to zero at the instant $t = t_0$, and that this change takes place over a very short period of time, t_r (see Fig. 2.57). Integrating eq. 2.271 with respect to time, we obtain:

$$R_k \int_{t_0}^{t_0 + t_r} I_k \, dt + L_k \int_{t_0}^{t_0 + t_r} \frac{\partial I_k}{\partial t} \, dt + \sum_{n=1}^{N} M_{nk} \int_{t_0}^{t_0 + t_r} \frac{\partial I_n}{\partial t} \, dt = \Phi_{0k}(t_0), \quad n \neq k$$

since $\Phi(t_0 + t_r) = 0$.

Taking into account that induced currents are absent at the first instant

Fig. 2.57. Example of a finite but short ramp time for the termination of a constant value of magnetic flux.

(t_0) and that the interval t_r is very short, this last equation can be approximated as follows:

$$L_k I_k(t_0 + t_r) + \sum_{n=1}^{N} M_{nk} I_n(t_0 + t_r) = \Phi_{0k}(t_0), \quad n \neq k$$

Introducing the notation:

$$t_0 = t_0^- \quad \text{and} \quad t_0 + t_r = t_0^+$$

we have:

$$L_k I_k(t_0^+) + \sum_{n=1}^{N} M_{nk} I_n(t_0^+) = \Phi_{0k}(t_0^-), \quad n \neq k \tag{2.272}$$

On the left-hand side of this expression we have a representation for the magnetic flux through any k-ring caused by the induced currents in all of the rings at the first instant after switching, while on the right-hand side, we have the expression for the primary flux, Φ_{0k}, before switching. Thus, we again observe a principal feature of electromagnetic induction when the primary flux changes as a step function:

$$\Phi_0 = \begin{cases} \Phi_0 & t \leqslant t_0 \\ 0 & t > t_0 \end{cases}$$

In fact, at the very first instant, the induced currents in every ring have exactly that magnitude so that the flux of the magnetic field caused by these currents through any ring is precisely equal to the primary flux.

Now we are prepared to describe the asymptotic behavior of the field in a conducting medium. We will assume that in the general case the medium is not uniform and that sources for the primary field can be located either outside or inside the conductor. Also, let us suppose that the primary flux Φ_0 at the instant $t = t_0$ disappears instantaneously. Before switching off the flux Φ_0, of the magnetic field has been constant in time, and therefore no induced currents are present in the conductor. Correspondingly, the circulation of the magnetic field inside the medium along an arbitrary path was zero:

$$\oint_L H_0 \cdot \mathrm{d}l \;=\; 0 \qquad\qquad (2.273)$$

providing the path of integration does not enclose a current from the primary source.

A conducting medium can be visualized as consisting of a system of current rings with arbitrary shapes and in this way, one can apply results which have been obtained above. Inasmuch as the flux piercing any ring at the instant $t = t_0^+$ remains the same as that for earlier times, the magnetic field at any point in the conducting medium does not change either. This conclusion stems from the fact that for an arbitrary surface inside the conductor:

$$\Phi(t_0^+) \;=\; \Phi_0$$

Thus, immediately after disappearance of the primary flux, Φ_0, we have:

$$H(t_0^+) \;=\; H_0 \qquad\qquad (2.274)$$

Let us emphasize that this relationship does not exist outside the conductor.

From eqs. 2.273 and 2.274, it follows that the circulation of the magnetic field for any path inside a conductor is zero at the instant t_0^+ and therefore there are no induced currents:

$$j \;=\; 0 \quad \text{and} \quad E \;=\; 0, \quad \text{if} \quad t \;=\; t_0^+ \qquad\qquad (2.275)$$

However, there must be sources of magnetic field which cause the primary field H_0 at the instant when the source is switched off. These sources are induced surface currents, which are situated closer to the source of the primary field if the primary field is situated outside the conductor.

If the source of the primary field is situated within the conductor, the induced currents initially exist only near the source. Induced currents concentrated on the surface or near the primary source decay with time, since the electromagnetic energy is converted to heat and appear at various points in the medium. It is obvious that the decay of the field takes place more rapidly in a highly resistive medium, while in a highly conducting medium, the field decreases more slowly.

Let us note that in solving many boundary problems related to the calculation of nonstationary fields, conditions 2.274 or 2.275 are extremely important, and it is usually called the "initial condition". It is, in essence, a modification Faraday's law, and any nonstationary field in a conducting medium must satisfy eq. 2.274 or 2.275.

It is clear that if the current in a primary source located outside a conductor changes very rapidly, the induced currents essentially remain on the surface of the conductor. This explains why the high-frequency asymptote coincides with that for the early stage of transient electromagnetic field.

114

2.5. ELECTROMAGNETIC FIELD EQUATIONS

In the previous paragraphs, by making use of Gauss's and Stoke's theorems, we have developed the basic laws for the electromagnetic fields in equation form. In accord with these laws the electromagnetic field must satisfy the following set of equations:

$$\oint_L E \cdot dl = -\frac{\partial}{\partial t} \int_S B \cdot dS \tag{2.276}$$

$$\oint_L H \cdot dl = \int_S j \cdot dS + \int_S \frac{\partial D}{\partial t} \cdot dS \tag{2.277}$$

$$\oint_S D \cdot dS = e \tag{2.278}$$

$$\oint_S B \cdot dS = 0 \tag{2.279}$$

where E and H are electric and magnetic field vectors, B and D are magnetic and electric induction vectors, e is a charge in a volume surrounded by a surface S, and j is conduction current density. The various vectors are related by a set of relationships known as the constitutive equations:

$$D = \epsilon E, \qquad B = \mu H, \qquad j = \sigma E$$

where ϵ, μ, and σ are the dielectric permeability, the magnetic permeability, and the electrical conductivity of the medium, respectively. The paths of integration, L, can be arbitrarily situated, and in some cases they can cross the boundaries between media having different properties. Equations 2.276–2.279 are called the Maxwell equations in integral form, and each one of them describes a specific physical law. For this reason, any distribution of an electromagnetic field must satisfy these equations. They define the field at any point in the medium, including points situated on interfaces. Maxwell's equations must describe the field everywhere, regardless of the nature of the change in electrical properties from one region to another.

The first equation (eq. 2.276) is in essence Faraday's law, while the second equation (eq. 2.277) follows from a combination of Ampere's law and the postulate of conservation of charge. The third equation (eq. 2.278) is obtained from Coulomb's law for a nonalternating electric field. However, it remains valid regardless of how quickly the field may change. In order to demonstrate this, we will use the postulate of conservation of charge (eq. 2.149):

$$\oint_S j \cdot dS = -\dot{e} \quad \text{or} \quad \text{div } j = -\frac{\partial \delta}{\partial t} \tag{2.280}$$

Applying the second Maxwell equation (eq. 2.277) twice along the contour L, once in one direction, and then in the opposite direction, and considering two surfaces S_1 and S_2 bounded by the same contour L (see Fig. 2.58) we have:

$$\oint_L H \cdot \mathrm{d}l = \int_{S_1} j \cdot \mathrm{d}S + \int_{S_1} \dot{D} \cdot \mathrm{d}S$$

$$\oint_L H \cdot \mathrm{d}l = \int_{S_2} j \cdot \mathrm{d}S + \int_{S_2} \dot{D} \cdot \mathrm{d}S$$

Adding the two equations and considering that the surfaces S_1 and S_2 form a closed surface, we obtain:

$$0 = \oint_S j \cdot \mathrm{d}S + \int_S \frac{\partial D}{\partial t} \cdot \mathrm{d}S$$

and in accord with eq. 2.280:

$$\oint_S \frac{\partial D}{\partial t} \cdot \mathrm{d}S = \dot{e}$$

whence

$$\oint_S D \cdot \mathrm{d}S = e$$

By analogy, using the first Maxwell equation, we also have:

$$\oint_S B \cdot \mathrm{d}S = 0$$

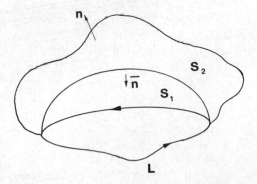

Fig. 2.58. Two surfaces S_1 and S_2 bounded by a common contour L.

The fourth equation (eq. 2.279) represents the fact that the magnetic flux through an enclosed surface is zero. This consideration demonstrates that the field equations can also be written as other sets of equations:

$$\oint_L E \cdot dl = -\frac{\partial}{\partial t} \int_S B \cdot dS$$

$$\oint_L H \cdot dl = \int_S j \cdot dS + \frac{\partial}{\partial t} \int_S D \cdot dS \tag{2.281}$$

$$\oint_S j \cdot dS = -\frac{\partial e}{\partial t}$$

inasmuch as eqs. 2.278 and 2.279 can be derived from the system given in eqs. 2.281. However, we will use the basic system of equations given in eqs. 2.276–2.279. It must be obvious that in any actual situation, the electromagnetic field has a finite value everywhere in space. However, in order to simplify the computation of fields, often some assumptions are made about the sources for the primary field. For example, in the place of an actual source, magnetic or electric dipoles may be considered. This type of approximation immediately leads to the existence of infinitely large values for the field at infinitesimal distance from near such sources. Therefore, eqs. 2.276–2.279 cannot be applied in the immediate vicinity of such idealized sources. For this reason a very small volume, in which the source is situated, is conceptually surrounded by a surface on which the field almost coincides with that caused by the currents and charges of such a primary source. In other words, near the source, the total field has to approach the primary field. One can say that this condition characterizes the type, intensity, and location of a primary field source.

On the other hand, with an unlimited increase in distance from the source the field must decrease in a proper way. This condition at infinity must be taken into account in the full description of a field. Finally, there is one more condition which appears when a transient field is being considered. For example, if the current or charges representing the source of the primary field change in the form of a step function at some moment $t = t_0$, eqs. 2.276 and 2.277 cannot be applied, since the derivatives with respect to time are not well defined at this instant. Therefore, at this instant, Maxwell's equations are replaced by an initial condition as described in paragraph 4 of this chapter.

Thus, a full description of the electromagnetic field includes not only Maxwell's equations as given in eqs. 2.276–2.279, but also conditions that must be met near the primary source and at infinity, along with an initial condition. Thus, the following series of steps can be recognized in defining an electromagnetic field through the use of eqs. 2.276–2.279:

(1) Determination of a set of functions satisfying the system of integral eqs. 2.276—2.279.

(2) Choice among these functions of those which satisfy the condition at infinity.

(3) Choice among the remaining functions of those which satisfy the condition near the source.

(4) Choice among the remaining functions of those which satisfy the initial condition, if a transient field is being considered.

From the physical point of view, it is apparent that a solution found in this way, represents in the electromagnetic field generated by the given distribution of sources. However, for the solution of a variety of problems, it is frequently preferable to apply differential equations. For this reason, we will consider a differential form of Maxwell's equations:

$$\text{curl } E = -\frac{\partial B}{\partial t} \qquad \text{div } D = \delta$$

$$\text{curl } H = \sigma E + \frac{\partial D}{\partial t} \qquad \text{div } B = 0$$

$$(2.282)$$

here δ is a free charge.

In contrast to the integral forms given in eq. 2.281, all the vectors that enter into each of the equations in 2.282 are considered at a single point. The essential feature of Maxwell's equations written in the differential form is that they describe the field only at points where the first derivatives of the field exist, that is, where the divergence and curl have meaning. Thus, unlike Maxwell's equations in the integral form, eqs. 2.282 can be applied only for so-called well behaved points. However, as is obvious there can be points, lines, and surfaces where some components of the electromagnetic field are discontinuous functions of the spatial variables. For example, the normal component of the electric field is usually a discontinuous function of the spatial variables at an interface separating two media with differing resistivity. As a consequence, we must make use of surface analogies to eqs. 2.282 at such interfaces. One must require the tangential components of electric and magnetic fields to be continuous functions, see eqs. 2.56 and 2.124, that is:

$$n \times (E_2 - E_1) = 0, \quad n \times (H_2 - H_1) = 0 \qquad (2.283)$$

where n is a unit normal vector to the interface, E_2 and E_1, and H_2 and H_1 are the electric and magnetic fields on either side of such an interface.

Thus, in essence, eqs. 2.283 are surface analogs to the corresponding Maxwell's equations given in differential form in eqs. 2.282. Thus, starting from the system of differential equations 2.282, the problem of defining the field consists of the following steps:

(1) Determination of a set of functions satisfying the differential equations 2.282.

(2) The choice among these functions of those satisfying the condition at infinity.

(3) The choice among the remaining functions of those having the given behavior near the source of the primary field.

(4) A choice among the still remaining functions of those that satisfy the boundary conditions given in eqs. 2.283.

(5) A choice among the remaining functions, E and H, of those that satisfy the initial conditions if a nonstationary field is being considered.

Considering that:

$$D = \epsilon E, \quad B = \mu H, \quad j = \sigma E$$

the system of equations in 2.282 usually contains two unknowns, namely the electric and magnetic field intensities. One can say that we have four differential equations in partial derivatives of the first order with respect to two unknown vectors, but more accurately, to six unknown components of the electromagnetic field.

Very frequently it is more convenient to derive equations in which the electric and magnetic fields are separated, than to make use of the simultaneous set of equations in 2.282. Let us consider points in the medium where the parameters σ, μ, and ϵ do not change:

$$\frac{\partial \sigma}{\partial l} = \frac{\partial \epsilon}{\partial l} = \frac{\partial \mu}{\partial l} = 0$$

and where dl is an arbitrarily oriented displacement. As has been shown previously (paragraph 2.3), electric charges are absent at such points, and therefore, Maxwell's equations take the form:

$$\operatorname{curl} E = -\mu \frac{\partial H}{\partial t} \qquad \operatorname{div} E = 0$$

$$\operatorname{curl} H = \sigma E + \epsilon \frac{\partial E}{\partial t} \qquad \operatorname{div} H = 0 \tag{2.284}$$

From the first Maxwell equation, we have:

$$\operatorname{curl} \operatorname{curl} E = -\mu \frac{\partial}{\partial t} \operatorname{curl} H$$

In making use of the vector identity:

$$\operatorname{curl} \operatorname{curl} E = \operatorname{grad} \operatorname{div} E - \nabla^2 E$$

and of the second Maxwell equation, we obtain:

$$\operatorname{grad} \operatorname{div} E - \nabla^2 E = -\mu \frac{\partial}{\partial t} \left(\sigma E + \epsilon \frac{\partial E}{\partial t} \right)$$

Taking into account Maxwell's third equation:

$$\operatorname{div} E = 0$$

we have:

$$\nabla^2 E - \mu\sigma \frac{\partial E}{\partial t} - \mu\epsilon \frac{\partial^2 E}{\partial t^2} = 0 \qquad (2.285)$$

where $\nabla^2 E = \Delta E$ is known as the Laplacian of the electric field.

Similarly, using the second equation in 2.284, we have:

$$\operatorname{curl} \operatorname{curl} H = \operatorname{grad} \operatorname{div} H - \nabla^2 H = \sigma \operatorname{curl} E + \epsilon \frac{\partial}{\partial t} \operatorname{curl} E$$

In making use of the first and fourth Maxwell equations, we obtain:

$$\nabla^2 H - \sigma\mu \frac{\partial H}{\partial t} - \mu\epsilon \frac{\partial^2 H}{\partial t^2} = 0 \qquad (2.286)$$

Thus, for points in the medium where the electric and magnetic properties do not vary spatially, we have obtained equations involving only the electric or magnetic fields, and the two equations are of identically the same form being of second order in partial derivatives. They are sometimes known as telegraph equations for conductive medium.

When these equations are used, the determination of the electromagnetic field can be done in almost the same sequence of steps as before:

(1) Definition of various functions that satisfy eqs. 2.285 and 2.286.

(2) The choice among these functions of those that satisfy the condition at infinity.

(3) The choice among the remaining functions of those whose behavior near the source corresponds to that for the primary field.

(4) The choice among the remaining functions of those that satisfy the surface conditions given in eqs. 2.283.

(5) The choice among the still remaining functions of those that satisfy the initial condition if a non-stationary field is being considered.

Now let us consider some special cases:

(A) First of all, assume that an electromagnetic field does not change with time, that is, all the derivatives with respect to time are zero, and that:

$$D = \epsilon_0 E \quad \text{and} \quad B = \mu_0 H$$

Then, in accord with eqs. 2.282 and 2.283, we have the following equations for well behaved points and for interfaces:

$$\operatorname{curl} E = 0 \qquad \operatorname{curl} H = j$$

$$\operatorname{div} E = \frac{\delta}{\epsilon_0} \qquad \operatorname{div} H = 0 \qquad (2.287)$$

and

$$n \times (E_2 - E_1) = 0 \qquad n \times (H_2 - H_1) = 0$$
$$E_n^{(2)} - E_n^{(1)} = \Sigma/\epsilon_0 \qquad H_n^{(2)} - H_n^{(1)} = 0 \qquad (2.288)$$

where Σ is the surface density of charge. In this case of a constant field ($\partial/\partial t = 0$), the system is split into two parts as follows:

$$\text{curl } E = 0 \qquad \text{div } E = \frac{\delta}{\epsilon_0}$$

and at interfaces:

$$n \times (E_2 - E_1) = 0 \qquad E_n^{(2)} - E_n^{(1)} = \Sigma/\epsilon_0 \qquad (A)$$

here δ and Σ are volume and surface charges, respectively, and:

$$\text{curl } H = j \qquad \text{div } H = 0$$

and at interfaces:

$$n \times (H_2 - H_1) = 0 \qquad H_n^{(2)} - H_n^{(1)} = 0 \qquad (B)$$

One part (A) defines the electric field and clearly shows that the sole source of the field is electric charge, which can exist at points where the conductivity changes, such as at interfaces. It is clear that the electric field can be found without any knowledge of the magnetic field, and that the electric field is governed only by Coulomb's law. When the electric field has been determined, current density can be calculated using Ohm's law, $j = \sigma E$, and making use of the second part (B), the magnetic field can be found as well. Also, it can be calculated using the Biot-Savart law.

At this point, we will consider a very important case, that of a quasi-stationary field, which is often also called a quasi-static field.

(B) Suppose that in the second equation describing the field, eq. 2.282, we can ignore the second term which represents displacement currents. Then, system 2.282 can be written as:

$$\text{curl } E = -\mu \frac{\partial H}{\partial t} \qquad \text{curl } H = j$$
$$\qquad (2.289)$$
$$\text{div } E = \frac{\delta}{\epsilon_0} \qquad \text{div } H = 0$$

and at interfaces as:

$$n \times (E_2 - E_1) = 0 \qquad n \times (H_2 - H_1) = 0$$
$$\qquad (2.290)$$
$$E_n^{(2)} - E_n^{(1)} = \frac{\Sigma}{\epsilon_0} \qquad H_n^{(2)} - H_n^{(1)} = 0, \quad \text{if } i = 0$$

From these expressions, it can be seen that the electric field has two sources:
(1) volume and surface charges, and
(2) change of the magnetic field with time.
Therefore, the electric field can be represented as a sum:

$$E = E^c + E^v$$

where E^c and E^v are both generally caused by charges and a change in magnetic field with time. However, when the field varies relatively slowly, E^c is very nearly the field caused only by charges, and it is defined by Coulomb's law, while E^v is the vortex electric field, generated by a change in the magnetic field with time.

In contrast to the behavior of the electric field, the quasi-stationary magnetic field has but one source, conduction currents. Comparing the equations for the magnetic field (eq. 2.289) with those for a constant magnetic field (eq. 2.287) we see that they are precisely the same. This means that the magnetic field at any point in the medium is defined by the instantaneous values of current density throughout the conducting medium, and can be calculated using only the Biot-Savart law. One can say that the quasi-stationary approximation means that we neglect propagation time for electromagnetic energy, that is, it is assumed that the field travels instantaneously from the transmitter to the receiver. In order to emphasize this, let us write down the field eqs. 2.285 and 2.286 for this approximation. Discarding terms involving displacement currents, we have:

$$\Delta E = \sigma\mu \frac{\partial E}{\partial t}, \qquad \Delta H = \sigma\mu \frac{\partial H}{\partial t} \tag{2.291}$$

These equations are known as the diffusion equations, that is, they describe the penetration of energy, but do not take into account wave propagation. They can be used provided that the time at which the signal is recorded or the period of the oscillations significantly exceeds the travel time for the field from the source to the observation point. It can be said that the quasi-stationary approximation is valid when conduction currents dominate over displacement currents in a conducting medium, and the arrival time for a signal in an insulator is much less than the duration of a measurement or the period of the observed oscillations. This assumption is equivalent to stating that the signal arrives instantly at all points where the field is measured.

It might be worth noting that even though the propagation effect is not considered in the quasi-stationary field approach, even this field contains some essential features of propagation.

At this point, we will examine a special case in which the electromagnetic field varies as a sinusoidal or a cosinusoidal function of time. This leads to some important simplifications in the presentation of Maxwell's equations (through the use of the so-called operator notation). Suppose that we have a sinusoidal oscillation:

$$M = M_0 \sin(\omega t + \phi) \tag{2.292}$$

where M_0 is the amplitude of the oscillation, ϕ is its phase, and ω is the radian frequency. Making use of Euler's formula:

$$e^{i(\omega t + \phi)} = \cos(\omega t + \phi) + i \sin(\omega t + \phi)$$

we can write eq. 2.292 as the imaginary part of an exponential term:

$$M_0 \sin(\omega t + \phi) = \operatorname{Im} \hat{M} e^{i\omega t} \tag{2.293}$$

where $\hat{M}$ is a complex amplitude given by:

$$\hat{M} = M_0 e^{i\phi} \tag{2.294}$$

Therefore, we have:

$$\hat{M} e^{i\omega t} = M_0 e^{i\phi} e^{i\omega t} = M_0 e^{i(\omega t + \phi)}$$

Whence

$$\operatorname{Im}(\hat{M} e^{i\omega t}) = \operatorname{Im}(M_0 e^{i(\omega t + \phi)})$$
$$= \operatorname{Im}[M_0\{\cos(\omega t + \phi) + i \sin(\omega t + \phi)\}]$$
$$= M_0 \sin(\omega t + \phi)$$

Similarly, a cosinusoidal oscillation can be represented by the real part of a complex function, that is:

$$M_0 \cos(\omega t + \phi) = \operatorname{Re} \hat{M} e^{i\omega t}$$

where $\hat{M}$ is again equal to $M_0 e^{i\phi}$.

Let us emphasize that the complex amplitude, $\hat{M}$, is defined by the real amplitude, M_0, and the phase, ϕ, of an oscillation.

Inasmuch as:

$$\hat{M} e^{i\omega t} = M_0 \cos(\omega t + \phi) + i M_0 \sin(\omega t + \phi)$$

and since both terms on the right-hand side of this equality are solutions of Maxwell's equations, one can operate using only the function $\hat{M} e^{i\omega t}$, and then after finding a function that satisfies this system of equations, one must take either the imaginary or real part. The representation of a solution by the form $\hat{M} e^{i\omega t}$ has a remarkable feature. In fact, it is actually the product of two functions, one being the complex amplitude, $\hat{M}$, which is a function of coordinates and the properties of the medium, as well as of frequency, but which does not depend on time. The second multiplier, $e^{i\omega t}$, depends on time in a simple manner and as is readily seen, after differentiation, still remains an exponential. This fact permits us to write Maxwell's equations in a form which does not contain the argument t, and this facilitates the solution. It is appropriate to note, that the sinusoidal function which is being considered has been in effect for such a long time that there is no need to take into account an initial condition.

Thus, representing a field and charges in the form:

$$H = \hat{H}e^{i\omega t}, \quad E = \hat{E}e^{i\omega t}, \quad \delta = \hat{\delta}e^{i\omega t} \tag{2.295}$$

and using these in Maxwell's equations (2.282) we obtain:

$$\text{curl } \hat{E} = -i\omega\mu\hat{H} \qquad \text{div } \hat{E} = \hat{\delta}$$
$$\text{curl } \hat{H} = \sigma\hat{E} + i\omega\epsilon\hat{E} \qquad \text{div } \hat{B} = 0 \tag{2.296}$$

inasmuch as:

$$\frac{\partial}{\partial t}e^{i\omega t} = i\omega e^{i\omega t}$$

Similarly, we have the following for eqs. 2.285 and 2.286:

$$\nabla^2\hat{H} - (i\sigma\mu\omega - \omega^2\epsilon\mu)\hat{H} = 0$$

and

$$\nabla^2\hat{E} - (i\sigma\mu\omega - \omega^2\epsilon\mu)\hat{E} = 0 \tag{2.297}$$

The quantity:

$$k^2 = i\sigma\mu\omega - \omega^2\epsilon\mu \tag{2.298}$$

is usually considered to be the square of a wave number, k. For quasi-stationary behavior of the field in which displacement currents are neglected, the wave number can be written as:

$$k^2 = (i\sigma\mu\omega)^{1/2} \tag{2.299}$$

Alternate forms are:

$$k = (i\sigma\mu\omega)^{1/2} = (\sigma\mu\omega/2)^{1/2}(1+i) = (1+i)/h \tag{2.300}$$

where h is the skin depth, defined as:

$$h = (2/\sigma\mu\omega)^{1/2} = (10^3/2\pi)(10\rho/f)^{1/2} \tag{2.301}$$

Here ρ is the resistivity in ohm meters, f is the frequency in hertz, and μ is the magnetic permeability, normally taken to be the value for free space, which is $4\pi \times 10^{-7}$ Henry/meter (Hm^{-1}).

Maxwell's equations can be written as follows for a harmonic quasi-stationary field behavior:

$$\text{curl } E = -i\omega\mu H \qquad \text{div } E = \frac{\delta}{\epsilon_0}$$
$$\text{curl } H = \sigma E \qquad \text{div } B = 0 \tag{2.302}$$

(Here, the "hat" previously used to indicate a complex amplitude has been omitted for simplicity.)

By algebraic recombination of these four equations we have Helmholtz's equations:

$$\nabla^2 H - i\sigma\mu\omega H = 0, \quad \nabla^2 E - i\sigma\mu\omega E = 0 \tag{2.303}$$

The system of equations in 2.302 is particularly simplified in the case in which a medium consists of parts within which conductivity is constant; that is, a piecewise uniform medium. In this case, electric charges can arise only in interfaces, and within the uniform pieces, the volume density of charge is zero. Therefore, in place of eq. 2.302, within each volume we have:

$$\text{curl } E = -i\omega\mu H \qquad \text{div } E = 0$$

$$\text{curl } H = \sigma E \qquad \text{div } H = 0$$

The piecewise uniform medium is the most widely used model for a geoelectric section, and it is appropriate here to formulate again the steps to use in determining the quasi-stationary harmonic field for this type of model. The steps are:

(1) Determination of solution functions that satisfy the systems of equations:

$$\begin{aligned} \text{curl } E &= -i\omega\mu H & \text{div } E &= 0 \\ \text{curl } H &= \sigma E & \text{div } B &= 0 \end{aligned} \tag{2.304}$$

or

$$\nabla^2 H - k^2 H = 0, \quad \nabla^2 E - k^2 E = 0, \quad \text{where } k^2 = i\sigma\mu\omega$$

(2) A choice among these functions of those which satisfy the source condition for the primary field.

(3) The choice among the remaining functions of those which satisfy the condition at infinity.

(4) The selection among the still remaining functions of those which satisfy various conditions at each interface, that is, the continuity of tangential components of the electromagnetic field.

As has been pointed out previously, inasmuch as the field under investigation must be stationary, there is no initial condition to be met, and because of this, the solution is made more simple. It might also be noted that a solution of eq. 2.303 will be in the form of a set of complex amplitudes for the electric and magnetic fields, and in accord with eq. 2.294, we obtain the amplitude M_0 and the phase ϕ of an oscillation in the basic form, eq. 292.

It should be apparent that when a solution has been obtained for harmonic fields, a solution can also be obtained for any arbitrary time dependence through the use of the Fourier transform.

Most frequently, the electric and magnetic field vectors cannot be completely described using but a single spatial component. For this reason, a solution can turn out to be very cumbersome. Some simplification can be obtained by making use of various auxiliary functions. There are two ways in which such auxiliary functions can be introduced. One approach follows from use of the third equation in the set 2.304:

$$\text{div } E = 0 \tag{2.305}$$

This is an approach which is commonly used when the field is energized using a non-grounded loop, as for example, represented by a magnetic dipole. In this case, only inductive excitation of the field takes place. In accord with eq. 2.305, the electric field can be defined as being a spatial derivation of the vector potential, A^*:

$$E = \text{curl } A^* \tag{2.306}$$

because the relationship:

$$\text{div curl } A^* = 0$$

always applies for any vector.

The function A^* is called a vector potential of the magnetic type. It should be obvious that the same electric field can be described by an infinite number of different functions A^*. For example, the gradient of any function can be added to some fixed potential A^* to provide the result:

$$\text{curl } (A^* + \text{grad } \phi) = \text{curl } A^* + \text{curl grad } \phi$$

Taking into account another vector identity:

$$\text{curl grad } \phi = 0$$

we can have:

$$\text{curl } (A^* + \text{grad } \phi) = \text{curl } A^* = E$$

This ambiguity in definition of A^* can be used to our advantage in simplifying equations when the vector potential is used, as well as to express both vectors of the field in terms of this single function.

To obtain a solution, we substitute eq. 2.306 into the second equation of 2.304 so that we have:

$$\text{curl } H = \sigma \text{ curl } A^* = \text{curl } \sigma A^*$$

since σ is considered to be a constant. This can all be written as:

$$\text{curl } (H - \sigma A^*) = 0$$

whence

$$H - \sigma A^* = \text{grad } \phi \tag{2.307}$$

where ϕ is some scalar function. Just as is the case with the vector potential, this function is ambiguous. Substituting the expressions for the vector quantities E and H in terms of the functions A^* and ϕ in the first eq. 2.304, we obtain:

$$\text{curl curl } A^* = -i\omega\mu(\sigma A^* + \text{grad } \phi)$$

Inasmuch as:

$$\text{curl curl } A^* = \text{grad div } A^* - \nabla^2 A^*$$

where ∇^2 is the vector Laplacian, we have the result:

$$\text{grad div } A^* - \nabla^2 A^* = -i\omega\mu\sigma A^* - i\omega\mu \text{ grad } \phi \tag{2.308}$$

The expressions which have been obtained for the functions A^* and ϕ are quite complicated. In order to simplify this last equation, we now choose a pair of functions A^* and ϕ that satisfy the condition:

$$\text{div } A^* = -i\omega\mu\phi \tag{2.309}$$

By using this so-called gauge condition, the differential equation becomes Helmholtz's equation for the vector potential A^*:

$$\nabla^2 A^* - k^2 A^* = 0 \tag{2.310}$$

This is precisely the same equation for either the electric or the magnetic field. However, in view of the use of condition 2.309, both vectors comprising an electromagnetic field are expressed in terms of a single vector potential quantity, A^*. In accord with eqs. 2.306, 2.307, and 2.309, we have the following representations for the two vectors quantities:

$$E = \text{curl } A^*$$

$$H = \sigma A^* - \frac{1}{i\omega\mu} \text{grad div } A^* \tag{2.311}$$

The behavior of the vector potential at interfaces follows from the required continuity for tangential components on the electric and magnetic fields at those boundaries. It is not particularly difficult to formulate conditions near the source of the primary field nor at infinity.

Let us examine another way for introducing the vector potential. From the fourth equation in the system 2.304, we can represent the magnetic field as being:

$$H = \text{curl } A \tag{2.312}$$

This function A is called a vector potential of the electrical type and this definition is normally used when the electromagnetic field is energized through the use of a grounded wire. Substituting this definition for the vector potential into the first equation of the set 2.304, we have:

curl E = $-i\omega\mu$ curl A

or

curl $(E + i\omega\mu A)$ = 0

whence

$E + i\omega\mu A$ = grad U

or $\qquad\qquad\qquad\qquad\qquad\qquad\qquad\qquad\qquad\qquad\qquad$ (2.313)

E = $-i\omega\mu A$ + grad U

Substituting 2.312 and 2.313 into the second Maxwell equation in the set 2.304, we have the result:

curl curl A = $-i\sigma\mu\omega A + \sigma$ grad U

or

grad div $A - \nabla^2 A$ = $-i\sigma\mu\omega A + \sigma$ grad U

Considering that there are an infinite number of functions A and U that will satisfy eqs. 2.312 and 2.313, we will seek a pair of them that simplifies the last equation. One such choice is:

$$U = \frac{1}{\sigma}\,\text{div}\,A \qquad\qquad\qquad\qquad\qquad\qquad (2.314)$$

With this gauge condition, we again obtain a Helmholtz equation for the electric vector potential:

$$\nabla^2 A - k^2 A = 0 \qquad\qquad\qquad\qquad\qquad\qquad (2.315)$$

And again, both vectors comprising the electromagnetic field are expressed in terms of a single vector potential function, A:

H = curl A

$\qquad\qquad\qquad\qquad\qquad\qquad\qquad\qquad\qquad\qquad\qquad$ (2.316)

E = $-i\omega\mu A + \dfrac{1}{\sigma}$ grad div A

As in the previous case, the behavior of this vector potential near the source and at infinity, as well as at interfaces, follows from the corresponding behavior of the electric and magnetic fields under these conditions.

In conclusion, let us review some of the results which are contained here. If one of the vector potentials is found, then the electric and magnetic fields can be determined by taking corresponding derivatives in accord with either eq. 2.311 or eq. 2.316. When an electromagnetic field is caused by both induced currents and charges, it may be necessary in most cases to make use of both vector potentials to determine the field, but there are some important exceptions. Just as is the case with harmonic fields, solutions for vector potentials can be extended to the case in which the functions depend arbitrarily on time.

2.6. RELATIONSHIPS BETWEEN VARIOUS RESPONSES OF THE ELECTROMAGNETIC FIELD

In this section we will explore some general relationships between the various responses of an electromagnetic field. First of all, we will start from a relationship between the quadrature and inphase components of the field. For example, representing the complex amplitude of the electric field as being the sum of two components:

$$E = \text{In}E + \text{i}QE$$

In substituting these into Helmholtz's equation (2.303) we have:

$$\nabla^2(\text{In}E + \text{i}QE) - \text{i}\sigma\mu\omega(\text{In}E + \text{i}QE) = 0$$

or

$$\nabla^2\text{In}E = -\omega\mu\sigma QE$$

$$\nabla^2 QE = \sigma\mu\omega\text{In}E \tag{2.317}$$

Thus, there is a relationship between the inphase and quadrature components of the spectrum. Let us examine this in more detail. We will make use of a solution in the form:

$$Me^{\text{i}\omega t}$$

where M is a complex amplitude of the spectrum. In obtaining an actual sinusoidal solution, one should take the imaginary part of this expression:

$$M_0 \sin(\omega t + \phi) = QMe^{\text{i}\omega t}$$

If the solution contains the complex amplitude term, from the physical point of view this means that there is a phase shift, and thus the field can be represented as being the sum of the quadrature (Q) and the inphase (In) components. We will have:

$$M = \text{In}M + \text{i}QM = M_0 \cos\phi + \text{i}M_0 \sin\phi \tag{2.318}$$

where M_0 and ϕ are the amplitude and phase of an oscillation, respectively.

Using the conventional symbols for representing a complex variable, we can write M as:

$$M(z) = U + \text{i}V \tag{2.319}$$

where U and V are the real and imaginary parts of the function $M(z)$ and z is an argument defined as:

$$z = x + iy$$

where x and y are coordinates on the complex plane z. Usually, the complex amplitude M of an electromagnetic field is an analytic function of frequency, ω. If this be the case, necessary and sufficient condition for analyticity of a function by the Cauchy-Reimann condition:

$$\frac{\partial U}{\partial x} = \frac{\partial V}{\partial y}, \quad \frac{\partial U}{\partial y} = -\frac{\partial V}{\partial x} \tag{2.320}$$

The Cauchy-Reimann conditions express the relationship that exists between the real and imaginary parts of an analytic function in the complex plane in differential form. In our case, the complex variable z is the frequency:

$$\omega = \mathrm{Re}\,\omega + i\,\mathrm{Im}\,\omega$$

and we will seek a relationship between the quadrature and the inphase components of the field for real values of ω, because the electromagnetic field is observed only at real frequencies. For this purpose, let us use the Cauchy formula which shows that if the function $M(z)$ is analytic within a contour C, as well as along this contour, and a is any point in the z-plane, then:

$$\oint_C \frac{M(z)}{z-a}\,dz = 2\pi i M(a) \cdot \begin{cases} 1 & \text{if } a \in C \\ \frac{1}{2} & \text{if } a \text{ is on } C \\ 0 & \text{if } a \notin C \end{cases} \tag{2.321}$$

This Cauchy formula permits us to evaluate $M(a)$ at any point within the contour C when the values for $M(z)$ are known along this contour. This relationship is a consequence of the close connection which exists among all values of an analytic function on the complex plane z.

Let us consider a path consisting of a semi-circle with an infinitely large radius, centered on the x-axis. The internal area of this contour includes the upper half plane as shown in Fig. 2.59. We will attempt to find a quadrature component for the function $M = U + iV$ by assuming that the inphase component U is known along the x-axis or vice versa. Using the Cauchy formula, we have:

$$M(\xi) = \frac{1}{i\pi} P \oint \frac{M(z)}{z - \xi}\,dz \tag{2.322}$$

The point $\xi = \epsilon + i\eta$ lies along the path of integration, and the symbol P indicates the principal value of the integral that is to be used. Inasmuch as the path of integration coincides with the x-axis ($\eta = 0$) we have:

$$M(\epsilon, 0) = \frac{1}{i\pi} P \int_{-\infty}^{\infty} \frac{M(x, 0)}{x - \epsilon}\,dx \tag{2.323}$$

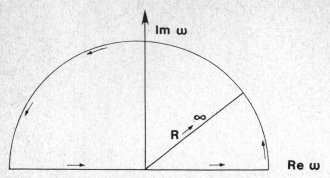

Fig. 2.59. Path of integration consisting of a semi-circle with an infinitely large radius and centered on the x-axis.

In developing eq. 2.323, it has been assumed that the value for the integral along the semi-circular part of the path of integration vanishes as the radius increases without limit. Because:

$$M(\epsilon, 0) = U(\epsilon, 0) + iV(\epsilon, 0)$$

and

$$M(x, 0) = U(x, 0) + iV(x, 0)$$

we obtain:

$$U(\epsilon, 0) = \frac{1}{\pi} P \int_{-\infty}^{\infty} \frac{V(x, 0)}{x - \epsilon} \, \mathrm{d}x \qquad (2.324)$$

$$V(\epsilon, 0) = -\frac{1}{\pi} P \int_{-\infty}^{\infty} \frac{U(x, 0)}{x - \epsilon} \, \mathrm{d}x \qquad (2.325)$$

The integrands in these expressions are characterized by a singularity which can readily be removed by making use of the identity:

$$P \int_{-\infty}^{\infty} \frac{\mathrm{d}x}{x - \epsilon} = 0$$

Now we can rewrite eqs. 2.324 and 2.325 in the form:

$$U(\epsilon, 0) = \frac{1}{\pi} \int_{-\infty}^{\infty} \frac{V(x, 0) - V(\epsilon, 0)}{x - \epsilon} \, \mathrm{d}x \qquad (2.326)$$

$$V(\epsilon, 0) = -\frac{1}{\pi} \int_{-\infty}^{\infty} \frac{U(x, 0) - U(\epsilon, 0)}{x - \epsilon} \, \mathrm{d}x \qquad (2.327)$$

inasmuch as:

$$V(\epsilon, 0) \int_{-\infty}^{\infty} \frac{dx}{x - \epsilon} = U(\epsilon, 0) \int_{-\infty}^{\infty} \frac{dx}{x - \epsilon} = 0$$

It can readily be seen that the integrands in eqs. 2.326 and 2.327 do not have singularities, and these expressions establish the relationship between the real and imaginary parts of some analytic function.

Let us return to consideration of the complex amplitude of the field:

$$M(\omega) = \text{In} \, M(\omega) + iQ \, M(\omega)$$

In accord with eqs. 2.326 and 2.327, the relationship between the quadrature and inphase components of the field are:

$$\text{In} M(\omega_0) = \frac{1}{\pi} \int_{-\infty}^{\infty} \frac{QM(\omega) - QM(\omega_0)}{\omega - \omega_0} \, d\omega \tag{2.328}$$

$$QM(\omega_0) = -\frac{1}{\pi} \int_{-\infty}^{\infty} \frac{\text{In} M(\omega) - \text{In} M(\omega_0)}{\omega - \omega_0} \, d\omega \tag{2.329}$$

Thus, when the spectrum of one of the components is known, the other component describing a field can be calculated by making use of either eq. 2.328 or 2.329.

It is now a simple matter to find the relationship between the amplitude and the phase responses of a field component. Taking the logarithm of the complex amplitude M, we have:

$$\ln M = \ln M_0 + i\phi \tag{2.330}$$

From this equation we see that the relationship between the amplitude and phase responses is the same as that for the quadrature and the inphase components. For example, for the phase response we have:

$$\phi(\omega_0) = -\frac{1}{\pi} \int_{-\infty}^{\infty} \frac{\ln M_0(\omega) - \ln M_0(\omega_0)}{\omega - \omega_0} \, d\omega \tag{2.331}$$

For most practical purposes, it is preferable to express the right-hand side of eq. 2.331 in another form. After some algebraic operations, we obtain:

$$\phi(\omega_0) = \frac{1}{\pi} \int_{-\infty}^{\infty} \frac{dL}{du} \left| \ln \coth \frac{u}{2} \right| du \tag{2.332}$$

where

$$L = \ln M_0, \quad u = \ln(\omega/\omega_0)$$

It can be seen from eq. 2.332 that the phase response depends on the slope of the amplitude response curve, when plotted on a logarithmic frequency scale. Inasmuch as the integration is carried out over the entire frequency range, the phase at any particular frequency ω_0 depends on the slope of the amplitude response curve over the entire frequency spectrum. However, the relative importance of the slope over various portions of the spectrum is controlled by a weighting factor $|\ln \coth u/2|$ which can also be written as $|\ln (\omega + \omega_0)/(\omega - \omega_0)|$.

The weighting factor is shown graphically in Fig. 2.60. It increases as the frequency is close to ω_0, and becomes logarithmic infinite at that point. Therefore, the slope of the amplitude response near the frequency for which the phase is to be calculated is much more important than the slope of the amplitude response curve at more distant frequencies.

It should be noted that calculation of the amplitude response from the phase can only be done with an uncertainty that amounts to a multiplicative scaling factor.

Equations 2.331 and 2.332 lead us to the following conclusions. First of all, measurements of the phase response does not provide additional information on the geoelectric section when the amplitude response is already known. However, it may well be that the shape of the phase response curve more clearly reflects some diagnostic features of this section than does the amplitude response curve.

It is important to stress that while there is in essence a unique relationship between the quadrature and the inphase responses, as well as between the amplitude and phase responses, this does not mean that there is a point by point relationship between them. In fact, measuring both amplitude and phase at one or a few frequencies provides two types of information characterizing the geoelectric section in a different manner. The same conclusion can be arrived at for the quadrature and the inphase measurements.

We will now investigate the relationship between frequency-domain and time-domain responses. In most cases considered in this section, a transient electromagnetic field is excited by a step function current in the source. For this reason, the relationship between frequency response and transient response corresponding to this single type of excitation will be our principal

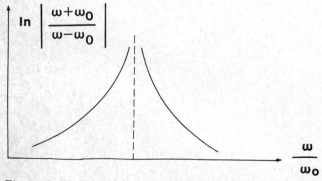

Fig. 2.60. The kernel function (weighting factor) for the transform given in eq. 2.332.

concern. The information we need is obtained through use of the Fourier transform, which takes the well known form:

$$F(t) = \frac{1}{2\pi} \int\limits_{-\infty}^{\infty} F^*(\omega)e^{-i\omega t}d\omega$$

$$F^*(\omega) = \int\limits_{-\infty}^{\infty} F(t)e^{i\omega t}dt$$

(2.333)

When the current in the source changes as a step function in time, the primary magnetic field accompanying this current does likewise:

$$H_0(t) = F_0(t) = \begin{cases} H_0 & t < 0 \\ 0 & t > 0 \end{cases}$$

(2.334)

According to eq. 2.333, the spectrum for the primary magnetic field is:

$$H_0(\omega) = F_0^*(\omega) = H_0/i\omega$$

(2.335)

The amplitude of this spectrum decreases inversely proportional to frequency, while the phase remains constant. Inasmuch as low frequencies prevail in the spectrum of the primary field when step function excitation is used, the use of this excitation is often preferable in practice.

In accord with eq. 2.333, the primary magnetic field can be written as:

$$H_0(t) = \frac{H_0}{2\pi} \int\limits_{-\infty}^{\infty} \frac{1}{i\omega} e^{-i\omega t}d\omega$$

(2.336)

where the path of integration is not permitted to pass through the point $\omega = 0$ (see Fig. 2.61). Let us write the right-hand integral as a sum:

$$\frac{1}{2\pi i} \int\limits_{-\infty}^{\infty} \frac{e^{-i\omega t}}{\omega} d\omega = \frac{1}{2\pi i} \int\limits_{-\infty}^{-\epsilon} \frac{e^{-i\omega t}}{\omega} d\omega + \frac{1}{2\pi i} \int\limits_{-\epsilon}^{+\epsilon} \frac{e^{-i\omega t}}{\omega} d\omega + \frac{1}{2\pi i} \int\limits_{\epsilon}^{\infty} \frac{e^{-i\omega t}}{\omega} d\omega$$

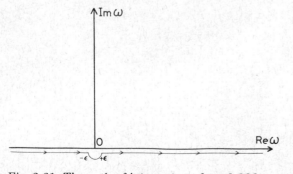

Fig. 2.61. The path of integration of eq. 2.336.

and integrate along a semi-circle around the origin whose radius tends to zero. In calculating the middle integral, we will introduce a new variable, ϕ:

$$\omega = \rho e^{i\phi}$$

Then we have:

$$d\omega = i\rho e^{i\phi} d\phi$$

and

$$\frac{1}{2\pi i} \int_{-\epsilon}^{\epsilon} \frac{e^{-i\omega t}}{\omega} d\omega = \frac{1}{2\pi i} \int_{\pi}^{2\pi} \frac{i\rho e^{i\phi} d\phi}{\rho e^{i\phi}} = \frac{1}{2}, \quad \text{if } \rho \to 0$$

Correspondingly, the second expression for the primary field when the variable of integration takes on only real values is:

$$H_0(t) = \frac{H_0}{2} + \frac{H_0}{2\pi} \int_{-\infty}^{\infty} \frac{e^{-i\omega t}}{i\omega} d\omega \qquad (2.337)$$

Now, making use of the principle of superposition, we obtain the following expressions for a nonstationary field:

$$H(t) = \frac{H_0}{2\pi i} \int_{-\infty}^{\infty} \frac{H(\omega)}{\omega} e^{-i\omega t} d\omega \qquad (2.338)$$

$$H(t) = \frac{H_0}{2} + \frac{H_0}{2\pi i} \int_{-\infty}^{\infty} \frac{H(\omega)}{\omega} e^{-i\omega t} d\omega \qquad (2.339)$$

where

$$H(\omega) = \operatorname{In}H(\omega) + iQH(\omega)$$

is the complex amplitude of the spectrum of the chosen component of the magnetic field, which is assumed to be known.

Let us write eq. 2.339 in the form:

$$H(t) = \frac{H_0}{2} + \frac{H_0}{2\pi} \int_{-\infty}^{\infty} \frac{QH(\omega)\cos\omega t - \operatorname{In}H(\omega)\sin\omega t}{\omega} d\omega$$

$$- \frac{i}{2\pi} H_0 \int_{-\infty}^{\infty} \frac{QH(\omega)\sin\omega t + \operatorname{In}H(\omega)\cos\omega t}{\omega} d\omega \qquad (2.340)$$

Inasmuch as:

$$\operatorname{In}H(\omega) = \operatorname{In}H(-\omega)$$

$$QH(\omega) = -QH(-\omega) \qquad (2.341)$$

the second integral in 2.340 is zero, and therefore:

$$H(t) = \frac{H_0}{2} + \frac{H_0}{\pi} \int_0^\infty \frac{QH(\omega)\cos\omega t - \mathrm{In}H(\omega)\sin\omega t}{\omega} \, d\omega \qquad (2.342)$$

For negative times, $H(t) = H_0$, and:

$$H_0 = \frac{H_0}{2} + \frac{H_0}{\pi} \int_0^\infty \frac{QH(\omega)\cos\omega t + \mathrm{In}H(\omega)\sin\omega t}{\omega} \, d\omega$$

or

$$0 = -\frac{H_0}{2} + \frac{H_0}{\pi} \int_0^\infty \frac{QH(\omega)\cos\omega t + \mathrm{In}H(\omega)\sin\omega t}{\omega} \, d\omega \qquad (2.343)$$

Let us take note that in these last two expressions, time is taken as positive.
Combining eqs. 2.342 and 2.343 we obtain:

$$H(t) = \frac{2}{\pi} H_0 \int_0^\infty \frac{QH(\omega)}{\omega} \cos\omega t \, d\omega$$

and $\qquad\qquad\qquad\qquad\qquad\qquad\qquad\qquad\qquad\qquad\qquad (2.344)$

$$H(t) = H_0 - \frac{2}{\pi} H_0 \int_0^\infty \frac{\mathrm{In}H(\omega)}{\omega} \sin\omega t \, d\omega$$

Correspondingly, for derivatives with respect to time of the magnetic induction, B, we have:

$$\dot{B}(t) = -\frac{2}{\pi} H_0 \int_0^\infty QB(\omega) \sin\omega t \, d\omega$$

and $\qquad\qquad\qquad\qquad\qquad\qquad\qquad\qquad\qquad\qquad\qquad (2.345)$

$$\dot{B}(t) = -\frac{2}{\pi} H_0 \int_0^\infty \mathrm{In}B(\omega) \cos\omega t \, d\omega$$

Equations 2.344 and 2.345 permit us to calculate the transient response when either the quadrature or the inphase component of the frequency-domain spectrum is known. It is often convenient to introduce new scale variables in these equations. Such a scale variable can be defined as:

$$y = \frac{1}{4\pi^2} \left(\frac{a}{h}\right)^2 = \frac{\sigma\mu a^2}{8\pi^2}\omega = \frac{\omega}{8\pi^2\alpha}$$

where h is skin depth, a is a linear dimension in the system, and:

$\alpha = 1/\sigma\mu a^2$

It should be obvious that:

$$\omega t = \left(\frac{\tau}{2\pi h}\right)^2 = \left(\frac{\tau}{a}\right)^2 y$$

where

$$\tau = 2\pi\sqrt{\frac{2t}{\sigma\mu}} \quad \text{and} \quad t = \frac{1}{8\pi^2\alpha}\left(\frac{\tau}{a}\right)^2$$

whence, for example, we will have:

$$\dot{B}(t) = -16\pi\alpha H_0 \int\limits_0^\infty \mathrm{In}B(8\pi^2\alpha y)\cos\left[\left(\frac{\tau}{a}\right)^2 y\right]dy$$

or $\hspace{8cm}$ (2.346)

$$\dot{B}(t) = -16\pi\alpha H_0 \int\limits_0^\infty QB(8\pi^2\alpha y)\sin\left[\left(\frac{\tau}{a}\right)^2 y\right]dy$$

Usually, because of the complexity of the expressions for the frequency spectrum, the only way to obtain numerical results from eqs. 2.344–2.346 is by numerical integration.

Up to this point, we have examined the relationship between frequency and transient responses, and have derived formulas for calculating the time-domain field for the case in which the primary field changes as a step function of time. In doing so, we have assumed that the frequency spectrum for the field is known. However, in practice, the use of this type of excitation meets with some practical difficulties. For example, due to the inductance in a transmitter loop, the current cannot be terminated instantly, and because of this, in place of a step current behavior, there is a gradual decrease of current in the transmitter. The time required for the current to vanish in the transmitter is usually called the ramp time. In order to investigate the effect caused by such behavior of the transmitter current, it is appropriate to use calculations of the transient field generated by a step function excitation, rather than refer to frequency domain fields by applying Fourier transform to the measurements. The further approach is based on the use of Duhamel's integral which is described below.

Assume that a primary field varies with time like the function shown in Fig. 2.62. It should be clear that this function can be thought of as being the sum of step functions with the amplitude $\Delta H_0(\tau)$, where τ is the instant at which the excitation occurs. Also, let us assume that the transient field caused by the unit step function is known, and is described by the function $A^*(t-\tau)$. It should be obvious that a step function with the amplitude $\Delta H_{0\tau}$ generates a transient field given by:

$$\Delta H_0(\tau)A^*(t-\tau)$$

Adding the actions of all such step functions occurring at various times, we find the expression for the total transient response for any component of the magnetic field:

$$H_i(t) = H_0(0)A_i^{*0}(t) + \sum_{\tau=0}^{\tau=t} \Delta H_0(\tau)A_i^*(t-\tau)$$

$$= H_0(0)A_i^{*0}(t) + \sum_{\tau=0}^{\tau=t} \frac{\partial H_0(\tau)}{\partial \tau} A_i^*(\tau - t)\Delta\tau$$

As can be seen from Fig. 2.62, the approximation:

$$\Delta H_0(\tau) = \frac{\partial H_0(\tau)}{\partial \tau} \Delta\tau$$

becomes more accurate with decreasing interval, $\Delta\tau$. In this expression, $A_i^*(t-\tau)$ is the response of the medium to the ith component of the magnetic field when the unit current step occurs at the instant $t = \tau$. For inductive excitation of the field, A_i^{*0} is identically zero, and therefore we have:

$$H_i(t) = \sum_{\tau=0}^{\tau=t} \frac{\partial H_0(\tau)}{\partial \tau} A_i^*(t-\tau)\Delta\tau$$

In the limiting case as $\Delta\tau$ approaches zero we obtain a convolution integral:

$$H_i(t) = \int_0^t \frac{\partial H_0(\tau)}{\partial \tau} A_i^*(t-\tau)\,d\tau \tag{2.347}$$

This integral is also called Duhamel's integral, and it permits us to find the

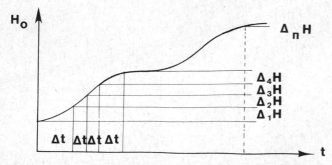

Fig. 2.62. Presentation of a primary field through Duhamel's integral.

138

transient response for an arbitrary shape of current excitation when the transient response of the medium of a step function excitation is already known.

Integrating the integral by parts, we obtain:

$$\int_0^t \frac{\partial H_0(\tau)}{\partial \tau} A_i^*(t-\tau)\mathrm{d}\tau = H_0(\tau)A_i^*(t-\tau)\Big|_0^t - \int_0^t H_0(\tau)\frac{\partial A_i^*(t-\tau)}{\partial \tau}\,\mathrm{d}\tau$$

$$= H_0(t)A_i^*(0) - H_0(0)A_i^*(t) - \int_0^t H_0(\tau)\frac{\partial A_i^*(t-\tau)}{\partial \tau}\,\mathrm{d}\tau \qquad (2.348)$$

This is known as the second form of Duhamel's integral. Similar expressions can be written for the electric field.

Let us now examine an example. Suppose that the behavior of the primary field as described by the linear function shown in Fig. 2.63, or:

$$H_0(\tau) = \begin{cases} H_0 & \tau < 0 \\ H_0(1-\tau/T) & 0 \leqslant \tau \leqslant T \\ 0 & \tau \geqslant T \end{cases} \qquad (2.349)$$

where T is the ramp time. Substituting eq. 2.349 into eq. 2.347 we obtain:

$$H_i(t) = -\frac{H_0}{T}\int_0^T A_i^*(t-\tau)\mathrm{d}\tau, \quad \text{if } t \geqslant T \qquad (2.350)$$

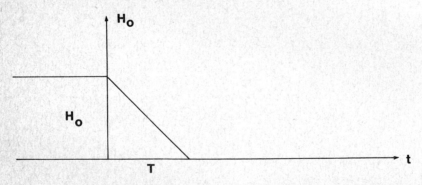

Fig. 2.63. Time variation of the primary field described by eq. 2.349.

In the limit, as T approaches zero and applying the central limit theorem, we have:

$$H_i(t) = -A_i^*(t)H_0 = A_i^{-1}H_0$$

where A_i^{-1} is the transient field caused by unit step function when the current is turned off.

Chapter 3

FREQUENCY AND TIME-DOMAIN BEHAVIOR OF THE FIELD CAUSED BY CURRENTS INDUCED IN A CONFINED CONDUCTOR

INTRODUCTION

In this chapter, we will consider in detail the behavior in both the frequency and time domain of currents induced in a confined conductive body embedded in an insulating full-space. We will pay particular attention to the relationship between the various components of the quasi-stationary field and the conductivity and geometric properties of the conductor. In addition, the relationships between various parts of the frequency spectrum or the transient response will be studied in some detail.

3.1. A CONDUCTING SPHERE IN A UNIFORM MAGNETIC FIELD (FREQUENCY DOMAIN)

Consider a sphere with a radius, a, conductivity, σ, and a magnetic permeability, μ, placed in a uniform harmonic magnetic field $H_0 e^{-i\omega t}$ directed along the z-axis as shown in Fig. 3.1. We will use a spherical coordinate system, R, θ, ϕ, with its origin at the center of this sphere.

As was pointed out in the preceding chapter, in general, a nearly uniformly magnetic field can be created over a limited range in space using various types of current sources. However, in the particular case we will consider at the moment, one will assume that the source for the primary field has only a j_ϕ component of the current density, which is independent of the angle ϕ, and that the center of the sphere is located on the axis of symmetry. In practice, a current ring of large diameter can be used as the source for an almost uniform magnetic field in the area occupied by such a sphere. Therefore, the vortex electric field, E_0, caused by the time rate of change of the primary field, $B_0 = \mu H_0$, is not uniform, but is equal to zero on the axis of symmetry. The field E_0 has but a single component in the spherical coordinate system, this being $E_{0\phi}$, which can be readily found from the first of Maxwell's equations expressed in integral form:

$$\oint_L E_0 \cdot dl = -\frac{\partial}{\partial t} \int_S B_0 \cdot dS \qquad (3.1)$$

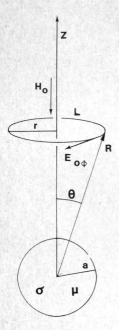

Fig. 3.1 Conducting sphere situated within a uniform magnetic field.

Applying this expression along a circular path L situated in a horizontal plane as shown in Fig. 3.1, and taking into account the axial symmetry and uniformity of the primary field, we obtain:

$$E_{0\phi} 2\pi r = i\omega\mu H_{0z} \pi r^2$$

or

$$E_{0\phi} = i\omega\mu \frac{H_{0z}}{2} R \sin\theta \qquad (3.2)$$

That is, the electric field, $E_{0\phi}$, increases linearly with distance from the z-axis. Inasmuch as the radial component of the primary field, E_0, is zero, that is, it does not cross the surface of the sphere, no electrical charge is formed, and therefore, the only sources for the secondary field are currents arising because of the vortex electrical field. It is obvious that current filaments within the sphere are located in horizontal planes, and have the form of circles with their centers situated on the z-axis.

Because of the interaction between induced currents (the skin effect), the current density at any point is a relatively complicated function of frequency, conductivity, and the radius of the sphere. The secondary electromagnetic field can be represented as being the sum of fields created by the individual elementary current rings, and for this reason, in spherical coordinates, the electromagnetic field is completely described by one electrical component,

$E_{1\phi}$, and two magnetic components, H_{1R} and $H_{1\theta}$. Since there is only a single component to the electric field, it is a simple matter to define the field $E_{1\phi}$ and then using Maxwell's equations to determine the magnetic field. In accord with Maxwell's equations for a harmonic quasi-stationary field (see eq. 2.297), we have:

$$\text{curl } E = i\omega\mu H \qquad \text{div } E = 0$$

$$\text{curl } H = \sigma E \qquad \text{div } H = 0 \qquad (3.3)$$

where E and H are the complex amplitudes of the field. From this, we have:

$$\text{curl curl } E = i\omega\mu \text{ curl } H = i\omega\mu \sigma E$$

that is, the electric field both outside and within the sphere satisfy the following equations respectively:

$$\text{curl curl } E^{e} = 0, \quad \text{if } R > a$$

and $\qquad (3.4)$

$$\text{curl curl } E^{i} - k^2 E^{i} = 0, \quad \text{if } R < a$$

where, as usual the wave number is defined as:

$$k = (i\sigma\omega\mu)^{1/2}$$

As always, the tangential components of the electromagnetic field must be continuous across the interface $R = a$, so that:

$$E_{\phi}^{e} = E_{\phi}^{i} \quad \text{and} \quad H_{\theta}^{e} = H_{\theta}^{i} \quad \text{if } R = a \qquad (3.5)$$

Using Maxwell's first equation:

$$H_{\theta} = -\frac{1}{i\omega\mu}\frac{1}{R}\frac{\partial}{\partial R}R E_{\phi}$$

the boundary conditions can be written as:

$$E_{\phi}^{e} = E_{\phi}^{i}, \quad \frac{1}{\mu_{e}}\frac{\partial}{\partial R}R E_{\phi}^{e} = \frac{1}{\mu_{i}}\frac{\partial}{\partial R}R E_{\phi}^{i}, \quad \text{if } R = a \qquad (3.6)$$

It is convenient in later calculations to represent the electric field observed outside the sphere as being the sum of primary and secondary parts:

$$E_{\phi}^{e} = E_{0\phi} + E_{1\phi}^{e} \qquad (3.7)$$

From the physical point of view, it should be clear that the field inside the sphere remains finite, while that outside the sphere, the secondary field tends to zero for large distances, R.

Determination of the electric field consists of the following steps:

(1) Solution of the set of equations 3.4.

(2) Choice among the solutions for eq. 3.4 of functions that satisfy the boundary conditions expressed in eq. 3.6.

(3) The choice among these last expressions of a function which will also satisfy the condition that in infinity the field goes to zero, and yet everywhere else is finite.

Before proceeding further, let us make some necessary changes in the algebraic form of the equations in set 3.4. In spherical coordinates:

$$\text{curl}_R E = \frac{1}{R \sin \theta} \frac{\partial}{\partial \theta} \sin \theta \, E_\phi = a_R$$

$$\text{curl}_\theta E = -\frac{1}{R} \frac{\partial}{\partial R} R E_\phi = a_\theta$$

and

$$\text{curl}_\phi E = 0, \quad \text{since } E_R = E_\theta = 0$$

Therefore,

$$\text{curl}_\phi \, \text{curl} \, E = \text{curl}_\phi (a_R R_0 + a_\theta \, \theta_0)$$

$$= -\frac{1}{R} \frac{\partial^2}{\partial R^2} (R E_\phi) - \frac{1}{R^2} \frac{\partial}{\partial \theta} \frac{1}{\sin \theta} \frac{\partial}{\partial \theta} \sin \theta \, E_\phi \tag{3.8}$$

where R_0 and θ_0 are unit vectors directed along the principal coordinates R and θ, respectively.

Substituting eq. 3.8 into eq. 3.4, we obtain equations in partial derivatives of the second order for the electric field both outside and inside the sphere, respectively:

$$\frac{1}{R} \frac{\partial^2}{\partial R^2} (R E_\phi^e) + \frac{1}{R^2} \frac{\partial}{\partial \theta} \frac{1}{\sin \theta} \frac{\partial}{\partial \theta} \sin \theta \, E_\phi^e = 0 \tag{3.9}$$

$$\frac{1}{R} \frac{\partial^2}{\partial R^2} (R E_\phi^i) + \frac{1}{R^2} \frac{\partial}{\partial \theta} \frac{1}{\sin \theta} \frac{\partial}{\partial \theta} \sin \theta \, E_\phi^i + k^2 E_\phi^i = 0 \tag{3.10}$$

Making use of the method of separation of variables, we solve eq. 3.10 with the axial symmetry ($\partial E_\phi / \partial \phi = 0$) being taken into account. The form of the solution will be:

$$E_\phi^i = T(R) \, \Phi(\theta) \tag{3.11}$$

Substituting this expression in eq. 3.10 and multiplying both sides by $R^2 / T\Phi$, we have:

$$\frac{R}{T} \frac{\partial^2}{\partial R^2} RT + \frac{1}{\Phi} \frac{\partial}{\partial \theta} \frac{1}{\sin \theta} \frac{\partial}{\partial \theta} \sin \theta \, \Phi + k^2 R^2 = 0 \tag{3.12}$$

As is always the case when a method of separation of variables is used, this equation separates into two ordinary differential equations which are:

$$\frac{R}{T} \frac{d^2 RT}{dR^2} + k^2 R^2 = m \tag{3.13}$$

$$\frac{1}{\Phi} \frac{d}{d\theta} \frac{1}{\sin \theta} \frac{d}{d\theta} \sin \theta \, \Phi = -m \tag{3.14}$$

where m is the separation constant, having dimensions of inverse distance.
After differentiation of the product RT, eq. 3.13 takes the form:

$$\frac{d^2 T}{dR^2} + \frac{2}{R} \frac{dT}{dR} + \left(k^2 - \frac{m}{R^2} \right) T = 0 \tag{3.15}$$

This is Bessel's equation, for which the solution can be written as:

$$T(k, R) = R^{-1/2} Z_{(\frac{1}{4} + m)^{1/2}} (ikR) \tag{3.16}$$

where

$$Z_{(\frac{1}{2} + m)^{1/2}} (ikR)$$

are modified Bessel's functions of the first and second kind. Carrying out the differentiation on the left-hand side of eq. 3.14, we have:

$$(1 - u^2) \frac{d^2 \Phi}{du^2} - 2u \frac{d\Phi}{du} - \frac{\Phi}{1 - u^2} + m \Phi = 0 \tag{3.17}$$

where

$$u = \cos \theta$$

This equation is Legendre's equation for associated functions with the index $l = 1$. From the behavior of these functions, we recognize that a non-trivial solution of eq. 3.17 is possible only if the parameter m has the values:

$$m = n(n + 1) \tag{3.18}$$

Thus, partial solutions of eqs. 3.15 and 3.17 are:

$$T_n(k, R) = \frac{1}{R^{1/2}} \left\{ A_n^1 I_{[\frac{1}{4} + n(n+1)]^{1/2}} (ikR) + B_n^1 K_{[\frac{1}{4} + n(n+1)]^{1/2}} (ikR) \right\} \tag{3.19}$$

and

$$\Phi_n(\theta) = C_n^1 P_n^{(1)}(u) + D_n^1 Q_n^{(1)}(u) \tag{3.20}$$

where

$$I_{[\frac{1}{4} + n(n+1)]^{1/2}} (ikR) \quad \text{and} \quad K_{[\frac{1}{4} + n(n+1)]^{1/2}} (ikR)$$

are modified Bessel functions of the first and second kind, $P_n^{(1)}(u)$ and $Q_n^{(1)}(u)$ are associated Legendre's functions, $u = \cos \theta$, and A_n^1, B_n^1, C_n^1, and D_n^1 are constants representing the amplitudes of the various radial and angular harmonics in the solutions. The coefficient for the function $Q_n^{(1)}(u)$ must be identically zero, because at all points along the z-axis ($\theta = 0$) both inside and outside the sphere this function will become infinitely large. At the center of the sphere, the Bessel function:

$$K_{[\frac{1}{4} + n(n + 1)]^{1/2}}(ikR)$$

also tends to be infinite, and for this reason, its coefficient must be zero in the general solution for the electric field inside the sphere:

$$E_\phi^i = R^{-1/2} \sum_{n=0}^{\infty} B_n I_{[\frac{1}{4} + n(n+1)]^{1/2}}(ikR)P_n^{(1)}(\cos\theta) \tag{3.21}$$

Equation 3.15 is considerably simpler in the non-conductive medium outside the sphere because the wave number is zero. Therefore:

$$\frac{d}{dR}\left(R^2 \frac{dT}{dR}\right) - n(n + 1)\,T = 0 \tag{3.22}$$

The solution of this equation is well known:

$$T_n(R) = C_n^1 R^n + F_n^1 R^{-n-1} \tag{3.23}$$

Inasmuch as the field created by current flow within the sphere tends to zero with increasing R, the expression for the secondary electric field outside the sphere must have the form:

$$E_{1\phi}^e = \sum_{n=0}^{\infty} F_n^1 R^{-n-1} P_n^{(1)}(\cos\theta) \tag{3.24}$$

Considering eqs. 3.7, 3.21 and 3.24, the expression for the electric field outside and inside the sphere can be written as:

$$E_\phi^e = \frac{i\omega\mu_e H_0}{2} R P_1^{(1)}(u) + \frac{i\omega\mu_e H_0}{2} \sum_{n=0}^{\infty} D_n R^{-n-1} P_n^{(1)}(u) \tag{3.25}$$

$$E_\phi^i = \frac{i\omega\mu_e H_0}{2} R^{-1/2} \sum_{n=0}^{\infty} C_n I_{[\frac{1}{4} + n(n+1)]^{1/2}}(ikR) P_n^{(1)}(u) \tag{3.26}$$

inasmuch as:

$$P_1^{(1)}(\cos\theta) = \sin\theta$$

In order to determine the coefficients D_n and C_n, the boundary conditions in eq. 3.6 can be applied.

Considering the orthogonality property of the associated Legendre functions instead of a system of an infinite number of equations with an infinite number of unknowns, we obtain two equations and two unknowns, D_n and C_n, for each harmonic with the index n. Considering that the primary electric field $E_{0\phi}$ is described by only the first harmonic ($n = 1$), we can readily see that all of the coefficients C_n and D_n, except for C_1 and D_1, are zero, and that for their determination we can write the following system of equations:

$$a^3 + D_1 = a^{3/2} C_1 I_{3/2}(ika)$$

$$\mu_i(2a^3 - D_1) = \mu_e a^{3/2} C_1[I_{3/2}(ika) + ika\, I'_{3/2}(ika)] \tag{3.27}$$

where

$$I'_{3/2}(x) = \frac{\partial}{\partial x} I_{3/2}(x)$$

Bessel's functions are related through recurrence relationships:

$$I_{3/2}(x) = I_{-1/2}(x) - \frac{1}{x} I_{1/2}(x)$$

$$x I'_{3/2}(x) = x I_{1/2}(x) - \frac{3}{2} I_{3/2}(x)$$

Substituting these expressions into eq. 3.27, after some straightforward algebraic operations, we have:

$$D_1 = \frac{(2\mu_i + \mu_e)\, x I_{-1/2}(x) - [\mu_e(1 + x^2) + 2\mu_i]\, I_{1/2}(x)}{(\mu_i - \mu_e)\, x I_{-1/2}(x) + [\mu_e(1 + x^2) - \mu_i]\, I_{1/2}(x)}\, a^3 \tag{3.28}$$

and

$$C_1 = \frac{3\mu_i x\, a^{3/2}}{(\mu_i - \mu_e)\, x I_{-1/2}(x) + [\mu_e(1 + x^2) - \mu_i]\, I_{1/2}(x)} \tag{3.29}$$

where $x = ika$, $k = (i\sigma\mu_i\omega)^{1/2}$; μ_i and μ_e are the magnetic permeability values for this sphere and for the surrounding medium, respectively.

Inasmuch as

$$C_n = D_n = 0 \quad \text{if } n \neq 1$$

we obtain a relatively simple set of expressions for the electric field outside and inside the sphere:

$$E_\phi^e = E_{0\phi} + \frac{i\omega\mu_e}{2} H_{0z} D_1 R^{-2} \sin\theta, \quad \text{if } R \geq a$$

$$E_\phi^i = \frac{i\omega\mu_e}{2} H_{0z} C_1 R^{-1/2} I_{3/2}(ikR) \sin\theta, \quad \text{if } R \leq a \tag{3.30}$$

From Maxwell's first equation:

$$\text{curl } E = -\partial B/\partial t$$

we have:

$$H_R = \frac{1}{i\omega\mu} \frac{1}{R \sin \theta} \frac{\partial}{\partial \theta} \sin \theta \, E_\phi$$

$$H_\theta = -\frac{1}{i\omega\mu} \frac{1}{R} \frac{\partial}{\partial R} RE_\phi \tag{3.31}$$

Thus, we can obtain the following expressions for the secondary magnetic field as observed outside the sphere:

$$H_{1R}^e = \frac{D_1}{R^3} H_{0z} \cos \theta, \qquad H_{1\theta}^e = \frac{D_1}{2R^3} H_{0z} \sin \theta, \qquad R \geqslant a \tag{3.32}$$

The electromagnetic field from induced currents is equivalent outside the sphere to the field that would be observed from a magnetic dipole situated at the center of the sphere and having the moment:

$$M = 2\pi D_1 H_0 \tag{3.33}$$

directed along the z-axis.

Because D_1 is in general complex, the secondary field differs from the primary field both in magnitude and phase, and for this reason, it can be written as the sum of inphase and quadrature components. The inphase component of the secondary field is either actually inphase or reversed inphase by $180°$ with respect to the primary field, while the quadrature component is shifted by $\pm 90°$ with respect to H_{0z}.

We will first investigate the behavior of the secondary field when the magnetic permeabilities μ_i and μ_e are equal, and assume the value for free space, $4\pi \times 10^{-7}\,\text{H m}^{-1}$.

In this case, the expression for the function D:

$$D_1 = Da^3 \tag{3.34}$$

is simplified markedly and we have:

$$D = \frac{3 x \cosh x - (3 + x^2) \sinh x}{x^2 \sinh x} = \frac{3 \coth x}{x} - \frac{3}{x^2} - 1 \tag{3.35}$$

because

$$I_{1/2}(x) = \left(\frac{2}{\pi x}\right)^{1/2} \sinh x, \qquad I_{-1/2}(x) = \left(\frac{2}{\pi x}\right)^{1/2} \cosh x$$

where

$$x = ika = i(i\sigma\omega\mu)^{1/2} a = (\sigma\omega\mu/2)^{1/2} a \, (1 - i) = p(1 - i) \tag{3.36}$$

The parameter p is the ratio between the radius of the sphere, a, and the thickness of the skin layer, h:

$$p = a/h \qquad (3.37)$$

In order to investigate the low-frequency part of the spectrum, assume that the skin layer thickness, h, is much greater than the radius of the sphere, a, then the function $\coth x$ can be written as a series:

$$\coth x = \frac{1}{x} + \frac{x}{3} - \frac{x^3}{45} + \frac{2x^5}{945} - \cdots = \frac{1}{x} + \sum_{k=1}^{\infty} \frac{2^{2k} B_{2k}}{(2k)!} x^{2k-1}, \quad \text{if } |x^2| < \pi^2$$

where B_{2k} are Bernoulli's numbers. Several of these numbers are listed in Table 3.I.

Table 3.I.

k	1	2	3	4	5	6	7
B_{2k}	$\dfrac{1}{6}$	$-\dfrac{1}{30}$	$\dfrac{1}{42}$	$-\dfrac{1}{30}$	$\dfrac{5}{66}$	$\dfrac{691}{2730}$	$\dfrac{7}{6}$

Substituting this series into eq. 3.35:

$$D = 3 \sum_{k=2}^{\infty} \frac{2^{2k} B_{2k}}{(2k)!} x^{2k-2}, \quad \text{if } |x^2| < \pi^2 \qquad (3.38)$$

or

$$D = 3 \sum_{k=2}^{\infty} \frac{(-\mathrm{i})^{k-1} 2^{2k} B_{2k}}{(2k)!} (\sigma\mu\omega a^2)^{k-1}$$

In accord with this last equation, the low-frequency part of the spectrum of the field contains only interger powers in ω (it is a MacLauren series) and as will be shown later, this feature of the spectrum is diagnostic for fields created by currents induced in confined conductors situated in insulating host rocks.

Shortening eq. 3.38 to the first few terms:

$$D = \mathrm{i}\frac{\sigma\mu a^2}{15} \omega - \frac{2}{315} (\sigma\mu a^2)^2 \omega^2 - \mathrm{i}\frac{1}{1575} (\sigma\mu a^2)^3 \omega^3 \qquad (3.39)$$

it is easily seen that the ratio of coefficients for these series tends to a constant value:

$$-\mathrm{i}\pi^2 /\sigma\mu a^2$$

defining the radius of convergence for a MacLauren series and also, as will be proved later, the character of the decay of a transient field at the late stage.

For small values of the parameter p (that is, at low frequency, for a large resistivity, or for a small radius), the quadrature component of the magnetic field is dominant, and increases in direct proportion to the frequency and the conductivity:

$$QH^e_{1R} \approx \frac{\sigma\mu\omega a^2}{15} H_{0z} \frac{a^3}{R^3} \cos\theta$$

$$\mathrm{In}H^e_{1R} \approx -\frac{2}{315} (\sigma\mu\omega a^2)^2 H_{0z} \frac{a^3}{R^3} \cos\theta$$

$$QH^e_{1\theta} \approx \frac{\sigma\mu\omega a^2}{30} H_{0z} \frac{a^3}{R^3} \sin\theta$$

$$\mathrm{In}H^e_{1\theta} = -\frac{1}{315} (\sigma\mu\omega a^2)^2 \frac{a^3}{R^3} \sin\theta$$

$$(3.40)$$

Now, consider the behavior of the currents at low frequencies ($p < 1$). From eq. 3.29 we have:

$$J_\phi = \sigma E_\phi = \frac{i\sigma\mu_e\omega}{2} H_{0z} C_1 R^{-1/2} I_{3/2}(ikR) \sin\theta \qquad (3.41)$$

where

$$C_1 = \frac{3a^{3/2}}{x I_{1/2}(x)}$$

and

$$I_{3/2}(ikR) = \left(\frac{2}{i\pi kR}\right)^{1/2} \left(\cosh ikR - \frac{1}{ikR} \sinh ikR\right)$$

$$I_{1/2}(ika) = \left(\frac{2}{i\pi ka}\right)^{1/2} \sinh ika$$

The expression for the current density can be written similarly as:

$$J_\phi = -\frac{3}{2R} \frac{x}{\sinh x} H_{0z} \left(\cosh ikR - \frac{1}{ikR} \sinh ikR\right) \sin\theta \qquad (3.42)$$

Expanding the hyperbolic functions in power series and retaining only the first two terms in each series, we obtain:

$$J_\phi \approx \frac{H_{0z}}{2} k^2 R \left[1 - k^2 \left(\frac{R^2}{10} - \frac{a^2}{6}\right)\right] \sin\theta \qquad (3.43)$$

The first term in eq. 3.43 is:

$$J_{0\phi} = H_{0z} \frac{i\sigma\mu\omega}{2} R \sin\theta = \sigma E_{0\phi} \tag{3.44}$$

Thus, for small values of p, the current density for the quadrature component of the induced current flow is defined only by the primary electric field $E_{0\phi}$. In other words, the interaction between current filaments can be neglected. In accord with eq. 3.44, the quadrature component of the current density increases linearly toward the surface of the sphere. However, the inphase component of j_ϕ caused by interaction of currents near the center of the sphere ($R \ll a$) increases almost linearly, reaches a maximum value and then decreases as it approaches its appropriate value on the surface of the sphere. The inphase component of j_ϕ for small values of p, in contrast to the behavior of the quadrature component, is proportional to k^4.

It should be clear that by neglecting the interaction of current filaments, we have markedly simplified the procedure for determining the quadrature component of the magnetic field, so that it consists merely of the calculation of current density:

$$J_\phi = \sigma E_{0\phi}$$

and the magnetic field contributed by current rings based on the use of the Biot-Savart law.

Now we can consider the high-frequency portion of the spectrum for which the parameter $p \gg 1$. Inasmuch as $\coth x \to 1$, when x tends to infinity, the function D tends to the value -1 and the appropriate formulas for the field are those for an ideal conductor, that is:

$$H_R^e = H_{0R} - \frac{a^3}{R^3} H_{0z} \cos\theta$$

$$H_0^e = H_{00} - \frac{a^3}{2R^3} H_{0z} \sin\theta \tag{3.45}$$

$$E_\phi^e = E_{0\phi} - \frac{i\omega\mu_e}{2} H_{0z} \frac{a^3}{R^2} \sin\theta, \quad \text{if } R \geqslant a$$

As a particular case, on the surface of the sphere we have:

$$H_R^e = 0, \quad H_\theta^e = -\frac{3}{2} H_{0z} \sin\theta, \quad E_\phi^e = 0 \tag{3.46}$$

Letting $\coth x = 1$ and keeping the rest of the terms in eq. 3.35, we obtain an asymptotic representation for the function D that describes the high-frequency portion of the spectrum:

$$D \approx \left(-1 + \frac{3}{2p}\right) + \frac{3i}{2p}\left(1 - \frac{1}{p}\right), \quad \text{if } p \gg 1 \tag{3.47}$$

As may be seen from this last equation, both the inphase and quadrature components approach their asymptotic values relatively slowly. It should be noted that the same result can be derived using approximate boundary conditions, as proposed by Leontovich (1948). At this point we can investigate some of the features of the distribution of induced currents for this portion of the spectrum. For large values of ka, the expression for current density in eq. 3.42 can be replaced by the following:

$$J_\phi \approx -\frac{3i}{2}\frac{ka}{R} H_{0z} e^{+ik(R-a)} \sin\theta \tag{3.48}$$

because

$$\sinh x = \cosh x = e^{+x}/2, \quad \text{as } |x| \to \infty$$

At all points inside the sphere, the current density tends to zero with increasing wave number, and if the skin depth is significantly less than the radius of the sphere, the major part of the current flows near the surface of the conductor. In accord with eq. 3.48, at $R = a$, the volume current density is:

$$J_\phi \approx -\frac{3}{2} ikH_{0z} \sin\theta, \quad |ka| \gg 1$$

and increases without limit as frequency increases. However, the cross-sectional area through which most of the current passes becomes smaller at the same time, and from the physical point of view, we can expect that the magnitude of the surface current in the limit is finite. In fact, the current passing through an elementary surface dS is:

$$dI = J_\phi\, dS = -\frac{3}{2} iH_{0z}\, ka\, e^{+ik(R-a)}\, dR\, d\theta \sin\theta$$

For the total current through a sector subtending an angle $d\theta$, we have:

$$dI = -\frac{3i}{2} ka H_{0z} \sin\theta\, d\theta \int_0^a e^{-ik(R-a)}\, dR$$

$$= -\frac{3}{2} aH_{0z} (1 - e^{ika}) \sin\theta\, d\theta$$

From this, we have:

$$dI = -\frac{3}{2} aH_{0z} \sin\theta\, d\theta, \quad \text{as } |k| \to \infty$$

Integrating this last expression with respect to θ over the limits from π to zero, we find that in the limiting case as frequency becomes infinite, the total surface current is:

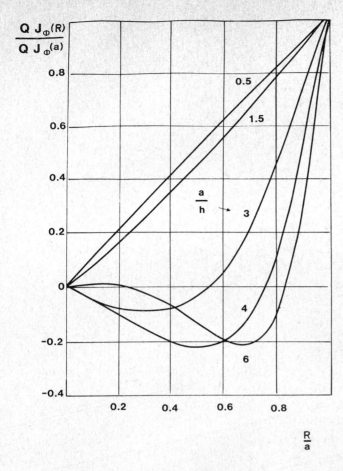

Fig. 3.2 Curves showing the distribution of the quadrature component of current density along the radius of the sphere for various values of the parameter p.

$$I = -3H_{0z}\,a \tag{3.49}$$

Curves showing the distribution of current density along the radius of the sphere for various values of the parameter p are shown in Figs. 3.2 and 3.3. The ratio of the current density at an arbitrary point, R, to that on the surface of the sphere is plotted along the ordinate.

It should be pointed out that as frequency increases and as the skin depth becomes much less than the radius of the sphere, the ratio of the tangential components of the electric and the magnetic fields on the surface of the sphere coincides with the expression for the impedance of a plane wave. In accord with eq. 3.47, we have the following for the total field:

$$E_\phi^e = \frac{i\omega\mu_e\,3H_0}{2k\,i}\sin\theta, \qquad H_\theta^e = \frac{3}{2}H_0\sin\theta$$

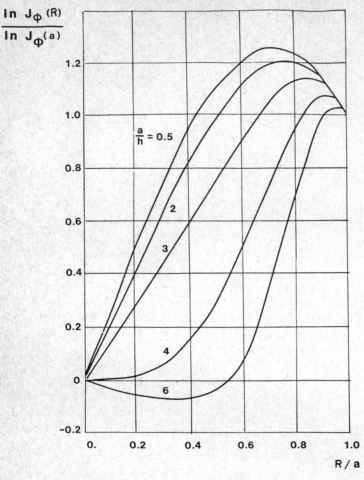

Fig. 3.3. Curves showing the inphase component of current density along the radius of the sphere for various values of the parameter p.

and

$$Z = \frac{E_\phi^e}{H_\theta^e} = \frac{\omega\mu}{k} = \left(\frac{\omega\mu}{i\sigma}\right)^{1/2} = \left(-\frac{i\omega\mu}{\sigma}\right)^{1/2} = \left(\frac{\omega\mu}{\sigma}\right)^{1/2} e^{-i\pi/4} \qquad (3.50)$$

We can now describe the whole spectrum of behavior for the electromagnetic field caused by currents induced in the sphere and described by the function D. The values for the real and imaginary parts of the function D for various values of μ_i/μ_e are given in Table 3.II. Curves for the functions $\mathrm{Re}\,D$ and $\mathrm{Im}\,D$ are shown in Fig. 3.4 for the case in which $\mu_i/\mu_e = 1$.

In considering these curves, we can recognize three diagnostic ranges for the frequency response of a field, namely:

TABLE 3.II

a/h	μ_i/μ_e 1		1.25		2.5		5	
	— ReD	ImD	ReD	ImD	ReD	ImD	ReD	ImD
0.200	0.406×10^{-4}	0.533×10^{-2}	0.154	0.710×10^{-2}	0.666	0.148×10^{-1}	0.114×10^{1}	0.245×10^{-1}
0.283	0.162×10^{-3}	0.107×10^{-1}	0.154	0.142×10^{-1}	0.666	0.296×10^{-1}	0.144×10^{1}	0.489×10^{-1}
0.400	0.649×10^{-3}	0.213×10^{-1}	0.153	0.284×10^{-1}	0.663	0.590×10^{-1}	0.113×10^{1}	0.970×10^{-1}
0.566	0.259×10^{-2}	0.425×10^{-1}	0.150	0.565×10^{-1}	0.653	0.117	0.111×10^{1}	0.189
0.800	0.102×10^{-1}	0.840×10^{-1}	0.138	0.111	0.615	0.224	0.102×10^{1}	0.341
1.13	0.390×10^{-1}	0.161	0.939×10^{-1}	0.209	0.491	0.385	0.804	0.509
1.60	0.132	0.274	-0.371×10^{-1}	0.339	0.227	0.513	0.496	0.584
2.26	0.326	0.353	—0.271	0.401	-0.698×10^{-1}	0.510	0.194	0.581
3.20	0.530	0.324	—0.485	0.353	—0.308	0.446	-0.808×10^{-1}	0.532
4.52	0.669	0.258	—0.633	0.283	—0.496	0.367	—0.313	0.456
6.40	0.766	0.198	—0.739	0.218	—0.637	0.288	—0.498	0.370
9.05	0.834	0.147	—0.815	0.163	—0.741	0.220	—0.638	0.290

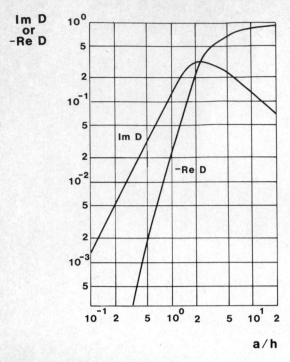

Fig. 3.4. Curves for the real and imaginary parts of D when the magnetic permeability contrast is unity.

(1) The low-frequency part of the spectrum, or the range for small values of the ratio a/h, where the quadrature component of the magnetic field increases almost in direct proportion to the frequency, and the conductivity, while the inphase component is considerably smaller.

(2) An intermediate-frequency range where the quadrature component reaches a maximum and then with further increases in frequency, decreases. This maximum for the quadrature component is observed when the radius of the sphere is about 2.5 times the thickness of the skin layer.

(3) A high-frequency part of the spectrum, or the range over which the parameter p is large. In this case, the induced currents are concentrated near the surface of the sphere, and the inphase component tends to become constant with increasing frequency, with the constant being equal to the field magnitude caused by currents on the surface of an ideally conducting sphere. At the same time, the quadrature component of the magnetic field tends to zero in inverse proportion to $\omega^{1/2}$.

It must be stressed that these particular features of the frequency characteristic of the electromagnetic field are inherent to quasi-stationary fields that are created by induced currents flowing in confined conductors of arbitrary shape, embedded in an insulating host medium. In this respect, it can be helpful to represent the function D characterizing the spectrum of the electromagnetic field in still another form. Making use of the identity:

$$\coth x = \frac{1}{x} + 2x \sum_{n=1}^{\infty} \frac{1}{\pi^2 n^2 + x^2}$$

along with eq. 3.35, we have:

$$D = 6 \sum_{n=1}^{\infty} \frac{1}{\pi^2 n^2 - i\sigma\mu a^2 \omega} - 1 \qquad (3.51)$$

Inasmuch as

$$1 = 6 \sum_{n=1}^{\infty} \frac{1}{\pi^2 n^2}$$

the function D can be represented as:

$$D = i\omega \frac{6}{\pi^2} \sum_{n=1}^{\infty} \frac{1}{n^2} \frac{1}{\pi^2 n^2 \alpha - i\omega} \qquad (3.52)$$

where

$$\alpha = 1/\sigma\mu a^2 \qquad (3.53)$$

Thus, the spectrum for electromagnetic field from currents flowing in the sphere can be described as a sum of simple fractions. Now let us consider the

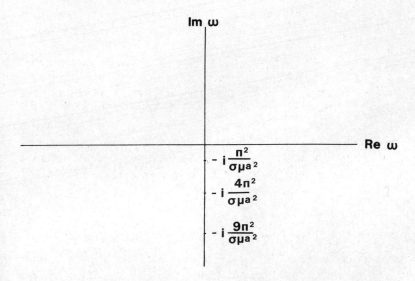

Fig. 3.5. Locations of zeros for the denominator in the fractions in eq. 3.52.

function D in the complex plane, ω, as shown in Fig. 3.5. It can readily be seen that the denominator in the fractions in eq. 3.52 are zero when:

$$\omega_n = -i\frac{\pi^2}{\sigma\mu a^2}n^2 \tag{3.54}$$

Therefore, the spectrum has singularities or poles, located along the imaginary axis as shown in Fig. 3.5. In accord with eq. 3.54, the distance from a pole and the origin $\omega = 0$ increases rapidly with larger values of n. For example:

$$\omega_1 = -i\frac{\pi^2}{\sigma\mu a^2}, \quad \omega_2 = -i\frac{4\pi^2}{\sigma\mu a^2}, \quad \omega_3 = -i\frac{9\pi^2}{\sigma\mu a^2}$$

and so on.

As is well known, the radius of convergence for MacLauren's series is defined by the distance from the origin ($\omega = 0$) to the first singularity. For this reason, we can say that the low-frequency expansion represented in eq. 3.39 converges when:

$$\omega < \pi^2/\sigma\mu a^2 \tag{3.55}$$

Now, consider the effect of the magnetic permeability, μ_i, on the frequency behavior of the field. In accord with eq. 3.28 we have:

$$D = \frac{(2K+1)x - [(1+2K)+x^2]\tanh x}{(K-1)x + [(1-K)+x^2]\tanh x} \tag{3.56}$$

because

$$I_{1/2}(x) = \left(\frac{2}{\pi x}\right)^{1/2}\sinh x, \quad I_{-1/2}(x) = \left(\frac{2}{\pi k}\right)^{1/2}\cosh x$$

where

$$K = \mu_i/\mu_e, \quad x = i(i\sigma\mu_i\omega)^{1/2}a$$

Letting $|x| < 1$ and replacing $\tanh x$ by the first few terms of its power series expansion:

$$\tanh x \approx x - \frac{x^3}{3} + \frac{2x^5}{15} - \ldots$$

we obtain:

$$D \approx \frac{2(K-1)}{2+K} + i\frac{3K\sigma\mu_i a^2}{5(2+K)^2}\omega - \ldots \tag{3.57}$$

Thus, in contrast to the behavior in the case of a non-magnetic sphere, at low frequencies, the inphase component of the magnetic field is dominant and with a decrease in frequency tends to a limit corresponding to a constant magentic field given by:

$$\lim_{\omega \to 0} D = \frac{2(K-1)}{2+K} = 2\frac{\mu_i - \mu_e}{\mu_i + 2\mu_e} \tag{3.58}$$

In accord with eq. 3.32, we have the following expressions for the magnetic field:

$$H^e_{1R} = 2\frac{\mu_i - \mu_e}{\mu_i + 2\mu_e}\frac{a^3}{R^3} H_{0z} \sin \omega t \cos \theta$$

and

$$H^e_{1\theta} = \frac{\mu_i - \mu_e}{\mu_i + 2\mu_e}\frac{a^3}{R^3} H_{0z} \sin \omega t \sin \theta, \quad \text{as } p \to 0 \tag{3.59}$$

where H_{0z} is the magnitude of the primary field.

At any instance, this magnetic field coincides with that when the primary field is constant and is equal to $H_{0z} \sin \omega t$. Equations 3.59 describe the field caused by a contrast in magnetic permeability between the sphere and the host medium. As a result of a change in this field with time, a vortex electric field is generated. In accord with eq. 3.30, we have:

$$E^e_{1\phi} = \omega \mu_e H_{0z}\frac{\mu_i - \mu_e}{\mu_i + 2\mu_e}\frac{a^3}{R^2} \sin \theta \cos \omega t, \quad \omega \to 0 \tag{3.60}$$

In order to derive corresponding expressions for the field inside the sphere, we can make use of asymptotic expressions for Bessel's functions:

$$I_{1/2}(x) \approx \left(\frac{2}{\pi x}\right)^{1/2}\left(x + \frac{x^3}{6}\right)$$

$$I_{-1/2}(x) \approx \left(\frac{2}{\pi x}\right)^{1/2}\left(1 + \frac{x^2}{2}\right)$$

Substituting these into eq. 3.29, we have:

$$C_1 \approx \left(\frac{\pi x}{2}\right)^{1/2}\frac{3Ka^{3/2}}{x^2\{1 + (K-1)/3\}}$$

Because:

$$I_{3/2}(\alpha x) \approx \left(\frac{2}{\pi x a}\right)^{1/2}\frac{1}{3}\alpha x$$

where

$$\alpha = R/2$$

158

we have:

$$E_\phi^i \approx i\omega B_{0z} \frac{3K}{2(2+K)} R \sin \theta \qquad (3.61)$$

where

$$B_{0z} = \mu_e H_{0z}$$

is the vector representing magnetic induction by the primary field.

Making use of Maxwell's equation:

$$\text{curl } E = -\partial B/\partial t$$

it can readily be shown that inside this sphere, the magnetic field is uniform, and equal to:

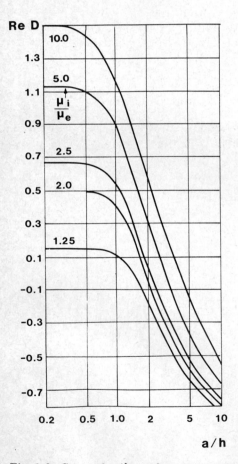

Fig. 3.6. Curves for the real part of the function D.

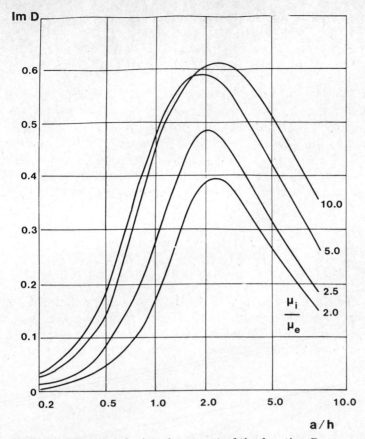

Fig. 3.7. Curves for the imaginary part of the function D.

$$B_i \quad \frac{3K}{2+K} B_0 \qquad\qquad (3.62)$$

For this reason, the induced current density at small values of p increases with an increase in μ_i, and for very large values of K ($K \gg 1$) it is almost three times greater than that for a non-magnetic sphere. This result follows directly from a consideration of the imaginary parts of the function D:

$$\frac{\operatorname{Im} D(K, p)}{\operatorname{Im} D(1, p)} = \frac{9K^2}{(2+K)^2}, \quad \text{if } p \ll 1 \qquad\qquad (3.63)$$

In particular, for the case with K much greater than 1, we have:

$$\frac{\operatorname{Im} D(K, p)}{\operatorname{Im} D(1, p)} = 9$$

The quadrature component of the magnetic field increases to a greater extent than does the current density as a consequence of the presence of the magnetic medium.

It should be clear that with an increase in magnetic permeability μ_i, the skin effect expresses itself at lower frequencies than would otherwise be the case.

However, at high frequencies, both the magnetic field and the induced currents vanish inside the sphere, and the sole source for the field is a surface current. Correspondingly, the influence of the magnetic permeability in the sphere disappears and in the limit, the function D tends to the value -1 (see Table 3.II). Curves for the real and imaginary parts of this function are shown in Figs. 3.6 and 3.7. It should be noted here that in the case of a magnetic sphere, the inphase component of the secondary magnetic field changes sign.

3.2. A CONDUCTIVE SPHERE IN A UNIFORM NON-STATIONARY MAGNETIC FIELD

We will now consider the electromagnetic field from currents induced in a sphere when the primary field has the form of step function excitation. Suppose that a sphere with radius a embedded in an insulating space is illuminated with a uniform magnetic field H_{0z} which at the moment $t = 0$ vanishes (see Fig. 3.8):

$$H_{0z} = \begin{cases} H_{0z} & t < 0 \\ 0 & t > 0 \end{cases} \tag{3.64}$$

In accord with the law for electromagnetic induction, at that instant, eddy currents appear on the surface of the sphere in such a way that they preserve the magnetic field unchanged inside the conductor, that is:

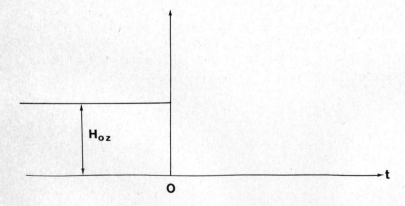

Fig. 3.8. Time variation of a magnetic field which vanishes at the moment $t = 0$.

$$H^i = H_{i0}, \quad \text{at } t = 0 \tag{3.65}$$

where H_{i0} is the constant magnetic field inside the conductive body. Shortly after this instant, a diffusion process begins, with induced currents developing within this sphere. Diffusion of the current is accompanied by a decrease in the field, since the electromagnetic energy is converted to heat. We will determine what the field is in two ways, namely, by solving the diffusion equation, and by applying the Fourier transform. The latter approach is the one more commonly used, and permits us to find what a non-stationary field will be in more complicated models of conducting media when the spectral response is known.

As has been shown in the previous chapter, the Maxwell equations for quasi-stationary field behavior have the form:

$$\operatorname{curl} E = -\frac{\partial B}{\partial t} \quad \text{(I)} \qquad \operatorname{div} E = 0 \quad \text{(III)}$$

$$\operatorname{curl} H = \sigma E \quad \text{(II)} \qquad \operatorname{div} B = 0 \quad \text{(IV)}$$

Making use of equation IV, we will introduce a vector potential of the electrical type satisfying the definition:

$$B = \operatorname{curl} A \tag{3.66}$$

Substituting this expression for B in Maxwell's equations I and II, we obtain:

$$\operatorname{curl} \operatorname{curl} A = \operatorname{grad} \operatorname{div} A - \nabla^2 A = \sigma \mu E$$

and

$$E = -\frac{\partial A}{\partial t} - \operatorname{grad} U$$

From this, we can write:

$$\operatorname{grad} \operatorname{div} A - \nabla^2 A = -\sigma \mu \frac{\partial A}{\partial t} - \sigma \mu \operatorname{grad} U$$

Using the gauge condition:

$$\operatorname{div} A = -\sigma \mu U$$

we have:

$$\nabla^2 A = \sigma \mu \frac{\partial A}{\partial t} \tag{3.67}$$

and

$$E = -\frac{\partial A}{\partial t} + \frac{1}{\sigma \mu} \operatorname{grad} \operatorname{div} A$$

Inasmuch as we have assumed that the primary field is caused by a horizontal current loop as shown in Fig. 3.1, we can make use of the axial symmetry of the problem and look for a solution with a single component A_ϕ. For this reason, in view of the axial symmetry, div $A = 0$, and:

$$B = \text{curl}\,A, \quad E = -\frac{\partial A}{\partial t} \tag{3.68}$$

As has already been pointed out, determination of the non-stationary field behavior is a somewhat more complicated task than determination of the harmonic field behavior and will usually consist of three principal steps:

(1) A solution of the diffusion equation:

$$\nabla^2 A = \sigma\mu \frac{\partial A}{\partial t}$$

(2) Determination of the conditions under which a solution of eq. 3.67 provides continuity of the tangential components of the field at the surface of the sphere:

$$E_\phi^i = E_\phi^e, \quad H_\theta^i = H_\theta^e$$

or in accord with eq. 3.68:

$$A_\phi^i = A_\phi^e, \quad \mu_e \frac{\partial}{\partial R} R A_\phi^i = \mu_i \frac{\partial}{\partial R} R A_\phi^e, \quad \text{when } R = a \tag{3.69}$$

(3) Determination of the field which satisfies the initial condition in eq. 3.65:

$$H^i(t) = H_{i0} \quad \text{or} \quad A_\phi^i(t) = A_{i0}, \quad \text{as } t = 0 \tag{3.70}$$

where A_{i0} is the vector potential for a constant magnetic field inside the sphere.

Of course, any solution has to be finite everywhere and must satisfy an appropriate condition at infinity.

Let us write a particular solution to the diffusion equation as the product:

$$A = X(R, \theta)T(t) \tag{3.71}$$

Substituting this solution into 3.67 as a trial, we have:

$$T\nabla^2 X(R, \theta) = \sigma\mu X\dot{T}$$

or

$$\frac{\nabla^2 X}{X} = \sigma\mu \frac{\dot{T}}{T} = -\sigma\mu\bar{q}_s = -k_s^2$$

where

$$k_s^2 = \sigma\mu\bar{q}_s$$

is the separation constant.

Thus, in place of eq. 3.67 we have obtained a single equation with partial derivatives in X and an ordinary diffential equation of the first order for the function $T(t)$:

$$\nabla^2 X(R, \theta) + k_s^2 X(R, \theta) = 0 \tag{3.72}$$

and

$$\frac{dT}{dt} + \bar{q}_s T = 0 \tag{3.73}$$

where

$$k_s^2 = \sigma\mu\bar{q}_s \tag{3.74}$$

It should be obvious that one solution for eq. 3.73 is the exponential function $e^{-\bar{q}_s t}$.

Equation 3.72 is the same as that for the electric field which was derived in the last paragraph except that in the place of the squared wave number, we have the quantity $k_s^2 = \sigma\mu\bar{q}_s$. As was shown under excitation of the sphere by a uniform magnetic field B_0 in the frequency domain, the electric field for the currents induced in the sphere can be described using only one spherical harmonic. Making use of this result, we will try to satisfy the boundary and initial conditions, assuming that the dependence of the vector potential on coordinates is also defined by a single harmonic. Proceeding from the equation $E = -\partial A/\partial t$ we will assume that the vector potential has only the component A_ϕ. Therefore, the general solution both inside and outside the sphere will be in the form:

$$A_\phi^i = R^{-1/2} \sin\theta \sum_s C_s J_{3/2}(k_s R) e^{-\bar{q}_s t}$$

$$\tag{3.75}$$

$$A_\phi^e = R^{-2} \sin\theta \sum_s D_s e^{-\bar{q}_s t}$$

In these expressions for the vector potential, the coefficients C_s, D_s, and the parameter $\bar{q}_s$ are unknown quantities.

It is clear that the functions A_ϕ^i and A_ϕ^e are solutions of the equations:

$$\nabla^2 A^i = \sigma\mu \frac{\partial A^i}{\partial t} \quad \text{and} \quad \nabla^2 A^e = 0$$

respectively.

Now we will seek a parameter k_s for which the boundary conditions in eqs. 2.6 are satisfied. Inasmuch as the primary field is zero for t greater than zero, we obtain from this boundary condition (eq. 3.69) a uniform system of equations as follows:

$$C_s a^{3/2} J_{3/2}(k_s a) = D_s$$

$$\mu_e a^2 C_s \frac{d}{da} [a^{1/2} J_{3/2}(k_s a)] = -\mu_i D_s$$

(3.76)

Eliminating the two coefficients C_s and D_s, we obtain an equation for k_s:

$$a^{1/2} \frac{d}{da} [a^{1/2} J_{3/2}(k_s a)] + K J_{3/2}(k_s a) = 0$$

After differentiation:

$$a \frac{d}{da} J_{3/2}(k_s a) + (K + \tfrac{1}{2}) J_{3/2}(k_s a) = 0$$

(3.77)

where $K = \mu_i/\mu_e$.

Thus, in order to satisfy the boundary conditions, the roots k_s for this equation must be found. The parameters $\bar{q}_s$, describing the field decay, are calculated from eq. 3.74 when the values for k_s^2 are known. Equation 3.77 can be simplified considerably by making use of the following relationships:

$$J'_{3/2}(x) = J_{1/2}(x) - \frac{3}{2x} J_{3/2}(x)$$

(3.78)

and

$$J_{1/2}(x) = \left(\frac{2}{\pi x}\right)^{1/2} \sin x$$

(3.79)

$$J_{3/2}(x) = \left(\frac{2}{\pi x}\right)^{1/2} \left(\frac{1}{x} \sin x - \cos x\right)$$

Substituting eq. 3.78 into eq. 3.77, we have:

$$x J_{1/2}(x) - \tfrac{3}{2} J_{3/2}(x) + (K + \tfrac{1}{2}) J_{3/2}(x) = 0$$

(3.80)

or

$$x J_{1/2}(x) + (K - 1) J_{3/2}(x) = 0$$

where $x = k_s a$. Using eq. 3.79, the last equation can be rewritten in the form:

$$x \sin x + (K - 1)\left(\frac{1}{x} \sin x - \cos x\right) = 0$$

Finally, we arrive at a transcendental equation for determining $k_s a$:

$$\tan k_s a = \frac{(K-1)k_s a}{k_s^2 a^2 + (K-1)} \tag{3.81}$$

If the sphere is non-magnetic, eq. 3.81 can be further simplified ($K = 1$) and we have:

$$\tan k_s a = 0 \tag{3.82}$$

or

$$k_s a = \pi s$$

and therefore:

$$\bar{q}_s = \frac{k_s^2}{\sigma \mu_i} = \frac{\pi^2 s^2}{\sigma \mu_i a^2} \tag{3.83}$$

where $s = 1, 2, 3, 4 \ldots$ and so on.

In order to find the unknown coefficients C_s and D_s, we will use the initial condition in eq. 3.70:

$$A_\phi^i(t) = A_{i0}$$

for $t = 0$. The vector potential A_{i0} is easily defined from the solution for frequency-domain behavior for the case where frequency tends to zero. In order to derive the asymptotic expression for A_{i0}, we will make use of eq. 3.61 and the second of the equations in 3.68:

$$A_{i0} = \frac{3KB_{0z}}{2(2+K)} R \sin \theta_{\varphi 0} \tag{3.84}$$

Therefore, in accord with eqs. 3.75 and 3.84, the initial condition 3.70 is written as:

$$\frac{3KB_{0z}}{2(2+K)} R^{3/2} = \sum_{s=1}^{\infty} C_s J_{3/2}(k_s R), \quad \text{if } R \leqslant a \tag{3.85}$$

This last expression is an expansion of the function:

$$f(R) = \frac{3KB_{0z}}{2(2+K)} R^{3/2}$$

in a series of Bessel functions with the index $n = 3/2$. It is known from the theory for Bessel functions that the coefficients in this equation are defined as:

$$C_s = \frac{\int_0^a R f(R) J_{3/2}(k_s R) dR}{\int_0^a R [J_{3/2}(k_s R)]^2 dR} \tag{3.86}$$

Replacing Bessel's functions by trigonometric functions after integration of eq. 3.79 we have:

$$C_s = \frac{3KB_{0z}a^{3/2}}{[k_s^2 a^2 + (K-1)(K+2)] J_{3/2}(k_s a)} \tag{3.87}$$

In accord with eq. 3.76, we also have:

$$D_s = \frac{3KB_{0z}a^3}{[k_s^2 a^2 + (K-1)(K+2)]} \tag{3.88}$$

Therefore, the expression for the vector potential outside this sphere is:

$$A_\phi^e = 3KB_{0z}a^3 R^{-2} \sin\theta \sum_{s=1}^{\infty} \frac{e^{-\bar{q}_s t}}{[k_s^2 a^2 + (K-1)(K+2)]} \tag{3.89}$$

Using eq. 3.68, we can obtain the following expression for the field:

$$E_\phi^e = 3KB_{0z}a^3 R^{-2} \sin\theta \sum_{s=1}^{\infty} \frac{\bar{q}_s e^{-\bar{q}_s t}}{[k_s^2 a^2 + (K-1)(K+2)]} \tag{3.90}$$

and

$$B_R^e = \frac{6KB_{0z}a^3}{R^3} \cos\theta \sum_{s=1}^{\infty} \frac{e^{-\bar{q}_s t}}{[k_s^2 a^2 + (K-1)(K+2)]}$$

$$B_\theta^e = \frac{3KB_{0z}a^3}{R^3} \sin\theta \sum_{s=1}^{\infty} \frac{e^{-\bar{q}_s t}}{[k_s^2 a^2 + (K-1)(K+2)]} \tag{3.91}$$

where $\bar{q}_s = k_s^2 / \sigma\mu_i$.

Let us first examine the transient response for induced currents in a non-magnetic sphere, that is for $K = 1$. In this case, in accord with eqs. 3.83 and 3.90–3.91, we have:

$$\bar{q}_s = \frac{\pi^2 s^2}{\sigma\mu a^2} = \frac{k_s^2}{\sigma\mu} \tag{3.92}$$

and

$$E_\phi^e = \frac{3B_{0z}a^3}{R^2} \alpha \sin\theta \sum_{s=1}^{\infty} e^{-\pi^2 s^2 \alpha t}$$

$$B_R^e = \frac{6B_{0z}a^3}{\pi^2 R^3} \cos\theta \sum_{s=1}^{\infty} \frac{1}{s^2} e^{-\pi^2 s^2 \alpha t} \tag{3.93}$$

$$B_\theta^e = \frac{3B_{0z}a^3}{\pi^2 R^3} \sin\theta \sum_{s=1}^{\infty} \frac{1}{s^2} e^{-\pi^2 s^2 \alpha t}$$

where $\alpha = 1/\sigma\mu a^2$.

Thus, the electromagnetic field outside the sphere is equivalent to that which would be observed for a magnetic dipole, just as was the case in the frequency domain. The moment of the dipole is directly proportional to the strength of the primary magnetic field, H_{0z}, and decreases with time

It will prove to be convenient in subsequent discussions to distinquish between three different parts of the time response of the electromagnetic field, namely: (1) the early stage; (2) an intermediate stage; and (3) a late stage.

During the early stage, in which $t \to 0$, in accord with eq. 3.93 we have the following expressions for the magnetic induction components:

$$B_R^e = \frac{6B_{0z}a^3}{\pi^2 R^3} \cos \theta \sum_{s=1}^{\infty} \frac{1}{s^2}$$

$$B_\theta^e = \frac{3B_{0z}a^3}{\pi^2 R^3} \sin \theta \sum_{s=1}^{\infty} \frac{1}{s^2}$$

Inasmuch as:

$$\sum_{s=1}^{\infty} \frac{1}{s^2} = \frac{\pi^2}{6}$$

we have:

$$B_R^e = B_{0z} \frac{a^3}{R^3} \cos \theta$$

$$B_\theta^e = B_{0z} \frac{a^3}{2R^3} \sin \theta$$

(3.94)

These expressions are exactly the same as those corresponding to the high-frequency part of the frequency-domain response; that is, during the early stage the induced currents flow on the surface of the sphere, and they create a magnetic field equal in amplitude to the field observed at high frequencies. At the same time, the magnetic field inside the sphere, according to the requirements for the initial condition, is H_0 and induced currents are absent.

Let us study the early-stage behavior in more detail. In accord with eq. 3.47, at high frequencies, the inphase and quadrature components of the magnetic field can be written as:

$$\text{In}H_{1R}^e = \left(-1 + \frac{3}{2p}\right)\frac{a^3}{R^3} H_{0z} \cos \theta, \qquad QH_{1R}^e = \frac{3}{2p}\frac{a^3}{R^3} H_{0z} \cos \theta$$

(3.95)

$$\text{In}H_{1\theta}^e = \left(-1 + \frac{3}{2p}\right)\frac{a^3}{2R^3} H_{0z} \sin \theta, \qquad QH_{1\theta}^e = \frac{3}{2p}\frac{a^3}{2R^3} H_{0z} \sin \theta$$

where

$$p = \frac{a}{h} = \left(\frac{\sigma\mu a^2}{2}\right)^{1/2} \omega^{1/2} = \left(\frac{1}{2\alpha}\right)^{1/2} \omega^{1/2} \tag{3.96}$$

Inasmuch as the early stage of the transient field is defined almost entirely by the high-frequency part of the frequency-domain response, we can make use of the limit theorem for Fourier's transformation. For example, starting from the quadrature component of H_{1R}^e, we have (see eq. 2.338):

$$\frac{\partial H_R^e(t)}{\partial t} = -\frac{2}{\pi} \int_0^\infty Q H_{1R}^e \sin \omega t d\omega \tag{3.97}$$

Substituting eq. 3.96 into this last equation, we obtain:

$$\frac{\partial H_R^e(t)}{\partial t} = -\frac{2}{\pi} \frac{3}{2} (2\alpha)^{1/2} \frac{a^3}{R^3} H_{0z} \cos\theta \int_0^\infty \frac{\sin \omega t}{\omega^{1/2}} d\omega, \quad \text{when } t \to 0$$

This integral can be written as:

$$\int_0^\infty \frac{\sin \omega t}{\sqrt{\omega}} d\omega = \frac{1}{t^{1/2}} \int_0^\infty \frac{\sin \omega t}{\sqrt{\omega t}} d\omega t = \frac{1}{t^{1/2}} \int_0^\infty \frac{\sin x}{\sqrt{x}} dx$$

Inasmuch as:

$$\int_0^\infty \frac{\sin x}{\sqrt{x}} dx = \left(\frac{\pi}{2}\right)^{1/2}$$

we have the following expression for $\partial H_R^e / \partial t$:

$$\frac{\partial H_R^e}{\partial t} = -\frac{3}{\sqrt{\pi}} \left(\frac{\alpha}{t}\right)^{1/2} \frac{a^3}{R^3} H_{0z} \cos\theta$$

or

$$\frac{\partial B_R^e}{\partial t} = -\frac{3}{\sqrt{\pi}} \left(\frac{\alpha}{t}\right)^{1/2} \frac{a^3}{R^3} B_{0z} \cos\theta \tag{3.98}$$

By analogy, we have:

$$\frac{\partial H_\theta^e}{\partial t} = -\frac{3}{\sqrt{\pi}} \left(\frac{\alpha}{t}\right)^{1/2} H_{0z} \frac{a^3}{2R^3} \sin\theta$$

and

$$\frac{\partial B_\theta^e}{\partial t} = -\frac{3}{\sqrt{\pi}} \left(\frac{\alpha}{t}\right)^{1/2} B_{0z} \frac{a^3}{2R^3} \sin\theta, \quad \text{for } t \to 0 \tag{3.99}$$

where $\alpha = 1/\sigma\mu a^2$.

Thus, during the early stage of time-domain behavior, the derivatives of the magnetic field with respect to time are directly proportional to $\alpha^{1/2}$, and decrease with time as $t^{-1/2}$.

Using eqs. 3.94 and 3.98, it is possible to derive expressions for the field itself. In fact, we have:

$$B_R^e(t) = B_R^e(0) + \int_0^t \frac{\partial B_R^e}{\partial t}\, dt$$

$$B_\theta^e(t) = B_\theta^e(0) + \int_0^t \frac{\partial B_\theta^e}{\partial t}\, dt$$

After integration:

$$B_R^e(t) = B_{0z} \frac{a^3}{R^3} \cos\theta \left[1 - \frac{6}{\sqrt{\pi}} (\alpha t)^{1/2} \right] \quad \text{for } t \to 0$$

$$B_\theta^e(t) = B_{0z} \frac{a^3}{2R^3} \sin\theta \left[1 - \frac{6}{\sqrt{\pi}} (\alpha t)^{1/2} \right]$$

$$(3.100)$$

It should be clear that with an increase in the conductivity or the radius of the sphere, the early-stage behavior persists over longer times. In particular, in the case of a perfectly conducting sphere, induced currents are present only on the surface of the sphere at all times, and do not decay. For this reason, the magnetic field of these currents is constant ($\alpha = 0$) with a value equal to that for $t = 0$. One of the most important characteristics of the magnetic field behavior during the early stage is a consequence of the fact that the eddy currents are almost entirely situated on the surface of the conductor. Therefore, the magnetic field is only weakly related to the conductivity, and mainly depends on the radius of the sphere and its location. However, the electric field caused by a change in the magnetic field with time, and therefore, with the diffusion of currents into the conductor, is directly proportional to the quantity $\alpha^{1/2}$. As can be seen from eqs. 3.98 and 3.99, the EMF in a receiving coil increases without limit as the time t tends to zero. This peculiarity is a consequence of the step function form of the excitation. In practice, there is always some range of time (a "ramp time") over which the current in the source drops to zero, and therefore, the EMF as well as the electric field will be finite at the initial instant.

As follows from eq. 3.99, with a decrease in the value for the parameter α, that is, with either an increase in the conductivity or the radius of the sphere, the electromotive force becomes smaller during the early stage as a consequence of the decrease in the rate of change of the magnetic field with time.

A comparison of this with the exact solution in eq. 3.93 shows that the asymptotic formulas in eqs. 3.99—3.100 can be used with an error of less than 10% when the condition:

$$\alpha t < 0.03 \tag{3.101}$$

is met.

In accord with eqs. 3.98—3.100, the parameter α can be determined during the early stage from the expression:

$$\frac{\dot{B}_R^e}{B_R^e} = \frac{\dot{B}_\theta^e}{B_\theta^e} = -\frac{3\alpha^{1/2}}{(\pi t)^{1/2}} \frac{1}{1 - \dfrac{6}{\sqrt{\pi}} (\alpha t)^{1/2}}, \quad \text{if } \alpha t < 0.03 \tag{3.102}$$

We will next consider the behavior at relatively large times, during the late stage of the time-domain field. As follows from eq. 3.93, no matter what the conductivity of this sphere, at sufficiently late times, the field is almost entirely determined by the first exponential of these sums, that is:

$$B_R^e = \frac{6B_{0z}}{\pi^2} \frac{a^3}{R^3} e^{-\pi^2 \alpha t} \cos \theta$$

$$\tag{3.103}$$

$$B_\theta^e = \frac{3B_{0z}}{\pi^2} \frac{a^3}{R^3} e^{-\pi^2 \alpha t} \sin \theta$$

and

$$E_\phi^e = \frac{3\alpha B_{0z}}{R^2} a^3 e^{-\pi^2 \alpha t} \sin \theta \tag{3.104}$$

It should be clear that the late-stage behavior is expressed when the condition:

$$t \geqslant 1/\pi^2 \alpha \tag{3.105}$$

holds.

Often the parameter $1/\pi^2 \alpha$ is called the time constant, τ_0, for a spherical conductor:

$$\tau_0 = \frac{1}{\pi^2 \alpha} = \frac{\sigma \mu a^2}{\pi^2} \tag{3.106}$$

Thus, the late-stage behavior for a transient field will be observed when measurements are made at times comparable to or exceeding the time constant τ_0. Correspondingly, the components of the electromagnetic field described in eqs. 3.103—3.104 can be rewritten as:

$$B_R^e = \frac{6B_{0z}}{\pi^2} \frac{a^3}{R^3} \cos\theta e^{-t/\tau_0}$$

$$B_\theta^e = \frac{3B_{0z}}{\pi^2} \frac{a^3}{R^3} \sin\theta e^{-t/\tau_0} \qquad \text{if } t \geqslant \tau_0 \qquad (3.107)$$

$$E_\phi^e = \frac{3B_{0z}}{\tau_0\pi^2} \frac{a^3}{R^2} \sin\theta e^{-t/\tau_0}$$

This remarkable simplicity for the expressions describing the field during the late stage of their transient response is a particularly important part of the behavior of the transient field from a practical point of view. The components for the magnetic field as well as those for the electric field decay exponentially with time, and the right-hand terms in the expressions for B_R and B_θ can be represented as the product of two functions, which are:

(1) A function which depends only on geometry and on the magnitude of the primary field. For example, for B_R, this function has the following representation

$$\frac{6B_{0z}}{\pi^2 R^3} a^3 \cos\theta$$

(2) A second function which is an exponential e^{-t/τ_0} that is the same for all of the components of the field, and which does not depend on the intensity of the primary field or the position of the observation point, but is only a function of the time t and the time constant τ_0. This remarkable behavior for the formulas simplifies the determination of the parameters describing this sphere, σ and a. For example, forming a logarithm of both sides of the equation for B_θ^e, we have:

$$\ln B_\theta^e = \ln \frac{3B_{0z}a^3}{\pi^2 R^3} \sin\theta - \frac{t}{\tau_0} \qquad (3.108)$$

This equation indicates that the slope of the function $\ln B_\theta^e(t)$ defines the time constant τ_0. This feature is also true for the other components B_R^e and E_ϕ^e. The time constant, τ_0, can also be determined from the ratio:

$$\tau_0 = -\frac{B_R^e}{\partial B_R^e/\partial t} = -\frac{B_\theta^e}{\partial B_\theta^e/\partial t} \qquad (3.109)$$

As will be discussed later, due to the dependence of the field at the late stage from parameter α, the transient method in exploration has considerable advantages with respect to resolution particularly in comparison with that available using the conventional amplitude and phase measurements.

Let us now consider an important feature of the behavior of induced currents during the late stage of the time-domain behavior. In accord with eqs. 3.68, 3.75 and 3.87:

$$J_\phi = \sigma E_\phi^i = \frac{3H_0}{a} \frac{a}{R} \left[\frac{a}{\pi R} \sin \frac{\pi R}{a} - \cos \frac{\pi R}{a} \right] e^{-\pi^2 \alpha t} \sin \theta \qquad (3.110)$$

where $0 \leqslant R \leqslant a, t \geqslant \tau_0$.

During the late stage, all induced currents decay in the same manner with time (e^{-t/τ_0}) and therefore, the relationship between current density at various points within the sphere is independent of time, and depends only on the distance from the center of the sphere ($\theta = $ const.) Near the center of the sphere, the current density varies in direct proportion to R, reaches a maximum, and then decreases approaching the proper value at the surface, as shown in Fig. 3.9. This behavior is similar to that for the inphase component of the current at low frequencies in the frequency domain (see Fig. 3.3). We should note that during the late stage of time-domain behavior, for regions near the surface of the sphere, we have:

$$J_\phi = \frac{3H_0}{a} e^{-\pi^2 \alpha t} \sin \theta \qquad (3.111)$$

It should be clear that at relatively late times, the current density at any point in the sphere will be in the same direction.

The values for the functions:

$$L_1 = \frac{6}{\pi^2} \sum_{s=1}^{\infty} \frac{1}{s^2} e^{-\pi^2 s^2 \alpha t}$$

and $\qquad (3.112)$

$$L_3 = -\tau_0 \frac{\partial L_1}{\partial t} = \frac{6}{\pi^2} \sum_{s=1}^{\infty} e^{-\pi^2 s^2 \alpha t}$$

permit us to calculate the transient magnetic field and the electromagnetic force; these are given in Table 3.III and are shown graphically in Figs. 3.11–3.12.

It is useful in understanding the relationship between the frequency-domain and time-domain responses to make use of the Fourier transform, to proceed from the simple expression for the function D (eq. 3.52) to derive an equation for the transient response. For example, for the component H_R we have:

$$H_R^e = \frac{1}{2\pi} \int_{-\infty}^{\infty} \frac{H_R^e(\omega)}{-i\omega} e^{-i\omega t} d\omega$$

$$= -\frac{H_{0z} a^3}{2\pi R^3} \cos \theta \frac{6}{\pi^2} \sum_{n=1}^{\infty} \frac{1}{n^2} \int_{-\infty}^{\infty} \frac{e^{-i\omega t}}{n^2 \pi^2 \alpha - i\omega} d\omega \qquad (3.113)$$

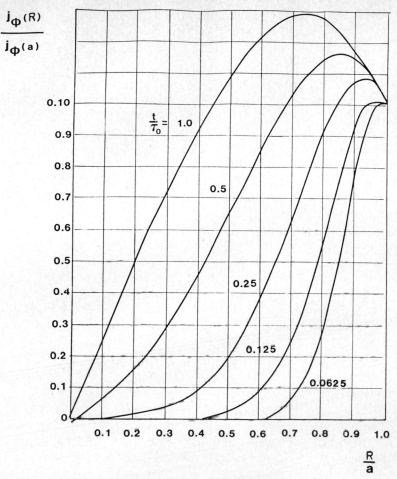

Fig. 3.9. Variation of current density as a function of distance from the center of the sphere.

As is well known, the values for the integrals written in eq. 2.43 can be determined in terms of their behavior around the poles in the integrand, which are situated on the imaginary ω-axis (eq. 3.54). Using the residue theorem, we have:

$$\int_{-\infty}^{\infty} \frac{e^{-i\omega t}}{\pi^2 n^2 \alpha - i\omega}\, d\omega = 2\pi i \sum_{n=1}^{\infty} \frac{e^{i\omega_n t}}{-i}$$

where ω_n are poles of the integrand. Finally, we obtain:

$$H_R^e = \frac{H_{0z} a^3}{R^3} \frac{6}{\pi^2} \cos\theta \sum_{n=1}^{\infty} \frac{1}{n^2} e^{-\pi^2 n^2 \alpha t}$$

TABLE 3.III

s	μ_i/μ_e							
	1		1.25		2.5		5.0	
	$k_s a$	$k_s^2 a^2$	$k_s a$	$k_s^2 a^2$	$k_s a$	$k_s^2 a^2$	$k_s a$	$k_s^2 a^2$
1	3.14	987	3.22	10.3	3.51	12.3	3.83	14.7
2	6.28	39.5	6.32	40.0	6.50	42.3	6.78	46.0
3	9.42	88.8	9.45	89.3	9.58	91.7	9.80	96.0
4	12.6	158	12.6	158	12.7	16.1	12.9	165
5	15.7	247	15.7	247	15.8	25.0	15.9	254
6	18.8	355	18.9	356	18.9	358	19.0	363
7	29.0	484	22.0	484	22.1	487	22.2	491
8	25.1	632	25.1	632	25.2	635	25.3	639
9	28.3	799	28.3	800	28.3	802	28.4	807
10	31.4	987	31.4	987	31.5	990	31.5	995

It is fundamental that these derivations show that the powers of the exponents describing the time-domain field behavior are defined by the behavior of the spectrum near its poles on the imaginary ω-axis.

In conclusion, let us examine the time-domain response for various ratios of magnetic permeability, $K = \mu_i/\mu_e$. Having solved the transcendental eq. 3.81, we find the poles for the spectrum:

$$\omega_{si} = -i\frac{q_{si}}{\sigma\mu_i a^2}, \quad \text{where } \bar{q}_{si} = q_{si}\alpha$$

Values for the parameters $k_s a$ and $k_s^2 a^2$, showing the pattern of poles in the spectrum as well as the powers of the exponents describing a transient response for the ratio μ_i/μ_e are given in Table 3.III. It should be observed that

$$q_{si} = k_s^2 a^2$$

As follows from Table 3.III, only the first poles in the spectrum and the corresponding powers for the exponentials in the first terms of the series describing the transient behavior are affected significantly by variations in magnetic permeability. Inasmuch as the first pole of the spectrum defines a time constant:

$$\tau_{0i} = \sigma\mu_i a^2/q_{1i}$$

we see from this table that over a relatively wide range of values for μ_i, the time constant is nearly directly proportional to the ratio μ_i/μ_e, as shown in Fig. 3.10.

In accord with eqs. 3.90 and 3.91, the magnetic field and electric field can be written as:

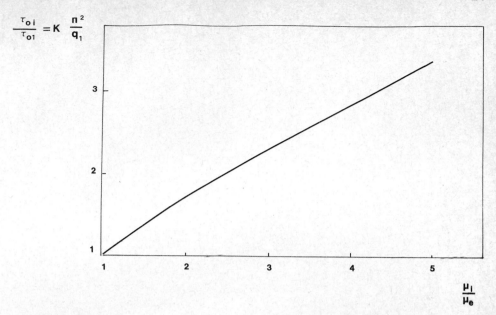

$$\frac{\tau_{oi}}{\tau_{o1}} = K \frac{n^2}{q_1}$$

Fig. 3.10. Curves showing that the time constant is nearly directly proportional to the ratio μ_i/μ_e.

$$H_R^e = H_{0z} \frac{a^3}{R^3} \cos\theta \, K\pi^2 L_1(K, t)$$

$$H_\theta^e = H_{0z} \frac{a^3}{2R^3} \sin\theta \, K\pi^2 L_1(K, t) \qquad (3.114)$$

$$E_\phi^e = B_{0z} \frac{a^3}{2R^2} \sin\theta \, \alpha K\pi^2 L_3(K, t)$$

and

$$L_1(K, t) = \frac{6}{\pi^2} \sum_{s=1}^{\infty} \frac{e^{-q_s \alpha t}}{q_s + (K-1)(K+2)}$$

$$L_3(K, t) = \frac{6}{\pi^2} \sum_{s=1}^{\infty} \frac{q_s e^{-q_s \alpha t}}{q_s + (K-1)(K+2)}$$

here

$$\alpha = \frac{1}{\sigma \mu_i a^2}, \quad q_s = k_s^2 a^2, \quad K = \mu_i/\mu_e \qquad (3.115)$$

values for the function $L_1(K, t)$ and $L_3(K, t)$ are given in Table 3.IV as functions of the ratio t/τ_{0i}, and are shown graphically in Figs. 3.11 and 3.12. The parameter τ_{0i} is defined as:

TABLE 3.IV

t/τ_{0i}	μ_i/μ_e							
	1		1.25		2.5		5.0	
	$L_1(K)$	$L_3(K)$	$L_1(K)$	$L_3(K)$	$L_1(K)$	$L_3(K)$	$L_1(K)$	$L_3(K)$
0.125	0.657	0.190×10^1	0.588	0.116×10^1	0.366	0.929	0.181	0.617
0.177	0.601	0.977	0.534	0.925	0.394	0.711	0.154	0.444
0.250	0.537	0.773	0.475	0.725	0.279	0.534	0.127	0.312
0.353	0.467	0.602	0.409	0.558	0.232	0.391	0.100	0.213
0.500	0.390	0.460	0.338	0.419	0.184	0.278	0.752	0.140
0.707	0.309	0.337	0.204	0.303	0.137	0.189	0.523	0.876×10^{-1}
1.000	0.226	0.235	0.191	0.208	0.924×10^{-1}	0.120	0.326×10^{-1}	0.509×10^{-1}
1.410	0.148	0.150	0.122	0.130	0.544×10^{-1}	0.686×10^{-1}	0.173×10^{-1}	0.261×10^{-1}
2.000	0.823	0.825	0.660×10^{-1}	0.694×10^{-1}	0.261×10^{-1}	0.326×10^{-1}	0.722×10^{-2}	0.107×10^{-1}
2.830	0.359×10^{-1}	0.359×10^{-1}	0.277×10^{-1}	0.290×10^{-1}	0.930×10^{-2}	0.116×10^{-1}	0.211×10^{-2}	0.313×10^{-2}
4.000	0.111×10^{-1}	0.111×10^{-1}	0.810×10^{-2}	0.849×10^{-2}	0.216×10^{-2}	0.269×10^{-2}	0.370×10^{-3}	0.549×10^{-3}
5.660	0.212×10^{-2}	0.212×10^{-2}	0.142×10^{-2}	0.149×10^{-2}	0.274×10^{-3}	0.342×10^{-3}	0.315×10^{-4}	0.468×10^{-4}
8.000	0.204×10^{-3}	0.204×10^{-3}	0.122×10^{-3}	0.128×10^{-3}	0.148×10^{-4}	0.185×10^{-4}	0.971×10^{-5}	0.144×10^{-5}

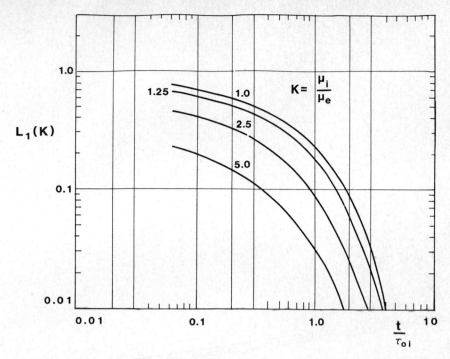

Fig. 3.11. Behavior of the function L_1 which is used in computing transient magnetic fields.

$$\tau_{0i} = \sigma \mu_i a^2 / \pi^2$$

and it is the time constant for a magnetic sphere. As is readily seen from the curves in Figs. 3.11 and 3.12, during the early stage, just as at high frequencies in the frequency domain, the effect of magnetic permeability vanishes and the function $KL_1(K, t)$ tends to unity. In contrast, during the late stage, with increase in time, the influence of changing values in μ_i becomes even more important, because the time constant depends strongly on magnetic permeability.

3.3. A CONDUCTIVE SPHERE ILLUMINATED BY THE MAGNETIC FIELD FROM A CURRENT LOOP, WITH AXIAL SYMMETRY

In this section we will consider the influence of non-uniformity in the primary magnetic field on the frequency- and time-domain behavior of induced currents in a conducting sphere. We will assume that the source of the primary magnetic field is a horizontal loop with a current I flowing in it as shown in Fig. 3.13. Using results from the first part of this chapter, we can assume that the electric field has a single component, E_ϕ, which satisfies

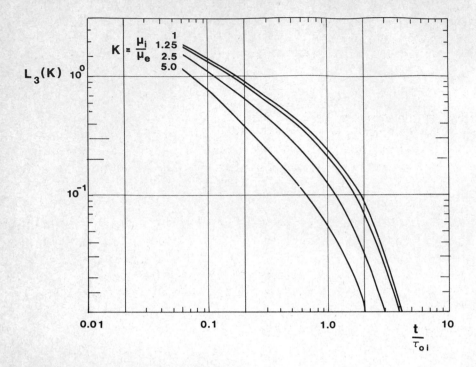

Fig. 3.12. Behavior of the function L_3 used in calculating the electromagnetic force.

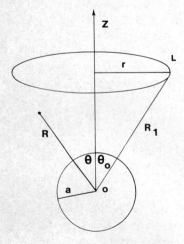

Fig. 3.13. The case in which the source of a primary magnetic field is a horizontal loop with a current I flowing in it is shown here.

eqs. 3.9 and 3.10 outside and inside the sphere, respectively, as well as the boundary conditions in eq. 3.6, and the appropriate condition at infinity. In accord with eq. 3.7, the total field outside this sphere can be written as the sum of two parts:

$$E_\phi^e = E_{0\phi} + E_{1\phi}^e \tag{3.116}$$

where $E_{0\phi}$ is the primary electric field caused by the change to primary magnetic field with time, and $E_{1\phi}^e$ is the secondary electric field caused by the change of the secondary magnetic field with time. As is well known, the primary electric field can be written as the sum of spherical harmonics in the form

$$E_{0\phi} = \frac{i\omega\mu I}{2} \sum_{n=1}^{\infty} \frac{\sin\theta_0}{n(n+1)} \left(\frac{R}{R_1}\right)^n P_n^1(\cos\theta_0) P_n^1(\cos\theta) \tag{3.117}$$

where I is the current flowing in the loop, and R, θ are spherical coordinates for an observation point, with the origin being at the center of the sphere. The parameters θ_0 and R_1 are as shown in Fig. 3.13. $P_n^1(x)$ is an associated Legendre function. For example:

$$P_1^1(x) = (1-x^2)^{1/2} \qquad P_3^1(x) = \tfrac{3}{2}(1-x^2)^{1/2}(5x^2-1)$$
$$P_2^1(x) = 3x(1-x^2)^{1/2} \qquad P_4^1(x) = \tfrac{5}{2}(1-x^2)^{1/2}(7x^3-3x)$$

There are recurrence relationships for these functions:

$$nP_{n+1}^1(x) - (2n+1)xP_n^1(x) + (n+1)P_{n-1}^1(x) = 0$$

Using eqs. 3.21, 3.24, and 3.117, the electric field outside and inside the sphere can be rewritten in the form:

$$E_\phi^e = \frac{i\omega\mu_e I}{2} \sum_{n=1}^{\infty} \frac{\sin\theta_0}{n(n+1)} \left(\frac{R}{R_1}\right)^n P_n^1(\cos\theta_0) P_n^1(\cos\theta)$$

$$+ \frac{i\omega\mu_e I}{2} \sum_{n=1}^{\infty} \frac{\sin\theta_0}{n(n+1)} \frac{D_n^1}{R^{n+1}} \frac{1}{R_1^n} P_n^1(\cos\theta_0) P_n^1(\cos\theta), \quad \text{if } R \geqslant a \tag{3.118}$$

$$E_\phi^i = \frac{i\omega\mu_e I}{2} R^{-1/2} \sum_{n=1}^{\infty} \frac{\sin\theta_0}{n(n+1)} \frac{1}{R_1^n} C_n I_{[\frac{1}{4}+n(n+1)]^{1/2}}(ikR) \times$$

$$P_n^1(\cos\theta_0) P_n^1(\cos\theta), \quad \text{if } R \leqslant a$$

where D_n^1 and C_n are coefficients whose values are yet to be determined. In order to find the values for these coefficients, we can use the boundary conditions in eq. 3.6 provided that $\mu_i = \mu_e$. In view of the orthogonality property of associated Legendre functions, it follows from the equality of sums of the type in eq. 3.118 that the spherical harmonics with the same index n can be equated. Thus, for each harmonic, we have two equations in two unknowns:

$$a^n + \frac{D_n^1}{a^{n+1}} = a^{-1/2} C_n I_{n+1/2}(ika)$$

$$(n+1)a^n - \frac{nD_n^1}{a^{n+1}} = \{\tfrac{1}{2}a^{-1/2}I_{n+1/2}(ika) + ika^{1/2}I'_{n+1/2}(ika)\}C_n$$

(3.119)

Hence:

$$C_n = \frac{(2n+1)a^{n+1/2}}{[(n+\tfrac{1}{2})I_{n+1/2}(x) + xI'_{n+1/2}(x)]}, \quad \text{where } x = ika$$

Inasmuch as:

$$I'_m(x) = I_{m-1}(x) - \frac{m}{x}I_m$$

we obtain:

$$C_n = \frac{(2n+1)a^{n+1/2}}{xI_{n-1/2}(x)}$$

(3.120)

Substituting this last expression into the first equation of 3.119, we have:

$$D_n^1 = a^{2n+1} \frac{\{(2n+1)I_{n+1/2}(x) - xI_{n-1/2}(x)\}}{xI_{n-1/2}(x)}$$

Since

$$\frac{2m}{x}I_m(x) = I_{m-1}(x) - I_{m+1}(x)$$

we have:

$$D_n^1 = -a^{2n+1} \frac{I_{n+3/2}(x)}{I_{n-1/2}(x)}$$

(3.121)

or

$$D_n^1 = -a^{2n+1}D_n$$

where:

$$D_n = \frac{I_{n+3/2}(ika)}{I_{n-1/2}(ika)}$$

(3.122)

and $I_\nu(x)$ is a modified Bessel function written as:

$$I_\nu(x) = \sum_{l=0}^{\infty} \frac{x^{\nu+2l}}{2^{2l+\nu}\Gamma(l+1)\Gamma(l+\nu+1)}$$

and $\Gamma(l+1)$ is the Γ function. Similarly, the expression for the secondary electric field has the form:

$$E_{1\phi}^e = -\frac{i\omega\mu_e I}{2} \sum_{n=0}^{\infty} \frac{\sin\theta_0}{(n+1)n} \left(\frac{a}{R_1}\right)^n D_n \left(\frac{a}{R}\right)^{n+1} P_n^1(\cos\theta_0) P_n^1(\cos\theta) \quad (3.123)$$

Similar expressions can be written for both components of the magnetic field, H_R^e and H_θ^e. From eq. 3.123 each spherical harmonic, n, in the primary field generates a spherical harmonic in a secondary field for the same index n. It should be obvious that the case of a uniform magnetic field corresponds to the first harmonic, $n = 1$.

The complex function D_n describes the spectrum for every harmonic of the field. At low frequencies when $|ka| < 1$, we can obtain the following expression by replacing Bessel's function by the first few terms in its series expansion:

$$D_n \approx -i \frac{\sigma\mu a^2 \omega}{(2n+3)(2n+1)} \left[1 + i\sigma\mu a^2 \frac{2\omega}{(2n+5)(2n+1)}\right], \quad \text{if } |ka| < 1 \tag{3.124}$$

Therefore, the ratio of the inphase component to the quadrature component of the field tends to the following limit for each of the harmonics:

$$\frac{\text{Re } D_n}{\text{Im } D_n} = -\sigma\mu a^2 \frac{2\omega}{(2n+5)(2n+1)}, \quad \text{as } \omega \to 0 \tag{3.125}$$

This means that at low frequencies, for spherical harmonics with larger number, n, the role of the quadrature component becomes more significant. In this part of the spectrum, the greater the non-uniformity of the primary field, the less will be the extent of the skin effect.

In particular, coincidence of the impedances for planar and spherical surfaces of the conductor occurs at high frequencies for still larger frequencies with larger n.

In order to consider the high-frequency part of the spectrum, one must make use of asymptotic expansions of the Bessel functions. One of these is:

$$I_{m+1/2}(x) \approx \frac{e^x}{(2\pi x)^{1/2}} \sum_{k=0}^{m} \frac{(-1)^k(m+k)!}{k!(m-k)!(2x)^k}$$

Keeping only the first two terms, we have:

$$I_{n-1/2}(x) \approx \frac{e^x}{(2\pi x)^{1/2}} \left[1 - \frac{n(n-1)}{2x}\right]$$

182

$$I_{n+3/2}(x) = \frac{e^x}{(2\pi x)^{1/2}}\left[1 - \frac{(n+2)(n+1)}{2x}\right]$$

In accord with eq. 3.122, we have:

$$D_n = \frac{I_{n+3/2}(x)}{I_{n-1/2}(x)} \approx 1 - \frac{2n+1}{ika} \tag{3.126}$$

Whence:

$$\mathrm{Re}D_n \approx 1 - \frac{2n+1}{2p}, \quad \mathrm{Im}D_n \approx -\frac{2n+1}{2p} \tag{3.127}$$

where

$$p = a/h = (\sigma\mu a^2/2)^{1/2}\omega^{1/2} = (1/2\alpha)^{1/2}\omega^{1/2}$$

These frequency responses for the function D_n are shown graphically in Fig. 3.14. Regardless of the number of the spherical harmonic, at high frequencies, the spectrum for the function D_n tends towards unity, as a consequence of the concentration of induced currents on the surface of the conductive sphere. At the same time, these responses clearly demonstrate that with an increase of n, the skin effect expresses itself at progressively higher frequencies.

Examples of the spectrum for the electromagnetic force induced in a small horizontal loop situated at the center of transmitter are shown in Fig. 3.15. From eqs. 3.123 and 3.124, follow the relative contribution of the first harmonic of the secondary field, which corresponds to a uniform primary field, increases in three cases, such as:

(1) When the distance from a current loop to the center of the sphere (R_1) increases. This occurs because the primary field becomes more uniform in the vicinity of the sphere.

(2) When the distance from the observation point to the center of the sphere increases. This can be explained in the following way. In accord with eq. 3.118, a real distribution of currents is represented as being the sum of various systems of currents, each of which corresponds to a specific spherical harmonic. Every such system except the one for the first harmonic consists of a group of currents having various directions. Due to the change in sign of these currents, it is clear that with an increase in the harmonic number, n, the corresponding harmonic of the field decreases more rapidly. Therefore, the effect of the first harmonic becomes dominant.

(3) In the third case the first harmonic regardless of the distance from the source or from the observation point to the center of the sphere plays the most important role.

Suppose that measurements are carried out at low frequencies. Using the expressions for Bessel's functions, one can show that for each spherical

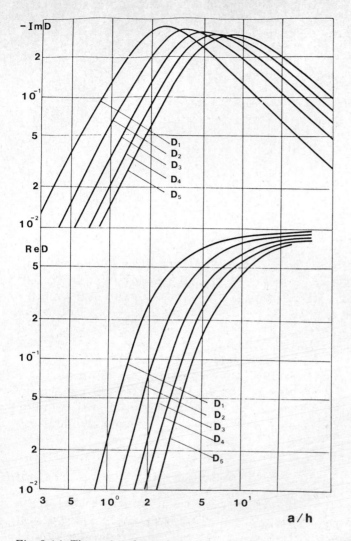

Fig. 3.14. The real and imaginary parts of the frequency response for the function D_n.

harmonic the function D_n can be expanded in a series containing only integer powers of ω,

$$D_n = a_{1n}\omega + a_{2n}\omega^2 + a_{3n}\omega^3 + a_{4n}\omega^4 + \tag{3.128}$$

Now suppose that in measuring a secondary field, that only the term proportional to ω is defined, following which only the term proportional to ω^2 is measured, and so on. It is obvious that the effect of spherical harmonics with various numbers n is mainly defined by the rate of decrease in the corresponding coefficients a_n. In accord with eq. 3.124, the ratio of the

184

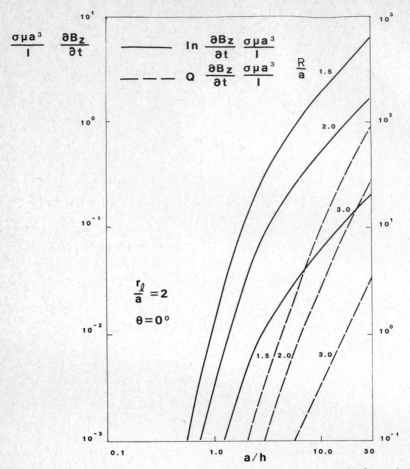

Fig. 3.15. Examples of the spectrum for the electromagnetic force induced in a small horizontal loop situated at the center of the transmitter loop.

coefficients a_{1n} and a_{11}, characterizing the contribution of a spherical harmonic with index n in measuring the part of the field proportional to ω is:

$$\frac{a_{1n}}{a_{11}} = \frac{15}{(2n+3)(2n+1)} \tag{3.129}$$

In particular:

$$\frac{a_{12}}{a_{11}} = \frac{3}{7}, \quad \frac{a_{13}}{a_{11}} = \frac{5}{21}, \quad \frac{a_{14}}{a_{11}} = \frac{5}{33}, \text{ i.e.,}$$

when the quadrature component of the magnetic field is observed at low frequencies, the magnitude of the complex amplitude D_n decreases relatively slowly with increasing n.

Now for a moment suppose that the secondary field proportional to ω^2 (the inphase component of the magnetic field at low frequencies) is measured. In this case, from eq. 3.124, we have:

$$\frac{a_{2n}}{a_{21}} = \frac{315}{(2n+5)(2n+3)(2n+1)^2} \tag{3.130}$$

In particular:

$$\frac{a_{22}}{a_{21}} = \frac{1}{5}, \quad \frac{a_{23}}{a_{21}} = \frac{5}{77}, \quad \frac{a_{24}}{a_{21}} = \frac{35}{1287}$$

Comparing eqs. 3.129 and 3.130 we see that the effect of the first spherical harmonic ($n = 1$), when the term proportional to ω^2 is measured, becomes greater. One can show that in measuring the terms in the spectrum containing higher powers in ω, the effect of the first harmonic, corresponding to a uniform primary field, increases gradually. Thus, no matter what the geometric factor, when the later terms in a series describing the field at low frequencies are measured, the effect of the non-uniformity of the primary field disappears.

We can now consider the time-domain response of the field caused by induction currents in the sphere. Using the Fourier transform for eq. 3.123, we can write an expression for the electric field:

$$E_\phi(t) = -\frac{\mu_e I}{2} \sum_{n=1}^{\infty} \frac{\sin\theta_0}{n(n+1)} \left(\frac{a}{R_1}\right)^n \dot{M}_n(t) \left(\frac{a}{R}\right)^{n+1} P_n^1(\cos\theta_0) P_n^1(\cos\theta) \tag{3.131}$$

where

$$\dot{M}_n(t) = \frac{1}{2\pi} \int_{-\infty}^{\infty} D_n(ika) e^{-i\omega t} d\omega \tag{3.132}$$

or by using eq. 3.122, we obtain:

$$\dot{M}_n(t) = \frac{1}{2\pi} \int_{-\infty}^{\infty} \frac{I_{n+3/2}(ika)}{I_{n-1/2}(ika)} e^{-i\omega t} d\omega \tag{3.133}$$

In these last three equations, the notation $\dot{M}_n(t)$ indicates the time derivative, that is:

$$\dot{M}_n(t) = \frac{d M_n(t)}{dt}$$

It must be emphasized that in deriving these equations we have assumed that the primary field has the form of a step function as shown in Fig. 3.8. In accord with eqs. 3.131—3.133, in order to find the transient field, E_ϕ, the integral of every spherical harmonic n, on the right-hand side of eq. 3.133

must be evaluated. To accomplish this, one can use the residue theorem. It is well known that the modified Bessel functions can be written in the form:

$$I_m(x) = i^{-m} J_m(ix)$$

From this, we have:

$$\frac{I_{n+3/2}(ika)}{I_{n-1/2}(ika)} = -\frac{J_{m+2}(-ka)}{J_m(-ka)}, \quad \text{where } m = n - 1/2$$

First, we will consider the behavior of the integrand:

$$F_1 = \frac{J_{m+2}(-ka)}{J_m(-ka)} e^{-i\omega t}$$

in the complex plane, ω. This function has singularities only at points which correspond to the roots of the equation:

$$J_m(-ka) = J_{n-1/2}(-ka) = 0 \qquad (3.134)$$

From the theory of Bessel functions, the roots of this equation are located on the imaginary axis in the ω-plane. It is important to note that for each spherical harmonic with any index n, there is an infinite number of roots. These roots can be written in the form:

$$\omega_{ns} = \pm i q_{ns}\alpha \qquad (3.135)$$

where $\alpha = 1/\sigma\mu a^2$, and q_{ns} are parameters characterizing the location of roots. The index s is the number of the root for the corresponding spherical harmonic, n. As an example, q_{11} defines the first (minimal) root for the first spherical harmonic, while q_{25} is the fifth root for the second harmonic, and so on. Values for various q_{ns} are listed in Table 3.V.

It should be clear that the roots given by eq. 3.135 are simple poles of complex amplitude in the spectrum, as was the case with the functions, $D_n e^{-i\omega t}$, which are integrands of eq. 3.123. Therefore, we can use the Cauchy formula using only those poles located on the negative imaginary axis in the complex plane ω inasmuch as for positive values of time, the power of the exponent $-i\omega t$ must be negative. Using the residue theorem:

$$\dot{M}_n(t) = \frac{1}{2\pi} \int_{-\infty}^{\infty} D_n(ika)e^{-i\omega t}d\omega = -\frac{1}{2\pi} \int_{-\infty}^{\infty} \frac{J_{n+3/2}(-ka)}{J_{n-1/2}(-ka)} e^{-i\omega t}d\omega$$

$$\qquad (3.136)$$

$$= -i \sum_{s=1}^{\infty} \text{Res}\left[\frac{J_{n+3/2}(-ka)}{J_{n-1/2}(-ka)} e^{-i\omega t}\right]$$

where

$$\text{Res}\left[\frac{J_{n+3/2}(-ka)}{J_{n-1/2}(-ka)} e^{-i\omega t}\right]$$

TABLE 3.V

Values of parameter q_{ns}

s	n						
	1	2	3	4	5	6	7
1	9.86960	20.1907	33.2173	48.8312	66.9543	87.5310	110.519
2	39.4784	59.6795	82.7192	108.516	137.005	168.130	201.850
3	88.8264	118.899	151.854	187.635	226.193	267.476	311.452
4	157.914	197.858	240.703	286.408	334.934	386.244	440.307
5	246.740	296.552	349.279	404.887	463.343	524.616	588.683
6	355.306	414.990	477.593	543.086	611.454	—	—
7	483.611	—	—	—	—	—	—

is the residue for the pole ω_{ns}. Inasmuch as the pole is simple, we have:

$$\text{Res } F_1 = \frac{J_{m+2}(x_{ns})e^{-i\omega_{ns}t}}{\dfrac{\partial}{\partial\omega}J_m(x)}, \quad \text{as } \omega \to \omega_{ns} \tag{3.137}$$

where

$$x = -ka, \quad x_{ns} = -k_{ns}a$$

and

$$k_{ns}\alpha = (i\omega_{ns}\mu\sigma a^2)^{1/2} = (i\omega_{ns}/\alpha)^{1/2}; \quad m = n - 1/2$$

It should be clear that

$$\frac{\partial}{\partial\omega}J_m(x) = J'_m(x)\frac{\partial x}{\partial\omega} = -\left(\frac{i}{\alpha}\right)^{1/2}\frac{J'_m(x)}{2(\omega_{ns})^{1/2}} = -\frac{1}{2}\left(\frac{i}{\omega_{ns}\alpha}\right)^{1/2}J'_m(x)$$

Making use of the relationship between the various Bessel functions and their derivations, such as:

$$J'_m(x) = -J_{m+1}(x) + \frac{m}{x}J_m(x)$$

$$\frac{2m}{x}J_m(x) = J_{m-1}(x) + J_{m+1}(x)$$

and considering that:

$$J_m(x_{ns}) = 0$$

we obtain:

$$\frac{J_{m+2}(x_{ns})}{\dfrac{\partial}{\partial x}J_m(x)} = -\frac{J_{m+2}(x_{ns})}{J_{m+1}(x_{ns})} = -\frac{2(m+1)}{x_{ns}}$$

Therefore

$$\text{Res } F_1 = -\frac{2(2n+1)}{i} \alpha e^{-q_{ns}\alpha t} \tag{3.138}$$

Thus, we obtain the following for the function $\dot{M}_n(t)$:

$$\dot{M}_n(t) = 2(2n+1)\alpha \sum_{s=1}^{\infty} e^{-q_{ns}\alpha t} \tag{3.139}$$

By using eqs. 3.131 and 3.139, the electric field, $E_\phi(t)$ can be calculated. The transient response for each spherical harmonic is given by an infinite sum of exponential terms. As an example, for the first and second harmonics, we have, respectively:

$$\dot{M}_1(t) = 6\alpha \sum_{s=1}^{\infty} e^{-q_{1s}\alpha t}$$

$$\dot{M}_2(t) = 10\alpha \sum_{s=1}^{\infty} e^{-q_{2s}\alpha t} \tag{3.140}$$

It is important to note that the power of the particular exponents in the expressions for $M_n(t)$ as indicated by eq. 3.132 is determined by the poles in the spectrum D_n. However, the power in the first exponent of each harmonic $q_{n1}\alpha$ is determined by the closest pole, and therefore, it specifies the radius of convergence for the series describing the low-frequency spectrum D_n. Therefore, it can be evaluated in an approximation from eq. 3.124, which can be rewritten as:

$$\left|\frac{\text{Im } D_n}{\text{Re } D_n}\right|_\omega \cong \frac{(2n+5)(2n+1)}{2}\alpha, \quad \text{i.e. } q_{n1} \cong \frac{(2n+5)(2n+1)}{2} \tag{3.141}$$

Comparing the results with values of q_{n1} listed in Table 3.V, it is apparent that eq. 3.141 provides reasonable accuracy, although it could be improved significantly if the ratio of the following terms in the various series with respect to ω were to be considered. The distribution of the parameter q_{ns} has three important features, as follows:

(1) For the first harmonic, q_{1s} coincides with the appropriate values of q_s found previously for excitation with a uniform primary field (eq. 3.83) and in accord with this equation, we have:

$$q_{1s} = \pi^2 s^2$$

(2) With increasing index s, q_{ns} increases, that is:

$$q_{n,s} < q_{n,s+1}$$

(3) With an increase in the number for the spherical harmonic, n, the corresponding values for q_{ns} will also increase, that is:

$$q_{n,s} < q_{n+1,s}$$

In particular, as can be seen from eq. 3.141:

$$q_{n,1} < q_{n+1,1}$$

Thus, one might conclude that with increase in time, the transient response will be defined almost entirely by the exponential term $e^{-q_{11}}$; that is at the late stage:

$$\dot{M}_1(t) > \dot{M}_2(t) > \dot{M}_3(t)\ldots$$

and

$$\dot{M}_1(t) \approx 6\alpha e^{-q_{11}\alpha t} \qquad (3.142)$$

or

$$\dot{M}_1(t) \approx 6\alpha e^{-\pi^2 \alpha t} = 6\alpha e^{-t/\tau_0} \qquad (3.143)$$

where $\tau_0 = \sigma\mu a^2/\pi^2$ is the time constant defined earlier in this chapter for the case of a uniform primary field excitation.

Similarly, we have the following expression for the electric field during the late stage of transient behavior:

$$E_\phi(t) = -\frac{\mu_e I}{2}\, 3\alpha \sin^2\theta_0 \sin\theta\, \frac{a^3}{R_1 R^2}\, e^{-\pi^2 \alpha t}, \quad \text{as } \frac{t}{\tau_0} > 1 \qquad (3.144)$$

or

$$E_\phi(t) = A\,e^{-t/\tau_0}$$

It must be obvious that this result does not depend on the position of the observation point nor on the location of the current loop, and it is of great practical importance. It means that the late-stage behavior of the secondary field, despite the non-uniformity of the primary field excitation, decays exactly in the same manner as would be the case for uniform primary field excitation. Correspondingly, in determining the time constant during the late-stage behavior, we do not have to take into account the geometry of the primary field excitation, and this markedly simplifies the procedures used in interpretation. It should also be noted that the same result holds for conductive bodies of arbitrary shape.

At this point, let us consider briefly the behavior of the electric field during the early stage. First, in accord with eq. 3.127, we have the following expression for the imaginary part of the complex amplitude D_n for large values of the parameter a/h:

$$\operatorname{Im} D_n = -\frac{(2n + 1)\sqrt{\alpha}}{\sqrt{2}\sqrt{\omega}}$$

Applying the Fourier transform to the function ωD_n, we have:

$$\dot{M}_n(t) = \frac{2}{\pi}\frac{(2n + 1)}{\sqrt{2}}\sqrt{\alpha}\int_0^\infty \frac{\sin \omega t}{\sqrt{\omega}}\,d\omega$$

Considering that:

$$\int_0^\infty \frac{\sin \omega t}{\sqrt{\omega}}\,d\omega = \left(\frac{\pi}{2}\right)^{1/2}\frac{1}{t^{1/2}}$$

we obtain:

$$\dot{M}_n(t) = \frac{2n + 1}{\sqrt{\pi}}\left(\frac{\alpha}{t}\right)^{1/2} = \frac{2n + 1}{\pi^{3/2}}\frac{1}{(t\tau_0)^{1/2}} \tag{3.145}$$

where τ_0 is the time constant.

During the early stage, the sperical harmonics decrease with time in a like manner, namely, in inverse proportion to the square root of time. Simultaneously, with an increase in the number, n, of a harmonic, its effect during the early stage becomes greater. The transient responses of the functions $\dot{M}_n(t)$ are shown in Fig. 3.16 for the first ten harmonics.

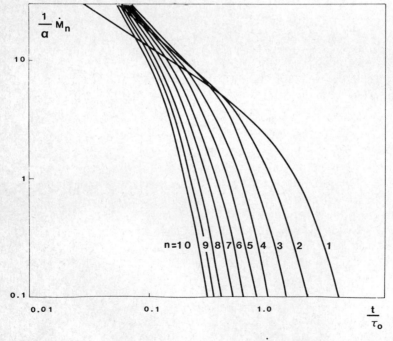

Fig. 3.16. The transient response for functions $\dot{M}_n(t)$ for the first 10 harmonics.

Up to this point, only the electric field has been considered. We will now derive expressions for the magnetic field and its derivatives with respect to time. Using the Maxwell equation:

$$\text{curl } E = -\frac{\partial B}{\partial t}$$

in spherical coordinates we have:

$$\frac{\partial B_R}{\partial t} = -\frac{1}{R \sin \theta} \frac{\partial}{\partial \theta} \sin \theta E_\phi$$

$$\frac{\partial B_\theta}{\partial t} = \frac{1}{R} \frac{\partial}{\partial R} R E_\phi$$

(3.146)

where B is the magnetic induction vector. Substituting eq. 3.131 into this last expression we have:

$$\frac{\partial B_R}{\partial t} = \frac{\mu_e I}{2R} \sum_{n=1}^{\infty} \frac{\sin \theta_0}{n(n+1)} \left(\frac{a}{R_1}\right)^n \dot{M}_n(t) \left(\frac{a}{R}\right)^{n+1} P_n^1(\cos \theta_0) \times$$

$$\frac{1}{\sin \theta} \frac{\partial}{\partial \theta} \sin \theta P_n^1(\cos \theta)$$

(3.147)

$$\frac{\partial B_\theta}{\partial t} = \frac{\mu_e I}{2R} \sum_{n=1}^{\infty} \frac{\sin \theta_0}{(n+1)} \left(\frac{a}{R_1}\right)^n \dot{M}_n(t) \left(\frac{a}{R}\right)^{n+1} P_n^1(\cos \theta_0) P_n^1(\cos \theta)$$

In order to derive expressions for the components of the induction vector, we have to replace the function $\dot{M}(t)$ in eq. 3.147 by the function $M(t)$, where

$$M(t) = \int_0^t \frac{\partial M}{\partial t} dt$$

This can be done because the magnetic field vanishes with increase in time.
Making use of eq. 3.139, we obtain:

$$M_n(t) = -2(2n+1) \sum_{s=1}^{\infty} \frac{1}{q_{ns}} e^{-q_{ns}\alpha t}$$

(3.148)

Transient behavior of the first ten harmonics of the function $M_n(t)$ are shown graphically in Fig. 3.17. As was the case with the electric field, the late stage of the magnetic field and its time derivative are described almost entirely by the first spherical harmonic, and in accord with eq. 3.147 we have:

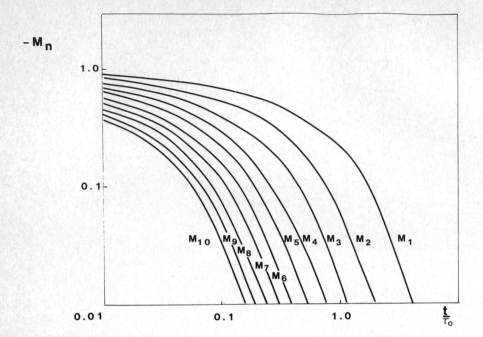

Fig. 3.17. Transient behavior of the first 10 harmonics of the function $M_n(t)$.

$$\frac{\partial B_R}{\partial t} = \frac{\mu_e I}{2R_1} \sin^2\theta_0 \, \frac{a^3}{R^3} \, \dot{M}_1(t)\cos\theta$$

$$\frac{\partial B_\theta}{\partial t} = \frac{\mu_e I}{4R_1} \sin^2\theta_0 \, \frac{a^3}{R^3} \, \dot{M}_1(t)\sin\theta$$

Inasmuch as:

$$\sin\theta_0 = r_1/R_1$$

where r_1 is the radius of the current loop source, the time rate of change of magnetic induction can be written as:

$$\dot{B}_R = \frac{\mu_e I r_1^2}{2R_1^3}\frac{a^3}{R^3}\,\dot{M}_1(t)\cos\theta$$

$$\text{if } t > \tau_0 \qquad\qquad (3.149)$$

$$\dot{B}_\theta = \frac{\mu_e I r_1^2}{4R_1^3}\frac{a^3}{R^3}\,\dot{M}_1(t)\sin\theta$$

where

$$I r_1^2/2R_1^3$$

is the primary magnetic field at the center of the sphere. In accord with eq. 3.143:

$$\dot{M}_1(t) = 6\alpha e^{-t/\tau_0} = \frac{6}{\pi^2} \frac{1}{\tau_0} e^{-t/\tau_0}$$

Comparing this expression with the result obtained in the previous paragraph, one can say that at the late stage, no matter what the degree of non-uniformity of the primary field excitation, the secondary field is exactly that which would be observed for a uniform primary field in the vicinity of this sphere, and with a value equal to the magnitude of the magnetic field at the center of the sphere. Thus, we have discovered that not only the character of the decay, but also the value of the field during the late stage is the same as though the primary field excitation were uniform. This is of great practical interest, since it remains valid for conductors of much more complicated shape, and markedly simplifies theoretical considerations and interpretation in mining inductive surveys.

As follows from eq. 3.149, the various components of the magnetic field are:

$$B_R = \frac{\mu_e I r_1^2}{2 R_1^3} \frac{a^3}{R^3} M_1(t) \cos \theta$$

$$B_\theta = \frac{\mu_e I r_1^2}{4 R_1^3} \frac{a^3}{R^3} M_1(t) \sin \theta \tag{3.150}$$

where

$$M_1(t) = -\frac{6}{\pi^2} e^{-\pi^2 \alpha t} = -\frac{6}{\pi^2} e^{-t/\tau_0}$$

As can be seen from the figures in Table 3.V, q_{21} is less than q_{12}, and therefore in approaching the late stage, the contribution from the second spherical harmonic is more important than the contribution of the term containing the exponential:

$$e^{-q_{12}\alpha t} = e^{-4\pi^2 \alpha t}$$

In this respect, it is important to note by comparing eqs. 3.139 and 3.148 that in measuring the magnetic field, late stage behavior is observed beginning at earlier times than is the case when the EMF is measured.

To complete our consideration of the magnetic field, consider its behavior during the early stage. From eqs. 3.145 and 3.148, we have:

$$M_n(t) = M_n(0) + \int_0^t \dot{M}_n(t)dt = -2(2n+1)\left[\sum_{s=1}^{\infty} \frac{1}{q_{ns}} - \frac{1}{\pi^{3/2}}\left(\frac{t}{\tau_0}\right)^{1/2} \right] \tag{3.151}$$

That is, the asymptote for each spherical harmonic approaches its limiting value as the square root of time.

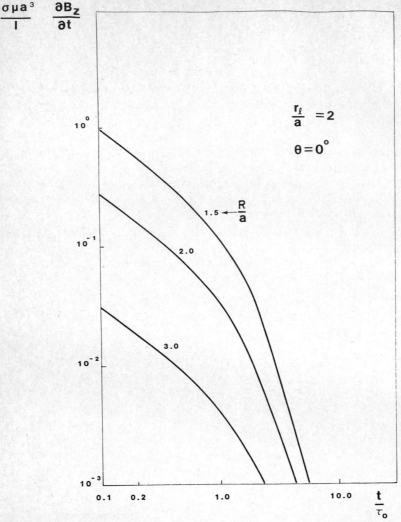

Fig. 3.18. Transient responses for the time rate of change of vertical magnetic induction at the center of the transmitting loop.

As an example, the transient responses for the function $\partial B_z/\partial t$ at the center of transmitting loop as shown graphically in Fig. 3.18.

In conclusion, we should point out that in the time domain, as in the frequency domain, there are three cases that can be recognized in the way in which the effect of the non-uniformity of the primary field disappears, namely:

(1) As the distance from the current loop to the center of the sphere increases.

(2) As the distance from the observation point to the sphere increases.

(3) With increasing time of measuring, regardless of the mutual position of the source loop, the spherical body and the observation point.

3.4. EQUATIONS FOR THE FREQUENCY- AND TIME-DOMAIN BEHAVIOR OF FIELDS CAUSED BY INDUCED CURRENTS IN CONDUCTIVE BODIES WITH AXIAL SYMMETRY

In the previous sections in this chapter, we have examined the frequency- and time-domain responses generated by induced currents in a conducting sphere. Now we can consider a more general case including electromagnetic fields of currents induced in a conductor or a system of conductors provided that the primary field and the conductive bodies are characterized by axial symmetry with a common axis. Example of the models for such cases are shown in Fig. 3.19.

In all possible cases, the primary magnetic field excitation is contributed by a current ring or a system of current rings, but the primary vortex electric field has only a single component, $E_{0\phi}$. In particular, the primary magnetic field excitation can be uniform in the neighborhood of the conductive

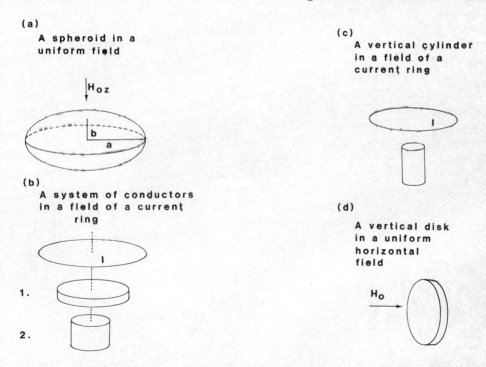

(a)

A spheroid in a uniform field

(b)

A system of conductors in a field of a current ring

1.

2.

(c)

A vertical cylinder in a field of a current ring

(d)

A vertical disk in a uniform horizontal field

Fig. 3.19. Various models in which a conductor or a system of conductors are characterized by axial symmetry with a common axis.

bodies. It is obvious that a primary vortex electric field arises as a consequence of the change of the magnetic field strength with time, and because
of the axial symmetry, this electric field does not intersect the surface of any
conductor. Because of this, no electric charges develop and due to the
existence of the vortex electric field at every point in the conductor, currents
will arise with a density given by:

$$j_\phi = \sigma(E_{0\phi} + E_{1\phi}^i) \tag{3.152}$$

where $E_{0\phi}$ is the primary vortex electric field strength, $E_{1\phi}^i$ is a secondary
vortex electric field caused by the magnetic field from induced currents in
the conductive bodies, and σ is the conductivity within the body. In a
cylindrical coordinate system, r, ϕ, z, it can be an arbitrary function of r
and z, but not of ϕ. It should be clear that the term $E_{1\phi}^i$ represents interaction
between currents, that is, the skin effect, and that its magnitude is not
known. However, we can be assured that because of the lack of surface
electrical charges, the induced currents as well as the primary field $E_{0\phi}$ have
only an azimuthal component j_ϕ. The interaction between current filaments
does not change the direction of current flow in this case.

In contrast to the model of the spherical conductor, the determination of
induced currents in these more general cases is a very complicated problem,
in spite of axial symmetry. Even when a conductor has a relatively simple
shape, as for example, spheroidal, the solution is still very cumbersome
regardless of the simplicity of the primary field excitation. In this case, the
method of separation of variables permits one to replace the equation:

$$\nabla^2 E + k^2 E = 0$$

in spheroidal coordinates as two ordinary differential equations, for which
the solutions are associated Legendre functions. The field inside and outside
the spheroid can be written as an infinite sum, where the amplitudes of the
spheroidal harmonics must be determined. In principle, these amplitudes can
be determined from an infinite system of equations in the unknowns which
describe the required continuity of the tangential components of the electric
and magnetic fields at the surface of the spheroid. However, in contrast to
the same type of problem with a sphere or a circular cylinder, the system
cannot be reduced to sets of two equations with two unknowns for each of
the spheroidal harmonics. This is a consequence of the fact that the spheroidal functions of the coordinates describing the field in the two media are
not orthogonal, inasmuch as the arguments of these functions contain the
various wave numbers. Moreover, a computational algorithm for the solution
of an infinite set of equations with complex amplitudes is in the general case
not available, and therefore, the application of the method of separation of
variables leads to an extremely cumbersome numerical problem. For this
reason, the field is usually defined using various numerical methods, such as,
the technique of integral equations. In particular, the frequency and transient

responses for the electromagnetic fields caused by currents induced in a conductive spheroid or an elliptical cylinder, as will be described below, were obtained making use of a system of two integral equations for the tangential components of the field along a path lying in the vertical plane. This method will be described in detail in the last chapter. However, before analyzing these responses, along with others, we should derive some general equations for the frequency and transient behavior which will prove useful for a better understanding of the frequency- and time-domain behavior of the fields, and the relationship between the two. With this purpose in mind, we will proceed from the equation for current density, eq. 3.152, which is actually the integral equation of interest. Applying the principle of super-position, it can be written as:

$$j_\phi(g) = j_{0\phi}(g) + i\omega\mu\sigma(g)\int_S G(p,g)j_\phi(p)dS \qquad (3.153)$$

where

$$j_{0\phi}(g) = \sigma(g)E_{0\phi}(g)$$

is the current density caused by the primary vortex electric field, and

$$i\omega\mu G(p,g)j_\phi(p)dS$$

is the vortex electric field at a point g caused by a circular current filament passing through an elementary current tube at the point p as shown in Fig. 3.20. It was shown earlier in Chapter 2 that we had the following expression for the function G:

$$G(p,g) = \frac{1}{\pi u}\left(\frac{r_p}{r_g}\right)^{1/2}[(1 - \tfrac{1}{2}u^2)K(u) - E(u)] \qquad (3.154)$$

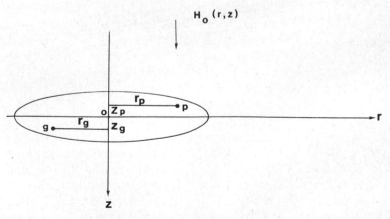

Fig. 3.20. Geometry for circular current filament passing through an elementary current tube at a point p.

198

where

$$u^2 = \frac{4r_p r_g}{(r_p + r_g)^2 + (z_p - z_g)^2}$$

and r_p and r_g are the distances from the points p and g to the z-axis, respectively, and $K(u)$ and $E(u)$ are the complete elliptical integrals of the first and second kinds.

Equation 3.153 is an integral equation of the second kind for the current density, j_ϕ. The first term on the right-hand side, $j_{0\phi}(g)$ is readily calculated.

Generally, integral equation can be reduced to a system of linear equations with constant coefficients, and it is of interest to examine this equivalence from the physical point of view. Let us represent a current distribution within a conductive mass as a sum of current rings flowing in elementary closed volumes, the centers of which, in view of axial symmetry, are situated on the z-axis. We can suppose that a cross-sectional area of such an elementary tube is small so that a change in current density, or in the field, or in the EMF across such a cross-section can be neglected. By so doing, the integral equation in 3.153 can be rewritten as:

$$j_\phi(g) = j_{0\phi}(g) + i\omega\mu\sigma(g) \sum_{p \neq g} G(p,g)I(p) + i\omega\mu\sigma(g) \int_{S(g)} G(p,g)j_\phi dS \quad (3.155)$$

In this equation, the summing is carried out over all the elementary tubes except the tube at the point g. Over the cross-section of this tube, the integration must still be performed (this comprises the last term in eq. 3.155). Here, $I(p)$, is the current flowing in the elementary tube in which the point p is located at the center of its cross-sectional area. Multiplying both sides of eq. 3.155 by the value $2\pi r_g/\sigma(g)$ and considering that:

$$j_\phi(g) \frac{2\pi r_g}{\sigma(g)} = \frac{I(g)}{dS(g)} \frac{2\pi r_g}{\sigma(g)} = I(g)R(g)$$

we obtain:

$$I(g)R(g) = \mathscr{E}_0(g) + i\omega\mu \sum_{p \neq g} \bar{G}(p,g)I(p) + i\omega\mu \int_{S(g)} \bar{G}(p,g)j_\phi dS \quad (3.156)$$

where $R(g)$, is the resistance of the elementary tube passing through the point g; and

$$\mathscr{E}_0(g) = \frac{2\pi r_g}{\sigma(g)} j_{0\phi}(g) = 2\pi r_g E_{0\phi}(g)$$

is the EMF around the ring g caused by the primary electric field:

$$\bar{G}(p,g) = 2\pi r_g G(p,g)$$

and finally:

$$i\omega\mu\bar{G}(p,g)\,I(p)$$

is the electromotive force in the ring g caused by the current flowing in the ring p, which is $I(p)$.

The last term in eq. 3.156 is the EMF in the ring g caused by the current flowing in it, $I(g)$, and this can be written as:

$$i\omega\mu \int_{S(g)} j_\phi \bar{G}\,dS = \frac{i\omega\mu I(g)}{S(g)} \int_{S(g)} \bar{G}\,dS$$

For convenience, we will make use of the following notation:

$$M_{pg} = -\mu\bar{G}(p,g) \text{ and } L_g = -\frac{\mu}{S(g)} \int_{S(g)} \bar{G}\,dS \tag{3.157}$$

where M_{pg} and L_g are the mutual inductance coefficients between coaxial rings passing through the points p and g, and the self-inductance of the ring at the point g, respectively. Substituting these definitions into eq. 3.156:

$$I(g)R(g) = \mathscr{E}_0(g) - i\omega \sum_{p \neq g} M_{pg}I(p) - i\omega L_g I(g)$$

or

$$(R_g + i\omega L_g)I(g) + i\omega \sum_{p \neq g} M_{pg}I(p) = \mathscr{E}_0(g) \tag{3.158}$$

Having written eq. 3.158 for every current ring, in place of the integral eq. 3.153, we have arrived at a system of linear equations with coefficients representing the various current filaments:

$$
\begin{aligned}
Z_{11}I_1 + i\omega M_{12}I_2 + \ldots + i\omega M_{1n}I_n + \ldots + i\omega M_{1N}I_N &= \mathscr{E}_{01} \\
i\omega M_{21}I_1 + Z_{22}I_2 + \ldots + i\omega M_{2n}I_n + \ldots + i\omega M_{2N}I_N &= \mathscr{E}_{02} \\
i\omega M_{n1}I_1 + i\omega M_{n2}I_2 + \ldots + Z_{nn}I_n + \ldots + i\omega M_{nN}I_N &= \mathscr{E}_{0n} \\
i\omega M_{N1}I_1 + i\omega M_{N2}I_2 + \ldots + i\omega M_{Nn}I_N + \ldots + Z_{NN}I_N &= \mathscr{E}_{0N}
\end{aligned}
\tag{3.159}
$$

where

$$Z_{nn} = R_n + i\omega L_n$$

is the impedance of the nth ring.

The right-hand side of eq. 3.159 is the EMF caused by the primary vortex electric field, $E_{0\phi}$, and it is directly proportional to $i\omega$.

The conversion from the integral equation to the system of equations in 3.159 is based on the equivalence of fields caused by a volume current distribution and a system of currents forming linear rings, with the resistance and the self and mutual inductances being characterized by the corresponding elementary tubes. The coefficients in the equations of system 3.159 can be evaluated using the well established expressions for R, L, and M, for a linear ring with a circular cross-section. For a thin circular ring:

$$R = \frac{2\pi r}{\sigma S}, \quad L = r\mu \left[\ln \frac{8r}{r_0} - 1.75 \right]$$

$$(3.160)$$

$$M_{12} = -\frac{2\mu}{u} (r_1 r_2)^{1/2} [(1 - \tfrac{1}{2} u^2) K - E]$$

where r is the radius of the ring, r_0 is the cross-sectional radius in that ring:

$$u^2 = \frac{4 r_1 r_2}{[(r_1 + r_2)^2 + (z_1 - z_2)^2]}, \quad S = \pi r_0^2$$

and μ is the magnetic permeability.

It should be clear that with an increase in the number of tubes with which the current distribution is approximated, the errors involved in replacing the cross-section of arbitrary shape with a circular shape becomes progressively smaller. It can readily be seen that the inductive resistance of an elementary ring must be significantly less than the ohmic resistance, R, or $R \gg \omega L$, since if this is not true, the radius of the cross-section of the ring r_0 becomes greater than a skin depth. When this happens, uniformity of current density and field behavior over the cross-section are not established. Therefore, in order to carry out the calculations at relatively high frequencies, it is necessary to increase the number of equations, and this is one reason why other approaches have been used for obtaining the frequency and transient responses for this more general case. However, as has been mentioned previously, an analysis of this system of equations in 3.159 permits one to establish some very fundamental characteristics for electromagnetic fields caused by induction currents in conductive bodies of arbitrary shapes surrounded by an insulating host rock.

According to eq. 3.158, the expression for current in a ring passing through any point p can be written as:

$$I(p) = i\omega \frac{P_{N-1}(p, i\omega)}{P_N(i\omega)}$$

$$(3.161)$$

where $P_N(i\omega)$ is a polynomial of order N with respect to $i\omega$, and equal to the determinant of the system. $P_{N-1}(p, i\omega)$ is a polynomial of $N - 1$ order with respect to $i\omega$. The multiplier $i\omega$ in eq. 3.161 is there because the primary EMF in each ring is directly proportional to $i\omega$. As is known from the theory of polynomials, the right-hand side of eq. 3.161 can be represented as a sum of simple fractions:

$$I(p) = i\omega \sum_{n=1}^{N} \frac{a_n}{\omega_n - i\omega} \tag{3.162}$$

where $-i\omega_n$ are the roots of the polynomial P_N, and ω_n are real numbers:

$$\omega_1 < \omega_2 < \omega_3 < \omega_4 < \dots$$

With an increasing number of elementary tubes, the accuracy in determining the current distribution in the conductor increases, and in the limit we have:

$$I(p) = i\omega \sum_{n=1}^{\infty} \frac{a_n}{\omega_n - i\omega} \tag{3.163}$$

As follows from eq. 3.159, the determinant of the system, that is the polynomial P_N, depends on frequency, conductivity, and the geometric shape of the conductor or a group of conductive bodies. Therefore, the roots of this polynomial, $-i\omega_n$ are functions of the conductivity structure and the shape as well as of the dimensions of the conductive body.

From eq. 3.163, let us represent the volume density of currents as follows:

$$j_\phi(p) = k^2 b^2 H_{0z} \sum_{n=1}^{\infty} \frac{\beta_n(p)}{q_n - k^2 b^2} \tag{3.164}$$

where $k = (i\sigma_0 \mu \omega)^{1/2}$ is the wave number defined as usual and σ_0 is the conductivity at a given point, with b being an arbitrary geometrical parameter for the conductive body. For example, it might be the radius of a sphere or the length of one of the semi-axes of a spheroid. The quantity H_{0z} is the vertical component of the primary magnetic field at any point within the conductor. This point might be located on a z-axis. In particular, H_{0z} could be a uniform field in the vicinity of the conductive bodies. However, it might also be the horizontal component of the primary field if the axis of symmetry is horizontal. The quantities $\beta_n(p)$ are functions of the coordinates of the point p within the conductor. Finally, the q_n are numbers that satisfy the following condition:

$$q_1 < q_2 < q_3 < q_4 < \dots \tag{3.165}$$

Proceeding from eq. 3.164 and applying the Biot-Savart and Faraday laws, we can write the complex amplitude for the electromagnetic field, for example, in cylindrical coordinates as:

$$H_z(a) = k^2 b^2 H_{0z} \sum_{n=1}^{\infty} \frac{d_{nz}(a)}{q_n - k^2 b^2}$$

$$H_r(a) = k^2 b^2 H_{0z} \sum_{n=1}^{\infty} \frac{d_{nr}(a)}{q_n - k^2 b^2} \tag{3.166}$$

$$E_\phi(a) = i\omega\mu k^2 b^3 H_{0z} \sum_{n=1}^{\infty} \frac{f_n(a)}{q_n - k^2 b^2}$$

(3.166)

This expression for the field is quite convenient inasmuch as the coefficients d_n and $f_n(a)$ are dimensionless. It should be noted that in accord with the results obtained earlier in Chapter 2, the calculation of the coefficients d_{nz} and d_{nr} involves the integration of complete elliptical integrals over the cross-section of the conductive body when the current density, J_ϕ is known. If we examine the behavior of the frequency responses of the current density and of the electromagnetic field on the complex plane ω as shown in Fig. 3.21, it may readily be seen that they all have the same singularities, in the form of simple poles, situated on the imaginary ω-axis. In fact, the denominators of the fractions in eq. 3.166 are zero as:

$$\omega_n = -i\frac{q_n}{\sigma_0\mu b^2}$$

(3.167)

While the poles of the spectrum, ω_n, depend on the resistivity of the conductor and on its shape and dimensions, the coefficients $d_n(a)$ and $f_n(a)$ are also functions of the position of the conductor with respect to the source and the location of the observation point.

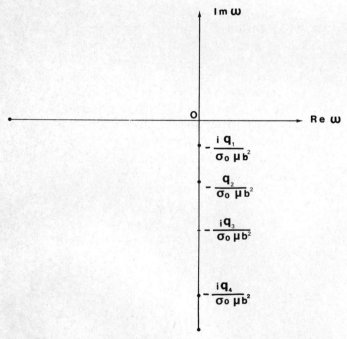

Fig. 3.21. Locations of singularities for the spectrums of current density and electromagnetic field strengths in the complex plane ω.

Applying the Fourier transform:

$$H(t) = \frac{1}{2\pi} \int\limits_{-\infty}^{\infty} \frac{H(i\omega)}{i\omega} e^{-i\omega t} d\omega$$

to each term in eq. 3.166 and repeating the same operations that were carried out in the previous section of this chapter, we obtain the following expressions for the transient field:

$$J_\phi(p) = H_{0z} \sum_{n=1}^{\infty} \beta_n(p) e^{-q_n \tau^2/8\pi^2 b^2}$$

$$H_z(a) = H_{0z} \sum_{n=1}^{\infty} d_{nz}(a) e^{-q_n \tau^2/8\pi^2 b^2}$$

$$\quad\quad\quad\quad\quad\quad\quad\quad\quad\quad\quad\quad\quad (3.168)$$

$$H_r(a) = H_{0z} \sum_{n=1}^{\infty} d_{nr}(a) e^{-q_n \tau^2/8\pi^2 b^2}$$

$$E_\phi(a) = -\frac{H_{0z}}{\sigma_0 b} \sum_{n=1}^{\infty} q_n f_n(a) e^{-q_n \tau^2/8\pi^2 b^2}$$

where $\tau = 2\pi(2t/\sigma_0\mu)^{1/2}$.

The primary field H_{0z} is a step function:

$$H_{0z} = \begin{cases} H_{0z} & t < 0 \\ 0 & t > 0 \end{cases}$$

In order to stress the relationship between the frequency and time domain responses caused by induced currents in the conductor when it has axial symmetry, we can rewrite eqs. 3.166 and 3.168 in a somewhat different form:

$$J_\phi(\omega, p) = i\omega H_{0z} \sum_{n=1}^{\infty} \frac{\beta_n(p)}{q_n/\sigma_0 \mu b^2 - i\omega}$$

$$J_\phi(t, p) = H_{0z} \sum_{n=1}^{\infty} \beta_n(p) e^{-q_n t/\sigma_0 \mu b^2}$$

$$H_z(\omega, a) = i\omega H_{0z} \sum_{n=1}^{\infty} \frac{d_{nz}}{q_n/\sigma_0 \mu b^2 - i\omega}$$

$$H_z(t, a) = H_{0z} \sum_{n=1}^{\infty} d_{nz} e^{-q_n t/\sigma_0 \mu b^2}$$

$$H_r(\omega, a) = i\omega H_{0z} \sum_{n=1}^{\infty} \frac{d_{nr}}{q_n/\sigma_0 \mu b^2 - i\omega}$$

$$H_r(t, a) = H_{0z} \sum_{n=1}^{\infty} d_{nr} e^{-q_n t/\sigma_0 \mu b^2}$$

$$E_\phi(\omega, a) = -\omega^2 H_{0z} \mu b \sum_{n=1}^{\infty} \frac{f_n}{q_n/\sigma_0 \mu b^2 - i\omega}$$

$$E_\phi(t, a) = -\frac{H_{0z}}{\sigma_0 b} \sum_{n=1}^{\infty} q_n f_n e^{-q_n t/\sigma_0 \mu b^2}$$

Comparing these frequency- and time-domain responses for the currents and components of the electromagnetic field, we can see that they are controlled by the same parameters including β_n, d_n, f_n, and $q_n/\sigma_0 \mu b^2$, along with H_{0z}.

Let us summarize the results of this section as follows:

(1) Equations that specify the frequency responses for induced currents and the electromagnetic field can be written as a sum of simple fractions.

(2) In the complex ω-plane, the frequency spectrum has singularities which are simple poles:

$$\omega_n = -i \frac{q_n}{\sigma_0 \mu b^2}$$

situated on the imaginary ω-axis. The distribution of these poles controls the principal features of the spectrum for real values of frequency.

(3) The spectrum of any of the electromagnetic field components, as for example, a magnetic component, is defined by two sets of parameters for a given frequency, and by the primary field; that is, by d_n and by $q_n/\sigma_0 \mu b^2$, where n has integer values.

(4) For a given primary field, the coefficients d_n and f_n are determined only by geometry, including such factors as the position, the dimension, and the shape of the conductive body and the coordinates of the observation site, while the coefficients q_n are functions only of the shape of the conductive body, providing the conductive body is uniform in properties.

(5) Equations describing the transient behavior of induced currents in a conductor or system of confined conductors can be written as a sum of exponential terms. This manner of presentation displays the most important features of the behavior of the transient field and it should not be considered that this presentation is the result of the approximation of functions describing the transient field as the sum of exponentials.

(6) The powers in the exponentials are terms involving the product of a time with the value of a pole in the spectrum in eq. 3.167 while the coefficients d_n and f_n satisfy the conditions:

$$d_{1z} > d_{2z} > d_{3z} > d_{4z} > \ldots$$
$$d_{1r} > d_{2r} > d_{3r} > d_{4r} > \ldots$$
$$f_1 > f_2 > f_3 > f_4 > \ldots$$

In this section we have derived equations describing the frequency and time domain behavior of induced currents in confined conductors of arbitrary shape but with axial symmetry. It will be shown later that this presentation remains valid for even much more general cases in both demains.

3.5. FREQUENCY AND TRANSIENT RESPONSES CAUSED BY CURRENT FLOW IN A CONDUCTIVE BODY WITH AXIAL SYMMETRY

In this section we will consider the principal features of the frequency and time domain responses of induced currents flowing in a confined conductive body characterized by axial symmetry, and illustrate these results considering the special case of a spheroidal conductive body. First, let us explore the low frequency part of the spectrum. In deriving the asymptotic expressions for the low-frequency part of the spectrum, we will begin from eqs. 3.164 and 3.165. The expression for current density from eq. 3.164 is:

$$J_\phi(p) = k^2 b^2 H_{0z} \sum_{n=1}^{\infty} \frac{\beta_n(p)}{q_n - k^2 b^2}$$

Each term in this series can be expanded in a MacLauren series if $|k^2 b^2| < q_n$. Carrying out such an expansion we obtain:

$$\frac{\beta_n}{q_n - k^2 b^2} = \frac{\beta_n}{q_n} \frac{1}{1 - \dfrac{k^2 b^2}{q_n}} \approx \frac{\beta_n}{q_n} \left\{ 1 + \frac{k^2 b^2}{q_n} + \frac{k^4 b^4}{q_n^2} + \right\}$$

$$= \frac{\beta_n}{q_n} \sum_{i=0}^{\infty} \frac{k^{2i} b^{2i}}{q_n^i} = \sum_{i=0}^{\infty} \frac{\beta_n}{q_n^{i+1}} k^{2i} b^{2i}, \quad \text{if } |k^2 b^2| < q_n$$

(3.169)

Similarly:

$$\frac{k^2 b^2 H_{0z} \beta_n}{q_n - k^2 b^2} = H_{0z} \sum_{i=0}^{\infty} \frac{\beta_n}{q_n^{i+1}} (kb)^{2i+2}$$

Replacing the index $i + 1$ with the index l, we have:

$$\frac{k^2 b^2 H_{0z} \beta_n}{q_n - k^2 b^2} = H_{0z} \sum_{l=1}^{\infty} \frac{\beta_n}{q_n^l} k^{2l} b^{2l} \tag{3.170}$$

Expanding every fraction in eq. 3.164 in the same manner and summing coefficients with the same power $k^{2l} b^{2l}$, one can represent the expression for current density in the form of a MacLauren series as follows:

$$J_\phi(p) = H_{0z} \sum_{l=1}^{\infty} \alpha_l k^{2l} b^{2l} \tag{3.171}$$

In accord with eq. 3.170, we have

$$\alpha_l = \frac{\beta_1}{q_1^l} + \frac{\beta_2}{q_2^l} + \frac{\beta_3}{q_3^l} + \frac{\beta_4}{q_4^l} + \dots \tag{3.172}$$

The radius of covergence for each fraction $\beta_n/(q_n - k^2 b^2)$ is defined by the corresponding value for q_n. It is obvious that the series in eq. 3.171 consisting of all possible fractions will converge when:

$$|k^2 b^2| < q_1 \tag{3.173}$$

inasmuch as q_1 is the minimum value for q_n, as may be seen from eq. 3.165.

Considering the definition of the wave number, we can rewrite eq. 3.171 in the form:

$$J_\phi(p) = H_{0z} \sum_{l=1}^{\infty} \alpha_l i^l (\sigma \mu b^2)^l \omega^l \tag{3.174}$$

where $i = \sqrt{-1} = e^{i\pi/2}$.

Then, we obtain the following expressions for the inphase and quadrature components of the induced currents at low frequencies, respectively:

$$\text{In} J_\phi(p) = H_{0z} \sum_{l=1}^{\infty} (-1)^l \alpha_{2l} (\sigma \mu b^2)^{2l} \omega^{2l} \tag{3.175}$$

$$Q J_\phi(p) = H_{0z} \sum_{l=1}^{\infty} (-1)^{l+1} \alpha_{2l-1} (\sigma \mu b^2)^{2l-1} \omega^{2l-1}$$

From eq. 3.173, we see that the series in eqs. 3.174 and 3.175 will converge when frequency satisfies the condition:

$$\omega < q_1/\sigma \mu b^2 \tag{3.176}$$

We see that the first pole in the spectrum defines the radius of convergence for these series.

Using the same approach for the various components of the electromagnetic field given by eq. 3.166, we obtain:

$$H_z = H_{0z} \sum_{l=1}^{\infty} C_{lz} k^{2l} b^{2l} \tag{3.177}$$

$$H_r = H_{0z} \sum_{l=1}^{\infty} C_{lr} k^{2l} b^{2l} \tag{3.178}$$

and

$$E_\phi = i\omega\mu b H_{0z} \sum_{l=1}^{\infty} e_l k^{2l} l^2 \tag{3.179}$$

where

$$C_{lz} = \frac{d_{1z}}{q_1^l} + \frac{d_{2z}}{q_2^l} + \frac{d_{3z}}{q_3^l} + \frac{d_{4z}}{q_4^l} + \dots \tag{3.180}$$

$$C_{lr} = \frac{d_{1r}}{q_1^l} + \frac{d_{2r}}{q_2^l} + \frac{d_{3r}}{q_3^l} + \frac{d_{4r}}{q_4^l} + \dots \tag{3.181}$$

and

$$e_l = \frac{f_1}{q_1^l} + \frac{f_2}{q_2^l} + \frac{f_3}{q_3^l} + \frac{f_4}{q_4^l} + \dots \tag{3.182}$$

As an example, we have the following expression for the quadrature and inphase components of the magnetic field:

$$\text{In } H_z = H_{0z} \sum_{l=1}^{\infty} (-1)^l C_{2l,z} (\sigma\mu b^2)^{2l} \omega^{2l}$$

$$Q H_z = H_{0z} \sum_{l=1}^{\infty} (-1)^{l+1} C_{2l-1,z} (\sigma\mu b^2)^{2l-1} \omega^{2l-1} \tag{3.183}$$

and

$$\text{In } H_r = H_{0z} \sum_{l-1}^{\infty} (-1)^l C_{2l,r} (\sigma\mu b^2)^{2l} \omega^{2l}$$

$$Q H_r = H_{0z} \sum_{l-1}^{\infty} (-1)^{l+1} C_{2l-1,r} (\sigma\mu b^2)^{2l-1} \omega^{2l-1} \tag{3.184}$$

It is clear that the series in eqs. 3.177 to 3.179, as well as those in eqs. 3.183 to 3.184 will converge when condition 3.176 is met. Therefore, for the low-frequency part of the spectrum, the various components of the electromagnetic field and of the induced current flow can be written in the form of a MacLauren series which contains only integer powers in ω. It should be stressed here that this behavior will not occur if any other medium, such as a layered medium is considered instead.

It can readily be seen by applying the Biot-Savart law for each term in eq. 3.171 that we will obtain corresponding terms in the series in eqs. 3.177 and 3.178 describing the electromagnetic field for these currents.

Returning to eq. 3.175, we write:

$$\text{In}J_\phi(p) = H_{0z}\{-\alpha_2(\sigma\mu b^2)^2\omega^2 + \alpha_4(\sigma\mu b^2)^4\omega^4 - \ldots\}$$

$$\text{if } \omega < \frac{q_1}{\sigma\mu b^2} \quad (3.185)$$

$$QJ_\phi(p) = H_{0z}\{\alpha_1(\sigma\mu b^2)\omega - \alpha_3(\sigma\mu b^2)^3\omega^3 + \ldots\}$$

Thus, the series for the inphase component contains only even integer powers of ω, while the series for the quadrature component consists of only odd integer powers of ω. Each term in these series is the product of three factors, namely:

(1) α_{2l} or α_{2l-1}

(2) $(\sigma\omega b^2)^{2l}$ or $(\sigma\mu b^2)^{2l-1}$

(3) ω^{2l} or ω^{2l-1}

However, the functions α depend on the coordinates of the point p in the conductive body and on its geometry. The factor $\sigma\mu b^2$ is determined by the conductivity, the magnetic permeability, and one of the characteristic dimensions of the body. With increasing order of the series, the power of ω increases markedly.

Assume for the moment that measurements are carried out at relatively low frequencies and that the values for the quadrature and inphase components will be determined mainly by the first terms. In this case, we can write:

$$QJ_\phi(p) \approx H_{0z}\alpha_1\sigma\mu b^2\omega$$

$$\text{In}J_\phi(p) \approx -H_{0z}\alpha_2(\sigma\mu b^2)^2\omega^2$$

$$(3.186)$$

Thus, at low frequencies, the quadrature component of the current density is dominant, and it is directly proportional to the conductivity and to the frequency. This behavior can be explained by the fact that the leading term in the series representing the function $Qj_\phi(p)$ is determined by the primary vortex electric field. However, the inphase component of the current density turns out to be more sensitive to a change in the characteristics of the conductive body, such as its conductivity. Strictly speaking, the coefficients α_l depend on the geometry of the conductive body in a different manner. However, as will be shown later, in many cases for practical purposes the distinction is not significant. We can illustrate this making use of the series which describes the low frequency part of the spectrum for the magnetic field (eqs. 3.177 and 3.178). From eqs. 3.180 and 3.181, we have:

$$C_l = \frac{d_1}{q_1^l} + \frac{d_2}{q_2^l} + \frac{d_3}{q_3^l} + \frac{d_4}{q_4^l} + \dots \tag{3.187}$$

where the second index on C, indicating the component of the magnetic field being considered, has been omitted for simplification. Inasmuch as the values for q_n increase with increasing n, the coefficients C_l for relatively large values of l, that is, for higher order terms in the series in eqs. 3.177 and 3.178, as well as in eq. 3.179, are defined mainly by the first term in eq. 3.187. Thus:

$$C_{l-1} \approx \frac{d_1}{q_1^{l-1}}, \quad C_l \approx \frac{d_1}{q_1^l}, \quad C_{l+1} \approx \frac{d_1}{q_1^{l+1}} \tag{3.188}$$

and so on.

These equations are better approximations for large values of the index l. From eq. 3.188, we can obtain:

$$C_l \approx \frac{C_{l-1}}{q_1}, \quad C_{l+1} \approx \frac{C_l}{q_1} \approx \frac{C_{l-1}}{q_1^2} \tag{3.189}$$

These expressions establish the relationship between the coefficient C_l with various values for the index l. We should remember that the quantity q_1 determines the minimum pole in the spectrum, but that d_1 is the coefficient multiplying the first exponential describing the transient field or the numerator of the first fraction of the sum representing the field in eq. 3.166.

Numerical calculations demonstrate that the relationships in eq. 3.188 or 3.189 are useful even for relatively small values of the index l, that is, for the first few terms of the series in eqs. 3.177 and 3.178, except for the first term which describes the quadrature component of the magnetic field. Assuming that the first coefficient is known, we can write an approximate expression for the field:

$$H_i \approx H_{0z} \left\{ C_{1i} k^2 b^2 + \sum_{l=2}^{\infty} C_{li} k^{2l} b^{2l} \right\}$$

$$\cong H_{0z} \left\{ C_{1i} k^2 b^2 + d_{1i} \sum_{l=2}^{\infty} \frac{k^{2l} b^{2l}}{q_1^l} \right\} \tag{3.190}$$

where the counter i indicates the r- or z-component of the magnetic field. A similar equation can be written for the electric field.

Now let us examine a method for obtaining the low frequency part of the spectrum, that is, for the determination of the coefficient C_l or e_l. First, we will proceed from eq. 3.171 which describes the low frequency spectrum for current density:

$$J_\phi(p) = H_{0z} \sum_{n=1}^{\infty} \alpha_n(kb)^{2n}$$

Substituting this into the integral equation of 3.153 and taking into account eq. 3.2 and equating like terms in powers of $(kb)^2$ we obtain a recurrence relationship for determining α_n:

$$\alpha_1 = \frac{1}{2b^2} r_p, \quad \alpha_n = \frac{1}{b^2} \int_S \alpha_{n-1} G(p,g)\mathrm{d}S \qquad (3.191)$$

In this expression, the notations are the same as those used in the previous paragraph and which were illustrated in Fig. 3.20. In particular, the function $G(p,g)$ represents the interaction between circular current filaments passing through the points p and g, p. The equation for determining α_1, describing the leading term for the quadrature component of the current density, does not need to be integrated. The first equation in 3.191 is written with the assumption that the primary field H_{0z} is uniform in the vicinity of the conductive body. It can readily be generalized for more complicated cases.

From eq. 3.191 we have:

$$\alpha_1(p) = \frac{1}{2b^2} r_p$$

$$\alpha_2(p) = \frac{1}{b^2} \int_S \alpha_1 G(p,g)\mathrm{d}S$$

$$\alpha_3(p) = \frac{1}{b^2} \int_S \alpha_2 G(p,g)\mathrm{d}S$$

and so on.

Knowing the low-frequency spectrum for circular current filaments in the conductive body and applying the Biot-Savart law, we can find the coefficients C_{ni} for the magnetic field at any point in space. As has been demonstrated previously (see Chapter 2), in the general case the magnetic field caused by a circular current filament is expressed in terms of elliptical integrals. In particular, for a point on the z-axis (see Fig. 3.21), we have:

$$C_{nz} = \frac{1}{2} \int_S \frac{\alpha_n(p) r_p^2 \mathrm{d}S}{[r_p^2 + (z_p - z_g)^2]^{3/2}} \qquad (3.192)$$

In the same manner, using the same coefficients $\alpha_n(p)$, we can derive the coefficients C_{nr} as well as e_ϕ.

It might be noted here that these derivations remain valid if the conductivity σ is the function of two coordinates, r and z. Suppose that the conductive body has the form of a spheroid with axes a and b (a greater than b) and that the primary uniform magnetic field is directed along the b-axis as shown

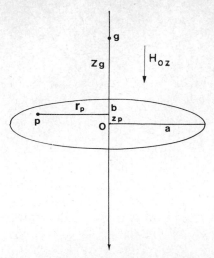

Fig. 3.22. Example of a conductive body with the form of a spheroid situated in a primary uniform magnetic system.

in Fig. 3.22. We will concentrate our attention on the field at the z-axis. Therefore, the low-frequency spectrum can be calculated using eq. 3.192. It is obvious that with an increase in the ratio of semi-axes, a/b, and of the depth, z/r, the secondary magnetic field will become as that for a vanishingly thin disc with varying longitudinal conductance, S, which is a function of the distance away from the b-axis. Let us note that we consider as an example the field on the z-axis as shown in Fig. 3.22. The expression for this conductance has the form:

$$S = 2b\sigma(1 - r^2/a^2)^{1/2} \tag{3.193}$$

where the product $2b\sigma$ is the conductance of the mid-part of the spheroid. The surface density of currents in the disc, I_ϕ, as follows from eq. 3.153, satisfies the equation:

$$I_\phi(p) = k_s^2 \frac{H_0}{2} \bar{r}_p (1 - \bar{r}_p^2)^{1/2} + k_s^2 (1 - \bar{r}_p^2)^{1/2} \int_0^1 G(p,g) I_\phi(g) d\bar{r}_g \tag{3.194}$$

where

$$k_s^2 = i\omega\mu S_0 a, \quad S_0 = 2b\sigma, \quad \text{and } \bar{r} = r/a$$

The current density function, j_ϕ, is related to the surface density through the expression:

$$I_\phi = j_\phi 2b(1 - \bar{r}^2)^{1/2}$$

Writing the surface current density at low frequency as a series:

$$I_\phi(p) = H_0 \sum_{n=1}^{\infty} \alpha_{ns}(p) k_s^{2n}$$

and substituting this in eq. 3.194, we again obtain the recurrence relationships needed to define α_{ns}:

$$\alpha_{1s} = \frac{1}{2} \bar{r}_p (1 - \bar{r}_p^2)^{1/2}$$

$$\alpha_{ns}(p) = (1 - \bar{r}_p^2)^{1/2} \int_0^1 G(p, g) \alpha_{n-1,s} \mathrm{d}\bar{r}_g$$

(3.195)

Considering that the disc is vanishingly thin, the calculation of the magnetic field through the use of the Biot-Savart law is much simplified. In particular, writing the low-frequency part of the magnetic field on the z-axis as:

$$H_z = H_0 \sum_{n=1}^{\infty} C_{nz}^s k_s^{2n}$$

we have

$$C_{nz}^s = \frac{1}{2} \int_0^1 \frac{\alpha_{ns}(p) \bar{r}_p^2}{(\bar{r}_p^2 + \bar{z}^2)^{3/2}} \mathrm{d}\bar{r}_p$$

(3.196)

where $\bar{z}$ is the distance from the center of the disc to the observation point, normalized to a, that is, $\bar{z} = z/a$.

Let us note that eqs. 3.194—3.196 become somewhat simpler for the vanishingly thin disc characterized by a constant longitudinal conductance $S = S_0$, since in place of the $(1 - \bar{r}_p^2)^{1/2}$, we have the factor unity in the previous equation.

From a comparison of the low frequency spectrum for the fields caused by currents flowing in a spheroid and those flowing in a disc, equivalent to that spheroid with varying conductance S, it can be seen that the coefficients C_{nz} and C_{nz}^s are related as:

$$C_{nz} \approx C_{nz}^s 2^n \left(\frac{a}{b}\right)^n$$

(3.197)

Numerical evaluations demonstrate that with an increase in the axial ratio a/b, the number n of coefficients C_{nz}^s for which this equality remains valid increases. For example, for a coefficient C_{1z} defining the leading term in the expression for the quadrature component, QH_z, at low frequencies, the equality in 3.197 remains valid for $a/b \geqslant 8$ with an error of no more than 5% for practically any z/b which is greater than unity. Within the same error, for the coefficient C_{2z} characterizing the leading term for the inphase component, this equality applies when $a/b > 32$, and finally, for the coefficients C_{3z} this equality holds when $a/b \geqslant 64$. As follows from eq. 3.197, the parameter q_1, characterizing the first pole in the spectrum for reasonably elongated spheroids is related to the corresponding parameter q_{1s} for the spectrum of currents flowing in the disc with a varying conductance S as follows:

$$q_1 = q_{1s} b/2a \qquad (3.198)$$

As calculations show, for a disc with

$$S = S_0 (1 - r^2/a^2)^{1/2}$$

the parameter q_{1s} is

$$q_{1s} = 7.70 \qquad (3.199)$$

If the disc were to have a constant conductance S_0, we have

$$q_{1s_0} = 5.51 \qquad (3.200)$$

The first seven coefficients C_{nz} for each of several points distributed along the z-axis are listed in Tables 3.VI and 3.VII. Table 3.VII contains values for the functions d_{1z} and q_1. These values illustrate the change in the field as a function of z for various axial ratios a/b. The values for the coefficients C_{1z}, d_{1z}, and q_1 listed in Tables 3.VI and 3.VII clearly demonstrate the accuracy in applying formulas 3.188 and 3.189. In accord with eq. 3.187, the first fraction in eq. 3.166 defines the coefficients C_{nz} with relatively minor error when n is greater than or equal to 2. If there is a necessity to calculate the first coefficient with even higher accuracy, one can use either the values listed in Table 3.VI or determine C_{1z} or C_{1r} by applying the Biot-Savart law. Under certain conditions, these calculations will permit one to derive an expression for this coefficient in explicit form. In particular, along the axes of a spheroid we have:

$$QH(z) = \omega\mu S_0 a C_{1z}^s = \frac{\omega\mu S_0 a H_{0z}}{4} \int_0^1 \frac{\bar{r}_p^3 (1 - \bar{r}_p^2)^{1/2}}{(\bar{r}_p^2 + \bar{z}^2)^{3/2}} \, d\bar{r}_p$$

$$= \frac{\omega\mu S_0 a H_{0z}}{8} \left[\left(1 + \frac{3z^2}{a^2}\right) \tan^{-1} \frac{a}{z} - \frac{3z}{a} \right], \quad \text{if } \frac{a}{b} \frac{z}{b} > 6 \qquad (3.201)$$

Whence

$$C_{1z} = \frac{a}{4b} \left[\left(1 + 3\frac{z^2}{a^2}\right) \tan^{-1} \frac{a}{z} - \frac{3z}{a} \right] \qquad (3.202)$$

and

$$C_{2z} \approx C_{1z} q_1, \quad C_{3z} \approx C_{1z} q_1^2, \ldots \text{ and so on.}$$

TABLE 3.VI

a/b	z_g/b	C_{1z}	C_{2z}	C_{3z}	C_{4z}	C_{5z}	C_{6z}	C_{7z}
1	1.1	0.501×10^{-1}	0.477×10^{-2}	0.477×10^{-3}	0.481×10^{-4}	0.487×10^{-5}	0.493×10^{-6}	0.499×10^{-7}
	1.5	0.197×10^{-1}	0.188×10^{-2}	0.188×10^{-3}	0.189×10^{-4}	0.192×10^{-5}	0.194×10^{-6}	0.197×10^{-7}
	2.0	0.833×10^{-2}	0.793×10^{-3}	0.793×10^{-4}	0.801×10^{-5}	0.811×10^{-6}	0.821×10^{-7}	0.831×10^{-8}
	3.0	0.247×10^{-2}	0.235×10^{-3}	0.235×10^{-4}	0.237×10^{-5}	0.240×10^{-6}	0.243×10^{-7}	0.246×10^{-8}
2	1.1	0.243	0.735×10^{-1}	0.217×10^{-1}	0.634×10^{-2}	0.184×10^{-2}	0.535×10^{-3}	0.155×10^{-3}
	1.5	0.144	0.422×10^{-1}	0.123×10^{-1}	0.359×10^{-2}	0.104×10^{-2}	0.302×10^{-3}	0.875×10^{-4}
	2.0	0.804×10^{-1}	0.229×10^{-1}	0.664×10^{-2}	0.193×10^{-2}	0.558×10^{-3}	0.162×10^{-3}	0.470×10^{-4}
	3.0	0.307×10^{-1}	0.852×10^{-2}	0.245×10^{-2}	0.708×10^{-3}	0.205×10^{-3}	0.595×10^{-4}	0.172×10^{-4}
4	1.1	0.832	0.686	0.531	0.400	0.299	0.222	0.165
	1.5	0.639	0.510	0.390	0.293	0.218	0.162	0.120
	2.0	0.463	0.358	0.270	0.202	0.150	0.111	0.828×10^{-1}
	3.0	0.252	0.187	0.139	0.103	0.764×10^{-1}	0.567×10^{-1}	0.421×10^{-1}
8	1.1	2.25	4.51	8.34	14.9	26.1	45.6	79.3
	1.5	1.97	3.88	7.10	12.6	22.1	38.5	66.9
	2.0	1.68	3.21	5.82	10.3	18.0	31.3	54.4
	3.0	1.22	2.23	3.97	6.97	12.1	21.1	36.7
16	1.1	0.530×10^{1}	0.237×10^{2}	0.979×10^{2}	0.389×10^{3}	0.152×10^{4}	0.586×10^{4}	0.225×10^{5}
	1.5	0.497×10^{1}	0.220×10^{2}	0.902×10^{2}	0.357×10^{3}	0.139×10^{4}	0.537×10^{4}	0.206×10^{5}
	2.0	0.458×10^{1}	0.200×10^{2}	0.814×10^{2}	0.321×10^{3}	0.125×10^{4}	0.482×10^{4}	0.185×10^{5}
	3.0	0.389×10^{1}	0.165×10^{2}	0.664×10^{2}	0.261×10^{3}	0.101×10^{4}	0.389×10^{4}	0.149×10^{5}
32	1.1	0.115×10^{2}	0.110×10^{3}	0.968×10^{3}	0.824×10^{4}	0.689×10^{5}	0.571×10^{6}	0.470×10^{7}
	1.5	0.112×10^{2}	0.106×10^{3}	0.929×10^{3}	0.789×10^{4}	0.660×10^{5}	0.546×10^{6}	0.450×10^{7}
	2.0	0.107×10^{2}	0.101×10^{3}	0.882×10^{3}	0.748×10^{4}	0.625×10^{5}	0.517×10^{6}	0.426×10^{7}
	3.0	0.989×10^{1}	0.914×10^{2}	0.795×10^{3}	0.672×10^{4}	0.560×10^{5}	0.463×10^{6}	0.381×10^{7}
64	1.1	0.241×10^{2}	0.473×10^{3}	0.866×10^{4}	0.154×10^{6}	0.269×10^{7}	0.466×10^{8}	0.804×10^{9}
	1.5	0.237×10^{2}	0.464×10^{3}	0.847×10^{4}	0.150×10^{6}	0.263×10^{7}	0.456×10^{8}	0.786×10^{9}
	2.0	0.232×10^{2}	0.453×10^{3}	0.826×10^{4}	0.146×10^{6}	0.256×10^{7}	0.444×10^{8}	0.765×10^{9}
	3.0	0.223×10^{2}	0.431×10^{3}	0.784×10^{4}	0.139×10^{6}	0.222×10^{7}	0.420×10^{8}	0.724×10^{9}

TABLE 3.VII

z_g/b	a/b						
	1	2	4	8	16	32	64
q_1:	9.87	3.45	1.35	0.575	0.261	0.122	0.058
1.1	0.457	0.900	0.135×10^1	0.165×10^1	0.186×10^1	0.195×10^1	0.178×10^1
1.5	0.180	0.508	0.988	0.139×10^1	0.170×10^1	0.188×10^1	0.174×10^1
2.0	0.767×10^{-1}	0.273	0.677	0.113×10^1	0.153×10^1	0.176×10^1	0.169×10^1
3.0	0.225×10^{-1}	0.100	0.344	0.762	0.123×10^1	0.158×10^1	0.160×10^1
4.0	0.951×10^{-2}	0.461×10^{-1}	0.191	0.523	0.100×10^1	0.141×10^1	0.151×10^1
6.0	0.282×10^{-2}	0.146×10^{-1}	0.727×10^{-1}	0.268	0.672	0.114×10^1	0.135×10^1
8.0	0.119×10^{-2}	0.632×10^{-2}	0.341×10^{-1}	0.149	0.463	0.923	0.121×10^1
10	0.609×10^{-3}	0.327×10^{-2}	0.184×10^{-1}	0.897×10^{-1}	0.326	0.753	0.109×10^1
15	0.180×10^{-3}	0.982×10^{-3}	0.576×10^{-2}	0.320×10^{-1}	0.150	0.464	0.833
20	0.761×10^{-4}	0.416×10^{-3}	0.248×10^{-2}	0.146×10^{-1}	0.785×10^{-1}	0.298	0.644
30	0.225×10^{-4}	0.124×10^{-3}	0.746×10^{-3}	0.457×10^{-2}	0.279×10^{-1}	0.137	0.396
40	0.951×10^{-5}	0.522×10^{-4}	0.316×10^{-3}	0.191×10^{-2}	0.127×10^{-1}	0.710×10^{-1}	0.253
60	0.282×10^{-5}	0.155×10^{-4}	0.941×10^{-4}	0.592×10^{-3}	0.398×10^{-2}	0.252×10^{-1}	0.115
80	0.119×10^{-5}	0.653×10^{-5}	0.397×10^{-4}	0.251×10^{-3}	0.171×10^{-2}	0.114×10^{-1}	0.599×10^{-1}
100	0.609×10^{-6}	0.334×10^{-5}	0.203×10^{-4}	0.129×10^{-3}	0.885×10^{-3}	0.607×10^{-2}	0.343×10^{-1}
150	0.180×10^{-6}	0.991×10^{-6}	0.603×10^{-5}	0.383×10^{-4}	0.265×10^{-3}	0.187×10^{-2}	0.115×10^{-1}
200	0.761×10^{-7}	0.418×10^{-6}	0.255×10^{-5}	0.161×10^{-4}	0.112×10^{-3}	0.798×10^{-3}	0.510×10^{-2}

It is easily seen that for sufficiently large values of the ratio of z/a, the magnetic field from currents induced in a spheroid is equivalent to that from a magnetic dipole with the moment

$$M = \frac{\sigma \mu b a^4}{15} \omega 2\pi H_{0z} \tag{3.203}$$

oriented in the direction of the primary magnetic field.

A comparison of these calculations with exact calculations based on the method of integral equations shows that the error involved in determining the field by series of approximation using seven terms is no more than 5—10% when:

$$\frac{b}{h_i} < 0.7 q_1^{1/2} \tag{3.204}$$

with h_i being the skin depth. For example, for $\rho = 0.1\,\Omega m$ meter, $b = 5\,m$, and $a = 40\,m$, we can apply eqs. 3.183 to 3.184 when the frequency is less than 300 Hz.

We will now consider the opposite case, that is, the high-frequency part of the spectrum. Let us make use of a flattened spheroidal system of coordinates, ξ, η and related with the cylindrical coordinates through the relationships shown in Fig. 3.23:

$$z = h\xi\eta, \quad r = h[(1 + \eta^2)(1 - \xi^2)]^{1/2} \tag{3.205}$$

where

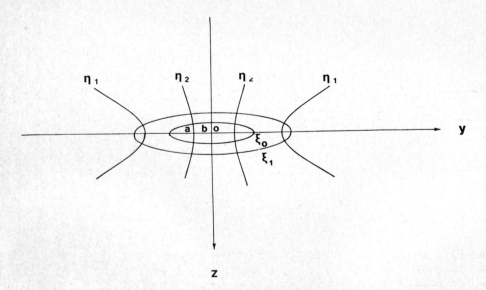

Fig. 3.23. Definition of a flattened spheroidal coordinate system.

$-1 \leqslant \xi \leqslant 1$, $\quad 0 \leqslant \eta < \infty$, $\quad$ and $\quad h = (a^2 - b^2)^{1/2}$

In accord with eq. 3.205, the primary vortex electric field has the form:

$$E_{0\phi} = \frac{i\omega\mu}{2} H_{0z} h [(1 + \eta^2)(1 - \xi^2)]^{1/2}$$

or

$$E_{0\phi} = \frac{\omega\mu}{2} H_{0z} h P_1^1(\xi) P_1^1(i\eta) \qquad (3.206)$$

where $P_1^1(x)$ is an associated Legendre function.

Let us write the secondary electric field as consisting of a single spatial harmonic satisfying the boundary condition at infinity:

$$E_s = \frac{\omega\mu}{2} H_{0z} h D_\infty Q_1^1(i\eta) P_1^1(\xi) \qquad (3.207)$$

where

$$Q_1^1(i\eta) = (1 + \eta^2)^{1/2} \left(\operatorname{ctg}^{-1}\eta - \frac{\eta}{1 + \eta^2} \right)$$

The total electric field outside the spheroid takes the form:

$$E_\phi^e = \frac{\omega\mu H_{0z} h}{2} [P_1^1(i\eta) + D_\infty Q_1^1(i\eta)] P_1^1(\xi) \qquad (3.208)$$

At very high frequencies, the electric field in a moderately conductive spheroid tends to zero and therefore the boundary condition at the surface has the same form as in the case of a perfectly conducting body, that is:

$$E_\phi^e(\eta_0) = E_{0\phi} + E_s = 0 \qquad (3.209)$$

where η_0 is the coodinate for the surface of the spheroidal conductor. Therefore, we obtain the following for the function D_∞:

$$D_\infty = -\frac{P_1^1(i\eta_0)}{Q_1^1(i\eta_0)} = -\frac{i}{\operatorname{ctg}^{-1}\eta_0 - \eta_0/(1 + \eta_0^2)}$$

and $\qquad\qquad\qquad\qquad\qquad\qquad\qquad\qquad$ as $\omega \to \infty \qquad (3.210)$

$$E_s = -\frac{\operatorname{ctg}^{-1}\eta - \eta/(1 + \eta^2)}{\operatorname{ctg}^{-1}\eta_0 - \eta_0/(1 + \eta_0^2)} E_{0\phi}$$

This expression represents only the asymptotic value for the quadrature component of the electric field. By using Maxwell's equations, it is a simple matter to arrive at the corresponding values for the inphase component of the magnetic field.

A more complete characterization of the high-frequency part of the spectrum, with some limitations, can be obtained by using the approximate set of boundary conditions known as Leontovich's conditions. As is well known, when the minimum radius of curvature of the surface of the conductive body is much greater than the skin depth, the ratio of the tangential components of the electric and magnetic fields on the inside of the conductor is equal to the impedance for a plane wave:

$$Z = (i\omega\mu/\sigma)^{1/2}$$

Taking into account the continuity of tangential components of the fields at the interface between media with different conductivities, we obtain the approximate boundary condition in the same manner as for a sphere:

$$\frac{E^e_{t_1}}{H^e_{t_2}} = \left(\frac{i\omega\mu}{\sigma}\right)^{1/2} = Z \tag{3.211}$$

where t_1 and t_2 are two unit vectors tangential to the surface of the spheroid and mutually perpendicular.

It is essential that both components of the field E^e and H^e describe the field at the outside of the surface of the conductor. In view of condition 3.211 the field outside the spheroid can be determined even when the field inside the conductor remains unknown. However, as in the general case, the field is described by an infinite sum of spheroidal harmonics. Near the z-axis, at distances from the surface of the spheroid which are significantly less than the horizontal semi-axis a $(a > b)$, the first spatial harmonic plays the leading role. For this reason, we will look for a representation for the tangential component of the electric field in terms of a single harmonic, as we did before:

$$E^e_\phi \approx \frac{\omega\mu H_{0z}}{2} h[P^1_1(i\eta) + DQ^1_1(i\eta)] P^1_1(\xi) \tag{3.212}$$

From the first of Maxwell's equations, we obtain the following expression for the tangential component of the magnetic field, H^e_ξ:

$$H^e_\xi = \frac{H_{0z}(1 - \xi^2)^{1/2}}{(\xi^2 + \eta^2)^{1/2}} [\eta - D(\eta \, \mathrm{ctg}^{-1}\eta - 1)] \tag{3.213}$$

Substituting these last two equations in the boundary condition of eq. 3.211 we have:

$$D = \frac{B\eta_0 - 1}{(1 - B\eta_0) \, \mathrm{ctg}^{-1}\eta_0 + B - \eta_0/(1 + \eta_0^2)} \tag{3.214}$$

where

$$B = \frac{2Z}{i\omega\mu h(1 + \eta_0^2)^{1/2}(\xi^2 + \eta_0^2)^{1/2}}$$

Result of calculations of the vertical component of the magnetic field along the z-axis based on the approximate formula:

$$H_z^s \approx H_{0z}D[\text{ctg}^{-1}\eta - \eta/(1 + \eta^2)] \tag{3.215}$$

and with the help of the integral equations, practically coincide when

$$b/h_i \geqslant 10 \quad \text{and} \quad z/a < 1$$

It is important to note here that in accord with eq. 3.215, because of the presence of the term proportional to the impedance, Z, with increase in frequency, the quadrature component of the magnetic field tends to zero as $1/\omega^{1/2}$. Also, the inphase component of the magnetic field approaches its asymptotic value for large frequency in the same way, that is, one can write:

$$\text{In}H_i \approx A_i - \frac{B_i}{\omega^{1/2}}, \quad QH_i \approx \frac{C_i}{\omega^{1/2}} \tag{3.216}$$

where A, B, and C are constants, and the index i indicates an arbitrary component of the field, either r or z. Taking the boundary conditions of eq. 3.211 into account, one can expect that this behavior of the magnetic field at high frequencies as a function of frequency will be observed for an arbitrarily shaped conductive body as well.

We have considered the asymptotic behavior of the harmonic field. In order to obtain a total response, calculations based on the use of the integral equation approach were performed. Examples of the spectral responses of the quadrature and inphase components of the vertical component of the magnetic field along the z-axis are shown in Figs. 3.24–3.29. Inasmuch as the field from currents induced in the spheroid is controlled by three parameters, namely b/h_i, a/h_i, and the position of the observation point, a comparison of the various frequency responses is shown in these illustrations reflects the effect of the ratio a/b. It manifests itself mainly in the fact that with an increase in the length of the horizontal axis, which is oriented perpendicular to the direction of the primary field, the field increases and the upper limit for the low-frequency spectrum is shifted towards lower frequencies. With an unlimited increase in the length of the horizontal axis, the secondary field also tends to become infinite, because the primary field remains uniform. However, if the source of the primary field is of limited dimensions and the skin depth is significantly less than the length of the semi-axis a, $(a/h_i \gg 1)$, the field from currents induced in this spheroid, approaches an asymptotic value corresponding to the secondary field for currents induced in a horizontal layer of unlimited extent with a thickness $2b$.

We can now consider the transient response for the same cases assuming

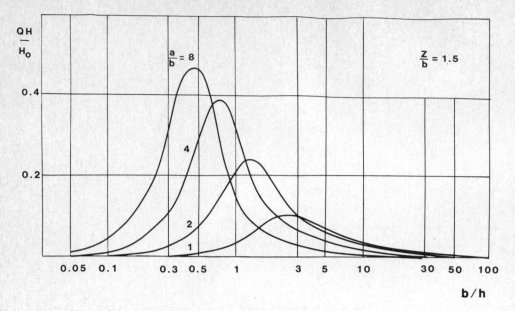

Fig. 3.24. Example of the behavior of the quadrature component of the vertical magnetic field along the z-axis for the case $z/b = 1.5$.

that a uniform source field vanishes abruptly (the asumption of a step-function excitation). In accord with eq. 3.168, we have:

$$H_i = H_0 \sum_{n=1}^{\infty} d_{ni} e^{-q_n t/\sigma\mu b^2} \tag{3.217}$$

where d_{ni} are coefficients depending on geometry. The coefficients q_n depend on the shape of the conductive body and meet the condition $q_1 < q_2 < q_3 < \ldots$. Let us emphasize that the functions d_{ni} and q_n are the same as those which have been used previously in describing frequency domain response. If in place of step-function excitation, another form of excitation is used, the form of eq. 3.49 remains the same, but there will be some combination of coefficients used in place of the coefficients d_{ni}.

We will first examine the late-stage behavior of the transient field. In accord with eq. 3.168, during the late stage, due to induced currents in the confined conductive body, we have:

$$J_\phi(p) = H_{0z}\beta_1(p)e^{-q_1 t/\sigma\mu b^2}$$

$$H_z(a) = H_{0z}d_{1z}(a)e^{-q_1 t/\sigma\mu b^2} \tag{3.218}$$

$$H_r(a) = H_{0z}d_{1r}(a)e^{-q_1 t/\sigma\mu b^2}$$

It is obvious that the late stage will begin only when the following condition is met:

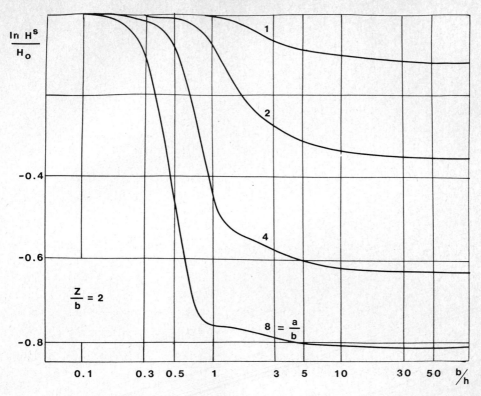

Fig. 3.25. Example of the spectral response of the inphase component of the vertical magnetic field along the z-axis for the case $z/b = 2$.

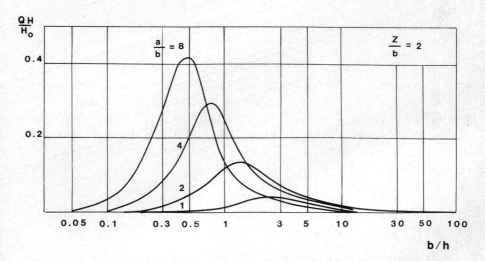

Fig. 3.26. Example of the spectral response of the quadrature component of the vertical magnetic field along the z-axis for the case $z/b = 2$.

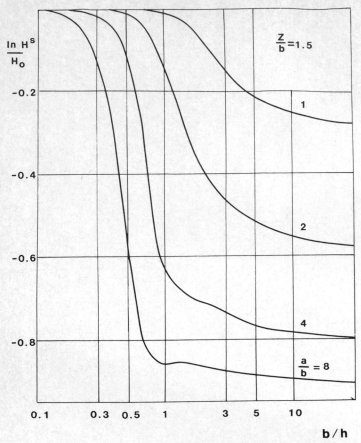

Fig. 3.27. Example of the spectral response of the inphase component of the vertical magnetic field along the z-axis for the case $z/b = 1.5$.

$$q_1 t / \sigma \mu b^2 > 1 \qquad (3.219)$$

In this case, all but the first terms in the series 3.168 are negligible due to the fact that $q_1 < q_2, < \ldots$, and we can restrict our attention to only the first term. It must be emphasized that the equations for the currents and the field have an especially simple form showing that the transient response during the late stage for confined conductive bodies decay exponentially with time.

With respect to the induced currents, it might be noted that during the late stage, the relative distribution does not change, and at all points the current density decreases in a like manner, as the exponential:

$$e^{-q_1 t / \sigma \mu b^2}$$

This is the reason why all of the components of the magnetic field decay in the same way. For example, when we take the ratio $H_z(a)/H_r(a)$, we obtain

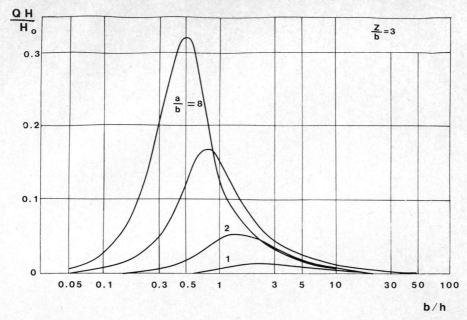

Fig. 3.28. Example of the spectral response for the quadrature component of the vertical magnetic field along the z-axis for the case $z/b = 3$.

a constant which is independent of time. Frequently, the parameter $\sigma\mu b^2/q_1$ is termed the time constant for the conductor:

$$\tau_0 = \sigma\mu b^2/q_1 \qquad (3.220)$$

The value q_1 depends on the shape of the conductive body. Correspondingly, the time constant τ_0 is the product of the conductivity with a function that depends on the geometry of the confined body:

$$\tau_0 = \sigma\mu F \qquad (3.221)$$

where

$$F = b^2/q_1 \qquad (3.222)$$

This time constant characterizes the time at which late-stage behavior manifests itself. From the physical point of view, with an increase in the conductivity or the dimensions of the conductive body, it is understable that the induced currents will decay more slowly, and therefore, the late-stage behavior will begin at greater time.

It is clear that the time constant τ_0, which does not depend on the primary field, the position of the conductive body, or the position of the observation site, but is only a function of the properties of the conductive body, is a highly important parameter that contains information of value about the conductor. If measurements are carried out during the late stage,

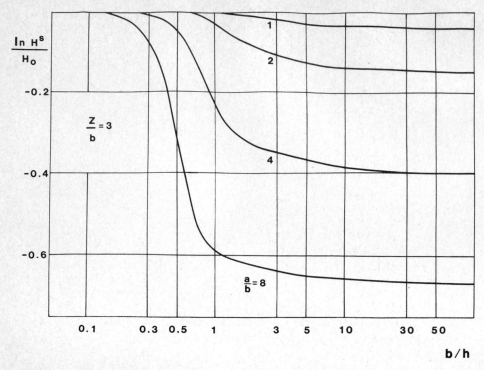

Fig. 3.29. Example of the spectral response for the inphase component of the vertical magnetic field along the z-axis for the case $z/b = 3$.

its definition is very simple. In fact, by taking a logarithm of the expression for H_z, we have:

$$\ln H_z(a, t) = \ln H_{0z} d_{1z}(a) - \frac{q_1 t}{\sigma \mu b^2} t = \ln H_{0z} d_{1z}(a) - \frac{t}{\tau_0} \qquad (3.223)$$

Thus, the slope of this function when the transient is plotted on a semi-logarithmic graph defines the time constant directly, as shown in Fig. 3.30. Moreover, the ordinate of the intersection of this straight line with the $t = 0$ axis contains information about the dimensions of the conductor and its location. Before we discuss the relationship between the late stage of time-domain behavior and the low frequency part of the frequency-domain behavior, let us consider an interesting feature of the equation for the field during the late stage. As follows from eq. 3.218, the expression for the magnetic field consists of two terms, which are:

(1) $H_{0z} d_{1i}(a)$, which is a function of the primary field and of geometry (the dimensions of the conductive body, its locations, and the location of the observation site), and

(2) An exponential term, e^{-t/τ_0} which depends on the measurement time,

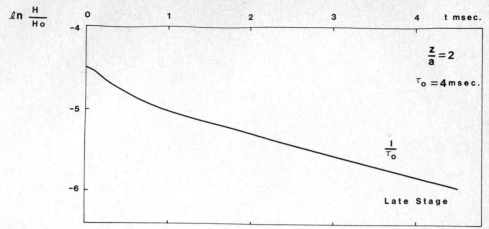

Fig. 3.30. Determination of the time constant from a semi-logarithmic plot of magnetic field strength as a function of time.

and the time contant τ_0. It is important to note that the last term does not depend on the strength of the primary field, or on the location of the conductive body or the coordinates of the observation site.

This particular structure for the equations markedly simplifies the interpretation of late-stage data, and due to the fact that the conductivity determines the power in the exponential term, the influence of geological noise can be subdued to a much greater extent in most cases in comparison with measurements of the quadrature or inphase components of the magnetic field. This fundamental concept will be considered in more detail later.

One comment is appropriate at this stage. Usually, the electromotive force is measured instead of the magnetic field. For example, the late stage EMF due to the vertical component H_z is:

$$\mathscr{E}(t) = -\mu H_{0z} d_{1z} d_{1z} M_R \frac{1}{\tau_0} e^{-t/\tau_0} \qquad (3.224)$$

where M_R is the moment of the receiver. It is obvious that the presence of the factor $1/\tau_0$ in front of the exponential term does not change the characteristic behavior during the late stage.

We can now discuss in more detail the relationship between the low frequency part of the spectrum and the late stage time domain behavior for the case in which eddy currents flow in a confined conductive body.

First of all, as has been shown previously, the functions d_{1i} and q_1 are precisely the same as those which comprise the first fractions in the equations for the spectral response (eq. 3.166). Moreover, let us remember that the parameter q_1 defines the minimum pole in the spectrum. Indeed, as follows from eq. 3.166, the first pole in the spectrum is situated on the imaginary ω axiz and is equal to:

$$\omega_1 = -1 \frac{q_1}{\sigma\mu b^2} = -\frac{i}{\tau_0} \qquad (3.225)$$

Its magnitude is equal to a quantity that is inversely proportional to time constant. Thus, the behavior of the field during the late stage of the time domain response is determined by the first pole in the spectral response. In order to emphasize this, we can write the eq. 3.218 for the horizontal component H_r as:

$$H_r(a, t) = H_{0z} d_{1r} e^{-|\omega_1|t} \qquad (3.226)$$

At the same time, the first pole defines the radius of convergence of the series in eqs. 3.177 to 3.179 which describe the low-frequency spectrum. Thus, the time constant τ_0 is inversely proportional to the radius of convergence for these series and therefore is related to the behavior of the coefficients in the series C_{li}. This phenomenon reflects the fundamental nature of the relationship between the low-frequency part of the spectrum and the late stage of time-domain behavior. But it does not mean that the quadrature and inphase components measured at low frequencies and the field observed at the late stage of time-domain behavior are related to the conductivity and geometry of the target body in the same manner. In fact, there is a fundamental difference between these functions. This can be illustrated as follows. A thorough understanding of this concept will assist us in investigating such important practical aspects as resolution, depth of investigation, and the effect of geological noise. For example, consider the vertical component of the magnetic field, H_z. First we will compare the late stage of transient-field behavior with the quadrature component at low frequencies. In accord with eqs. 3.183 and 3.218, we have:

$$QH_z \approx H_{0z} C_{1z} \sigma\mu\omega b^2 \approx H_{0z} C_{1z} q_1 \tau_0 \omega, \quad \text{if } \omega \to 0 \qquad (3.227)$$

$$H_z(t) \approx H_{0z} d_{1z} e^{-q_1 t/\sigma\mu b^2} = H_{0z} d_{1z} e^{-t/\tau_0}, \quad \text{if } \frac{t}{\tau_0} > 1$$

Of course, both field components are proportional to the primary field strengths, H_{0z}. As follows from an examination of eq. 3.187, there is a significant difference between the functions $C_{1z} q_1$ and d_{1z}, and therefore the influence of the geometry, including the shape of the conductive body, its dimensions, its location, and the location of the observation site all manifest themselves in a different way on the quadrature component than they do on the transient field during the late stage. However, there are cases when this dependence of the function $QH(\omega)$ and $H(t)$ is practically the same. For example, this occurs for conductive bodies which are not overly elongate, or for measurements made at points quite far from the conductive body. The greatest differences between these functions are observed when the dependence on time constant is considered, that is, the influence of the conductivity and the dimensions of the body. At low frequencies, the quadrature component QH_z, is directly proportional to the time constant, τ_0, regardless of how low the frequency is. At the same time, the influence of the parameter τ_0 on the field during late stage behavior is more profound and with increase in time, it becomes greater and greater. This is a consequence of the fact that the time constant, τ_0, controls the power of the exponential term, and the ratio t/τ_0 is greater than unity. This difference in sensitivity to a change in τ_0 is even apparent when the time constant varies only slightly. As an example, suppose that τ_0 is changed by an amount $\Delta\tau_0$ with $\Delta\tau_0/\tau_0$ being less than unity. It is clear that the quadrature component will vary exactly in direct proportion to the value of the ratio $\Delta\tau_0/\tau_0$, and for the transient behavior we will have:

$$\exp\left(-\frac{t}{\tau_0 + \Delta\tau_0}\right) = \exp\left(-\frac{t}{\tau_0}\frac{1}{1 + \Delta\tau_0/\tau_0}\right) \approx \exp\left(-\frac{t}{\tau_0}(1 - \Delta\tau_0/\tau_0)\right)$$

$$= \exp\left(-\frac{t}{\tau_0}\right)\exp\left(\frac{\Delta\tau_0}{\tau_0}\frac{t}{\tau_0}\right); \quad \text{if } \frac{\Delta\tau_0}{\tau_0} < 1$$

Moreover, assuming that

$$\frac{\Delta\tau_0}{\tau_0}\frac{t}{\tau_0} < 1$$

but still with t/τ_0 greater than unity, we obtain:

$$e^{-t/\tau_0 + \Delta\tau_0} = e^{-t/\tau_0}\left(1 + \frac{t}{\tau_0}\frac{\Delta\tau_0}{\tau_0}\right)$$

Inasmuch as the ratio t/τ_0 is greater unity during the late stage, we will observe a stronger change in the field than is the case when we measure the quadrature component in the frequency domain. As the time constant varies over a wider range of values, the difference between the behavior of the two field components becomes more important. For example, if τ_0 were to be changed by a factor of two, the quadrature component of the magnetic field will also change by a factor of two, but the transient field will change by almost an order of magnitude ($e^{\pm 2}$) as $t = \tau_0$. This example clearly demonstrates the fundamental fact which can be formulated to state that the low-frequency part of the spectrum representing the quadrature component in the frequency domain is considerably less sensitive to variations in the conductivity and dimensions of a conductive body than is the transient field during the late stage of time-domain behavior. It should be noted here that a variation in frequency, so long as one remains within the low-frequency part of the spectrum, does not change the relationship between the field and the time constant, that is, the conductivity or the dimensions of the conductive body, while an increase in time markedly increases the sensitivity of the transient response to a change in the time constant during the late-time behavior.

We can now compare the late stage of the transient field behavior with the behavior of the inphase component of the field, H_z, in the frequency domain. As follows from eqs. 3.183 and 3.218, we can write:

$$\text{In} H_z \approx -H_{0z} C_{2z} (\sigma \mu \omega b^2)^2 = -H_{0z} C_{2z} q_1^2 \tau_0^2 \omega^2$$

$$H_z(t) \approx H_{0z} d_{1z} e^{-q_1 t / \sigma \mu b^2} = H_{0z} d_{1z} e^{-t / \tau_0} \tag{3.228}$$

In accord with eq. 3.187, the difference between the functions $C_{2z} q_1^2$ and d_{1z} is smaller than in the case of the quadrature component. In other words, a dependence on geometrical factors for the inphase component and for the transient field is closer to each other than in measuring the quadrature component.

According to eq. 3.228, in contrast to the behavior of the quadrature component, the inphase component is more sensitive to a change in the time constant, τ_0. In fact, in place of the first power of τ_0 in the quadrature component, we have the factor τ_0^2, that is, the inphase component is relatively more sensitive to a change in the conductivity and dimensions of a conductive body than is the quadrature component. As will be demonstrated later, this has led to field techniques in which the inphase component is the preferred measurement in order to reduce geological noise and therefore to increase the depth of investigation when frequency-domain measurements are being made. However, returning to the comparison of the behavior of the inphase component with that of the transient field during the late stage, we can see that the transient measurement has a higher level of sensitivity to a change in time constant. Moreover, with increase in time the effect of the parameter τ_0, that is, of the conductivity and dimensions of the conductive body, becomes even stronger, whereas its effect on the inphase component or on the quadrature component does not depend on the frequency being used.

Use of the inphase component rather than of the quadrature component of the magnetic field in the frequency domain does increase the influence of conductivity on measurements, but even so, the relationship is not so strong as holds during the late stage of transient behavior. We are now prepared to compare the late-stage behavior with higher order terms in this series representation for the low-frequency spectrum. First, in accord with eq. 3.187 through 3.189 we can see that with an increase in the order of the term l, the equations:

$$C_{lz} = \frac{d_{1z}}{q_1^l}, \quad C_{lr} = \frac{d_{1r}}{q_1^l} \tag{3.229}$$

can be applied with progressively higher precision. In other words, the dependence on geometry during late stage and the high-order terms in these series is practically the same. Taking eq. 3.229 into account, for each of the terms in the series 3.181 or 3.182, when l is large, one can write the equation in the following manner:

$$\pm C_{li}(\sigma\mu b^2)^l \omega^l = \cong \pm d_{1i}\left(\frac{\sigma\mu b^2}{q_1}\right)^l \omega^l \cong \pm d_{1i}\tau_0^l \omega^l \tag{3.230}$$

We can say that with increasing order of the term, l, as well as with increase in time, the sensitivity of the field to the time constant becomes progressively greater. This clearly demonstrates the close relationship between the behavior of the field during the late stage of transient response and the part of the low-frequency field in the frequency domain which is represented by the high-order terms in the series 3.181 and 3.182. Therefore, even completely accepting the validity of the Fourier transform theory, one can say that all the information obtainable from late-stage behavior is contained also in the low-frequency spectrum. However, this information is contained primarily in the high-order terms of the series, and therefore, their contribution to the value of the quadrature and inphase components is relatively small. This analysis has been carried out in such a way as to demonstrate once again that the leading terms in the series representing the quadrature and inphase components which define the main part of the signal in the frequency domain are related to the properties of the conductive body in quite a different way than is the transient field during the late stage of time-domain behavior.

Considering further the relationship between the low-frequency part of spectrum and the late-stage behavior, let us consider one additional factor concerning the magnitude of the signals in both cases. First of all, we can consider the leading terms in the low-frequency spectrum. For simplicity and with some degree of approximation, let us assume that for all of the terms in eq. 3.189 that the condition:

$$C_{li} \approx d_{1i}/q_1^l$$

is valid. Then, the ratio of the quadrature component of the field to the field strength during the late-stage transient behavior is:

$$\frac{QH_i(\omega)}{H_i(t)} \approx \frac{C_{1i}H_0\sigma\mu\omega b^2}{C_{1i}q_1 e^{-t/\tau_0}H_0} = \omega\tau_0 e^{t/\tau_0} \tag{3.231}$$

In accord with eq. 3.225, which defines the upper limit of frequencies corresponding to the low-frequency spectrum, the product $\omega\tau_0$ must satisfy the condition:

$$\omega\tau_0 \leqslant 1 \tag{3.232}$$

However, in order to derive the time constant, it is sufficient to place an upper bound on the measurement of time based on the condition:

$t/\tau_0 < 2$

Correspondingly, a reasonable range of times for our consideration will be:

$$\tau_0 < t < 2\tau_0 \tag{3.233}$$

For example, if the frequency were such that $\omega\tau_0$ approximately equals one-half, the ratio in eq. 3.231 changes from a value of approximately unity to approximately 5. In accord with eq. 3.228, we have the following expression for the ratio of the strength of the inphase component of the magnetic field to that during the late stage of transient behavior:

$$\frac{\mathrm{In}\,H_i(\omega)}{H_i(t)} \cong \frac{C_{2i}H_0}{d_{1i}H_0} \frac{(\sigma\mu\omega b^2)^2}{\mathrm{e}^{-t/\tau_0}} = (\omega\tau_0)^2\,\mathrm{e}^{t/\tau_0} \tag{3.234}$$

This ratio will change from approximately 0.5 to 2.5 providing that condition 3.233 is met and that $\omega\tau_0$ is approximately one-half. It is clear that under the same conditions, this ratio for the next following terms in these series will be smaller.

These two numerical examples illustrate to a limited extent the relationship between the signal strengths which will be observed at low frequencies in the frequency domain or during the late stage in the time domain. However, one must note that in practice the transient field is commonly stacked, with signals from many successive transmissions being added together in order to increase the signal to noise ratio.

We have considered the late-stage behavior in a time domain and its relationship with the low-frequency part of the spectrum in the frequency domain in detail. Next, we should briefly consider the contrary case, that is, the early stage behavior of the transient response. First, at the instant when current in the source is turned off, induced currents will appear only on the surface of any conductive bodies. This surface distribution is defined by a single condition, namely, at the instant at which the magnetic field is switched off, the magnetic field inside the conductor must be equal to the primary field, H_0. As a consequence, at the switching instant, the transient response does not depend on the time constant. In fact, in accord with eq. 3.217 we have the following expression for any of the components of the magnetic field, H_i:

$$H_i = H_0 \sum_{n=1}^{\infty} d_{ni} \tag{3.235}$$

It is clear that from a physical point of view that this series must converge. Also, it is clear that eq. 3.235 defines the magnitude of the field at the high frequency limit in the frequency domain, since in both cases, induced currents are absent within the conductor, flowing only on its surface. This relationship can be written as the Fourier limit theorem:

$$\lim_{\omega \to \infty} H(\omega) \; = \; \lim_{t \to 0} H(t) \tag{3.236}$$

Thus, there is an particularly simple relationship between the high frequency part of the spectrum and the initial stage of the transient field, a condition which is quite unlike the relationship between the low-frequency spectrum and the late stage behavior.

In accord with eq. 3.236, one way of determining the strength of the transient field at time zero is by solution of the boundary problem for a perfectly conducting body, with the time variation of the field being in the form of a sinusoid. Usually, this problem is much simpler than the more general problem, and it can be solved either analytically or numerically.

Next let us examine the way in which the transient response departs from its limit at the early stage. In order to derive the appropriate formulas we will make use of our understanding of the frequency-domain behavior of the field at high frequencies. As was pointed out above, when at high frequencies the skin depth is less than the radius of curvature of the surface of the conductive body, the ratio of tangential components of the electric and magnetic fields which are mutually perpendicular is equal to the impedance of a plane wave on this surface, Z,

$$Z \; = \; (i\omega\mu/\sigma)^{1/2}$$

This approximate boundary condition leads to the result that each component of the field can be considered as a function of $\omega^{1/2}$, and can be written in the following manner:

$$\mathrm{In}\, H_i \; \approx \; \left[A_i - \frac{B_i}{\sqrt{\omega}} \right] H_0, \quad QH_i \; = \; \frac{C_i}{\sqrt{\omega}} \, H_0 \tag{3.237}$$

where A_i, B_i, and C_i are coefficients that depend on the geometry and the conductivity of the conductive body. This behavior has been clearly demonstrated earlier in this chapter when we considered the high frequency spectrum for currents induced in this spheroid. The product $A_i H_0$ is the field for the case in which the body is perfectly conducting.

Let us find an asymptotic expression for the function $\partial H_i(t)/\partial t$. In accord with eq. 2.338, we have:

$$\frac{\partial H_i(t)}{\partial t} \; = \; -\frac{2}{\pi} \int_0^\infty QH_i(\omega) \sin \omega t \, d\omega \; = \; -\frac{2}{\pi} C_i H_0 \int_0^\infty \frac{\sin \omega t}{\sqrt{\omega}} \, d\omega, \quad \text{as } t \to 0$$

Inasmuch as:

$$\int_0^\infty \frac{\sin \omega t}{\sqrt{\omega}} \, d\omega \; = \; \frac{1}{\sqrt{t}} \int_0^\infty \frac{\sin \omega t}{\sqrt{\omega t}} \, d\omega t \; = \; \frac{1}{\sqrt{t}} \int_0^\infty \frac{\sin x}{\sqrt{x}} \, dx \; = \; \left(\frac{\pi}{2t} \right)^{1/2}$$

232

we obtain:

$$\dot{H}_i(t) = -\left(\frac{2}{\pi}\right)^{1/2} C_i H_0 \frac{1}{t^{1/2}} \tag{3.238}$$

Integrating this last equation with respect to time, we find

$$H_i(t) = H_i(0) - \frac{2^{3/2}}{\pi^{1/2}} C_i H_0 t^{1/2}$$

or

$$H_i(t) \approx \left(A_i - \frac{2^{3/2}}{\pi^{1/2}} C_i t^{1/2}\right) H_0 \tag{3.239}$$

It is possible to show that $B_i = C_i$ by making use of another form of the Fourier transform.

Thus, in accord with eqs. 3.238 and 3.239, the magnetic field departs from its asymptotic value at zero time relatively slowly, varying as $t^{1/2}$, and it depends on the properties of a conductor in the same manner as the field in the frequency domain at high frequencies. However, the derivatives of the magnetic field with respect to time increases without limit as time decreases. Correspondingly, this same behavior is characteristic for the EMF observed during the early stage.

Up to this point, we have examined in detail both the late and early stages of transient behavior. Now, for examples, let us study the total transient response assuming that the conductive body has the form of a spheroid and that it is illuminated by a uniform magnetic field, H_0 which has the form of a step function. In the quasi-stationary approximation, the primary field propagates at infinite velocity through an insulator interacting simultaneously at all points on the surface of a conductive body. As has been pointed out earlier, at time zero, the surface currents create a field inside the conductor which is equal to the primary field:

$$H^i = H_0, \quad \text{as } t = 0 \tag{3.240}$$

This provides an initial condition. Then, the process of diffusion begins, and for sufficiently large times, the intensity of the currents and of the fields decreases exponentially.

Determination of the behavior of the nonstationary field, based on a solution of the diffusion equation:

$$\nabla^2 E = \frac{1}{\sigma\mu} \frac{\partial E}{\partial t}$$

by making use of the method of separation of variables is a much more complicated problem than is the solution for the case of a sinusoidal field.

Kaufman II Chapter 3 page 94

This is a consequence of the fact that it is necessary not only to assure continuity of tangential components of the field on the surface of the spheroid, but the initial condition must be satisfied as well. One can show that the most difficult part of this problem is the determination of the values for q_n from the boundary conditions, because this requires the solution of an infinite system of equations with an infinite number of unknowns. Therefore, we have calculated the transient responses by making use of the Fourier transform:

$$H_i(t) = \frac{2}{\pi} \int\limits_0^\infty \frac{QH_i(\omega)}{\omega} \cos \omega t \, d\omega, \quad \dot{H}_i = -\frac{2}{\pi} \int\limits_0^\infty QH_i \sin \omega t \, d\omega$$

Similar relationships can be written with the inphase component of the magnetic field. In addition, some use has been made of asymptotic expressions for the late stage (see eq. 3.218) which are of the form

$$H_i \approx H_0 d_{1i} e^{-q_1 \alpha t}, \quad \mathscr{E} \approx - \Phi_0 d_{1i} q_1 \alpha e^{-q_1 \alpha t}$$

where $\alpha = 1/\sigma \mu b^2$ and where Φ_0 is the magnetic flux representing the coupling of the primary field through the coil used as a receiver.

The transient responses of the magnetic field on the z-axis and values for the functions $\mathscr{E}(t)/\Phi_0 \alpha$ are shown graphically in Figs. 3.31—3.35. A dimensionless time τ/b, where $\tau = (2\pi\rho t \times 10^7)^{1/2}$ is used as the abscissa. During the initial stage, the curves for the magnetic field depart from a horizontal asymptote, the value of which is the inphase component at very high frequencies as given in eq. 3.236. With an increase in the length of the horizontal semi-axis of the conductive body, there is coincidence between this asymptote and the curves at progressively longer times. Inasmuch as the primary magnetic field switches instantaneously, the electromotive force at the initial instant tends to be infinite, and decreases with time, but in contrast to the behavior of the magnetic field component, during the early stage, the electromotive force is less than for the more elongate spheroids. During the late stage, the electromotive force is a function of the ratio of axial lengths in the same manner as the magnetic field, inasmuch as with an increase in the length of the horizontal axis, which is perpendicular to the primary field excitation, the currents in the conductive body decay progressively more slowly with the parameter q_1 decreasing.

The minimum values for the parameter τ/b which are the initial limit for which one can make use of the late-stage formulas with an error of no more than 5—10% are given in Table 3.VIII.

These values permit us to write an approximate condition that defines the range of late stage behavior for both magnetic field and the electromotive force:

$$\frac{\tau}{a} \geqslant 1.2 \quad \text{and} \quad \frac{\tau}{a} \geqslant 2, \quad \text{if} \frac{a}{b} \geqslant 1 \tag{3.241}$$

234

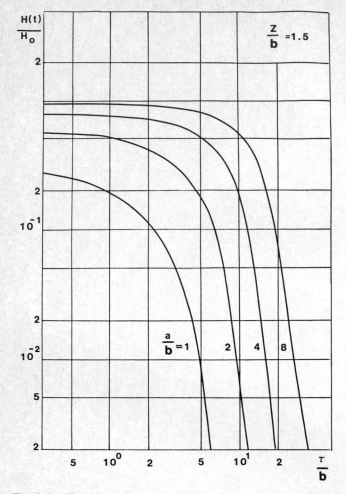

Fig. 3.31. Transient response of the magnetic field along the z-axis.

TABLE 3.VIII

	a/b						
	1	2	4	8	16	32	64
H	1.3	2.0	5.6	12	25	35	50
$\mathscr{E}$	2.5	3.0	11	18	30	40	60

respectively, where τ is the scale time as defined earlier.

In conclusion the following comments are appropriate:

(1) Regardless of the fact that the primary field can be nonuniform in any manner, the late stage will be described accurately by eq. 3.218, where H_0 is

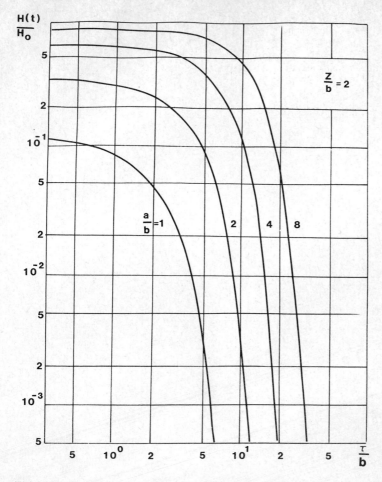

Fig. 3.32. Transient response of the magnetic field along the z-axis for the case $z/b = 3$.

that part of the primary field which is uniform in the vicinity of the conductive body.

(2) When the ratio of axial lengths for the spheroid, a/b is large enough, the exponential term describing the late stage behavior can be written as:

$$e^{-q_1 \frac{t}{\sigma \mu b^2}} = e^{-7.70 \frac{b}{2a} \frac{t}{\sigma \mu b^2}} = e^{-7.70 \frac{t}{\mu S a}} \tag{3.242}$$

where S is the conductance of the spheroid at its center ($S = 2\sigma b$). Thus, the time constant:

$$\tau_0 = \frac{\mu S a}{7.70} \tag{3.243}$$

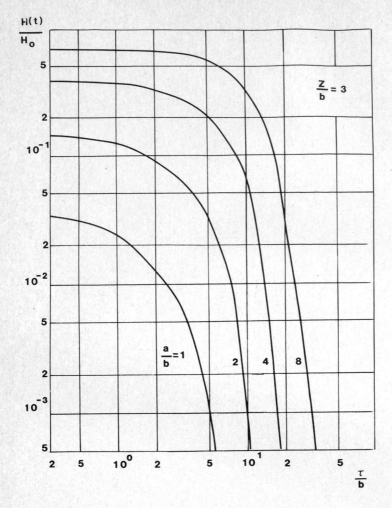

Fig. 3.33. Transient response for the magnetic field along the z-axis for the case $z/b = 3$.

is defined as the product of the conductance, S, and the semiaxial length, a, in a direction perpendicular to the primary field excitation. From eq. 3.199, if the spheroid were to be replaced by a disc with constant thickness, the time constant will be:

$$\tau_0 = \frac{\mu Sa}{5.51} \qquad\qquad (3.244)$$

The ratio of the time constant for the spheroid to that for a disc is shown as a function of a/b in Fig. 3.36.

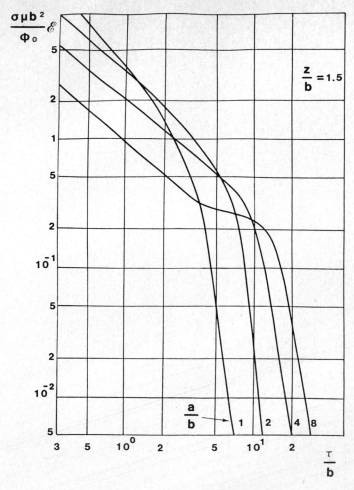

Fig. 3.34. Transient response for the electromotive force along the z-axis for the case $z/b = 1.5$.

3.6. A RIGHT CIRCULAR CYLINDER IN A UNIFORM MAGNETIC FIELD

In this case, let us assume that a conductive body consists of a markedly elongate body in the direction of the primary electrical field. Electrical charges will arise on the edges and lateral surface near the edges, with the electric field from these being zero for the normal component of current density on the surface of a conductive body which is embedded in an otherwise insulating medium. Vortex currents will form closed loops within the conductor and are almost parallel to a generating surface.

From a physical point of view, it is clear that with an increase in the

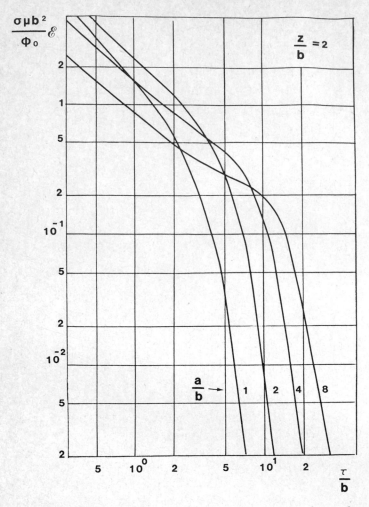

Fig. 3.35. Transient response for the electromotive force along the z-axis for the case $z/b = 2$.

length of the conductor in the direction of the primary electrical field, E_0, in comparison to the cross-section of the dimensions and the distance from the body to an observation point, the influence from the edge effect will decrease progressively. The contribution to the total magnetic field from currents which are not parallel to the generating surface for the conductor become smaller and smaller. Therefore, under certain conditions, the electromagnetic field from the currents in the confined conductive body is almost exactly the same as that from currents in an infinitely long cylinder. A similar conclusion can apply when the primary electromagnetic field vanishes near the edges of an elongate conductive body. We will investigate this problem in a later

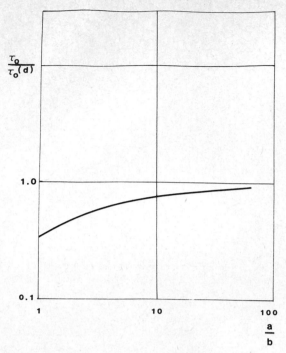

Fig. 3.36. The ratio of the time constant for a spheroid to that for a disc as a function a/b.

section in more detail, but at this point, we will start with a description of the behavior of the field caused in a cylindrical conductive body.

Assume that a right circular conductive cylinder is situated in a uniform magnetic field $H_{0z}e^{-i\omega t}$ which is directed perpendicularly to the axis of the cylinder. Also assume that the primary field is caused by two systems of linear current filaments, which are oriented parallel to the axis of the conductor, but with the currents flowing in opposite directions as shown in Fig. 3.37. In this case, the primary electric field has only a single component, E_{0x}, and by making use of Maxwell's first equation in its integral form:

$$\oint E \cdot dl = -\frac{\partial \Phi}{\partial t}$$

we find the following for an arbitrary path, L, lying in the plane xy as shown in Fig. 3.37:

$$E_{0x} = -i\omega\mu H_{0z} r \sin\phi \qquad (3.245)$$

where

$$\sin\phi = y/r \quad \text{and} \quad r = (z^2 + y^2)^{1/2}$$

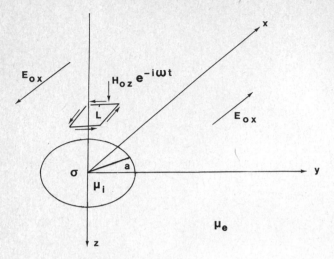

Fig. 3.37. A primary field caused by two systems of linear current filaments which are oriented parllel to the axis of a conductor but with currents flowing in opposite directions.

We see from this equation that the primary electric field is directed oppositely on either sides of the plane $z0x$. This field causes vortex currents to flow in the conductor, with the current filaments being oriented parallel to the axis and closing on themselves at infinite. The secondary field can be represented as being the sum of fields caused by the linear source currents, and in particular, in a cylindrical system of coordinates, the secondary field can be described completely by the three components H_r, H_ϕ, E_x.

First let us find the vortex electric field, as shown earlier, the electric field inside the cylinder will be described by the equation:

$$\nabla^2 E_x^i + k^2 E_x^i = 0 \qquad (3.246)$$

where k is the wave number as previously defined. Outside the cylinder, the electric field satisfies the equation:

$$\nabla^2 E_x^e = 0 \qquad (3.247)$$

Both the fields E_x^e and E_x^i consist of two parts, a primary electric field, E_{0x}, and a vortex secondary field, contributed by the time rate of change of the secondary magnetic field. In accord with appropriate boundary conditions, we must insist that the tangential components are continuous at the surface of the cylinder:

$$E_x^i = E_x^e, \quad H_\phi^i = H_\phi^e$$

Making use of Maxwell's equation:

Making use of Maxwell's equation:

$$\text{curl } E = -\frac{\partial B}{\partial t}$$

this condition can be written as:

$$E_x^i = E_x^e, \quad \frac{1}{\mu_e}\frac{\partial E_x^e}{\partial r} = \frac{1}{\mu_i}\frac{\partial E_x^i}{\partial r} \tag{3.248}$$

Inasmuch as the field does not depend on the coordinate x, eqs. 3.246 and 3.247 can be rewritten as:

$$\frac{\partial^2 E_x^i}{\partial r^2} + \frac{1}{r}\frac{\partial E_x^i}{\partial r} + \frac{1}{r^2}\frac{\partial^2 E_x^i}{\partial \phi^2} + k^2 E_\phi^i = 0 \tag{3.249}$$

and

$$\frac{\partial^2 E_x^e}{\partial r^2} + \frac{1}{r}\frac{\partial E_x^e}{\partial r} + \frac{1}{r^2}\frac{\partial^2 E_x^e}{\partial \phi^2} = 0 \tag{3.250}$$

We will seek a solution for these equations in the form of the product of two functions $R(r)\Phi(\phi)$. Substituting this assumed solution into eq. 3.249 and multiplying both sides by $r^2/R\Phi$, we obtain:

$$\frac{r^2}{R}R'' + \frac{r}{R}R' + \frac{\Phi''}{\Phi} + k^2 r^2 = 0$$

This equation can easily be separated into two ordinary differential equations:

$$\frac{d^2 R}{dr^2} + \frac{1}{r}\frac{dR}{dr} - \left[(ik)^2 + \frac{n^2}{r^2}\right]R = 0 \tag{3.251}$$

and

$$\frac{d^2 \Phi}{d\phi^2} + n^2 \Phi = 0 \tag{3.252}$$

The second of these two equations is a very common one, being the equation for a simple harmonic oscillator, which has the well known solution:

$$\Phi_n(\phi) = A_n^1 \cos n\phi + B_n^1 \sin n\phi \tag{3.253}$$

Equation 3.251 is a form of the Bessel equation, for which the solutions are modified Bessel functions:

$$R_n^i = C_n^1 I_n(ikr) + D_n^1 K_n(ikr) \tag{3.254}$$

In the area outside the cylinder, the form for eq. 3.251 is much simpler:

$$\frac{d^2R}{dr^2} + \frac{1}{r}\frac{dR}{dr} - \frac{n^2}{r^2}R = 0 \tag{3.255}$$

Particular solutions to this equation are:

$$R_n^e = C_n''r^n + D_n''r^{-n}, \quad \text{if } n \neq 0$$

and $\hspace{10cm}$ (3.256)

$$R_0^e = C_0''\ln r + D_0'', \quad \text{if } n = 0$$

Inasmuch as the magnetic field from induced currents in a cylinder observed at distances that are considerably greater than the radius of the cylinder is equivalent to the field from a simple dipole, that is, $H = 1/r^2$ as $r \to \infty$, the secondary electric field will tend to zero as a limit as $1/r$ with increase in r. Thus, we can write the expression for the electric field as follows:

$$E_x^e = -i\omega\mu_e H_0 r \sin\phi - i\omega\mu_e H_0 \sum_{n=1}^{\infty} r^{-n}(B_n \cos n\phi + T_n \sin n\phi) \tag{3.257}$$

$$E_x^i = -i\omega\mu_e H_0 \sum_{n=1}^{\infty} I_n(ikr)(D_n \cos n\phi + C_n \sin n\phi)$$

The Bessel function $K_n(ikr)$ cannot be used to describe the field inside the cylinder because:

$$K_n(ikr) \to \infty, \quad \text{as } r \to 0$$

In determining the as yet unknown coefficients in the solution, we apply the boundary conditions of eq. 3.250, and making use of the orthogonality property of the trigonometric functions, we obtain a system of equations for each harmonic:

$$a + a^{-1}T_1 = I_1(ika)C_1$$
$$\hspace{4cm} \text{if } n = 1 \hspace{3cm} (3.258)$$
$$\frac{1}{\mu_e}(1 - a^{-2}T_1) = \frac{1}{\mu_i}ikI_1'(ika)C_1,$$

and

$$a^{-n}T_n - I_n(ika)C_n = 0$$
$$\hspace{4cm} \text{if } n \neq 1 \hspace{3cm} (3.259)$$
$$\frac{n}{\mu_e}a^{-n-1}T_n + \frac{1}{\mu_i}ikI_n'(ika)C_n = 0,$$

The determinant for the uniform system of equations of the form 3.259 is nonzero, and therefore $T_n = C_n = 0$ providing $n \neq 1$. In the same way, we can demonstrate that all the coefficients B_n and D_n are zero. Thus, the field is completely described by a single cylindrical harmonic, $n = 1$. Solving the system of eq. 3.258 and making use of the recurrence relationships for modified Bessel functions which are as follows:

$$I_n'(x) = I_{n-1}(x) - \frac{n}{x} I_n(x)$$

$$\frac{2n}{x} I_n(x) = I_{n-1}(x) - I_{n+1}(x)$$

after some algebraic operations we arrive at the result:

$$T_1 = a^2 \frac{(\mu_i - \mu_e)I_0(ika) - (\mu_i + \mu_e)I_2(ika)}{(\mu_i + \mu_e)I_0(ika) - (\mu_i - \mu_e)I_2(ika)} \tag{3.260}$$

and

$$C_1 = \frac{4\mu_i}{ik[(\mu_i + \mu_e)I_0 - (\mu_i - \mu_e)I_2]} \tag{3.261}$$

From eq. 3.257, we see that the electromagnetic field from currents induced in a right circular cylinder is the same as that generated by a simple dipole situated in free space:

$$E_{1x}^e = -i\omega\mu_e H_0 T_1 \frac{\sin\phi}{r}$$

$$H_{1r}^e = -H_0 T_1 \frac{\cos\phi}{r^2} \tag{3.262}$$

$$H_{1\phi}^e = -H_0 T_1 \frac{\sin\phi}{r^2}$$

These equations describe a secondary field which is characterized by unusually simple geometrical structure. In particular, the current density in the half planes $0 < \phi < \pi$ and $\pi < \phi < 2\pi$ has opposite directions.

The function T_1 is particularly simple when the magnetic permeability in both media has the same value. Then:

$$T_1 = Ta^2 = -\frac{I_2(ika)}{I_0(ika)} a^2, \quad \text{if } \mu_i = \mu_e \tag{3.263}$$

The function $I_0(ika)$ has simple roots; that is, the singularities of the spectrum for the complex amplitude of T are simple poles situated on the imaginary ω axis, and the first of them defines the radius of convergence for the low frequency series.

For the low-frequency case when the magnitude of ka is less than unity, replacing the Bessel functions by the first few terms in their series expansion, we have:

$$T \approx i \frac{\sigma\mu\omega a^2}{8} - \frac{1}{48}(\sigma\mu\omega a^2)^2 + \dots \tag{3.264}$$

The spectrums for the imaginary and real parts of the function T are shown in Figs. 3.38 and 3.39 for various ratios μ_i/μ_e. The behavior is similar to that

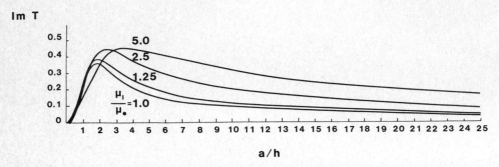

Fig. 3.38. Spectrum for the imaginary part of the function T for various ratios μ_i/μ_e.

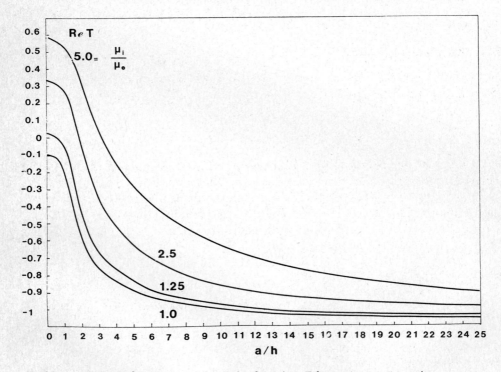

Fig. 3.39. Spectrum for the real part of the function T for various ratios μ_i/μ_e.

for the cases of conductive bodies characterized by axial symmetry in that an increase in frequency causes the quadrature component to increase, to pass through a maximum, and then to tend to zero in the limit. However, the inphase component increases monotonically, and approaches an asymptotic value corresponding to that for a perfectly conducting body. Expanding the Bessel functions in an asymptotic form:

$$I_n(x) \approx \frac{e^x}{(2\pi x)^{1/2}} \left[I - \frac{4n^2 - 1}{1!8x} + \frac{(4n^2 - 1)(4n^2 - 3^2)}{2!(8x)^2} - \cdots \right] \qquad (3.265)$$

and keeping only terms up to the second order, we have:

$$T \approx -1 + \frac{2}{ika}, \quad \text{if } |ka| \gg 1$$

Both components, the quadrature and inphase, approach their limit in inverse proportion to $\omega^{1/2}$.

Values for the function T which controls the spectral response of the field are listed in Table 3.IX. According to eq. 3.262, and by applying the Fourier transform, one can affirm that the time-domain behavior is also completely described by a single cylindrical harmonic of order $n = 1$.

In studying the transient field, we will make use of a vector potential of the electrical type:

$$B = \text{curl } A \qquad (3.266)$$

As follows from Maxwell's equation, the electric field is derivable from the vector potential A as follows:

$$E = -\frac{\partial A}{\partial t} \qquad (3.267)$$

The vector A has only a single component, A_x, which satisfies the diffusion equation:

$$\nabla^2 A_x^i = \sigma\mu \frac{\partial A_x}{\partial t} \qquad (3.268)$$

inside the cylinder, the Laplace's equation outside the cylinder:

$$\nabla^2 A_x^e = 0 \qquad (3.269)$$

Using the method of separation of variables as has been done previously, the component A_x can be written as:

$$A_x^i = \sin\phi \sum_{s=1}^{\infty} C_s J_1(k_s r) e^{-q_s \alpha t} \qquad (3.270)$$

TABLE 3.IX

| a/h | μ_i/μ_e | | | | | | | |
| | 1.0 | | 1.25 | | 2.5 | | 5.0 | |
	ImT	ReT	ImT	ReT	ImT	ReT	ImT	ReT
0.100	0.250×10^{-2}	-0.833×10^{-5}	0.247×10^{-2}	0.111	0.204×10^{-2}	0.429	0.139×10^{-2}	0.667
0.141	0.500×10^{-2}	-0.333×10^{-4}	0.494×10^{-2}	0.111	0.408×10^{-2}	0.428	0.278×10^{-2}	0.667
0.200	0.100×10^{-1}	-0.133×10^{-3}	0.987×10^{-2}	0.111	0.816×10^{-2}	0.428	0.555×10^{-2}	0.667
0.283	0.200×10^{-1}	-0.533×10^{-3}	0.197×10^{-1}	0.111	0.163×10^{-1}	0.428	0.111×10^{-1}	0.666
0.400	0.399×10^{-1}	-0.213×10^{-2}	0.394×10^{-1}	0.109	0.326×10^{-1}	0.427	0.222×10^{-1}	0.666
0.566	0.791×10^{-1}	-0.843×10^{-2}	0.782×10^{-1}	0.103	0.649×10^{-1}	0.424	0.443×10^{-1}	0.664
0.800	0.153	-0.325×10^{-1}	0.152	0.814×10^{-1}	0.128×10^{-1}	0.410	0.878×10^{-1}	0.657
1.13	0.270	-0.114	0.273	0.504×10^{-2}	0.240	0.360	0.169	0.631
1.60	0.370	-0.308	0.391	-0.187	0.388	0.210	0.297	0.544
2.26	0.347	-0.539	0.387	-0.439	0.461	-0.543×10^{-2}	0.415	0.349
3.20	0.263	-0.686	0.304	-0.613	0.420	-0.298	0.449	0.109
4.52	0.196	-0.778	0.232	-0.725	0.354	-0.480	0.438	-0.113
6.40	0.144	-0.843	0.173	-0.805	0.285	-0.622	0.398	-0.317
9.05	0.104	-0.889	0.127	-0.862	0.221	-0.729	0.340	-0.489
12.8	0.750×10^{-1}	-0.922	0.920×10^{-1}	-0.902	0.167	-0.806	0.276	-0.626
18.1	0.537×10^{-1}	-0.945	0.662×10^{-1}	-0.931	0.124	-0.862	0.216	-0.730
25.6	0.383×10^{-1}	-0.960	0.474×10^{-1}	-0.951	0.902×10^{-1}	-0.903	0.164	-0.807

$$A_x^e = \frac{\sin \phi}{r} \sum_{s=1}^{\infty} T_s e^{-q_s \alpha t} \qquad (3.270)$$

where

$$k_s^2 a^2 = q_s, \quad \alpha = 1/\sigma \mu a^2 \qquad (3.271)$$

Using the Maxwell equation:

$$B = \text{curl } A$$

we obtain the following expression for the vector potential characterizing the primary field:

$$A_{0x} = -B_0 r \sin \phi \qquad (3.272)$$

Assuming a step-function form of excitation, the initial condition at $t = 0$ is:

$$B_0 r = - \sum_{s=1}^{\infty} C_s J_1(k_s r), \quad \text{if } r < a \qquad (3.273)$$

Making use of the boundary conditions at the surface of the cylinder, we obtain a uniform system of equations:

$$aC_s J_1(k_s a) = T_s$$
$$a^2 k_s C_s J_1'(k_s a) = -T_s \qquad (3.274)$$

which has a non-zero solution for C_s and T_s only if the determinant for system is zero, that is:

$$J_1(k_s a) + k_s a J_1'(k_s a) = 0$$

Using the recurrence relationship for Bessel functions, this equation can also be written as:

$$J_0(k_s a) = 0 \qquad (3.275)$$

Therefore, continuity of the tangential components of the electric and magnetic field in the time domain on the surface of a cylinder is satisfied providing that the values for $k_s a$ are roots of the Bessel function order zero. The first five values for these roots are listed in Table 3.X.

TABLE 3.X

s				
1	2	3	4	5
$k_s a$ 2.40483	5.52008	8.65373	11.79153	14.93092
q_s 5.78321	30.47128	74.88704	139.04018	222.93237

From the initial condition expressed in eq. 3.273, making use of formulas for the determination of the coefficients of the expansion of the function $B_0 r$ in a series in terms of the function $J_1(k_s a)$ we find (Smythe, 1950):

$$C_s = -\frac{2B_0}{k_s} \frac{J_2(k_s a)}{[J_1(k_s a)]^2}, \quad T_s = -\frac{4B_0 a^2}{(k_s a)^2} \tag{3.276}$$

Whence, for the transient field outside the cylinder, we obtain the following expressions:

$$E_x^e = -\mu H_0 \frac{4a^2}{r} \alpha \sin \phi \sum_{s=1}^{\infty} e^{-q_s \alpha t}$$

$$H_r^e = -\frac{H_0 a^2}{r^2} 4 \cos \phi \sum_{s=1}^{\infty} \frac{1}{q_s} e^{-q_s \alpha t} \tag{3.277}$$

$$H_\phi^e = -\frac{H_0 a^2}{r^2} 4 \sin \phi \sum_{s=1}^{\infty} \frac{1}{q_s} e^{-q_s \alpha t}$$

From these expressions it is clear that during a very early stage of the time domain behavior of the magnetic field, there is no dependence on the parameter α and we have:

$$H_r^e = -4H_0 \frac{a^2}{r^2} \cos \phi \sum_{s=1}^{\infty} \frac{1}{q_s}$$

$$H_\phi^e = -4H_0 \frac{a^2}{r^2} \sin \phi \sum_{s=1}^{\infty} \frac{1}{q_s} \tag{3.278}$$

At the same time the electromotive force tends to be infinite. Applying the Fourier transform to the asymptotic expression of the function T as given earlier on page 245 we can see again that with decrease in time, the electromotive force increases in inverse proportion to the square root of time, while the magnetic field approaches the asymptote in eq. 3.278 as a function which is proportional to $t^{1/2}$. At the late stage, the field is described by a single exponential term, just as was the case for conductive bodies with axial symmetry. In fact, in accord with eq. 3.277, we have:

$$E_x^e = -4\mu H_0 \frac{a^2}{r} \alpha e^{-q_1 \alpha t} \sin \phi$$

$$H_r^e = -4H_0 \frac{a^2}{r^2} \frac{1}{q_1} e^{-q_1 \alpha t} \cos \phi \tag{3.279}$$

$$H_\phi^e = -4H_0 \frac{a^2}{r^2} \frac{1}{q_1} e^{-q_1 \alpha t} \sin \phi, \quad \text{if } t/\tau_0 > 1$$

where τ_0 is a time constant given by:

$$\tau_0 = \sigma \mu a^2 / q_1$$

and where q_1 is approximately 5.78. In comparison with the late stage of the field caused by currents induced in a spherical body, we can see that if $\alpha_{sph} = \alpha_c$ the field generated by linear current filaments decays more slowly than that caused by circular current filaments.

The functions:

$$L_1 = 4 \sum_{s=1}^{\infty} \frac{1}{q_s} e^{-q_s \alpha t} \quad \text{and} \quad L_2 = \frac{4}{\tau_0 q_1} \sum_{s=1}^{\infty} e^{-q_s \alpha t} \tag{3.280}$$

which describe the transient response of magnetic field and the electromotive force respectively are shown graphically in Fig. 3.40. Values for these functions are listed in Table 3.XI.

In conclusion, it is necessary to make one particular comment. As follows from eq. 3.264, the ratio of the inphase component to the quadrature component at the low-frequency part of the spectrum defines the position of

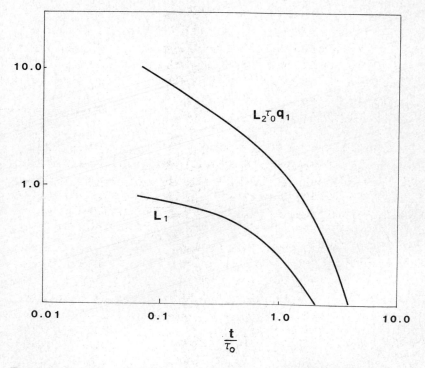

Fig. 3.40. Behavior of the functions L_1 and L_2 which describe the transient response of the magnetic field and the electromotive force, respectively.

TABLE 3.XI

t/τ_0	L_1	$L_2\tau_0 q_1$
0.625×10^{-1}	0.776	0.967×10^1
0.884×10^{-1}	0.737	0.806×10^1
0.125	0.690	0.662×10^1
0.177	0.637	0.539×10^1
0.250	0.576	0.435×10^1
0.353	0.507	0.347×10^1
0.500	0.429	0.272×10^1
0.707	0.344	0.207×10^1
0.100×10^1	0.255	0.149×10^1
0.141×10^1	0.168	0.975
0.200×10^1	0.936×10^{-1}	0.541
0.283×10^1	0.409×10^{-1}	0.236
0.400×10^1	0.127×10^{-1}	0.733×10^{-1}
0.566×10^1	0.242×10^{-2}	0.140×10^{-1}
0.800×10^1	0.232×10^{-3}	0.134×10^{-2}

the first pole in the spectrum within reasonable accuracy as well as the corresponding time constant. We have:

$$\left| \frac{\mathrm{Re}\ T}{\mathrm{Im}\ T} \right| \approx \frac{\sigma\mu a^2}{6}\omega \approx \tau_0\omega$$

3.7. A RIGHT CIRCULAR CYLINDER IN A FIELD CREATED BY AN INFINITELY LONG CURRENT-CARRYING LINE

In order to understand better the conditions under which the solution obtained in the preceding section can be applied, we can consider the secondary field of currents induced in the conductor when the source of the primary field is current flowing in an infinitely long filament parallel to the axis of the cylinder as shown in Fig. 3.41. Here, the current line passes through the point 0. The magnetic field caused by this current is:

$$H^0_{\phi_1} = \frac{I_0}{2\pi r_1}e^{-i\omega t} \tag{3.281}$$

Defining a vector potential A^0 which has a single component A^0_x, we obtain:

$$H^0_{\phi_1} = \mathrm{curl}_{\phi_1}\ A^0$$

or

$$H^0_{\phi_1} = -\frac{\partial A^0_x}{\partial r_1} = \frac{I_0}{2\pi r_1}e^{-i\omega t}$$

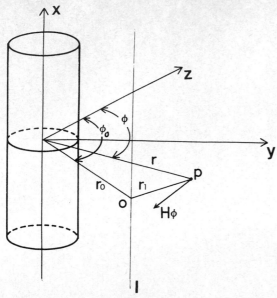

Fig. 3.41. A right circular cylinder in a field created by an infinitely long current-carrying line.

The magnitude of the vector potential A_x^0 is:

$$A_x^0 = -\frac{I_0}{2\pi} \ln r_1 \tag{3.282}$$

To define the field caused by currents induced in the cylinder, we can write the vector potential in terms of the cylindrical harmonics. As is well known, we have:

$$A_x^0 = \frac{I_0}{2\pi} \left[\sum_{n=1}^{\infty} \frac{1}{n} \left(\frac{r}{r_0} \right)^n (\cos n\phi_0 \cos n\phi + \sin n\phi_0 \sin n\phi) - \ln r_0 \right],$$

$$\text{if } \frac{r}{r_0} < 1 \tag{3.283}$$

where r_0, ϕ_0 are the coordinates of the current filament, and r, ϕ are the coordinates of the observation point.

In determining the secondary field, we can formulate a boundary problem with respect to a vector potential of the electrical type, A. From the first of Maxwell's equations, we have

$$\text{curl } E = -\mu \frac{\partial}{\partial t} \text{curl } A = \text{curl } \mu \frac{\partial A}{\partial t}$$

or

$$\text{curl}\left(E + \mu\frac{\partial A}{\partial t}\right) = 0$$

The primary vortex electric field is:

$$E_x^0 = -\mu\frac{\partial A_x^0}{\partial t} - \text{grad}_x U \tag{3.284}$$

Inasmuch as the field is independent of the x-coordinate, we have:

$$E_x^0 = -\mu\frac{\partial A_x^0}{\partial t} \tag{3.285}$$

Therefore, from eq. 3.283, we have the following expression for the primary electric field:

$$E_x^0 = i\omega\mu\frac{I_0}{2\pi}\left[\sum_{n=1}^{\infty}\frac{1}{n}\left(\frac{r}{r_0}\right)^n(\cos n\phi_0 \cos n\phi + \sin n\phi_0 \sin n\phi) - \ln r_0\right] \tag{3.286}$$

Under the action of this field, induction currents arise within the cylinder, directed along the x-axis. Interaction between these current filaments will not change their direction, and therefore, the secondary electric field, like the primary field, has but a single component, E_x. Therefore, eq. 3.285 can be applied for the total field as well:

$$E_x = -\mu\frac{\partial A_x}{\partial t} \tag{3.287}$$

Let us use the usual notation for the electric field outside and inside the cylinder:

$$\begin{aligned} E_x^e &= E_x^0 + E_x^s, && \text{if } r > a \\ E_x^i, && \text{if } r < a \end{aligned} \tag{3.288}$$

In accord with the results described in the previous paragraph, the electric field will satisfy the equations:

$$\begin{aligned} \nabla^2 E_x^e &= 0 &&, \quad \text{if } r > a \\ \nabla^2 E_x^i + k^2 E_x^i &= 0, && \text{if } r < a \end{aligned} \tag{3.289}$$

as well as the following conditions at the surface of the cylinder:

$$E_x^e = E_x^i, \quad \frac{1}{\mu_e}\frac{\partial E_x^e}{\partial r} = \frac{1}{\mu_i}\frac{\partial E_x^i}{\partial r} \tag{3.290}$$

As follows from eq. 3.257 and 3.286, the field can be written as:

$$E_x^e = i\omega\mu_e \frac{I_0}{2\pi} \left[\sum_{n=1}^{\infty} \frac{1}{n} \left(\frac{r}{r_0}\right)^n \cos n(\phi - \phi_0) - \ln r_0 \right]$$

$$+ i\omega\mu_e \frac{I_0}{2\pi} \sum_{n=1}^{\infty} \frac{1}{r^n} \{T_n^1 \cos n(\phi - \phi_0) + B_n \sin n(\phi - \phi_0)\} \qquad (3.291)$$

and

$$E_x^i = i\omega\mu_e \frac{I_0}{2\pi} \sum_{n=1}^{\infty} I_n(ikr)\{C_n \cos n(\phi - \phi_0) + D_n \sin n(\phi - \phi_0)\} \qquad (3.292)$$

It should be noted that the first term in eq. 3.291 requires that r be less than r_0.

Making use of the boundary conditions expressed in eq. 3.290 as well as the orthogonality of trigonometric functions as was done previously, we will find that:

$$B_n = D_n = 0$$

The following system of equations is derived for determining the coefficients T_n:

$$\frac{1}{n}\left(\frac{a}{r_0}\right)^n + \frac{1}{a^n} T_n^1 = C_n I_n(ika)$$

$$\frac{1}{\mu_e}\left(\frac{1}{r_0^n} a^{n-1} - \frac{n}{a^{n+1}} T_n^1\right) = \frac{C_n}{\mu_i} ikI_n'(ika) \qquad (3.293)$$

Solving this system for the function T_n^1, we have:

$$T_n^1 = \frac{KnI_n(x) - xI_n'(x)}{KnI_n(x) + xI_n'(x)} \frac{a^{2n}}{nr_0^n} \qquad (3.294)$$

where $K = \mu_i/\mu_e$, $x = ika$.

In particular, for a non-magnetic cylinder:

$$T_n^1 = \frac{nI_n(x) - xI_n'(x)}{nI_n(x) + xI_n'(x)} \frac{a^{2n}}{nr_0^n} \qquad (3.295)$$

Using the recurrence relationships:

$$I_n'(x) = I_{n-1}(x) - \frac{n}{x} I_n(x)$$

$$\frac{2n}{x} I_n(x) = I_{n-1}(x) - I_{n+1}(x)$$

equation 7.15 is considerably simplified so that we have:

$$T_n^1 = -\frac{I_{n+1}(ika)}{I_{n-1}(ika)}\frac{a^{2n}}{nr_0^n} = -T_n\frac{a^{2n}}{nr_0^n} \tag{3.296}$$

where

$$T_n = \frac{I_{n+1}(ika)}{I_{n-1}(ika)}$$

Thus, we have the following expression for the secondary electric field:

$$E_x^s = -i\omega\mu_e\frac{I_0}{2\pi}\sum_{n=1}^{\infty}T_n\left(\frac{a}{r_0}\right)^n\left(\frac{a}{r}\right)^n\frac{1}{n}\cos n(\phi-\phi_0) \tag{3.297}$$

Using Maxwell's equation:

$$\mathrm{curl}\,E = -\frac{\partial B}{\partial t}$$

we obtain the following expression for the magnetic field components:

$$i\omega\mu_e H_r = \frac{1}{r}\frac{\partial E_x}{\partial\phi}, \quad i\omega\mu_e H_\phi = -\frac{\partial E_x}{\partial r}$$

Thus:

$$H_r = \frac{I_0}{2\pi r}\sum_{n=1}^{\infty}T_n\left(\frac{a}{r_0}\right)^n\left(\frac{a}{r}\right)^n\sin n(\phi-\phi_0)$$

$$H_\phi = -\frac{I_0}{2\pi r}\sum_{n=1}^{\infty}T_n\left(\frac{a}{r_0}\right)^n\left(\frac{a}{r}\right)^n\cos n(\phi-\phi_0) \tag{3.298}$$

In accord with these last two equations, the secondary field is the sum of fields caused by linear multipoles, and the first harmonic represents the secondary field when the primary field is uniform near the cylinder. The examples of current distributions for several multipoles are shown in Fig. 3.42.

The complex amplitude of each spatial harmonic of the secondary field is described by the function T_n. Let us first investigate the low frequency part of the spectrum. Making use of the asymptotic expansions:

$$I_{n+1}(x) \approx \frac{(\frac{1}{2}x)^{n+1}}{\Gamma(n+2)} + \frac{(\frac{1}{2}x)^{n+3}}{\Gamma(n+3)}$$

and

$$I_{n-1}(x) \approx \frac{(\tfrac{1}{2}x)^{n+1}}{\Gamma(n)} + \frac{(\tfrac{1}{2}x)^{n+1}}{\Gamma(n+1)}$$

with

$$\Gamma(n) = (n-1)!$$

we have:

$$T_n = -\frac{i\sigma\mu\omega a^2}{4n(n+1)} + \frac{(\sigma\mu\omega a^2)^2}{8n^2(n+1)(n+2)} \qquad (3.299)$$

Considering that the functions $I_{n+1}(x)$ and $I_{n-1}(x)$ can be represented in terms of a power series in x, it is clear that in the low-frequency part of the spectrum, $|ka| < 1$, the functions T_n are expandable in a MacLauren series containing only integer powers in ω. The right-hand side of eq. 3.299 contains the leading terms of these series for the quadrature and inphase components respectively at very low .frequencies. As follows from this equation, the ratio between the inphase component and the quadrature component for each cylindrical harmonic is:

$$\frac{\text{In } T_n}{Q T_n} \approx -\frac{\sigma\mu a^2}{2n(n+2)} \qquad (3.300)$$

Thus, just as was the case for a spherical conductive body, with an increase in the order n of the spatial harmonic, the skin effect manifests itself to a progressively lesser degree. In other words, the interaction between current begins to come into play at progressively higher frequencies. This means that the radius of convergence for each spatial harmonic at low frequencies increases with increasing n. Making use of Cauchy's rule, eq. 3.300 permits us to evaluate the radius of convergence of this series for each harmonic, at least approximately. This radius ω_n, is:

$$\omega_n \approx \frac{2n(n+2)}{\sigma\mu a^2} = 2n(n+2)\alpha \qquad (3.301)$$

where $\alpha = 1/\sigma\mu a^2$.

Inasmuch as the first cylindrical harmonic, which corresponds to excitation by a uniform primary field, has the minimum radius of convergence, a series describing the low-frequency spectrum of the field caused by all of the harmonics is characterized by this radius. From eq. 3.301, it is approximately:

$$\omega_1 \approx 6\alpha \qquad (3.302)$$

Let us consider the high frequency part of the spectrum briefly. Making use of the asymptotic expressions for the Bessel functions:

$$I_{n+1}(x) \approx \frac{e^x}{(2\pi x)^{1/2}}\left[1 - \frac{4(n+1)^2 - 1}{8x}\right]$$

$$\qquad\qquad\qquad\qquad\qquad\qquad\qquad \text{for } x \gg 1$$

$$I_{n-1}(x) \approx \frac{e^x}{(2\pi x)^{1/2}}\left[1 - \frac{4(n-1)^2 - 1}{8x}\right],$$

we obtain:

$$T_n \approx 1 - 2n/ika \tag{3.303}$$

Therefore, the components of the complex amplitude are:

$$\text{In } T_n \approx 1 - n\sqrt{2}\sqrt{\alpha}/\sqrt{\omega}$$
$$QT_n \approx -n\sqrt{2}\sqrt{\alpha}/\sqrt{\omega} \tag{3.304}$$

Thus, these components approach their asymptotic value just as they did in the case of spherical symmetry, that is, as the inverse square root of frequency.

On the basis of numerical calculations, the frequency spectrums for the first five functions T_n have been obtained and are shown in Figs. 3.43 and 3.44. Inasmuch as the complex amplitudes for the various T_n are different and the extent of the contribution of a cylindrical harmonic depends

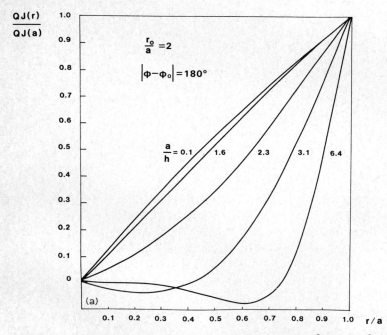

Fig. 3.42.a. Distribution of the quadrature component of current density.

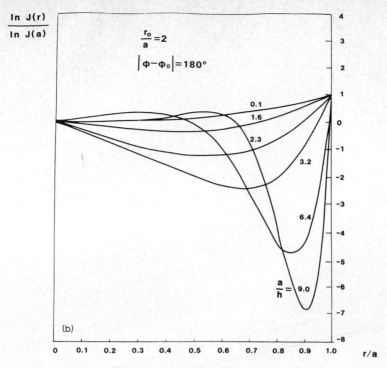

Fig. 3.42.b. Distribution of the inphase component of current density.

strongly on geometrical factors, one can generally expect that the frequency spectrums for the secondary field at different points will differ one from another. It is clear that these differences will increase as either the source of the primary field or the observation point is placed closer to the cylinder.

Examples of frequency spectrum for the secondary field are shown in Figs. 3.45 and 3.46 for various distances from the cylinder.

As follows from eqs. 3.297—3.299, it is possible to formulate three conditions similar to those expressed for spherical conductive bodies for the case in which the effect of the first cylindrical harmonic increases and becomes dominant. These conditions are:

(1) As the distance from the source, r_0, increases and the primary magnetic field becomes almost uniform in the vicinity of the cylinder,

(2) As the distance from the observation point to the cylinder, r, increases, and because cylindrical harmonics with larger n decrease more rapdily, the relative contribution of the first harmonic prevails.

(3) As seen from eq. 3.209 in measuring the second term in this series, the effect of the following cylindrical harmonics becomes smaller. It is possible to show that by measuring later terms for the lower frequency part of the spectrum of the complex amplitude, T_n the effect of all the cylindrical

In T_n

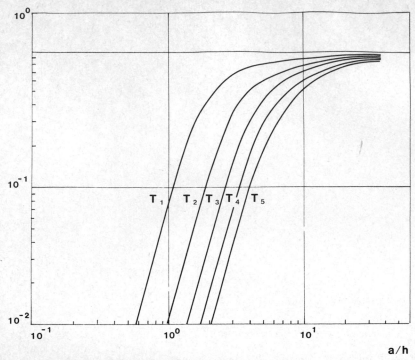

Fig. 3.43. Frequency spectrums for the inphase component of the first 5 functions T_n.

harmonics except the first one can be reduced to any degree, regardless of the size of r and r_0.

In all three cases, the secondary field caused by current induced in a cylinder approaches that for the case in which the primary field is uniform and therefore, in accord with eqs. 3.297 and 3.298 we have:

$$E_x^s = -i\omega\mu_e \frac{I_0}{2\pi r_0} T_1 \frac{a^2}{r} \cos(\phi - \phi_0)$$

$$H_r = \frac{I_0}{2\pi r_0} T_1 \frac{a^2}{r^2} \sin(\phi - \phi_0) \qquad (3.305)$$

$$H_\phi = -\frac{I_0}{2\pi r_0} T_1 \frac{a^2}{r^2} \cos(\phi - \phi_0)$$

The term $I_0/2\pi r_0$ represents the primary magnetic field contributed by the source current filament.

We can now consider the transient behavior of the field contributed by currents induced in the cylinder. Applying the Fourier transform to eq. 3.297 and assuming step-function excitation, we have:

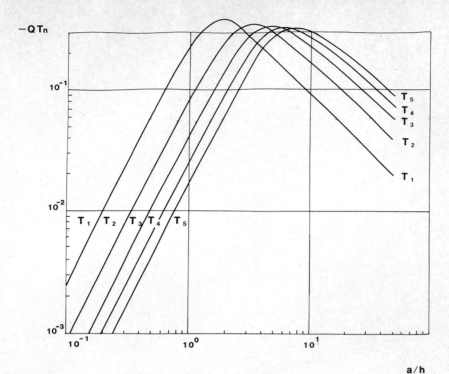

Fig. 3.44. Frequency spectrums for the quadrature component of the first 5 functions T_n.

$$E_x(t) = -\frac{\mu_e I_0}{2\pi} \sum_{n=1}^{\infty} \frac{1}{n} \dot{N}_n(t) \left(\frac{a}{r_0}\right)^n \left(\frac{a}{r}\right)^n \cos n(\phi - \phi_0) \qquad (3.306)$$

where

$$\dot{N}_n(t) = \frac{1}{2\pi} \int_{-\infty}^{\infty} T_n(\mathrm{i}ka) \mathrm{e}^{-\mathrm{i}\omega t} \mathrm{d}\omega \qquad (3.307)$$

or

$$\dot{N}_n(t) = \frac{1}{2\pi} \int_{-\infty}^{\infty} \frac{I_{n+1}(\mathrm{i}ka)}{I_{n-1}(\mathrm{i}ka)} \mathrm{e}^{-\mathrm{i}\omega t} \mathrm{d}\omega \qquad (3.308)$$

and

$$\dot{N}_n(t) = \frac{\mathrm{d}N_n}{\mathrm{d}t}$$

From eq. 3.306, in order to determine the electric field $E_x(t)$ the integral $\dot{N}_n(t)$ has to be calculated for each cylindrical harmonic. By analogy with

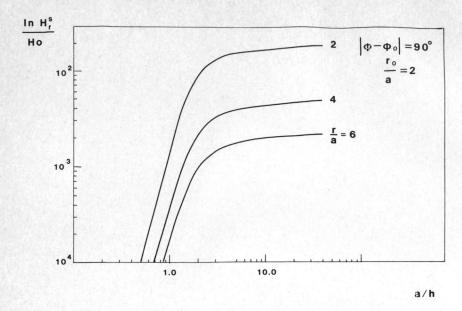

Fig. 3.45. Frequency spectrum for the inphase part of the secondary field for various distances from the cylinder.

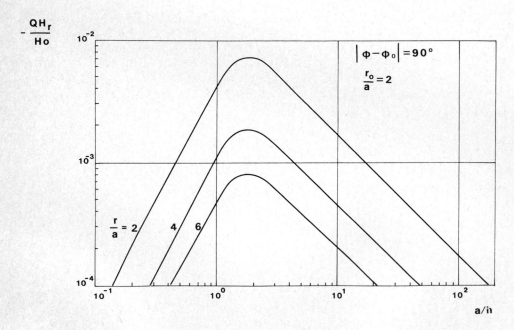

Fig. 3.46. Frequency response for the quadrature component of the secondary field for various distances from the cylinder.

previously obtained results (see Paragraph 3 of this chapter), the residue theorem will be used. Inasmuch as:

$$I_m(x) = i^{-m} J_m(ix)$$

we have

$$\frac{I_{n+1}(ika)}{I_{n-1}(ika)} = -\frac{J_{n+1}(-ka)}{J_{n-1}(-ka)}$$

It is clear that the function:

$$F_2 = \frac{J_{n+1}(-ka)}{J_{n-1}(-ka)} e^{-i\omega t}$$

has singularities along the imaginary ω-axis with coordinates given by:

$$J_{n-1}(-ka) = 0 \tag{3.309}$$

For each cylindrical harmonic, n, there is an infinite number of roots to eq. 3.309. They can be written in the form:

$$\omega_{ns} = \pm i q_{ns} \alpha \tag{3.310}$$

where $\alpha = 1/\sigma\mu a^2$, and where q_{ns} are numbers characterizing the position of the roots. The index s also defines the number of the root for the corresponding cylindrical harmonic. Examples of values for q_{ns} are listed in Table 3.XII. The roots, ω_{ns}, given in eq. 3.310 can be considered to be simple poles of the complex amplitude spectrum, and therefore of functions $T_n e^{-i\omega t}$, being integrands in eq. 3.307. Applying the residue theorem and taking into account only the poles situated on the negative portion of the imaginary ω-axis, we obtain:

$$\dot{N}_n(t) = -i \sum_{s=1}^{\infty} \operatorname{Res}\left[\frac{J_{n+1}(-ka)}{J_{n-1}(-ka)} e^{-i\omega t} \right] \tag{3.311}$$

Because the poles are simple, we have:

$$\operatorname{Res} F_2 = \frac{J_{m+2}(x_{ns}) e^{-i\omega_{ns} t}}{\dfrac{\partial}{\partial \omega} J_m(x)}, \quad \text{as } \omega \to \omega_{ns}$$

where

$$x = -ka, \quad x_{ns} = -k_{ns}a, \quad k_{ns}a = (i\omega_{ns}\mu\sigma a^2)^{1/2} = (i\omega_{ns}/\alpha)^{1/2};$$
$$m = n - 1$$

Repeating the process done earlier on page 187, we have:

TABLE 3.XII

Values of parameters q_{ns}

s \ n	1	2	3	4	5	6	7	8	9
1	0.578×10^1	0.147×10^2	0.264×10^2	0.407×10^2	0.576×10^2	0.769×10^2	0.987×10^2	0.123×10^3	0.149×10^3
2	0.305×10^2	0.492×10^2	0.708×10^2	0.953×10^2	0.122×10^3	0.152×10^3	0.185×10^3	0.220×10^3	0.257×10^3
3	0.749×10^2	0.103×10^3	0.135×10^3	0.169×10^3	0.207×10^3	0.246×10^3	0.289×10^3	0.334×10^3	0.382×10^3
4	0.139×10^3	0.178×10^3	0.219×10^3	0.263×10^3	0.310×10^3	0.360×10^3	0.413×10^3	0.468×10^3	0.526×10^3
5	0.223×10^3	0.271×10^3	0.323×10^3	0.377×10^3	0.434×10^3	0.494×10^3	0.556×10^3	0.622×10^3	0.690×10^3
6	0.327×10^3	0.385×10^3	0.446×10^3	0.510×10^3	0.577×10^3	0.647×10^3	0.719×10^3	0.795×10^3	0.873×10^3
7	0.450×10^3	0.518×10^3	0.589×10^3	0.663×10^3	0.740×10^3	0.819×10^3	0.902×10^3	0.987×10^3	0.108×10^4
8	0.593×10^3	0.671×10^3	0.752×10^3	0.836×10^3	0.922×10^3	0.101×10^4	0.110×10^4	0.120×10^4	0.130×10^4
9	0.756×10^3	0.844×10^3	0.934×10^3	0.103×10^4	0.112×10^4	0.122×10^4	0.133×10^4	0.141×10^4	0.154×10^4
10	0.938×10^3	0.104×10^4	0.114×10^4	0.124×10^4	0.135×10^4	0.146×10^4	0.157×10^4	0.168×10^4	0.180×10^4

$$\frac{\partial}{\partial \omega} J_m(x) = -\tfrac{1}{2}\left(\frac{i}{\omega_{ns}\alpha}\right)^{1/2} J'_m(x)$$

Making use of the recursive relationships between the Bessel functions and taking into account that:

$$J_m(x_{ns}) = 0$$

we have:

$$\frac{J_{n+1}(x_{ns})}{\dfrac{\partial}{\partial x} J_{n-1}(x)} = -\frac{2n}{x_{ns}}$$

Therefore:

$$\text{Res } F_2 = -\frac{4n}{i}\,\alpha e^{-q_{ns}\alpha t} \tag{3.312}$$

We obtain the following for the function $\dot{N}_n(t)$:

$$\dot{N}_n(t) = 4n\alpha \sum_{s=1}^{\infty} e^{-q_{ns}\alpha t} \tag{3.313}$$

Substituting this result into eq. 3.306, the electric field can be calculated. In accord with eq. 3.313, the transient response of each cylindrical harmonic is described by an infinite sum of exponential terms, just as was the case with the spherical harmonics. It is important to note that:

$$q_{n1} < q_{n2} < q_{n3} < q_{n4} < \ldots \tag{3.314}$$

As has been shown earlier, the power of the first exponent in each cylindrical harmonic $q_{n1}\alpha$ is determined by the minimum pole, which also defines the radius of the convergence for the series describing the low frequency part of the spectrum. Applying Cauchy's rule, we can evaluate the parameter q_{n1} in an approximation, proceeding from eq. 3.300:

$$q_{n1} \approx 2n(n+2) \tag{3.315}$$

As is seen from Table 3.XII, eq. 7.34 can be applied with reasonable accuracy in determining the first power in the exponent for each harmonic. Comparing the values for q_{ns} in Tables 3.I and 3.XII, we conclude that they have three important features which are the same, and which were pointed out in Paragraph 3. Thus, during the late stage of transient response, the electric field is defined by the first harmonic and it can be presented as:

$$E_x(t) \approx -\frac{2\mu_e I_0\alpha}{\pi r_0}\frac{a^2}{r}\,e^{-t/\tau_0}\cos(\phi - \phi_0), \quad \text{if } t > \tau_0 \tag{3.316}$$

264

where

$$\tau_0 = \sigma\mu a^2 / q_{11}$$

is a time constant that was defined by assuming uniformity of the primary field.

Just as was the case with the spherical conductive body, this result is independent of the position of the source current line and of the observation point.

At this point, let us consider briefly the behavior of the electric field during the very early stage. Applying the Fourier transform to eq. 304, we have:

$$\dot{N}_n(t) = \frac{2n\sqrt{2\alpha}}{\pi} \int_0^\infty \frac{\sin \omega t}{\sqrt{\omega}} \, d\omega$$

or (3.317)

$$\dot{N}_n(t) = \frac{2n}{\sqrt{\pi}} \left(\frac{\alpha}{t}\right)^{1/2} = \frac{2n}{(\pi q_{11})^{1/2}} \frac{1}{(t\tau_0)^{1/2}}$$

where $\tau_0 = 1/q_{11}\alpha$ is a time constant. Thus, we can state that all of the cylindrical harmonics decay in inverse proportion to the square root of time during the early stage. The transient behavior of the functions $\dot{N}(t)$ are shown in Fig. 3.47 for the first nine harmonics.

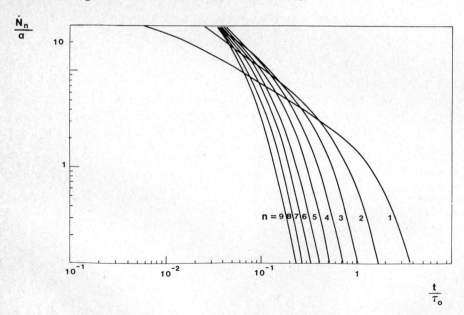

Fig. 3.47. The transient behavior for the first nine harmonics of the function $\dot{N}_n(t)$.

Up to this point, we have considered only the behavior of the electric field. In order to obtain expressions for the magnetic field and its time derivative, we will use the first of Maxwell's equations written in cylindrical coordinates:

$$\frac{\partial B_r}{\partial t} = -\frac{1}{r}\frac{\partial E_x}{\partial \phi}, \qquad \frac{\partial B_\phi}{\partial t} = \frac{\partial E_x}{\partial r}$$

As follows from eq. 3.306:

$$\frac{\partial B_r(t)}{\partial t} = -\frac{\mu_e I_0}{2\pi r}\sum_{n=1}^{\infty}\dot{N}_n(t)\left(\frac{a}{r_0}\right)^n\left(\frac{a}{r}\right)^n \sin n(\phi - \phi_0)$$

$$\frac{\partial B_\phi(t)}{\partial t} = \frac{\mu_e I_0}{2\pi r}\sum_{n=1}^{\infty}\dot{N}_n(t)\left(\frac{a}{r_0}\right)^n \cos n(\phi - \phi_0)$$

$$(3.318)$$

In order to arrive at expressions for the magnetic field functions, the $N_n(t)$ must be found. Using eq. 3.313 and considering that:

$$N_n(t) = \int_{\infty}^{t}\dot{N}_n(t)dt$$

we have:

$$N_n(t) = -4n\sum_{n=1}^{\infty}\frac{1}{q_{ns}}e^{-q_{ns}\alpha t} \qquad (3.319)$$

The transient behavior for $N_n(t)$ is shown in Fig. 3.48 for the first nine harmonics.

In accord with eq. 3.318, we have:

$$B_r(t) = -\frac{\mu_e I_0}{2\pi r}\sum_{n=1}^{\infty}N_n(t)\left(\frac{a}{r_0}\right)^n\left(\frac{a}{r}\right)^n \sin n(\phi - \phi_0)$$

$$B_\phi(t) = \frac{\mu_e I_0}{2\pi r}\sum_{n=1}^{\infty}N_n(t)\left(\frac{a}{r_0}\right)^n\left(\frac{a}{r}\right)^n \cos n(\phi - \phi_0)$$

$$(3.320)$$

Inasmuch as the parameter q_{11} is minimal, we have for the late stage:

$$\dot{B}_r(t) \approx -\frac{\mu_e I_0}{2\pi r}4\alpha e^{-t/\tau_0}\frac{a^2}{rr_0}\sin(\phi - \phi_0)$$

$$= -\frac{\mu_e I_0}{2\pi r_0}4\alpha\left(\frac{a}{r}\right)^2 e^{-t/\tau_0}\sin(\phi - \phi_0) \qquad (3.321)$$

and

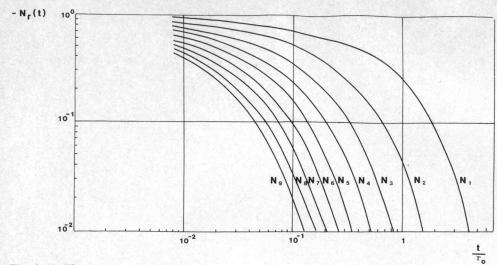

Fig. 3.48. The transient behavior for the first nine harmonics of the function $N_n(t)$.

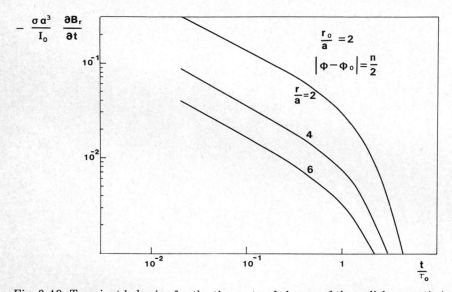

Fig. 3.49. Transient behavior for the time rate of change of the radial magnetic induction.

$$\dot{B}_\phi(t) = \frac{\mu_e I_0}{2\pi r_0} \, 4\alpha \left(\frac{a}{r}\right)^2 e^{-t/\tau_0} \cos(\phi - \phi_0)$$

The term $\mu_e I_0/2\pi r_0$ is the magnetic induction vector representing the primary field. Comparing this with results obtained in a preceding paragraph, we can see that during the late stage, the field will be the same as that observed when the primary field is uniform in the vicinity of the cylinder.

Similar expressions can be written for the magnetic field during late-stage behavior.

Applying precisely the same approach as in the case of the spherical conductor, we have the following result for the functions $N_n(t)$ at the early stage:

$$N_n(t) = N_n(0) + \int_0^t \dot{N}_n(t)\,dt = -4n\left[\sum_{s=1}^{\infty} \frac{1}{q_{ns}} - \frac{1}{(\pi q_{11})^{1/2}}\left(\frac{t}{\tau_0}\right)^{1/2}\right] \quad (3.322)$$

Thus, each cylindrical harmonic of the magnetic field appoaches its limit as the square root of time. As an example, the transient behavior of the function $\partial B_r/\partial t$ is shown in Fig. 3.49 for various locations.

In conclusion, it should be noted that in the case of a cylindrical conductor, just as in the case of a spherical conductor, there are three similar conditions under which the nonuniformity of the primary field almost vanishes.

3.8. EQUATIONS FOR THE FREQUENCY AND TRANSIENT BEHAVIOR CAUSED BY INDUCED CURRENTS IN A CYLINDRICAL CONDUCTOR WITH AN ARBITRARY CROSS-SECTION

In this paragraph, we derive equations for the field caused by induced linear currents in a cylindrical conductor or a system of cylindrical conductors. Some examples of these cylindrical forms are shown in Fig. 3.50.

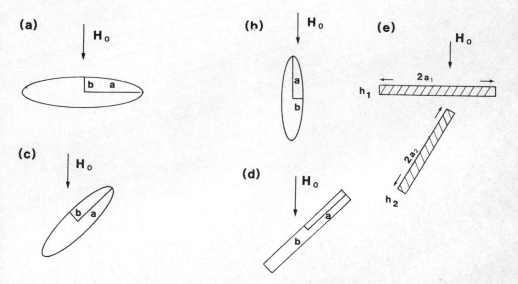

Fig. 3.50. Typical examples of cylindrical conductors with arbitrary cross-sections.

It is assumed that the conductors are infinitely elongate and that the primary vortex electric field is directed along the cylindrical axis, and is invariant in this direction. The results will be applicable for cases where the conductivity varies in an arbitrary manner over the cross-section of the cylinders. It is obvious that due to the existence of a primary vortex electric field, induced currents will arise within the conductive cylinder. In turn, each current filament creates a magnetic field which, because it varies with time, generates a vortex electric field. From a physical point of view it is clear that the inter-action between currents, that is, the effect of the secondary vortex electric field, does not change the direction of current flow. Regardless of the frequency, the conductivity, or the dimensions of the cylinder, the induced currents will have but a single component, that being directed along the axis of the cylinder. Inasmuch as the surrounding medium is insulating, the induced currents must form closed flow lines within the conductors, and therefore, we can always find a pair of current elements which are oppositely directed and equal in strength.

In order to investigate the general features for the frequency and transient behavior of the field, we will make use of an integral equation with respect to current density which is similar to that applied previously for models with axial symmetry. We start as before from Ohm's law:

$$j_x = \sigma(E_{0x} + E^i_{1x})$$ (3.323)

where E_{0x} is the primary vortex electric field caused by the time rate of change of the primary magnetic field, while E^i_{1x} is the secondary vortex electric field arising as a consequence of the time rate of change of the magnetic fields generated by induced currents. Applying the principle of superposition, eq. 3.323 can be written as

$$j_x(g) = j^0_x(g) + i\omega\mu\sigma \int_S G(p,g)j_x(p)dS$$ (3.324)

where $i\omega\mu G(p,g)j_x(p)dS$ is the vortex electric field at the point g caused by a linear current filament passing through the point p (see Fig. 3.51), j^0_x is the current density caused by the primary vortex electric field at the point g.

Inasmuch as the currents which provide the source of the primary field are oriented along the x-axis, in general, the vortex electric field can be written as:

$$E_{0x}(g) = -i\omega\mu H_0(g)G_0$$ (3.325)

where $H_0(g)$ is the magnitude of the primary magnetic field, and G_0 is a function that depends on the distances from the current sources to the point g. If for example, there is a single current filament passing through point a as shown in Fig. 3.51, we have (see page 251):

$$G_0 = r_{ag} \ln r_{ag}$$ (3.326)

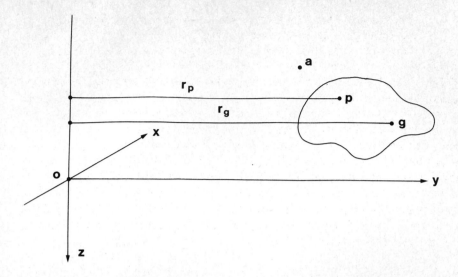

Fig. 3.51. Geometry of the points referred to in eq. 3.324.

where r_{ag} is the distance between the points a and g. Applying the principle of superposition, it is a simple matter to derive expressions for the function G_0 for an arbitrary distribution of linear current sources.

As follows from eq. 3.326, we have the following expression for the function $G(p, g)$ characterizing the interaction between currents:

$$G(p, g) = \frac{\ln r_{pg}}{2\pi} \tag{3.327}$$

where r_{pg} is the distance between the points p and g.

In place of the function G given by eq. 3.327, sometimes it is more convenient to use another function, namely:

$$G = \frac{1}{2\pi} \ln \frac{r_{p_2, g}}{r_{p_1, g}} \tag{3.328}$$

which reflects the effect of two currents passing through points p_2 and p_1, respectively, and which have equal amplitudes but opposite directions.

Following the approach described in Paragraph 4 of this chapter, we will represent the current field in the cylinder as being the sum of elementary volume current tubes, with axes parallel to each other and to the cylindrical axis. We can assume that the cross-section of each elementary tube is small, so that we can neglect any variation in the current density within the tube, the field, or the electromagnetic force within the cross-section of the tube.

Multiplying both parts of eq. 3.324 by the factor dx/σ, and replacing the integration sign by the summation sign, we have:

$$\mathscr{E}(g) = \mathscr{E}^0(g) + i\omega\mu \sum_{p \neq g} \overline{G}(p,g) I(p) + i\omega\mu \int_{S(g)} \overline{G} j_x(g) \mathrm{d}S \qquad (3.329)$$

where $\mathscr{E}(g)$ is the electromotive force within the interval $\mathrm{d}x$ given by $E_x(g)\mathrm{d}x$, and $\mathscr{E}_0(g)$ is the electromotive force within this interval caused by the primary electric field. In the further development of this section, the length of the tubular element will be assumed to be one unit ($\mathrm{d}x = 1$). We will use the notation:

$$\overline{G}(p,g) = G(p,g)\mathrm{d}x$$

and recognize that the summation includes all the elementary tubes except the tube within which the point g is situated. In accord with eq. 3.329, integration has to be carried out within the cross-sectional area of this particular tube. $I(p)$ is the current through the tube p. The third term in eq. 3.329 can be written in the form:

$$i\omega\mu \int_{S(g)} j_x \overline{G} \mathrm{d}S = \frac{i\omega\mu I(g)}{S(g)} \int_{S(g)} \overline{G} \mathrm{d}S$$

and we can make use of the following abbreviated notation:

$$M_{pg} = -\mu\overline{G}(p,g) \quad \text{and} \quad L_g = -\frac{\mu}{S(g)} \int_{S(g)} \overline{G} \mathrm{d}S$$

where M_{pg} and L_g are the coefficients of mutual inductance between current tubes and of self-inductance of the tube g, respectively. Considering that from Ohm's law:

$$\mathscr{E}(g) = I(g) R(g)$$

where $R(g)$ is the resistance in an element of the tube g, having the length $\mathrm{d}x$, that is:

$$R(g) = \frac{\mathrm{d}x}{\sigma S(g)}$$

In place of the integral equation 3.329, we obtain a system of linear equations with constant coefficients with respect to current. This system can be written as:

$$Z_{11}I_1 + i\omega M_{12}I_2 + i\omega M_{13}I_3 + \ldots + i\omega M_{1n}I_n = \mathscr{E}_1^0$$
$$i\omega M_{21}I_1 + Z_{22}I_2 + i\omega M_{23}I_3 + \ldots + i\omega M_{2n}I_n = \mathscr{E}_2^0 \qquad (3.330)$$
$$i\omega M_{N1}I_1 + i\omega M_{N2}I_2 + i\omega M_{N3}I_3 + \ldots + Z_{NN}I_N = \mathscr{E}_N^0$$

where $Z_{ii} = R_i + i\omega L_i$ is the impedance of the filament i. It is important to note that the right-hand side of the system in 3.330 represents the electromotive force in each linear tube caused by the primary vortex electric field,

and in accord with eq. 3.325, this EMF is directly proportional to the frequency, and is known. The transition from the integral equation form in 3.324 to the system of algebraic equations in 3.330 requires equivalence of the fields caused by the volumetric current flow with the system of linear currents, resistances, inductances, and mutual inductances which characterize the corresponding current tubes. As is well known, the expressions for mutual and self-inductance for unit length are quite simple. These are:

$$M_{pg} = -\mu \ln r_{pg}/2\pi, \quad L = \mu/8\pi \tag{3.331}$$

It is intuitively obvious that as the number of tubes used to replace the volume of the subsurface is increased, the error associated with this approximation decreases.

Comparing equations for the currents, eqs. 3.159 and 3.330, we see that they differ from each other only by constant coefficients. For this reason, the principal conclusions that were derived for models with axial symmetry are valid for cylindrical conductors, in the case in which the field does not change along the axis of the cylinder.

We can write the following expression for the current at any point in a conductor:

$$I(g) = i\omega \frac{P_{N-1}(g, i\omega)}{P_N(i\omega)} \tag{3.332}$$

where $P_N(i\omega)$ is a polynomial in $i\omega$, and is the same as the system determinant. The multiplier $i\omega$ appears as a consequence of the fact that the primary vortex EMF is directly proportional to $i\omega$. As is well known for the behavior of polynomials, the right-hand side of eq. 3.332 can be written as a sum of simple fractions, as follows:

$$I(g) = i\omega \sum_{n=1}^{N} \frac{a_n}{\omega_n - i\omega} \tag{3.333}$$

where $-i\omega_n$ are the roots of the polynomial $P_N(i\omega)$, and ω_n are real numbers such that:

$$\omega_1 < \omega_2 < \omega_3 < \omega_4 < \dots$$

Going to the limit in which the cross-section of the tubular elements becomes vanishingly small, from eq. 3.333 we have:

$$I(g) = i\omega \sum_{n=1}^{\infty} \frac{a_n}{\omega_n - i\omega} \tag{3.334}$$

or

$$j_x(g) = k^2 b^2 H_{0i} \sum_{n=1}^{\infty} \frac{\beta_n(g)}{q_n - k^2 b^2}$$

Using the Biot-Savart law, the various components of the secondary magnetic field can be expressed in terms of the distribution of currents, and in a cylindrical coordinate system this expression takes the form:

$$H_{1z}(a) = k^2 b^2 H_{0i} \sum_{n=1}^{\infty} \frac{d_{nz}(a)}{q_n - k^2 b^2}$$

$$(3.335)$$

$$H_{1r}(a) = k^2 b^2 H_{0i} \sum_{n=1}^{\infty} \frac{d_{nr}(a)}{q_n - k^2 b^2}$$

where

$$q_1 < q_2 < q_3 < q_4 < \dots \quad \text{and} \quad k^2 = i\sigma\mu\omega$$

with b being an arbitrary dimension representing a cross-section of one of the conductive elements, and H_{0i} is any component of the primary field at some point. Therefore, just as in the case for axial symmetry, the spectrums for the current are functions with singularities in the form of simple poles situated on the imaginary ω-axis:

$$\omega = -i\frac{q_n}{\sigma\mu b^2} = -iq_n\alpha = -i\omega_n$$

The pole ω_1 located closest to the origin defines the radius of convergence for the MacLauren series describing the low-frequency part of the spectrum. While the poles in the spectrum depend on the conductivity, the shape of the conductor and its dimensions, the coefficients $d_n(a)$ and $\beta_n(g)$ are functions of the coordinates of the observation point and of the conductor as well.

Applying the Fourier transform and taking into account that the primary field is a step function, we obtain:

$$j_x(g) = H_{0i} \sum_{n=1}^{\infty} \beta_n(g) e^{-q_n\alpha t}$$

$$H_{1z}(a) = H_{0i} \sum_{n=1}^{\infty} d_{nz}(a) e^{-q_n\alpha t}$$

$$(3.336)$$

$$H_{1r}(a) = H_{0i} \sum_{n=1}^{\infty} d_{nr}(a) e^{-q_n\alpha t}$$

We see again that the transient behavior can be described as a sum of exponential terms in which the exponents are determined by the poles in the spectrum.

Up to this point we have derived expressions for the frequency in transient responses caused by induction currents for two types of models:

(1) For conductors characterized by axial symmetry and with the primary field possessing the same symmetry, and

(2) For cylindrical conductors which are infinitely elongate, with a primary field that is constant along the cylindrical axis.

In both cases, the sources of the secondary field are only vortex currents within the conductors, inasmuch as the electric field does not intersect any surfaces between zones with different resistivities, and therefore no electrical charges can appear, and because the surrounding medium is assumed to be an insulator. Also, it is important that in both cases, the direction of the induction currents does not change with frequency, and can be specified independently.

Finally, the similarity of the expressions for frequency and transient responses for the various models that have been considered results from the fact that in the case of infinitely long conductors, the field and the induction currents do not change along the axial direction.

In the later sections of this chapter, more complicated situations will be examined.

3.9. FREQUENCY AND TRANSIENT RESPONSES OF THE FIELD CAUSED BY LINEAR INDUCED CURRENTS

Proceeding from the results obtained in the previous paragraph, we will now investigate the principal characteristics of the field caused by currents induced in cylindrical conductors in the case in which the primary vortex electric field is parallel to the axis of the conductor. Frequency and transient responses caused by currents in the elliptical cylinders will be considered as an example of this case. Inasmuch as the expressions for the induced currents and the field for the case of a cylindrical conductor are essentially the same as those for the field when the conductor is characterized by axial symmetry along the same axis, we might expect that the behavior in the frequency and time domain will be similar in the two cases.

First let us investigate the low-frequency portion of the spectrum. In order to illustrate some transformations, at this point we will consider as an example the vertical component of the secondary magnetic field, H_z. From eq. 3.335, we have:

$$H_{1z} = H_{0i} k^2 b^2 \sum_{n=1}^{\infty} \frac{d_{nz}}{q_n - k^2 b^2} \tag{3.337}$$

Assuming that the following inequality:

$$|k^2 b^2| < q_n \tag{3.338}$$

holds for every fraction, they can be expanded as a power series:

$$\frac{d_{1z}}{q_1 - k^2 b^2} = \frac{d_{1z}}{q_1} \frac{1}{1 - \frac{k^2 b^2}{q_1}} \approx \frac{d_{1z}}{q_1} \left\{ 1 + \frac{k^2 b^2}{q_1} + \frac{(k^2 b^2)^2}{q_1^2} + \frac{(k^2 b^2)^3}{q_1^3} + \cdots \right\}$$

$$\frac{d_{2z}}{q_2 - k^2 b^2} = \frac{d_{2z}}{q_2} \frac{1}{1 - \frac{k^2 b^2}{q_2}} \approx \frac{d_{2z}}{q_2} \left\{ 1 + \frac{k^2 b^2}{q_2} + \frac{(k^2 b^2)^2}{q_2^2} + \frac{(k^2 b^2)^3}{q_2^3} + \cdots \right\}$$

$$\frac{d_{lz}}{q_l - k^2 b^2} = \frac{d_{lz}}{q_l} \frac{1}{1 - \frac{k^2 b^2}{q_l}} \approx \frac{d_{lz}}{q_l} \left\{ 1 + \frac{k^2 b^2}{q_l} + \frac{(k^2 b^2)^2}{q_l^2} + \frac{(k^2 b^2)^3}{q_l^3} + \right\}$$

$$(3.339)$$

and so on.

It is obvious that each successive term in eq. 3.337 when expanded in a power series in terms of $k^2 b^2$ has a progressively greater radius of convergence, because the following relation holds between the parameters q_n:

$$q_1 < q_2 < q_3 <$$

Collecting the terms in these series with the same power in $k^2 b^2$ and remembering the multiplying term $k^2 b^2$ in front of the summation sign in eq. 3.337, we obtain:

$$H_{1z} = H_{0i} \sum_{n=1}^{\infty} C_{nz} (k^2 b^2)^n \tag{3.340}$$

where

$$C_{nz} = \frac{d_{1z}}{q_1^n} + \frac{d_{2z}}{q_2^n} + \frac{d_{3z}}{q_3^n} + \frac{d_{4z}}{q_4^n} + \cdots \tag{3.341}$$

Note that we have written the expression for the vertical component of the magnetic field as a power series with respect to $k^2 b^2$, or in other words, with respect to the frequency ω. As follows from these simple transformations, the radius of convergence for this series is the minimum value for q_n, that is, the value for q_1. By analogy, similar expressions can be written for the other components of the field and of the current. In particular, we have:

$$J_x = H_{0i} \sum_{n=1}^{\infty} \alpha_n (k^2 b^2)^n$$

$$(3.342)$$

$$H_{1r} = H_{0i} \sum_{n=1}^{\infty} C_{nr} (k^2 b^2)^n$$

where

$$\alpha_n = \frac{\beta_1}{q_1^n} + \frac{\beta_2}{q_2^n} + \frac{\beta_3}{q_3^n} + \frac{\beta_4}{q_4^n} + \ldots$$

$$C_{nr} = \frac{d_{1r}}{q_1^n} + \frac{d_{2r}}{q_2^n} + \frac{d_{3r}}{q_3^n} + \frac{d_{4r}}{q_4^n} + \ldots$$

(3.343)

It is clear that the radius of the convergence of the series 3.342 is also q_1. Writing the series in eqs. 3.340 and 3.343 in terms of powers of ω, we have:

$$J_x = H_{0i} \left\{ \sum_{n=1}^{\infty} (-1)^{n+2} \alpha_{2n} (\sigma\mu b^2)^{2n} \omega^{2n} \right.$$

$$\left. + i \sum_{n=1}^{\infty} (-1)^{n+1} \alpha_{2n-1} (\sigma\mu b^2)^{2n-1} \omega^{2n-1} \right\}$$

$$H_{1z}^e = H_{0i} \left\{ \sum_{n=1}^{\infty} (-1)^{n+2} C_{2n,z} (\sigma\mu b^2)^{2n} \omega^{2n} \right.$$

(3.344)

$$\left. + i \sum_{n=1}^{\infty} (-1)^{n+1} C_{2n-1,z} (\sigma\mu b^2)^{2n-1} \omega^{2n-1} \right\}$$

$$H_{1r}^e = H_{0i} \left\{ \sum_{n=1}^{\infty} (-1)^{n+2} C_{2n,r} (\sigma\mu b^2)^{2n} \omega^{2n} \right.$$

$$\left. + i \sum_{n=1}^{\infty} (-1)^{n+1} C_{2n-1,z} (\sigma\mu b^2)^{2n-1} \omega^{2n-1} \right\}$$

The radius of convergence for these series is given by the inequality:

$$\omega < q_1/\sigma\mu b^2 \quad \text{or} \quad \omega < 1/\tau_0$$

(3.345)

where τ_0 is a time constant characterizing the conductor.

As in previous cases with models possessing coaxial symmetry, it is appropriate to consider the range of frequencies for which the inequality in eq. 3.345 is satisfied, as that corresponding to the low-frequency part of the spectrum. Starting from eq. 3.344, we see again that the series representing the quadrature component of the induction currents and the magnetic field contain only odd powers of the frequency ω, while the series describing the inphase component contain even powers of ω.

A similar expression can be written for the secondary electric field which is directed along the strike of the conductor:

$$E_{1x}^e = H_{0i} \left\{ \sum_{n=1}^{\infty} (-1)^{n+2} e_{2n} (\sigma\mu b^2)^{2n-1} \omega^{2n} \right.$$

$$\left. + i \sum_{n=1}^{\infty} (-1)^{n+2} e_{2n+1} (\sigma\mu b^2)^{2n} \omega^{2n+1} \right\} \qquad (3.346)$$

As follows from a comparison of results of calculations of frequency responses for a circular cylinder, the low-frequency range, corresponding to condition 3.345, includes a relatively large part of the response curve. For example, the quadrature component nearly reaches its maximum value when $\omega = 1/\tau_0$. The same behavior will be observed in considering the frequency behavior of induced currents in elliptical conductors. It is relevant to note that the ascending branch of the spectral curve for the quadrature component arising from induction currents in spheroidal conductors also corresponds to low frequency portion of the spectrum. It is obvious that the same consideration can be applied for the inphase components. In other words, a significant part of the frequency responses of the quadrature and inphase components can be calculated using eqs. 3.344 and 3.346, provided that the inequality:

$$\omega < 1/\tau_0$$

can be satisfied. On occasion, it is more convenient to write eq. 3.345 in the form:

$$b/h < (q_1/2)^{1/2}$$

h is the skin depth.

Expression of the field in terms of a MacLauren series over a large part of the frequency-response curve allows us to see very clearly the important features of its behavior within this range, being what we term the low frequency part of the spectrum.

Let us briefly examine this portion of the spectrum, since its principle features are precisely the same as those of the field caused by conductors with axial symmetry:

(1) At very low frequencies, when we can discard all the terms in the series 3.344 but the first, the quadrature component of the magnetic field is directly proportional to both the conductivity and frequency, while the inphase component is smaller, but more closely dependent on conductivity (σ^2). For example, in this case we have:

$$Q H_{1z}^e \approx C_{1z} \sigma\mu b^2 \omega H_{0i}$$
$$\text{In } H_{1z}^e \approx -C_{2z} (\sigma\mu b^2)^2 \omega^2 H_{0i} \qquad (3.347)$$

In the general case, the coefficients C_{1z} and C_{2z} depend on the geometry in different manner. However, quite frequently this difference is relatively small. Then, in accord with eq. 3.341, we have:

$$QH_{1z}^e \approx i\frac{d_{1z}}{q_1}\sigma\mu b^2\omega H_{0i} = d_{1z}\omega\tau_0 H_{0i}$$

$$\text{In}H_{1z}^e \approx -\frac{d_{1z}}{q_1^2}(\sigma\mu b^2)^2\omega^2 H_{0i} = -d_{1z}(\omega\tau_0)^2 H_{0i}$$

(3.348)

In these approximations, the dependence of both components on the geometrical parameters is the same, and therefore, by taking the ratio of the two, we can determine the time constant from measurements made at a single frequency. In fact, we have:

$$\left|\frac{\text{In}H_{1z}^e}{QH_{1z}^e}\right| \approx \omega\tau_0 \quad\text{or}\quad \tau_0 \approx \frac{1}{\omega}\left|\frac{\text{In}H_{1z}^e}{QH_{1z}^e}\right|, \quad\text{as } \omega \to 0$$

(3.349)

Similar expressions can be written for the horizontal component of the magnetic field and for the electric field. Equation 3.349 can be considered to be the relationship that we can use to evaluate the time constant characterizing the late-state behavior of the transient response.

In the approximation considered here, the dependence of all the terms in the series 3.344 or 3.346 on the coordinates of the observation point is the same. This is a consequence of the fact that in eqs. 3.341 and 3.343, all terms except the first have been discarded. For this reason, the low-frequency spectrum can be written in terms only of the first fraction in eqs. 3.335. Thus, as an approximation we have:

$$H_{1z}^e \approx H_{0i}k^2b^2\frac{d_{1z}}{q_1 - i\sigma\mu b^2\omega}$$

$$H_{1r}^e \approx H_{0i}k^2b^2\frac{d_{1r}}{q_1 - i\sigma\mu b^2\omega} \quad\text{if } \omega < 1/\tau_0$$

(3.350)

$$j_x \approx H_{0i}k^2b^2\frac{\beta_1}{q_1 - i\sigma\mu b^2\omega}$$

(2) Returning to the series describing the low frequency part of the spectrum, it should be noted that there are two main features, namely, with an increase in the number, n, the dependence of the terms on the conductivity of the conductor and its dimensions increases without limit, and the dependence of the coefficients C_n on the coordinates of the observation point becomes essentially the same. In accord with eqs. 3.341 and 3.343, we have:

$$C_{n-1,z} \to \frac{d_{1z}}{q_1^{n-1}} \quad C_{n,r} \to \frac{d_{1z}}{q_1^n} \quad C_{n+1,z} \to \frac{d_{1z}}{q_1^{n+1}}$$

$$C_{n-1,r} \to \frac{d_{1r}}{q_1^{n-1}} \quad C_{n,z} \to \frac{d_{1r}}{q_1^n} \quad C_{n+1,r} \to \frac{d_{1r}}{q_1^{n+1}}$$

(3.351)

Correspondingly, the time constant, τ_0, can be obtained from the following equation:

$$\tau_0 = \sigma\mu b^2 \lim_{n \to \infty} \frac{C_{n,z}}{C_{n-1,z}} = \sigma\mu b^2 \lim_{n \to \infty} \frac{C_{n,r}}{C_{n-1,r}} \tag{3.352}$$

However, for all practical purposes, in order to calculate τ_0 and from that the radius of convergence of these series, it is sufficient to consider only the first few terms in eqs. 3.340 and 3.342.

(3) Inasmuch as these series describe a significant part of the frequency response, and in particular to the range of frequencies where measurements are usually carried out, it is appropriate at this point to describe a relatively simple approach to calculating the coefficients C_{nz} and C_{nr}. Applying an approach similar to that which was used in the case of coaxial symmetry, we will start with the integral equation for the current density, j_x. Suppose that a cross-section of a conductor is symmetrical with respect to plane $y = 0$ and primary electrical field has the same magnitude and opposite sign at points with coordinates y, z and $-y, z$ (see Fig. 3.51). The vector potential of the electrical type A_x, created by currents flowing in two parallel lines but in opposite directions, $+I$ and $-I$ can be represented as:

$$A_x = \frac{I}{2\pi} \ln \frac{r_2(p, g)}{r_1(p, g)} \tag{3.353}$$

where

$$r_1 = [(z_p - z_g)^2 + (y_p - y_g)^2]^{1/2}, \quad r_2 = [(z_p - z_g)^2 + (y_p + y_g)^2]^{1/2}$$

In the integral equation 3.324:

$$J_x(p) = J_{0x}(p) + \frac{k^2}{2\pi} \int_S J_x(g) G_1(p, g) \mathrm{d}S$$

where $k^2 = i\sigma\mu\omega$, $G_1(p, g) = \ln r_2/r_1$ in this case.

S is a half cross-section of the conductor. Substituting the first equation of the set 3.342:

$$J_x = H_{0i} \sum_{n=1}^{\infty} \alpha_n (k^2 b^2)^n$$

into the integral equation for j_x, we can again find the recurrence relationship between the coefficients α_n:

$$\alpha_1 k^2 b^2 H_{0i} = J_{0x} = \sigma E_{0x} = i\sigma\mu\omega H_{0i} G_0$$

or
$$\tag{3.354}$$

$$\alpha_1 = \frac{1}{b^2} G_0$$

and

$$\alpha_n = \frac{1}{b^2} \int_S \alpha_{n-1} G_1 \, dS \qquad (3.355)$$

where G_0 is a function which depends on the type of primary field, that is, on the geometry of the source. In accord with the Biot-Savart law, we can write the following expressions for the coefficients C_{nz} and C_{nr}:

$$C_{nz} = \frac{1}{2\pi} \int_S \frac{\alpha_n(p)(y_g - y_p)}{(y_p - y_g)^2 + (z_p - z_p)^2} \, dS$$

$$(3.356)$$

$$C_{nr} = \frac{1}{2\pi} \int_S \frac{\alpha_n(p)(z_g - z_p)}{(y_p - y_g)^2 + (z_p - z_g)^2} \, dS$$

S is whole cross section of the conductor.

In particular, on the z-axis:

$$C_{nz} = -\frac{1}{2\pi} \int_S \frac{\alpha_n(p) y_p}{y_p^2 + (z_p - z_g)^2} \, dS \qquad (3.357)$$

Thus, the method of successive approximations, which requires evaluation only of single integrals, permits us to find all components of the electromagnetic field at low frequencies.

Next, we will examine very briefly the high-frequency part of the spectrum. As the frequency tends to infinity, because of the skin effect, induction currents will concentrate near the surface of the conductive body, and the field behaves in the same manner as though it were perfectly conducting. Therefore, in the limit, we have the following condition on the surface:

$$E_x = 0 \qquad (3.358)$$

At the same time the electromagnetic field inside the conductor vanishes. This arises because the primary and secondary magnetic fields cancel each other. This condition defines a distribution of induction currents over the surface of the conductive body. However, the induced currents are located within a very thin layer near the surface, and the total induced current flowing in this infinitesimal layer is practically independent of conductivity and of frequency. Therefore, in the limit, the magnetic field caused by this current is defined only by geometric factors. As a particular case, outside the conductor, we have:

$$\lim |H^s(\omega)| \to |A(g)H_{0i}|, \quad \text{as } \omega \to \infty \qquad (3.359)$$

where $A(g)$ is a real function; that is, the quadrature component at the high frequency limit disappears.

Evaluation of the secondary field in this approximation, that is, of the function $A(q)$ usually is not a particularly difficult problem numerically. This is a consequence of the following facts. First of all, outside the conductor the magnetic field satisfies Laplace's equation:

$$\nabla^2 H^e = 0$$

Also, in view of eq. 3.358, the normal component of the total field is zero:

$$H_n^s = -H_n^0 \quad \text{on } S \tag{3.360}$$

where H_n^0 is the normal component of the primary field, which is known. This information is sufficient to define the components of the magnetic field. We should note that the asymptotic values of the field can be expressed in terms of the functions d_{ni}. In accord with eq. 3.335, we obtain:

$$H_{1z}(a) \to -H_{0i} \sum_{n=1}^{\infty} d_{nz}(a)$$

$$\tag{3.361}$$

$$H_{1r}(a) \to -H_{0i} \sum_{n=1}^{\infty} d_{nr}(a)$$

A significant improvement on this asymptotic expression can be obtained by again making use of the impedance relationship for the tangential components of the electromagnetic field at the surface of the conductor. As is well known:

$$\frac{E_x^e}{H_t^e} = Z = \left(\frac{\omega\mu}{i\sigma}\right)^{1/2}, \quad \text{on } S \tag{3.362}$$

provided that the skin depth is less than the radius of curvature. Here, H_t^e is the tangential component of the total magnetic field, perpendicular to E_x.

Making use of the boundary condition 3.362, we can represent the magnetic field at high frequencies in the form:

$$\text{In} H_{1r} \approx H_{0i} \left\{ A_r - \frac{a_r}{(\omega\mu\sigma)^{1/2}} \right\}$$

$$QH_{1r} \approx H_{0i} \frac{a_r}{(\omega\mu\sigma)^{1/2}}$$

$$\tag{3.363}$$

$$\text{In} H_{1z} \approx H_{0i} \left\{ A_z - \frac{a_z}{(\omega\mu\sigma)^{1/2}} \right\}$$

$$QH_{1z} \approx H_{0i} \frac{a_z}{(\omega\mu\sigma)^{1/2}}$$

where

$$A_r = - \sum_{n=1}^{\infty} d_{nr}, \quad A_z = - \sum_{n=1}^{\infty} d_{nz}$$

and where a_r and a_z are coefficients which depend only on geometry.

It is important to note that the magnetic field approaches its asymptotic value in inverse proportion to the square root of $\omega\sigma$. This behavior of the field is a consequence of the boundary conditions in eq. 3.362, and it is inherent for all conductive bodies regardless of their shape or dimensions.

We will now consider some general features of the transient responses of the electromagnetic field from linear currents in cylindrical conductors. As follows from eqs. 3.336:

$$J_x(g) = H_{0i} \sum_{n=1}^{\infty} \beta_n(g) e^{-g_n \alpha t}$$

$$H_{1z}(a) = H_{0i} \sum_{n=1}^{\infty} d_{nr}(a) e^{-g_n \alpha t} \tag{3.364}$$

$$H_{1r}(a) = H_{0i} \sum_{n=1}^{\infty} d_{nr}(a) e^{-g_n \alpha t}$$

the general features of the transient responses are the same as those for fields caused by induced currents in conductive bodies characterized by axial symmetry. For this reason, we can restrict our consideration to the most important characteristics of these responses. Assuming that the wave form of the current in the source is a step function, we arrive at the following results:

(1) At the instant that current switches, the magnetic field is independent of conductivity, and we have

$$H_{1z}(a) \to H_{0i} \sum_{n=1}^{\infty} d_{nz}(a)$$

$$\tag{3.365}$$

$$H_{1r}(a) \to H_{0i} \sum_{n=1}^{\infty} d_{nr}(a)$$

Comparing eqs. 3.361 and 3.365, we see that at high frequencies and during the very early stage the asymptotic values for the field amplitude coincide.

(2) Applying the Fourier transform to eqs. 3.363 in the same manner as was done for conductive bodies with axial symmetry, we obtain:

$$H_{1z}(t) \approx - H_{0i} \left\{ A_z - 2 \left(\frac{2}{\pi\sigma\mu} \right)^{1/2} a_z \sqrt{t} \right\} \tag{3.366}$$

$$H_{1r}(t) \approx -H_{0i}\left\{A_r - 2\left(\frac{2}{\pi\sigma\mu}\right)^{1/2} a_r\sqrt{t}\right\} \qquad (3.366)$$

and

$$\frac{\partial H_{1z}(t)}{\partial t} \approx \left(\frac{2}{\pi\sigma\mu}\right)^{1/2} a_z \frac{H_{0i}}{\sqrt{t}}$$

$$(3.367)$$

$$\frac{\partial H_{1r}(t)}{\partial t} \approx \left(\frac{2}{\pi\sigma\mu}\right)^{1/2} a_r \frac{H_{0i}}{\sqrt{t}}$$

Therefore, the magnetic field during the early stage of transient coupling approaches its zero-time asymptote as a function which is directly proportional to square root of time, and therefore, the electromotive force increases without limit.

The dependent of the second term in eq. 3.366 on the conductivity and on the geometry is precisely the same as that for the second term in eq. 3.363, the equation which describes the inphase component of the magnetic field, or of the leading term of the quadrature component, which is much smaller. For this reason, the electromotive force of the transient field during the early stage is related with the conductivity and the geometry in the same manner as is the electromotive force caused by the quadrature component of the magnetic field at high frequencies. As follows from eq. 3.363, the electromotive force caused by the inphase component of the magnetic field is directly proportional to the factor $H_{0i}A_z\omega$ or the factor $H_{0i}A_r\omega$, and therefore, it is independent of conductivity and its dependence on geometry is fundamentally different than that of the electromotive force in time domain. These considerations show that while the leading term of the magnetic field at high frequencies behaves in the same manner as the early stage of the transient magnetic field, the behavior of the electromotive force of the transient field and the leading term in the expression for EMF at high frequencies is different.

(3) During the late stage, the behavior of induced currents and of the electromagnetic field as functions of time is remarkably simple. In accord with eq. 3.364, we have:

$$J_x(g) \approx H_{0i}\beta_1(g)e^{-t/\tau_0}$$

$$H_{1z}(a) \approx H_{0i}d_{1z}(a)e^{-t/\tau_0} \qquad \text{if } t > \tau_0 \qquad (3.368)$$

$$H_{1r}(a) \approx H_{0i}d_{1r}(a)e^{-t/\tau_0}$$

Thus, during the late state, the linear induced currents and the electromagnetic field decay exponentially, in exactly the same manner as did the currents in the field for models characterized by axial symmetry. Moreover, the change with time is defined a time constant, τ_0, given by:

$$\tau_0 = \sigma \mu b^2 / q_1$$

As was pointed out earlier in this chapter, a remarkable feature of the equations describing the late stage is that every one of them can be considered as a product of two terms. The first term is $H_{0i}d_{1z}$ of the term $H_{0i}d_{1r}$, which is dependent only on the primary field and on the geometric parameters such as the position of the conductive body, its shape, and its dimensions, as well as the coordinates of the observation point. The second term, e^{-t/τ_0}, is a function of the time constant, τ_0, only. That is, it depends on the conductivity, the shape, and the dimensions of the conductive body. It is quite important to recognize that this term is independent of the strength of the primary field in the vicinity of the conductive body, that is, it is independent of its position and of the coordinates of the observation point. As will be demonstrated in a later chapter, this behavior for the transient response during the late stage is a decisive factor which defines the maximum depth of the investigation and the greatest resolution that can be obtained with the transient method in the presence of geological noise.

In accord with eq. 3.368, during the late stage, the time constant τ_0 can be easily defined. Moreover, it follows from eq. 3.352, that information about the time constant resides in higher order terms of the series that describe the low frequency behavior in the frequency domain. In other words, in order to calculate the time constant, τ_0, in the frequency domain it is necessary to invent a system for measuring only the higher order terms in these series. However, in order to obtain an approximate evaluation of the time constant, it is adequate to measure the leading terms for the quadrature and inphase components as follows from eq. 3.348:

$$QH_z^e \approx d_{1z}\omega\tau_0 H_{0i}, \quad InH_{1z}^e \approx -d_{1z}(\omega\tau_0)^2 H_{0i}$$

or
$$(3.369)$$

$$QH_r^e \approx d_{1r}\omega\tau_0 H_{0i}, \quad InH_{1r}^e \approx -d_{1r}(\omega\tau_0)^2 H_{0i}$$

and the time constant can be evaluated from the ratio of quadrature and inphase components. It is appropriate to note here that such a determination of the time constant will not involve any particularly large error in most cases. Moreover, in comparing eq. 3.368 with eq. 3.369, we see that the dependence on geometry when the quadrature and inphase components are measured is practically the same as for measurements of the transient response during the late stage. Therefore, one might say that in the cases in which geological and ambient noise can be neglected, the presently used frequency and transient methods are equivalent from the point of view of their capabilities for detecting a cylindrical conductive body and determination of its parameters. Of course, this conclusion holds both for models with axial symmetry and for models with more complicated geometry. However, when geological noise must be taken into account, there will be a different sensitivity of the field to the time constant τ_0 at low frequencies in the frequency domain and during the late stage in the time domain, which is a fundamental factor leading to differing depths of investigation for the frequency- and time-domain methods now in wide use. This will be the subject of a detailed investigation in a later chapter.

Let us now consider as an example the frequency- and time-domain responses caused by currents induced in an elliptical cylinder. Assume that a cylinder with semi-axes a and b is placed in a uniform magnetic field, $H_0 e^{i\omega t}$, which can be, as a special case, generated by two linear currents, $+I$ and $-I$, oriented parallel to the x-axis and located at the points $(0, -y', 0)$ and $(0, y', 0)$ at a relatively large distance from the conductor (see Fig. 3.52). Assuming that the primary magnetic field is uniform, its vortex electric field is non-uniform, and applying Faraday's law, we have:

$$E_{0x} = -i\omega\mu H_{0z}y, \quad \text{if } |y| \ll |y'| \tag{3.370}$$

Inasmuch as the electric field E_{0x} does not intersect the surface of the conductor, the only source for a secondary field will be vortex currents which are parallel to the axis of the cylinder. In the plane $y = 0$, the current density is zero, and at points located symmetrically with respect to this plane the current density will differ only in sign. For this reason, it is convenient to assume that currents are closed at infinity. As was mentioned

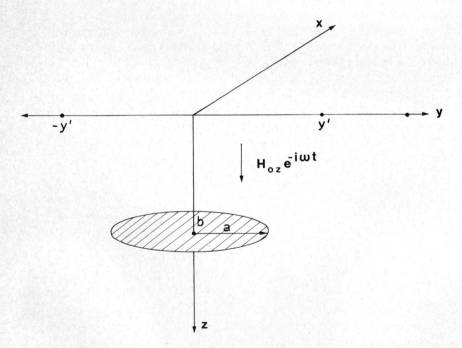

Fig. 3.52. Geometry of a cylinder with semi-axes a and b in a uniform magnetic field.

previously, the method of separation of variables cannot be used successfully in our case, and numerical methods are the principal approach that permits us to calculate the frequency and transient responses. In particular, the numerical results contained in this section were obtained by using the method of integral equations. We will consider two special cases, namely:

(1) that in which the primary magnetic field H_{0z} is directed along the minor semi-axis b (see Fig. 3.53a); and

(2) in which the primary magnetic field H_{0z} is directed along the major and semi-axis a (see Fig. 3.53b).

In the more general case in which a uniform primary magnetic field is oriented arbitrarily with respect to the axial orientation of the elliptical cylinder (as shown in Fig. 3.53c), the secondary field can be considered to be the sum of two fields corresponding to each of the previous two cases. Let us start at the low frequency range of the spectrum. In accord with eqs. 3.340 and 3.342 we have:

$$J_x = H_{0z} \sum_{n=1}^{\infty} \alpha_n (kb)^{2n}$$

$$H_z = H_{0z} \sum_{n=1}^{\infty} C_{nz} (kb)^{2n} \qquad (3.371)$$

$$H_r = H_{0z} \sum_{n=1}^{\infty} C_{nr} (kb)^{2n}$$

Using successive approximations to the integral equation with respect to the current density j_x and applying the Biot-Savart law, we find the coefficients C_{nz} and C_{nr} describing the low frequency portion of the spectrum for the quadrature and inphase components. Examples of values for the coefficients C_{nz} for points situated on the z-axis are given in Tables 3.XIII and 3.XIV for the case in which the primary magnetic field H_0 is directed either along the minor axis or the major axis of the cylinder. Corresponding values for the parameter q_1 are also given. As may be seen from the values listed in these tables, the ratios of the coefficients C_{nz} rapidly approach the value q_1 which in turn defines the radius of convergence for the power series 3.371. A curve showing the behavior of the parameter q_1 as a function of the ratio of the two semi-axes for the elliptical cylinder is shown in Fig. 3.54.

As the calculated values demonstrate, when the ratio of the axial lengths is great enough, that is when the ratio a/b is much greater than unity, and when the primary field is directed along the minor axis as shown in Fig. 3.53a, we have the following approximate expression for q_1:

$$q_1 \approx 5.49 \, b/2a \qquad (3.372)$$

If in place of the elliptical plate we have a plate with a constant thickness, $2b$, the coefficient q_1 is:

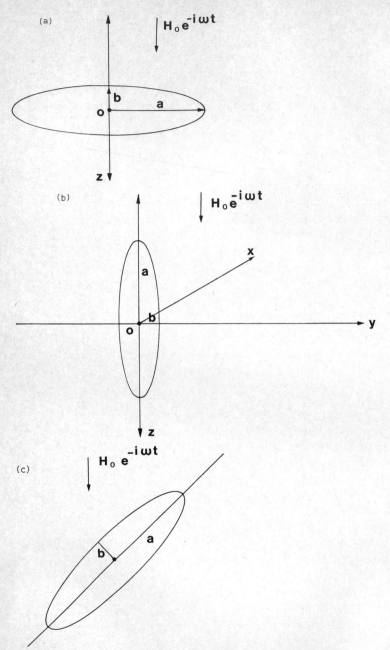

Fig. 3.53. Special cases for which numerical results have been obtained using the integral equation method. a. The case in which the primary magnetic field is directed along the minor semi-axis of the cylinder. b. The case in which the primary magnetic field is directed along the major semi-axis of the cylinder. c. The more general case in which the primary magnetic field is oriented arbitrarily with respect to the axial orientation of the elliptical cylinder.

$$q_1 \approx 4.01\, b/2a \qquad (3.373)$$

In the case when the primary magnetic field is directed along the major axis, the parameter q_1 rapidly approaches its limiting value:

$$q_1 \approx 3.0, \quad \text{if } a/b \gg 1 \qquad (3.374)$$

In accord with eqs. 3.341 and 3.342, we have the following approximations:

$$C_{1z} \approx \frac{d_{1z}}{q_1}, \quad C_{2z} \approx \frac{d_{1z}}{q_1{}^2}, \quad C_{3z} \approx \frac{d_{1z}}{q_1{}^3}$$

$$\qquad (3.375)$$

$$C_{1r} \approx \frac{d_{1r}}{q_1}, \quad C_{2r} \approx \frac{d_{1r}}{q_1{}^2}, \quad C_{3r} \approx \frac{d_{1r}}{q_1{}^3}$$

and so on.

As follows from the data listed in Tables 3.XIII and 3.XIV, the accuracy with which these approximations can be applied depends on the ratio of the two semi-axes a/b. If the elliptical cylinder is not strongly elongate, the error in calculating the coefficient C_{1z} and C_{1r} will not exceed 10% and becomes even smaller for subsequent terms in the low-frequency series. However, for strongly elongated cylinders, eq. 3.375 could be used with the same accuracy

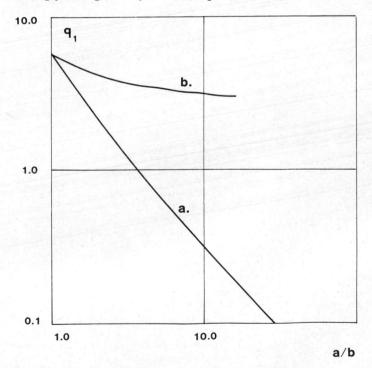

Fig. 3.54. Curves showing the behavior of the parameter q_1 as a function of the ratio of the semi-axes for an elliptical cylinder.

TABLE 3.XIII

a/b	q_1	z_a/b	C_{1z}	C_{2z}	C_{3z}	C_{4z}	C_{5z}	C_{6z}	C_{7z}
1	5.78	1.1	0.103	0.172×10^{-1}	0.296×10^{-2}	0.511×10^{-3}	0.884×10^{-4}	0.153×10^{-4}	0.264×10^{-5}
		1.5	0.556×10^{-1}	0.924×10^{-2}	0.159×10^{-2}	0.275×10^{-3}	0.475×10^{-4}	0.822×10^{-5}	0.142×10^{-5}
		2.0	0.312×10^{-1}	0.520×10^{-2}	0.895×10^{-3}	0.154×10^{-3}	0.267×10^{-4}	0.462×10^{-5}	0.797×10^{-6}
		3.0	0.139×10^{-1}	0.231×10^{-2}	0.398×10^{-3}	0.687×10^{-4}	0.119×10^{-4}	0.206×10^{-5}	0.354×10^{-6}
2	2.16	1.1	0.402	0.190	0.889×10^{-1}	0.412×10^{-1}	0.191×10^{-1}	0.883×10^{-2}	0.409×10^{-2}
		1.5	0.278	0.129	0.600×10^{-1}	0.278×10^{-1}	0.129×10^{-1}	0.595×10^{-2}	0.275×10^{-2}
		2.0	0.185	0.848×10^{-1}	0.392×10^{-1}	0.181×10^{-1}	0.839×10^{-2}	0.388×10^{-2}	0.180×10^{-2}
		3.0	0.957×10^{-1}	0.431×10^{-1}	0.199×10^{-1}	0.919×10^{-2}	0.425×10^{-2}	0.197×10^{-2}	0.910×10^{-3}
4	0.889	1.1	1.22	1.46	1.67	1.89	2.13	2.40	2.70
		1.5	1.00	1.18	1.34	1.52	1.71	1.92	2.16
		2.0	0.791	0.911	1.03	1.16	1.31	1.47	1.66
		3.0	0.513	0.575	0.647	0.728	0.818	0.920	1.03
8	0.395	1.1	3.08	8.55	22.4	57.3	146	369	934
		1.5	2.79	7.63	19.9	50.8	129	327	827
		2.0	2.47	6.65	17.2	43.9	111	282	714
		3.0	1.94	5.09	13.1	33.2	84.1	213	539
16	0.185	1.1	7.00	42.3	239	1320	7180	38900	211000
		1.5	6.65	39.8	224	1230	6710	36400	197000
		2.0	6.25	37.1	208	1140	6200	33600	182000
		3.0	5.53	32.2	179	981	5330	28900	156000

TABLE 3.XIV

a/b	q_1	$z_{a/b}$	C_{1z}	C_{2z}	C_{3z}	C_{4z}	C_{5z}	C_{6z}	C_{7z}
2	4.33	3.0	0.337×10^{-1}	0.722×10^{-2}	0.163×10^{-2}	0.375×10^{-3}	0.864×10^{-4}	0.199×10^{-4}	0.460×10^{-5}
		4.0	0.173×10^{-1}	0.376×10^{-2}	0.856×10^{-3}	0.197×10^{-3}	0.454×10^{-4}	0.105×10^{-4}	0.242×10^{-5}
		6.0	0.725×10^{-2}	0.159×10^{-2}	0.363×10^{-3}	0.836×10^{-4}	0.193×10^{-4}	0.445×10^{-5}	0.103×10^{-5}
		8.0	0.440×10^{-2}	0.881×10^{-3}	0.201×10^{-3}	0.464×10^{-4}	0.107×10^{-4}	0.247×10^{-5}	0.570×10^{-6}
4	3.57	6.0	0.179×10^{-1}	0.448×10^{-2}	0.121×10^{-2}	0.332×10^{-3}	0.925×10^{-4}	0.259×10^{-4}	0.724×10^{-5}
		8.0	0.889×10^{-2}	0.229×10^{-2}	0.622×10^{-3}	0.172×10^{-3}	0.480×10^{-4}	0.134×10^{-4}	0.377×10^{-5}
		10.0	0.541×10^{-2}	0.141×10^{-2}	0.384×10^{-3}	0.107×10^{-3}	0.297×10^{-4}	0.833×10^{-5}	0.233×10^{-5}
		15.0	0.230×10^{-2}	0.602×10^{-3}	0.165×10^{-3}	0.459×10^{-4}	0.128×10^{-4}	0.359×10^{-5}	0.107×10^{-5}
8	3.17	10.0	0.155×10^{-1}	0.403×10^{-2}	0.117×10^{-2}	0.355×10^{-3}	0.110×10^{-3}	0.344×10^{-4}	0.108×10^{-4}
		15.0	0.520×10^{-2}	0.146×10^{-2}	0.438×10^{-3}	0.135×10^{-3}	0.421×10^{-4}	0.132×10^{-4}	0.418×10^{-5}
		20.0	0.272×10^{-2}	0.776×10^{-3}	0.235×10^{-3}	0.726×10^{-4}	0.227×10^{-4}	0.715×10^{-5}	0.226×10^{-5}
		30.0	0.115×10^{-2}	0.332×10^{-3}	0.101×10^{-3}	0.313×10^{-4}	0.982×10^{-5}	0.309×10^{-5}	0.977×10^{-6}
16	3.04	20.0	0.779×10^{-2}	0.206×10^{-2}	0.610×10^{-3}	0.190×10^{-3}	0.605×10^{-4}	0.196×10^{-4}	0.641×10^{-5}
		30.0	0.261×10^{-2}	0.748×10^{-3}	0.231×10^{-3}	0.732×10^{-4}	0.237×10^{-4}	0.773×10^{-5}	0.254×10^{-5}
		40.0	0.136×10^{-2}	0.398×10^{-3}	0.124×10^{-3}	0.395×10^{-4}	0.128×10^{-4}	0.420×10^{-5}	0.138×10^{-5}
		60.0	0.576×10^{-3}	0.170×10^{-3}	0.533×10^{-4}	0.171×10^{-4}	0.556×10^{-5}	0.182×10^{-5}	0.600×10^{-6}

in calculating both the coefficients C_{2z} and C_{2r} and the following terms. Correspondingly, for all the cases that had been considered, the coefficients C_{nz} and C_{nr} describing the low-frequency part of the spectrum depend on the coordinates of the observation point in almost the same way, except the first coefficients C_{nz} and C_{nr}, when the ratio of semi-axes is large enough. It should be remembered here that the coefficients C_{1z} and C_{1r} as well as the coefficients C_{2z} and C_{2r} characterize the leading terms of the series that describe the quadrature and inphase components, respectively. In accord with eq. 3.350, the low-frequency portion of the spectrum can be approximated as follows:

$$H_{1z}(a) \approx H_{0z}k^2b^2 \frac{d_{1z}}{q_1 - i\sigma\mu\omega b^2}$$

$$\text{if } \omega < 1/\tau_0 \qquad (3.376)$$

$$H_{1r}(a) \approx H_{0z}k^2b^2 \frac{d_{1r}}{q_1 - i\sigma\mu\omega b^2}$$

As follows from eq. 3.356, in evaluating the field we can make use of the following expressions for C_{1z} and C_{1r}:

$$C_{1z} = \frac{1}{2\pi} \int_0^\infty \frac{\alpha_1(p)(y_a - y_p)}{(y_p - y_a)^2 + (z_p - z_a)^2} \, dS$$

$$ (3.377)$$

$$C_{1r} = \frac{1}{2\pi} \int_0^\infty \frac{\alpha_1(p)(z_a - z_p)}{(y_p - y_a)^2 + (z_p - z_a)^2} \, dS$$

where $\alpha_1(p)$ is defined as follows. Considering that the current density due to the primary vortex electric field E_{0x} is:

$$j_{0x} = -i\sigma\mu\omega H_{0z}y_p = -k^2b^2 \frac{H_{0z}}{b^2} y_p$$

we will have:

$$\alpha_1 = -\frac{1}{b^2} y_p \qquad (3.378)$$

Thus, computation of the coefficients C_{1z} and C_{1r} does not require the use of the method of successive approximations, while for determination of the coefficients C_{2z} and C_{2r}, it is necessary to carry out only one successive approximation. Considering that the parameter q_1 is known, the coefficients C_{nz} and C_{nr} corresponding to later terms in the series 3.371 can be found from eq. 3.351.

At this point we will examine the high frequency behavior. As has been mentioned earlier, when the skin depth in the conductor is significantly less than its radius of curvature, the field inside the conductor, near its surface, is

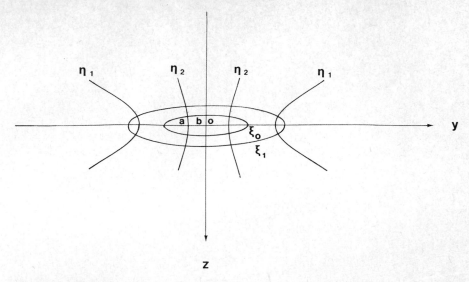

Fig. 3.55. Definition of an elliptical coordinate system related to the dimensions of an elliptical cylinder.

close to that for a plane wave, and the tangential components of the electric and magnetic fields are related by the equation:

$$E_x^i/H_t^i = Z = \left(-\frac{i\omega\mu}{\sigma}\right)^{1/2} \tag{3.379}$$

where H_t^i is the tangential component of the magnetic field perpendicular to E_x. In view of the required continuity of the tangential components, this relationship is valid for the external components of the field as well, that is:

$$E_x^e/H_t^e = Z, \quad \text{on } S \tag{3.380}$$

For this reason, at high frequencies, determination of the field outside the cylinder is markedly simplified under certain conditions. Let us use an elliptical coordinate system, related with the elliptical cylinder. Suppose for example that the horizontal semi-axis a is longer than the vertical semi-axis b, as shown in Fig. 3.55. Then

$$y = \alpha \cosh \xi \cos \eta$$

$$z = \alpha \sinh \xi \sin \eta$$

where $0 \leqslant \eta \leqslant 2\pi$ and $0 \leqslant \xi < \infty$.

The foci for the system are situated at the points $y = \pm \alpha, z = 0$. The coordinate ξ_0 corresponds to the surface of the elliptical cylinder. Therefore:

$$a = \alpha \cosh \xi_0 \quad \text{and} \quad b = \alpha \sinh \xi_0$$

that is $\alpha = (a^2 - b^2)^{1/2}$.

Outside the cylinder, in the insulating medium, the electric field satisfies Laplace's equation:

$$\frac{\partial^2 E_x}{\partial \xi^2} + \frac{\partial^2 E_x}{\partial \eta^2} = 0 \tag{3.381}$$

In the external medium, we will seek an approximate solution using a single spatial harmonic:

$$E_x = -i\omega\mu H_{0z}\alpha(\cosh \xi + Te^{-\xi})\cos \eta \tag{3.382}$$

where T is some unknown function. From Maxwell's equation that states that the curl $E = -\dot{B}$, we obtain:

$$H_\eta \approx \frac{\alpha H_{0z}}{h}\{\sinh \xi - Te^{-\xi}\}\cos \eta \tag{3.383}$$

where h is a metric factor:

$$h = \alpha (\cosh^2\xi - \cos^2\eta)^{1/2}$$

Applying the boundary condition expressed in 3.380, we obtain:

$$\frac{E_x(\xi_0)}{H_\eta(\xi_0)} \approx -i\omega\mu h(\xi_0, \eta)\frac{\cosh \xi_0 + Te^{-\xi_0}}{\sinh \xi_0 - Te^{-\xi_0}} \cong Z \tag{3.384}$$

where ξ_0 is the coordinate on the surface of the conductor. Therefore:

$$T = \frac{\sinh \epsilon_0 + \sigma Zh \cosh \epsilon_0}{1 - \sigma Zh} e^{\xi_0} \tag{3.385}$$

With increasing frequency, the function T becomes a constant, equal to $-\cosh \xi_0 e^{\xi_0}$, and the total electric field on the conductor surface approaches zero, at the same time that the ratio $H_\eta/H_{0\eta}$ approaches a constant value equal to $1 + a/b$.

It should be noted that application of the approximate boundary condition leads to representation of the field in the internal medium in terms of a sum of spatial harmonics, and therefore, in a strict sense, it will be necessary to solve an infinite system of equations with an infinite number of unknowns. However, a comparison of calculated results obtained with the integral equation method indicates that representation of the field using only the first harmonic (as in eqs. 3.382–3.383) provides reasonable accuracy for the conditions that $h/a < 0.1$ and $b/a < 0.25$, and when the observation point is situated near the central part of the cylinder.

Having expanded the function T in a series of powers in $\omega^{-1/2}$, we obtain asymptotic expressions for the magnetic field component H_ξ and H_η, which correspond to eq. 3.363:

$$\ln H_\xi \approx H_{0z} \left\{ A_\xi - \frac{a_\xi}{(\omega\mu\sigma)^{1/2}} \right\}$$

$$QH_\xi \approx -H_{0z} \frac{a_\xi}{(\omega\mu\sigma)^{1/2}}$$

$$\ln H_\eta \approx H_{0z} \left\{ A_\eta - \frac{a_\eta}{(\omega\mu\sigma)^{1/2}} \right\}$$

(3.386)

$$QH_\eta \approx -H_{0z} \frac{a_\eta}{(\omega\mu\sigma)^{1/2}}$$

where the coefficients A_ξ, a_ξ and A_η, a_η can be readily obtained from eqs. 3.382, 3.383 and 3.384. Examples of the frequency responses of the tangential components of the magnetic and electric field observed on the surface of the elliptical cylinder obtained by evaluation of the integral equations are shown in Figs. 3.56—3.59. The parameter specifying the individual curves is the value for the coordinate r:

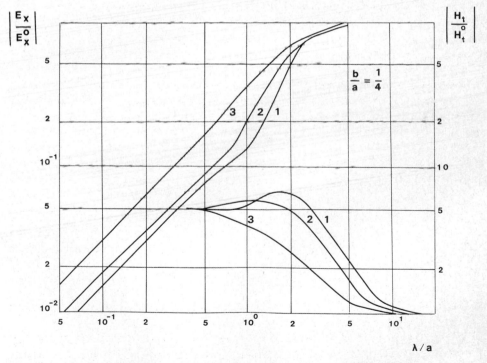

Fig. 3.56. Curves for the frequency response of the tangential component of the magnetic and electric field at the surface of an elliptical cylinder for the case $b/a = 1/4$, $\lambda = 2\pi h$.

$$\Phi\!\left(\frac{E_x}{E_x^o}\right),\Phi\!\left(\frac{H_t}{H_t^o}\right)$$

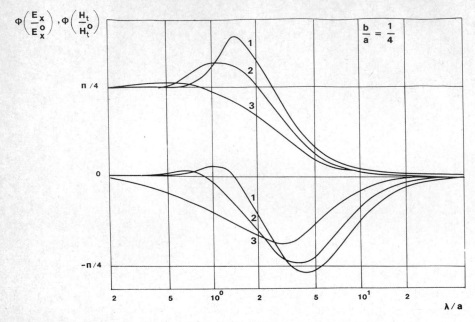

Fig. 3.57. Examples of the phase of the tangential components of the magnetic and electric fields observed on the surface of an elliptic cylinder for the case $b/a = 1/4$.

$$r_1 = 0.23a, \quad r_2 = 0.65a, \quad r_3 = 0.92a$$

As may be seen from these curves, the frequency response of the field near the conductor depends strongly on the coordinates of the observation point.

At this point we can examine the transient response of the magnetic field caused by currents induced in the elliptical cylinder. From eq. 3.364, during the late stage we have:

$$H_{1z}(a) \approx H_{0z}d_{1z}(a)e^{-t/\tau_0}$$
$$H_{1r}(a) \approx H_{0z}d_{1r}(a)e^{-t/\tau_0} \quad , \text{if } t > \tau_0 \tag{3.387}$$

where $\tau_0 = \sigma\mu b^2/q_1$ is a time constant.

As follows from the values listed in Table 3.XIII, with increasing length of the major semi-axis a (remembering that the primary field is directed along the minor semi-axis b), the time constant increases also and in correspondence with this, the late-stage behavior begins to manifest itself at a later time. From eq. 3.372, when the ratio a/b is large enough, we have the following value for the time constant:

$$\tau_{0b}^p = \frac{\sigma\mu b^2}{5.49}\frac{2a}{b} = 0.182\mu S_0 a, \quad \text{if } \frac{a}{b} \gg 1 \tag{3.388}$$

where $S_0 = 2\sigma b$ is the longitudinal conductance at the central plane of the cylinder.

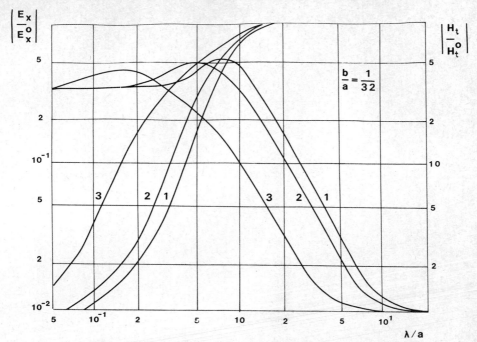

Fig. 3.58. Examples of the amplitude of the frequency response for the tangential compo·
nents of the magnetic and electric fields observed on the surface of an elliptic cylinder for
the case $b/a = 1/32$.

If the conductive body has a rectangular cross-section, that is, if the
longitudinal conductance is constant and a is much larger than b, by using
eq. 3.373 we have:

$$\tau_{0b}^{p(1)} = \frac{\sigma\mu b^2}{4.01} \frac{2a}{b} = 0.250 \, \mu S_0 a \tag{3.389}$$

Thus, in both of the two cases, the time constant is defined as the product of
longitudinal conductance and the length of the major semi-axis a, oriented
perpendicular to the direction of the primary magnetic field. If the primary
field H_{0z} is oriented along the major axis and the ratio a/b is large, in accord
with eq. 3.374 the time constant τ_{0a}^p is:

$$\tau_{0a}^p = 0.33 \, \sigma\mu b^2 = 0.165 \, \mu S_0 b, \quad \text{if } a/b \gg 1 \tag{3.390}$$

and therefore, it does not depend on the length of the major semi-axis a.

It is clear that when the primary field is arbitrarily oriented with respect
to the principal axes of the cylinder, the secondary field can be represented
as being the sum of two fields, each caused by induced currents in a case
when the primary field is parallel to one of the principle axes of the cylinder.
As an example, at the late stage we would have:

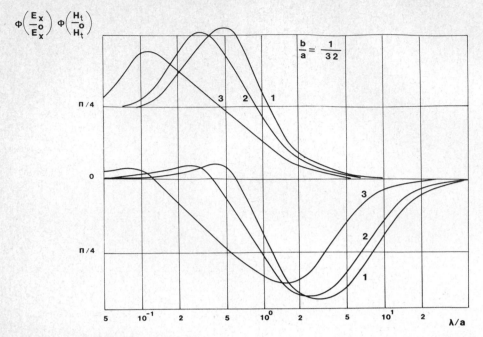

Fig. 3.59. Examples of the phase of the tangential components of the magnetic and electric fields observed on the surface of an elliptical cylinder for the case $b/a = 1/32$.

$$H_z = H_{0b} d_{1z}^{(b)} e^{-t/\tau_{0b}} + H_{0a} d_{1z}^{(a)} e^{-t/\tau_{0a}}$$
$$H_r = H_{0b} d_{1r}^{(b)} e^{-t/\tau_{0b}} + H_{0a} d_{1r}^{(a)} e^{-t/\tau_{0a}} \tag{3.391}$$

where H_{0b} and H_{0a} are the components of the primary magnetic field along the axes b and a respectively, while d_{1z}^b, d_{1r}^b and d_{1z}^a, d_{1r}^a are coefficients corresponding to the two principle directions, as well as τ_{0b} and τ_{0a}.

In accord with the values listed in Tables 3.XIII and 3.XIV, the following inequality is observed:

$$\tau_{0b} > \tau_{0a}, \quad \text{if } a/b > 1 \tag{3.392}$$

Therefore, during the late stage of a transient field, the field is controlled primarily by the first term in eq. 3.391. However, if the primary field is directed mainly along the major semi-axis so that H_{0a} is much greater than H_{0b}, there can be a range of times at the beginning of the late stage in which the second term of eq. 3.391 is dominant and the decay is defined by the time constant τ_{0a}. Behavior of the time constant as a function of the ratio of semi-axes is presented in Figs. 3.60. Along axis of ordinates, the ratios τ_{0b}/τ_{0b}^p and τ_{0a}/τ_{0a}^p are plotted, respectively. In accord with eqs. 3.388 and 3.90, we have:

$$\tau_{0b}^p = 0.182\,\mu S_0 a, \quad \tau_{0a}^p = 0.165\,\mu S_0 b$$

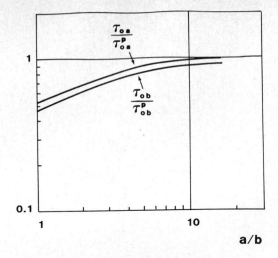

Fig. 3.60. Time constants observed for cylindrical conductors.

It might be noted here that the ratio of the time constant τ_{0b}^p and τ_{0a}^p for a relatively thin cylinder in which a is much larger than b is practically equal to the ratio of semi-axes, that is:

$$\tau_{0b}^p/\tau_{0a}^p \approx a/b, \quad \text{if } a/b \gg 1 \tag{3.393}$$

In earlier paragraphs, the effect of non-uniformity in the primary magnetic field has been considered. Some of these results can be applied in the case of the elliptical cylinder. In particular, during the late stage, the effect of non-uniformity becomes insignificant.

We have considered the late stage behavior in detail. Now, applying the Fourier transform to eqs. 3.386 and taking condition 3.366 into account, we can write asymptotic formulations for the early stage:

$$H_\xi = H_{0z}\left\{ A_\xi - 2\left(\frac{2}{\pi\sigma\mu}\right)^{1/2} a_\xi \sqrt{t} \right\}$$

$$H_\eta = H_{0z}\left\{ A_\eta - 2\left(\frac{2}{\pi\sigma\mu}\right)^{1/2} a_\eta \sqrt{t} \right\} \tag{3.394}$$

In conclusion, let us consider one more topic related with the behavior of the field caused by induction currents in infinitely long conductors. A comparison of the frequency- and time-domain responses caused by currents induced in conductors with axial symmetry and with cylindrical conductors provided that current density does not change along the axis demonstrates the complete similarity. For example, in both cases we have:

(1) Singularities in the spectrum which are simple poles located on the imaginary ω-axis.

(2) Spectrums which can be written as the infinite sum of simple fractions.

(3) The low-frequency part of the spectrum can be expanded as a MacLauren series containing only integer powers in ω.

(4) The transient response is described by an infinite sum of exponential terms.

(5) During the late stage the transient field delays exponentially, e^{-t/τ_0}, where τ_0 is a time constant determined by the position of the closest pole in the spectrum:

$$\omega = -i\,\frac{1}{\tau_0}$$

However, this coincidence of the principal characteristics of the field caused by currents in a confined conductor and in an infinitely long cylinder occurs if the current density does not change along the strike in the conductor, and it will not hold in the general case of cylindrical conductors. In order to illustrate this, let us consider a relatively simple example. Suppose that a vertical magnetic dipole with the moment $M_0 e^{i\omega t}$ is situated on the axis of a thin cylindrical shell with a radius a and having a conductance S as shown in Fig. 3.61. By definition, the longitudinal conductance S is the product of the thickness of the shell and its conductivity. In view of the

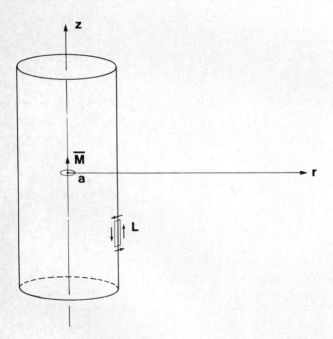

Fig. 3.61. A vertical magnetic dipole source situated on the axis of a thin cylindrical shell with a radius a and having a conductance S.

axial symmetry and the fact that the primary vortex electric field has only a single component $E_{0\phi}$, the currents induced in this shell also have only the component I_ϕ and therefore they are situated in horizontal planes. In order to find the field caused by surface currents with a density I_ϕ, we can start from Maxwell's equations. It is clear that inside and outside the shell we have:

$$\text{curl } E = -i\omega\mu H \quad \text{div } E = 0$$

$$\text{curl } H = 0 \qquad\qquad \text{div } H = 0$$

Assuming that the shell is vanishingly thin, we can introduce approximate boundary conditions on the surface of the conductor. In fact, applying Maxwell's equations in the integral form to the contour L as shown in Fig. 3.61, we have

$$\oint_L E \cdot dl = -i\omega\mu \int_S H \cdot dS$$

$$\int_L H \cdot dl = I$$

Taking the thickness of the shell to be vanishingly small, we can use approximate boundary conditions at the surface of the conductor. Using the above listed integral expressions of Maxwell's equations, we have:

$$\begin{aligned} E_\phi^e - E_\phi^i &= 0 \\ &\qquad\qquad , \quad \text{if } r = a \\ H_z^e - H_z^i &= -I_\phi \end{aligned} \qquad (3.395)$$

where the indices e and i indicate the field outside and inside the shell, respectively. First, we will introduce a vector potential of the magnetic type. As follows from Maxwell's third equation, we can write:

$$E = \text{curl } A^* \qquad (3.396)$$

where A^* is the vector potential. From the equation:

$$\text{curl } H = 0$$

we have:

$$H = -\text{grad } U^* \qquad (3.397)$$

where U^* is a scalar. Substituting eqs. 3.396 and 3.397 into the first equation of Maxwell's system, we obtain:

$$\text{curl curl } A^* = i\omega\mu \text{ grad } U^*$$

or

$$\text{grad div } A^* - \nabla^2 A^* = i\omega\mu \text{ grad } U^*$$

Making use of the gauge condition:

$$\mathrm{div}\, A^* = i\omega\mu U^* \tag{3.398}$$

we have:

$$\nabla^2 A^* = 0 \tag{3.399}$$

and both components of the electromagnetic field are expressed in terms of a single vector potential A^*:

$$E = \mathrm{curl}\, A^*$$
$$i\omega\mu H = -\,\mathrm{grad}\,\mathrm{div}\, A^* \tag{3.400}$$

As we know, the primary electromagnetic field can be described using only the vertical component of the vector potential, $A_z^{*(0)}$ which in free space is:

$$A_z^{*(0)} = i\omega\mu\,\frac{M}{4\pi}\frac{1}{R} \tag{3.401}$$

where $R = (r^2 + z^2)^{1/2}$.

For the same reason, we will seek a secondary field using the z component of the vector potential A^*. In view of the axial symmetry, eq. 3.399 can be written as:

$$\frac{\partial^2 A_z^*}{\partial r^2} + \frac{1}{r}\frac{\partial A_z^*}{\partial r} + \frac{\partial^2 A_z^*}{\partial z^2} = 0 \tag{3.402}$$

Using the method of separation of variables and considering the symmetry with respect to the z-axis, the solution can be assumed to be the product of a modified Bessel's function, either I_0 or K_0 and a cosinusoid. Inasmuch as the function $A_z^{*(0)}$ can be written as:

$$A_z^{*(0)} = i\omega\mu\,\frac{M}{2\pi^2}\int_0^\infty K_0(mr)\cos mz\,dm$$

we will seek a general solution in the form:

$$A_z^{*i} = i\omega\mu\,\frac{M}{2\pi^2}\int_0^\infty [K_0(mr) + C_m I_0(mr)]\cos mz\,dm$$

$$A_z^{*e} = i\omega\mu\,\frac{M}{2\pi^2}\int_0^\infty D_m K_0(mr)\cos mz\,dm \tag{3.403}$$

where C_m and D_m are unknown coefficients.

It is a simple matter to check that the functions A_z^{*i} and A_z^{*e} satisfy eq. 3.402. We will now derive the boundary conditions for this vector potential. In accord with eq. 3.400, in cylindrical coordinates we have:

$$E_\phi = -\frac{\partial A_z^*}{\partial r}, \quad H_z = -\frac{1}{i\omega\mu}\frac{\partial^2 A_z^*}{\partial z^2}$$

Therefore, in order to satisfy condition 3.395, it will be sufficient if the vector potential A_z^* satisfies the following conditions:

$$\frac{\partial A_z^{*(i)}}{\partial r} = \frac{\partial A_z^{*(e)}}{\partial r}$$

$$\frac{\partial^2 A_z^{*(i)}}{\partial z^2} - \frac{\partial^2 A_z^{*(e)}}{\partial z^2} = -i\omega\mu S\frac{\partial A_z^{*(e)}}{\partial r}$$

(3.404)

inasmuch as:

$$I_\phi = SE_\phi$$

Substituting expression 3.403 into the boundary conditions 3.404 and taking the orthogonality of the trigonometric functions into account, we obtain two equations with two unknown for each value of m:

$$-K_1(ma) + C_m I_1(ma) = -D_m K_1(ma)$$

$$m\{K_0(ma) + C_m I_0(ma) - D_m K_0(ma)\} = i\omega\mu SD_m K_1(ma)$$

inasmuch as:

$$K_0' = -K_1, \quad \text{and} \quad I_0' = I_1$$

Eliminating C_m, we obtain:

$$D_m = \frac{1}{1 + i\omega\mu SaI_1(ma)K_1(ma)}$$

In this process, the identity:

$$I_1(x)K_0(x) + I_0(x)K_1(x) = \frac{1}{x}$$

has been utilized. Thus, the expression for the vector potential outside the shell can be written in the form:

$$A_z^{*(e)} = i\omega\mu\frac{M}{2\pi^2}\int_0^\infty \frac{K_0(mr)\cos mz}{1 + i\omega\mu SaI_1(ma)K_1(ma)}\,dm$$

(3.405)

A similar expression can be obtained for the function A_z^{*i}. By examining eq. 3.405, we can describe some interesting properties of the frequency- and time-domain responses that differ from those that have been considered in earlier sections of this chapter.

First, we will recall the asymptotic behavior for the functions $I_1(ma)$ and $K_1(ma)$. These asymptotic conditions are:

$$I_1(x) \to \frac{x}{2} \quad \text{and} \quad K_1(x) \to \frac{1}{x} \quad,$$

and

$$I_1(x) \to \frac{e^x}{(2\pi x)^{1/2}} \quad \text{and} \quad K_1(x) \to \left(\frac{\pi}{2x}\right)^{1/2} e^{-x} \quad, \text{as } x \to \infty$$

Whence, we have:

$$I_1(ma)K_1(ma) \to \frac{1}{2} \qquad \text{as } m \to 0$$

$$\text{(3.406)}$$

$$I_1(ma)K_1(ma) \to \frac{1}{2ma} \qquad \text{as } m \to \infty$$

In accord with eq. 3.405 the spectrum for the vector potential as well as the field has singularities in the complex plane in ω when the denominator of the integrand is zero. Correspondingly, the distribution of singularities is defined by the equation:

$$\omega(m) = i \frac{1}{\mu Sa I_1(ma)K_1(ma)} \tag{3.407}$$

Thus, $\omega(m)$ is a continuous function of the variable of integration, m, and in contrast to the previous cases, the singularities in the spectrum which are situated on the imaginary ω-axis are not poles. In eq. 3.406, with an increase in m, the function $\omega(m)$ gradually increases from:

$$\omega_{min} = i \frac{2}{\mu Sa}$$

to infinity. This continuous distribution of singularities in the spectrum is shwon in Fig. 3.62. As a consequence the frequency spectrum cannot be represented as being the sum of simple fractions. However, the low-frequency part of the spectrum can be expanded using a MacLauren series containing only integer powers in ω. Actually, by expanding the part of the integrand of eq. 3.405:

$$\frac{1}{1 + i\omega\mu Sa I_1(ma)K_1(ma)}$$

into a series in terms of $i\omega\mu Sa$ and integrating each term, we can obtain expressions for the quadrature and inphase components containing only odd and even powers of ω, respectively.

We now obtain an expression for the transient response for step function excitation. Applying the Fourier transform, we have:

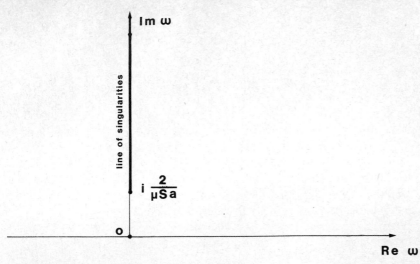

Fig. 3.62. The continuous distribution of singularities in the spectrum in accord with eq. 3.407.

$$A_z^{*(e)}(t) = \frac{\mu M}{4\pi^3} \int\limits_0^\infty K_0(mr) \cos mz \, dm \int\limits_0^\infty \frac{e^{i\omega t}}{1 + i\omega\mu Sa I_1(ma) K_1(ma)} \, d\omega$$

$$(3.408)$$

The interior integral is of a form that has been tabulated and which is well known:

$$\int\limits_0^\infty \frac{e^{i\omega t} d\omega}{1 + i\omega\mu Sa I_1 K_1} = \frac{2\pi e^{-t/\mu Sa I_1 K_1}}{\mu Sa I_1(ma) K_1(ma)}$$

Whence

$$A_z^{*(e)}(t) = \frac{M}{Sa \, 8\pi^2} \int\limits_0^\infty e^{-t/\mu Sa I_1 K_1} \frac{K_0(mr) \cos mz}{I_1(ma) K_1(ma)} \, dm \qquad (3.409)$$

In accord with eqs. 3.405 and 3.409, the high-frequency asymptote coincides with the expression for the early stage as time approaches zero.

Inasmuch as the function as $1/I_1 K_1$ increases without limit as m increases, we can state that for large times, the integral 3.409 is defined principally by small values of m, when the product $I_1 K_1$ is almost equal to one-half. Numerical calculations show that at the late stage, the vector potential A_z^* decays as:

$$\frac{1}{t^{2/3}} e^{-t/\tau_0}$$

where

$$\tau_0 = \mu S a / 2,$$

that is, the behavior is not that of a simple exponential.

3.10. A CONDUCTING SPHERE IN THE FIELD OF AN ARBITRARILY ORIENTED MAGNETIC DIPOLE SOURCE

In the previous sections of this chapter, we have examined the frequency- and time-domain behavior of currents induced in conductors under the condition that the primary vortex electric field did not intersect the surface of the conductor. From the physical point of view, this means that no surface charges are present, and that the electromagnetic field is contributed only by induced currents. Now we will consider the more general case in which the primary vortex electric field intersects the surface of the confined conductor. As an example, let us assume that a spherical conductor is situated in the field of a magnetic dipole source as shown in Fig. 3.63. In deriving formulas, we will perceive from the results which will be obtained in Chapter 6. In the general case, the field will be described using two potentials Π_1 and Π_2. In accord with these results, we have:

$$\Pi_1 = \frac{i\omega\mu}{k_e b} \frac{\sin\phi}{4\pi} M \sum_{n=1}^{\infty} \frac{\rho^{2n+1}}{[(2n-1)!!]^2 n^2} \frac{m^2 - 1}{m^2 + \dfrac{n+1}{n}} \xi_n(k_e b)$$

$$\frac{\xi_n(k_e r)}{r} P_n^{(1)}(\cos\theta) \quad \text{if } |k_i a| < 1 \tag{3.410}$$

$$\Pi_2 = \frac{\cos\phi}{4\pi b} M \sum_{n=1}^{\infty} \frac{\rho^{2n+1} D_n \xi_n'(k_e b)}{[(2n-1)!!]^2 n(n+1)} \frac{\xi_n(k_e r)}{r} P_n^{(1)}(\cos\theta)$$

$$\text{if } |k_e a| \ll 1$$

where k_e is the wave number in the host medium surrounding the conductive body, b is the distance from the center of the dipole to the center of the sphere, $\rho = k_e a$, and $m = k_i / k_e$, k_i is the wave number for the material of the sphere, and $\xi(x)$ and $\xi'(x)$ are the spherical Bessel function and its first derivative. D_n is the complex amplitude of the spectrum described in detail in the consideration of coaxial models in Paragraph 3 of this chapter.

In order to obtain expressions for the field in an insulator ($k_e = 0$) we will assume that $k_e a$ approaches zero. In this case the function $\xi_n(x)$, $\xi_n'(x)$ can be replaced by their asymptotic expansions:

$$\xi_n(x) = (2n-1)!! \frac{i}{x^n} e^{-ix}$$

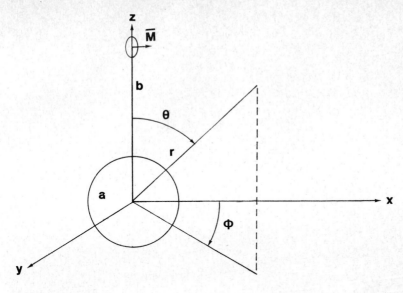

Fig. 3.63. A spherical conductor situated in the field from a magnetic dipole source.

$$\xi'_n(x) = -(2n-1)!!n \frac{ie^{-ix}}{x^{n+1}}$$

In a nonconducting medium we have the following:

$$\Pi_1 = -i\omega\mu \frac{\sin\phi}{4\pi r} M \sum_{n=1}^{\infty} \frac{1}{n^2} \left(\frac{a}{b}\right)^{n+1} \left(\frac{a}{r}\right)^n P_n^{(1)}(\cos\theta)$$

$$\Pi_2 = \frac{\cos\phi}{4\pi b} \frac{M}{r} \sum_{n=1}^{\infty} \frac{D_n}{(n+1)} \left(\frac{a}{b}\right)^{n+1} \left(\frac{a}{r}\right)^n P_n^{(1)}(\cos\theta)$$

(3.411)

inasmuch as m approaches infinity.

The potentials Π_2 and Π_1 are related to the field as follows:

$$E_r^{(2)} = 0, \qquad E_\theta^{(2)} = \frac{-i\omega\mu}{r\sin\theta} \frac{\partial}{\partial x} (r\Pi_2), \quad E_\phi^{(2)} = \frac{i\omega\mu}{r} \frac{\partial(r\Pi_2)}{\partial\theta}$$

$$H_r^{(2)} = \frac{\partial^2(r\Pi_2)}{\partial r^2}, \quad H_\theta^{(2)} = \frac{1}{r} \frac{\partial^2(r\Pi_2)}{\partial r\partial\theta}, \quad H_\phi^{(2)} = \frac{1}{r\sin\theta} \frac{\partial^2(r\Pi_2)}{\partial r\partial\phi}, \quad (3.412)$$

and

$$E_r^{(1)} = \frac{\partial^2 (r\Pi_1)}{\partial r^2}, \quad E_\theta^{(1)} = \frac{1}{r} \frac{\partial^2 (r\Pi_1)}{\partial r \partial \theta},$$

$$E_\phi^{(1)} = \frac{1}{r \sin \theta} \frac{\partial^2 (r\Pi_1)}{\partial r \partial \phi}$$

$$H_r^{(1)} = H_\theta^{(1)} = H_\phi^{(1)} \equiv 0, \quad \text{for } \sigma_e = 0$$

Thus, the secondary field outside the sphere can be represented as being the sum of two fields, each behaving quite differently.

Inasmuch as the spectrum of the field is described in terms of the functions D_n, we can conclude that the basic characteristics of the frequency- and time-domain behavior are the same as those for the field response which was obtained in the models with axial symmetry. Next let us examine the second part of the field described by the potential Π_1. First, in accord with eq. 3.412, this part of the quasi-stationary field (displacement currents are neglected) consists only of an electric field, and the magnetic field is absent. The source of this part of the field are electrical charges distributed over the surface of this sphere. These charges are shown schematically in Fig. 3.64. They are of opposite sign on opposite sides of the surface; $y > 0$ and $y < 0$. Their density increases on approach to the axis y. The normal component of

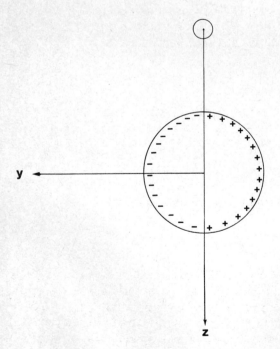

Fig. 3.64. The charge distribution which will give rise to an electric field.

the electric field and the corresponding normal component of the current density are zero at points on the inside surface of the conductor, as a consequence of the presence of these charges. With time, the polarity of the charges changes periodically. It is important to note that the charge density does not depend on the conductivity of the sphere, but that it is directly proportional to ω as well as to the strength of the primary vortex electric field.

From eqs. 3.411 and 3.412, the normal component of the electric field on the outside surface of the sphere is

$$E_r = -i\omega\mu \frac{M}{4\pi a^2} \sin\phi \sum_{n=-\infty}^{\infty} \frac{n+1}{n} \left(\frac{a}{b}\right)^{n+1} P_n^{(1)}(\cos\theta)$$

Inasmuch as the radial component of the field within the conductor is zero, and the difference of normal components of the induction vector, D_r, on the surface is equal to the charge density, we have:

$$\Sigma = -i\omega\mu \frac{\epsilon_0 M}{4\pi a^2} \sin\phi \sum_{n=-\infty}^{\infty} \frac{n+1}{n} \left(\frac{a}{b}\right)^{n+1} P_n^{(1)}(\cos\theta) \tag{3.413}$$

where ϵ_0 is the dielectric constant. These charges contribute a quasi-static electric field outside the sphere which obeys Coulomb's law. To illustrate this, consider a very simple case. Suppose that the dipole is located far from the sphere, that is, b is much greater than a. In this case, we need only consider the first term in the series in eq. 3.411, that is, we have:

$$\Pi_1 \approx -i\omega\mu \frac{Ma^3}{4\pi r^2 b^2} \sin\theta \sin\phi$$

Correspondingly, we obtain the following for the electric field:

$$E_r^{(1)} = \frac{2a^3}{r^3} E_0 \sin\phi \sin\theta$$

$$E_\theta^{(1)} = -\frac{a^3}{r^3} E_0 \sin\phi \cos\theta \tag{3.414}$$

$$E_\phi^{(1)} = -\frac{a^3}{r^3} E_0 \cos\phi$$

where E_0 is the primary vortex electric field at the center of the sphere. It is readily seen that eqs. 3.414 coincide with those for a static field when the primary electric field is uniform in the vicinity of the conductor.

Summarizing the results of this paragraph, we can say the following:

(1) Deviations from axial symmetry result in the appearance of surface electrical charges which do not depend on the conductivity of the sphere and

which change with frequency in the same manner as the primary electric field. In the case which was considered, these charges were found to be directly proportional to frequency.

(2) The total electric field outside this sphere consists of two parts, one being a vortex part which is caused by a change in the magnetic field with time and is exactly the same as that in the case we examined previously with axial symmetry and a galvanic part contributed by the electric charges and satisfying Coulomb's law, that is

$$E = E^v + E^c$$

Considering the effect of the galvanic part contributed by the charges in the time domain a constant is added to the transient response.

In both the frequency and time domains, the development of electrical charges does not produce any profound change in the behavior of the electric field when the surrounding medium is insulating.

(3) Inasmuch as in the quasi-stationary approximation, the magnetic components of the fields associated with the charge accumulations are zero, deviation from the condition of axial symmetry does not change the basic characteristics of the frequency and transient responses derived in the previous paragraphs.

(4) In conclusion, it is relevant to note that the effect of electric charges on both electric and magnetic field components can be significant when the surrounding medium is conductive. This problem will be examined in more detail in the last chapter.

3.11. A CIRCULAR CYLINDER IN THE FIELD OF A MAGNETIC DIPOLE FOR THE CASE IN WHICH THE SURROUNDING MEDIUM IS INSULATING

Up to this point, we have considered only those electromagnetic fields caused by currents induced in cylindrical conductors for cases in which the primary vortex electrical field does not intersect the surface of the cylinder. In order to consider what happens when such is the case, let us assume that the source of the primary field is a magnetic dipole with its moment oriented arbitrarily with respect to the axis of the cylinder, as shown in Fig. 3.65. In this case, the primary vortex electrical field intersects the surface of the cylinder, and as a consequence, electric charges develop. This means that currents created within the cylinder are driven by two types of electric field, namely that due to a change in the strength of the magnetic field with time (the inductive part) and a second part caused by the surface electrical charges (the galvanic part). It should be obvious that the distribution of these charges must be such as to make the normal component of current density zero near the surface of the cylinder, that is:

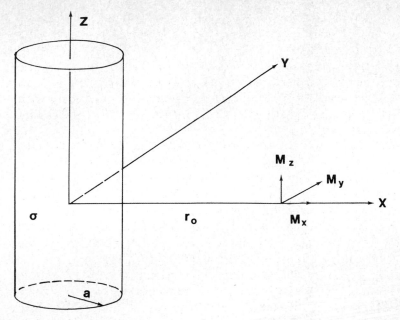

Fig. 3.65. A circular cylinder illuminated by the field from a magnetic dipole source.

$$j_n = 0 \quad \text{or} \quad E_n^i = 0 \tag{3.415}$$

where E_n^i is the normal component of the electric field near the surface of the cylinder on the inside.

From the physical point of view, one can expect that with increase in distance along the z-axis, the primary vortex electrical field becomes smaller and in correspondence with the surface charge density decreases.

First, let us write a solution to the boundary value problem in a way that has been described in detail by Wait. Maxwell's equations written outside and inside the cylinder are:

$$\operatorname{curl} E^e = i\omega\mu H^e \quad \operatorname{div} E^e = 0$$
$$\operatorname{curl} H^e = 0 \quad\quad \operatorname{div} H^e = 0 \tag{3.416}$$

and

$$\operatorname{curl} E^i = i\omega\mu H^i \quad \operatorname{div} E^i = 0$$
$$\operatorname{curl} H^i = \sigma E^i \quad\quad \operatorname{div} H^i = 0 \tag{3.417}$$

In accord with the equation:

$$\operatorname{curl} H^e = 0$$

the magnetic field outside the cylinder can be represented as being the gradient of some scalar function U:

$$H^e = - \operatorname{grad} U \tag{3.418}$$

Substituting this expression into the fourth of Maxwell's equations:

$$\operatorname{div} H^e = 0$$

we find that the function U satisfies Laplace's equation:

$$\nabla^2 U = 0 \tag{3.419}$$

The field inside the conductor cannot be written using such a scalar potential function inasmuch as the curl of the magnetic field is non-zero. We will seek a solution for the magnetic field within the conductor as the sum of two terms:

$$E^i = E^i_1 + E^i_2$$
$$H^i = H^i_1 + H^i_2 \tag{3.420}$$

where each term is expressed in terms of vector potentials of the magnetic and electric types, respectively:

$$E^i_1 = i\omega\mu \operatorname{curl} \Pi^*$$
$$H^i_2 = \operatorname{curl} \Pi \tag{3.421}$$

We will seek a solution under the assumption that both vector potentials have only z-components, that is:

$$\Pi^* = (0,0,\Pi^*_z) \quad \text{and} \quad \Pi = (0,0,\Pi_z) \tag{3.422}$$

We will now demonstrate that both parts of the electromagnetic field can be expressed in terms of a corresponding vector potential, and that both of them satisfy Helmholtz's equation. From the equation:

$$\operatorname{curl} H_1 = \sigma E_1$$

we have

$$\operatorname{curl} H_1 = i\sigma\mu\omega \operatorname{curl} \Pi^*$$

or

$$H_1 = k^2 \Pi^* - \operatorname{grad} \Phi^* \tag{3.423}$$

where $k^2 = i\sigma\mu\omega$ and Φ^* is the scalar potential of the magnetic type.

Substituting the first of the equations in 3.421 and 3.423 into equation:

$$\operatorname{curl} E = i\omega\mu H$$

we obtain:

$$\operatorname{curl} \operatorname{curl} \Pi^* = k^2 \Pi^* - \operatorname{grad} \Phi^*$$

or $\tag{3.424}$

$$\text{grad div } \Pi^* - \nabla^2 \Pi^* = k^2 \Pi^* - \text{grad } \Phi^*$$

In order to simplify this equation, we make use of the following guage condition:

$$\Phi^* = -\text{div } \Pi^* \tag{3.425}$$

Then, in place of eq. 3.424 we have:

$$\nabla^2 \Pi^* + k^2 \Pi^* = 0 \tag{3.426}$$

In accord with eqs. 3.421 and 3.423, we have:

$$E_1^i = i\omega\mu \text{ curl } \Pi^*$$

$$H_1^i = k^2 \Pi^* + \text{grad div } \Pi^* \tag{3.427}$$

The second part of the field, that which is related to the vector potential of the electric type, can be obtained as follows. From the first of Maxwell's equations we have:

$$\text{curl } E_2^i = i\omega\mu H_2^i = i\omega\mu \text{ curl } \Pi$$

or

$$E_2^i = i\omega\mu \Pi - \text{grad } \Phi \tag{3.428}$$

where Φ is the scalar potential of the electric type. Substituting the second of eqs. 3.421 and 3.428 into the following relationship:

$$\text{curl } H_2 = \sigma E_2$$

we obtain

$$\text{curl curl } \Pi = k^2 \Pi - \sigma \text{ grad } \Phi$$

or

$$\text{grad div } \Pi - \nabla^2 \Pi = k^2 \Pi - \sigma \text{ grad } \Phi \tag{3.429}$$

Making use of still another gauge condition:

$$\Phi = -\frac{1}{\sigma} \text{div } \Pi \tag{3.430}$$

we obtain a Helmholtz equation in place of eq. 3.429:

$$\nabla^2 \Pi + k^2 \Pi = 0 \tag{3.431}$$

Both the electric and magnetic vectors for this part of the field are expressed in terms of the vector potential Π:

$$H_2 = \text{curl } \Pi, \quad E_2 = i\omega\mu \Pi + \frac{1}{\sigma} \text{grad div } \Pi \tag{3.432}$$

Thus, the total field inside the cylinder can be represented using the two vector potentials as follows:

$$E^i = E_1^i + E_2^i = i\omega\mu(\text{curl } \Pi^* + \Pi) + \frac{1}{\sigma}\text{grad div } \Pi$$

$$H^i = H_1^i + H_2^i = k^2\Pi^* + \text{grad div } \Pi^* + \text{curl } \Pi$$

(3.433)

Before we formulate the boundary conditions at the surface of the cylinder for the vector potentials, it will be convenient to express them in terms of cylindrical harmonics. With this object in mind, let us start from the scalar potential U describing the magnetic field outside the cylinder (eq. 3.148). It can be written as the sum of two terms:

$$U = U_0 + U_s$$

(3.434)

where U_0 is the scalar potential of the magnetic dipole in free space, and U_s is the scalar potential for the magnetic field caused by induced currents. As is well known, the potential U_0 can be written as:

$$U_0 = \frac{M}{4\pi}\frac{\partial}{\partial l}\frac{1}{R}$$

(3.435)

where M is the moment of the dipole which is oriented in the l direction. We should note that the function $M/4\pi R$ can be interpreted as being the potential of a fictitious magnetic charge. To simplify the procedure deriving the formulas, we will assume that the primary scalar potential is:

$$U_0^p = \frac{M}{4\pi}\frac{1}{R}$$

(3.436)

and then taking a derivative with respect to l, the final solution corresponding to the magnetic dipole excitation is obtained. In eq. 3.436, R is the distance between the source and the observation point. Making use of the identity:

$$\frac{1}{R} = \frac{1}{\pi}\int_{-\infty}^{\infty} K_0(\lambda\bar{r})e^{-i\lambda(z-z_0)}d\lambda$$

(3.437)

where

$$R = [\bar{r}^2 + (z-z_0)^2]^{1/2}$$

and

$$\bar{r} = [r_0^2 + r^2 - 2rr_0\cos(\phi-\phi_0)]^{1/2}$$

and the additional relationship:

$$K_0(\lambda\bar{r}) = \sum_{n=0}^{\infty} \epsilon_n K_n(\lambda r_0)I_n(\lambda r)\cos n(\phi-\phi_0), \quad \text{of } r < r_0$$

where

$\epsilon_0 = 1,$ if $n = 0$; and $\epsilon_n = 2,$ if $n \geqslant 1$

eq. 3.436 can be written as:

$$U_0^p = \frac{M}{8\pi^2} \sum_{n=-\infty}^{\infty} \int_{-\infty}^{\infty} K_n(\lambda r_0) I_n(\lambda r) e^{-i\lambda(z-z_0)} d\lambda \, e^{-in(\phi-\phi_0)} \qquad (3.438)$$

or

$$U_0^p = \frac{M}{2\pi^2} \sum_{n=0}^{\infty} \epsilon_n \int_0^{\infty} K_n(\lambda r_0) I_n(\lambda r) \cos \lambda(z-z_0) d\lambda \cos n(\phi-\phi_0)$$

As follows from eq. 3.438, the potential representing the primary field is the sum of cylindrical harmonics.

For convenience in further algebraic operations, let us represent eq. 3.438 in the form:

$$U_0^p = \Gamma \, I_n(\lambda r) \qquad (3.439)$$

where Γ is an operator, which in accord with eq. 3.438 indicates the operations which must be carried out in order to derive the function U_0^p.

It is convenient to represent the secondary potential in a form similar to that used for the primary potential by assuming that every cylindrical harmonic in the primary potential causes a corresponding harmonic in the secondary potential. This means that the function U_s should be written as:

$$U_s = \frac{M}{8\pi^2} \sum_{n=0}^{\infty} \int_0^{\infty} K_n(\lambda r_0) A_n K_n(\lambda r) e^{-i\lambda(z-z_0)} d\lambda \, e^{-in(\phi-\phi_0)} \qquad (3.440)$$

or

$$U_s = \Gamma \, A_n K_n(\lambda r) \qquad (3.441)$$

The function $I_n(\lambda r)$ is replaced by $K_n(\lambda r)$ because the secondary field must decrease with increase in distance from the induced currents and surface charges. $A_n(\lambda)$ is some unknown function which will be determined from boundary conditions. Thus, the expression for the total field is:

$$U^p = \Gamma \, [I_n(\lambda r) + A_n K_n(\lambda r)] \qquad (3.442)$$

where U^p is the potential outside the cylinder for the case in which the source of the primary field is a fictitious magnetic charge.

Now we will find expressions for both of the vector potentials inside the cylinder. Eq. 3.426 can be written as:

$$\nabla^2 \Pi_z^* + k^2 \Pi_z^* = 0$$

In the cylindrical coordinate system, it has the form:

$$\frac{1}{r} \frac{\partial}{\partial r} \left(r \frac{\partial \Pi_z^*}{\partial r} \right) + \frac{1}{r^2} \frac{\partial^2 \Pi_z^*}{\partial \phi^2} + \frac{\partial^2 \Pi_z^*}{\partial z^2} + k^2 \Pi_z^* = 0$$

Performing the standard operation of separating variables and considering that the field inside the conductor must have only finite values, the function Π_z^* can be written as:

$$\Pi_z^* = \Gamma\, a_n(\lambda)\, I_n(ur) \tag{3.443}$$

where $a_n(\lambda)$ is an unknown function, and

$$u = (\lambda^2 - k^2)^{1/2} \tag{3.444}$$

By analogy, the vector potential of the electric type, Π_z, has the form:

$$\Pi_z = \Gamma\, b_n(\lambda)\, I_n(ur) \tag{3.445}$$

The unknown coefficients $A_n(\lambda)$, $a_n(\lambda)$ and $b_n(\lambda)$ can be determined from boundary conditions, i.e., continuity of tangential components of the electric and magnetic fields at the surface of the cylinder. Inasmuch as the continuity of tangential components of the electric field requires the continuity of normal components of the magnetic field when the magnetic permeability is the same inside and outside the cylinder, it will be convenient to formulate the boundary conditions in terms of the continuity of all components of the magnetic field, that is:

$$H_\phi^e = H_\phi^i, \quad H_z^e = H_z^i, \quad H_r^e = H_r^i, \quad \text{at } r = a \tag{3.446}$$

In accord with eq. 3.418, the magnetic field outside the cylinder for the case in which the primary field source is a fictitious charge is expressed in terms of the scalar potential as follows:

$$H_\phi^e = -\frac{1}{r} \frac{\partial U^p}{\partial \phi}, \quad H_z^e = -\frac{\partial U^p}{\partial z}, \quad H_r^e = -\frac{\partial U^p}{\partial r} \tag{3.447}$$

Making use of eq. 3.433 and considering that:

$$\operatorname{div} \Pi^* = \frac{\partial \Pi_z^*}{\partial z}$$

we have:

$$H_\phi^i = \frac{1}{r} \frac{\partial^2 \Pi_z^*}{\partial \phi \partial z} - \frac{\partial \Pi_z}{\partial r}$$

$$H_z^i = k^2 \Pi_z^* + \frac{\partial^2 \Pi_z^*}{\partial z^2} \tag{3.448}$$

$$H_r^i = \frac{\partial^2 \Pi_z^*}{\partial r \partial z} + \frac{1}{r} \frac{\partial \Pi_z}{\partial \phi}$$

Therefore, the boundary conditions for the potentials are:

$$-\frac{1}{a}\frac{\partial U^{\mathrm{p}}}{\partial \phi} = \frac{1}{a}\frac{\partial^2 \Pi_z^*}{\partial z \partial \phi} - \frac{\partial \Pi_z}{\partial r}$$

$$-\frac{\partial U^{\mathrm{p}}}{\partial z} = k^2 \Pi_z^* + \frac{\partial^2 \Pi_z^*}{\partial z^2} \qquad , \text{at } r = a \qquad (3.449)$$

$$-\frac{\partial U^{\mathrm{p}}}{\partial r} = \frac{\partial^2 \Pi_z^*}{\partial r \partial z} + \frac{1}{a}\frac{\partial \Pi_z}{\partial \phi}$$

Making use of eqs. 3.440—3.445, the first equation in the set 3.449 can be written as:

$$\Gamma\frac{in}{a}[I_n(\lambda a) + A_n K_n(\lambda a)] = \Gamma\left[-ub_n I_n' - \frac{\lambda n}{a} a_n I_n(ua)\right]$$

Taking into account the orthogonality of the functions $e^{-in(\phi-\phi_0)}$, which are multipliers in the operator Γ, this last equation can be rewritten as:

$$-\frac{in}{a}[I_n(\lambda a) + A_n K_n(\lambda a)] = ub_n I_n'(ua) + \frac{\lambda n}{a} a_n I_n(ua) \qquad (3.450)$$

Correspondingly, the second equation in the set 3.449 has the form:

$$\Gamma i\lambda[I_n(\lambda a) + A_n K_n(\lambda a)] = -\Gamma u^2 a_n I_n(ua)$$

or $\qquad\qquad\qquad\qquad\qquad\qquad\qquad\qquad\qquad\qquad\qquad (3.451)$

$$i\lambda[I_n(\lambda a) + A_n K_n(\lambda a)] = -u^2 a_n I_n(ua)$$

By analogy, for the last equation in the set 3.449, we have:

$$\lambda[I_n'(\lambda a) + A_n K_n'(\lambda a)] = i\lambda u a_n I_n'(ua) + \frac{in}{a} b_n I_n(ua) \qquad (3.452)$$

At this point, we have arrived at the following system of equations:

$$-\frac{in}{a}[I_n(\lambda a) + A_n K_n(\lambda a)] = ub_n I_n'(ua) + \frac{\lambda n}{a} a_n I_n(ua)$$

$$i\lambda[I_n(\lambda a) + A_n K_n(\lambda a)] = -u^2 a_n I_n(ua) \qquad (3.453)$$

$$\lambda[I_n'(\lambda a) + A_n K_n'(\lambda a)] = i\lambda u a_n I_n'(ua) + \frac{in}{a} b_n I_n(ua)$$

Eliminating the coefficients $a_n(\lambda)$ and $b_n(\lambda)$, we arrive at the following expression for A_n:

$$A_n = \frac{\tilde{I}_n(m) - \tilde{I}_n(m_1) + \frac{in^2 p}{m^2 m_1^4} \frac{1}{\tilde{I}_n(m_1)} \frac{I_n(m)}{K_n(m)}}{\tilde{K}_n(m) - \tilde{I}_n(m_1) + \frac{in^2 p}{m^2 m_1^4} \frac{1}{\tilde{I}_n(m_1)}} \qquad (3.454)$$

where

$$m = \lambda a, \quad m_1 = ua = (\lambda^2 a^2 - k^2 a^2)^{1/2}, \quad k^2 a^2 = ip, \quad p = \sigma\mu\omega a^2,$$

$$\tilde{I}_n(m) = I_n'(m)/mI_n(m), \quad \tilde{K}_n(m) = K_n'(m)/mK_n(m),$$

and where $I_n'(m)$ and $K_n'(m)$ are the first derivatives with respect to m.

The secondary magnetic field due to the fictitious charge is:

$$H_s^p = -\nabla U_s^p$$

or

$$H_\phi^s = \frac{1}{r} \Gamma i n A_n K_n(\lambda r)$$

$$H_r^s = -\Gamma \lambda A_n K_n'(\lambda r) \qquad (3.455)$$

$$H_z^s = \Gamma i\lambda A_n K_n(\lambda r)$$

We will now derive expressions for the magnetic field for the case in which the source of the primary field is a magnetic dipole. First of all, we will assume that the dipole moment is directly along the z-axis.

Axial magnetic dipole

In this case, in accord with eq. 3.435, the potential contributed by the axial dipole U_0^a is:

$$U_0^a = \frac{\partial}{\partial z_0} U_0^p$$

and correspondingly:

$$U_s^a = \frac{\partial}{\partial z_0} U_s^p$$

As follows from eq. 3.440, we have:

$$U_s^a = \frac{M}{2\pi^2} \sum_{n=0}^{\infty} \epsilon_n \int_0^{\infty} A_n K_n(\lambda r_0)\lambda K_n(\lambda r) \sin \lambda (z - z_0)dz \cos n(\phi - \phi_0)$$

$$(3.456)$$

Writing the secondary magnetic field as:

$$H_r^s = \frac{M}{4\pi r^3} h_r^a, \quad H_\phi^s = \frac{M}{4\pi r^3} h_\phi^a \quad \text{and} \quad H_z^s = \frac{M}{4\pi r^3} h_z^a$$

we obtain:

$$h_r^a = -\frac{2}{\pi} \beta^3 \sum_{n=0}^{\infty} \epsilon_n \int_0^{\infty} m^2 A_n K_n(\alpha m) K_n'(\beta m) \sin m\bar{z} \, dm \cos n(\phi - \phi_0)$$

$$(3.457)$$

$$h_\phi^a = \frac{4}{\pi} \beta^2 \sum_{n=1}^{\infty} n \int_0^{\infty} m A_n K_n(\alpha m) K_n(\beta m) \sin m\bar{z} \, dm \sin n(\phi - \phi_0) \quad (3.458)$$

$$h_z^a = -\frac{2}{\pi} \beta^3 \sum_{n=0}^{\infty} \epsilon_n \int_0^{\infty} m^2 A_n K_n(\alpha m) K_n(\beta m) \cos m\bar{z} \, dm \cos n(\phi - \phi_0)$$

$$(3.459)$$

here

$$m = \lambda a, \quad \alpha = r_0/a, \quad \beta = r/a, \quad \bar{z} = (z - z_0)/a \tag{3.460}$$

Radial magnetic dipole

In this case, the moment of the dipole is oriented along the radius r_0, and therefore the primary potential is:

$$U_0^r = \frac{\partial}{\partial r_0} U_0^p$$

As follows from eq. 3.440, we have

$$U_s^r = \frac{M}{2\pi^2} \sum_{n=0}^{\infty} \epsilon_n \int_0^{\infty} \lambda A_n K_n'(\lambda r_0) K_n(\lambda r) \cos \lambda(z - z_0) d\lambda \cos n(\phi - \phi_0)$$

$$(3.461)$$

Correspondingly, for the components of the magnetic field written as:

$$H_r = \frac{M}{4\pi r^3} h_r^r, \quad H_\phi = \frac{M}{4\pi r^3} h_\phi^r, \quad H_z = \frac{M}{4\pi r^3} h_z^r$$

we obtain:

$$h_r^r = -\frac{2}{\pi} \beta^3 \sum_{n=0}^{\infty} \epsilon_n \int_0^{\infty} m^2 A_n K_n'(\alpha m) K_n'(\beta m) \cos m\bar{z} \, dm \cos n(\phi - \phi_0)$$

$$(3.462)$$

$$h_\phi^r = \frac{4}{\pi} \beta^2 \sum_{n=1}^{\infty} n \int_0^{\infty} m A_n K_n'(\alpha m) K_n(\beta m) \cos m\bar{z} \, dm \sin n(\phi - \phi_0)$$

$$(3.463)$$

318

and

$$h_z^r = -\frac{2}{\pi} \beta^3 \sum_{n=0}^{\infty} \epsilon_n \int_0^{\infty} m^2 A_n K_n'(\alpha m) K_n(\beta m) \sin m\bar{z}\, dm \cos n(\phi - \phi_0)$$

(3.464)

Angular magnetic dipole

Finally, we will derive expressions for the secondary field for the case in which the moment of the dipole is oriented along a line, ϕ_0. The expression for the potential of the primary field in this case is:

$$U_0^{an} = \frac{1}{r_0} \frac{\partial U_0^p}{\partial \phi_0}$$

and therefore:

$$U_s^{an} = \frac{M}{r_0 \pi^2} \sum_{n=1}^{\infty} n \int_0^{\infty} A_n(\lambda) K_n(\lambda r_0) K_n(\lambda r) \cos \lambda (z - z_0) d\lambda \sin n(\phi - \phi_0)$$

(3.465)

Correspondingly, when the components of the magnetic field are normalized to the function $M/4\pi r^3$, we have:

$$h_r^{an} = -\frac{4\beta^3}{\pi\alpha} \sum_{n=1}^{\infty} n \int_0^{\infty} A_n K_n(m\alpha) m K_n'(m\beta) \cos m\bar{z}\, dm \sin n(\phi - \phi_0) \quad (3.466)$$

$$h_\phi^{an} = -\frac{4\beta^2}{\pi\alpha} \sum_{n=1}^{\infty} n^2 \int_0^{\infty} A_n K_n(m\alpha) K_n(m\beta) \cos m\bar{z}\, dm \cos n(\phi - \phi_0) \quad (3.467)$$

$$h_z^{an} = -\frac{4\beta^3}{\pi\alpha} \sum_{n=0}^{\infty} n \int_0^{\infty} A_n K_n(m\alpha) m K_n(m\beta) \sin m\bar{z}\, dm \sin n(\phi - \phi_0) \quad (3.468)$$

These various equations permit us to determine the magnetic field caused by inducted currents when the magnetic dipole is arbitrarily oriented with respect to the axis of the cylinder. However, it should be noted that this magnetic field defines only the vortex electric field, and that there is another, or galvanic, part of the electric field which is contributed by surface charges. This arises as a consequence of the fact that displacement currents have been neglected, and that the corresponding equations for the magnetic field do not contain information about the other part of the electric field.

Proceeding from the equations for vector potentials and magnetic field strengths, let us examine some of the general features of the field. First of all, in contrast to the case of an infintely long source parallel to the axis of the cylinder, the field is a function of all three coordinates, r, ϕ, and z. Consequently, both the frequency-domain and time-domain responses of the field in general depend on the coordinates of the observation site.

In the solution which was obtained, all the components of the field and therefore all the currents induced in the cylinder are represented as being the sum of cylindrical harmonics of the azimuthal angle ϕ regardless of the type of excitation. For example, in accord with eq. 3.440, the potential of the secondary field for the case in which the source is a fictitious charge is:

$$U_s = \frac{M}{2\pi^2} \sum_{n=0}^{\infty} \epsilon_n L_n \cos n(\phi - \phi_0) \tag{3.469}$$

where

$$L_n = \int\limits_0^\infty A_n(\lambda, a, h) K_n(\lambda r_0) K_n(\lambda r) \cos \lambda(z - z_0) d\lambda$$

or $\tag{3.470}$

$$L_n = \int\limits_0^\infty \Phi(\lambda, n, a, h, r, r_0) \cos \lambda(z - z_0) d\lambda$$

The function L_n is an amplitude coefficient expressed in terms of azimuthal harmonic with the index n, and it is a function of all the geometrical parameters except ϕ, and of the thickness of the skin layer:

$$h = (2/\sigma\mu\omega)^{1/2}$$

In accord with eq. 3.470, the amplitude of the azimuthal harmonic L_n is expressed in terms of a Fourier integral where the variable of integration can be considered to be a spatial frequency for harmonics along the z-axis, while the functions $\Phi = A_n K_n(\lambda r_0) K_n(\lambda r)$ describe their amplitudes.

Strictly speaking, the vector potentials as well as the field are defined by all the spatial harmonics with respect to the variables λ and n. However, the relative contribution of the harmonic with a given spatial frequency λ depends primarily on geometric parameters and the thickness of a skin layer. Also, the effect of each harmonic with a frequency λ and n varies with a change in the orientation of the magnetic dipole as will be shown later.

It is appropriate to note here that the dependence on the thickness of a skin layer, regardless of the orientation of the source of the primary field, manifests itself only in terms of the function $A_n(\lambda, a, h)$. Therefore, the principal features characterizing the electromagnetic spectrum can be derived from analysis of this function, and these results can be applied in obtaining an understanding of the behavior of transient responses during the early and late stages.

Analysis of the function A_n

In accord with eq. 3.454, we have:

$$A_n = \frac{\tilde{I}_n(m) - \tilde{I}_n(m_1) + \dfrac{in^2 p}{m^2 m_1^4} \dfrac{1}{\tilde{I}_n(m_1)}}{\tilde{K}_n(m) - \tilde{I}_n(m_1) + \dfrac{in^2 p}{m^2 m_1^4} \dfrac{1}{\tilde{I}_n(m_1)}} \frac{I_n(m)}{K_n(m)} \tag{3.471}$$

where

$$m = \lambda a, \quad m_1 = (m^2 - ip)^{1/2}, \quad p = \sigma\mu\omega a^2$$

$$\tilde{I}_n(m) = I_n'(m)/mI_n(m), \quad \tilde{K}_n(m) = K_n'(m)/mK_n(m)$$

Let us first examine the behavior of the function A_n when the spatial frequency m is small and the condition:

$$m^2 \ll p \tag{3.472}$$

applies. This assumption about the value for m means that we will consider here only that part of the field caused by induced currents with a density that differs only slightly with a change in the coordinate z. The argument for the harmonic functions $m(z - z_0)$ is small over relatively wide ranges of values for z and therefore the induced currents along with the field caused by them is practically independent of z.

It is convenient to write the function A_n in the form:

$$A_n = \tilde{A}_n \frac{I_n(m)}{K_n(m)} \tag{3.473}$$

where

$$\tilde{A}_n = \frac{\tilde{I}_n(m) - \tilde{I}_n(m_1) + \dfrac{in^2 p}{m^2 m_1^4} \dfrac{1}{\tilde{I}_n(m_1)}}{\tilde{K}_n(m) - \tilde{I}_n(m_1) + \dfrac{in^2 p}{m^2 m_1^4} \dfrac{1}{\tilde{I}_n(m_1)}} \tag{3.474}$$

Inasmuch as the angular dipole is considered, the zero harmonic is absent.

For small values of m we have:

$$I_n(m) \approx \frac{m^n}{2^n n!}, \quad I_n'(m) \approx \frac{m^{n-1}}{2^n (n-1)!}, \qquad n \geqslant 0$$

$$K_n(m) \approx \frac{(n-1)! \, 2^{n-1}}{m^m}, \quad K_n'(m) \approx -\frac{n! \, 2^{n-1}}{m^{n+1}} \qquad \text{if } n \geqslant 1$$

Whence

$$\tilde{I}_n(m) = \frac{n}{m^2}, \quad \tilde{K}_n(m) \approx -\frac{n}{m^2}$$

Substituting these expressions into eq. 3.474, w obtain:

$$\tilde{A}_n \approx -\frac{\dfrac{n}{m^2} - \dfrac{I_n'(m_1)}{m_1 I_n(m_1)} + \dfrac{in^2 p}{m^2 m_1^{\,4}} \dfrac{m_1 I_n(m_1)}{I_n'(m_1)}}{\dfrac{n}{m^2} + \dfrac{I_n'(m_1)}{m_1 I_n(m_1)} - \dfrac{in^2 p}{m^2 m_1^{\,4}} \dfrac{m_1 I_n(m_1)}{I_n'(m_1)}} \quad \text{if } n \geqslant 1 \tag{3.475}$$

Replacing m_1^2 by $-ip$, we obtain:

$$\frac{in^2 p}{m^2 m_1^{\,4}} \frac{m_1 I_n(m_1)}{I_n'(m_1)} \approx -\frac{n^2}{m^2} \frac{I_n(\sqrt{-ip})}{\sqrt{-ip}\, I_n'(\sqrt{-ip})}$$

Therefore, we obtain the following for the numerator in eq. 3.475:

$$\frac{n}{m^2} - \frac{I_n'(m_1)}{m_1 I_n(m_1)} - \frac{n^2}{m^2 m_1} \frac{I_n(m_1)}{I_n'(m_1)} \approx \frac{n[\sqrt{-ip}\, I_n'(\sqrt{-ip}) - n I_n(\sqrt{-ip})]}{m^2 \sqrt{-ip}\, I_n'(\sqrt{-ip})}$$

$$\tag{3.476}$$

It should be noted at this point that the term $m^2 (I_n')^2$ has been neglected.
   Making use of the identity:

$$I_n'(\sqrt{-ip}) = \frac{n}{\sqrt{-ip}}\, I_n(\sqrt{-ip}) + I_{n+1}(\sqrt{-ip})$$

the numerator in the function $\tilde{A}_n$ becomes:

$$\frac{n I_{n+1}(\sqrt{-ip})}{m^2 I_n'(\sqrt{-ip})} \tag{3.477}$$

In the same manner, we can obtain the following approximation for the denominator of $\tilde{A}_n$:

$$\frac{n[\sqrt{-ip}\, I_n'(\sqrt{-ip}) + n I_n(\sqrt{-ip})]}{m^2 \sqrt{-ip}\, I_n'(\sqrt{-ip})} = \frac{n[2n I_n(\sqrt{-ip}) + \sqrt{-ip}\, I_{n+1}(\sqrt{-ip})]}{m^2 \sqrt{-ip}\, I_n'(\sqrt{-ip})}$$

Making use of the recurrent relationship:

$$I_n(\sqrt{-ip}) = \sqrt{-ip}\, \frac{I_{n-1}(\sqrt{-ip}) - I_{n+1}(\sqrt{-ip})}{2n}$$

the last expression is considerably simplified, so that we have:

$$\frac{n I_{n-1}(\sqrt{-ip})}{m^2 I_n'(\sqrt{-ip})} \tag{3.478}$$

Taking into account the condition 3.472, we finally obtain an expression for the function A_n which is valid for small values of m:

$$A_n \approx -\frac{I_{n+1}(ika)}{I_{n-1}(ika)} \frac{m^{2n}}{2^{2n-1}n!(n-1)!} = -T_n \frac{m^{2n}}{2^{2n-1}n!(n-1)!} \qquad (3.479)$$

where

$$T_n = I_{n+1}(ika)/I_{n-1}(ika) \qquad (3.480)$$

From a comparison of eq. 3.480 and eq. 3.296, we see that the field corresponding to small spatial frequencies m such that $m < |ka|$ coincides with that caused by linear induced currents. Therefore, we may draw the following conclusions about this spectrum for this part of the field:

(1) On the complex ω-plane, the spectrum has singularities located on the imaginary axis,

(2) Regardless of the value for n, if condition 3.472 applies, poles for the spectrum are defined by the equation $I_{n-1}(ika) = 0$.

(3) Each harmonic n has an infinite number of poles, and the minimum pole corresponds to the first harmonic when $n = 1$.

This analysis leads directly to the conclusion that the transient response of this field is described by a sum of exponential terms, and that during the late stage, the transient field decays as a single exponential term with the time constant $\tau_0 = 1/\omega_0$, where $i\omega_0$ is the minimum pole characterizing the spectrum.

Now let us consider the behavior of the function A_n for large spatial harmonics, m.

As is well known, we have the identity:

$$I_n(x) = \frac{e^x}{(2\pi x)^{1/2}} \quad \text{and} \quad K_n(x) = e^{-x}(\pi/2x)^{1/2}, \quad \text{if } x \gg 1$$

Thus:

$$I_n'(x) = \frac{e^x}{(2\pi x)^{1/2}} \quad \text{and} \quad K_n'(x) = -e^{-x}(\pi/2x)^{1/2}$$

Whence

$$\tilde{I}_n(m) = 1/m, \quad \tilde{I}_n(m_1) = 1/m_1, \quad \tilde{K}_n(m) = -1/m$$

Correspondingly, for the function $\tilde{A}_n$, we have:

$$\tilde{A}_n = -\frac{m_1^3 m + in^2 p - m^2 m_1^2}{m_1^3 m - in^2 p + m^2 m_1^2} \qquad (3.481)$$

and the singularities characterizing the spectrum on the complex ω-plane are defined by the equation:

$$m^2 m_1^2 + m_1^3 m - i n^2 p = 0$$

In the limit as m approaches infinity, we have on replacing m_1 by m:

$$\omega = -i \frac{2m^4}{\sigma \mu a^2 n^2}, \quad \text{for } m \gg 1, \, m > n \tag{3.482}$$

In other words, the singularities being functions of m are continuously distributed along the imaginary ω-axis.

In the general case for an arbitrary value of m, the equation describing the singularities according to eq. 3.474 is:

$$\tilde{K}_n(m) = \tilde{I}_n(m_1) - \frac{i n^2 p}{m^2 m_1^4} \frac{1}{\tilde{I}_n(m_1)} \tag{3.483}$$

Thus, this analysis has shown that the behavior of the electromagnetic spectrum for every harmonic with spatial frequency m depends on the value of m except for very small m. Inasmuch as the contribution of these harmonics is determined by geometrical factors as well as the frequency ω and the orientation of the source, we can expect various types of behavior for the frequency- and time-domain responses when these factors change. However, we can draw some conclusions about the frequency-domain response for the total field. Inasmuch as the singularities in the spectrum are situated at some distance from the origin of the ω-axis on the complex ω-plane, the low-frequency spectrum can be expanded in a MacLauren series containing only integer powers of ω. In accord with eq. 3.474, at high frequencies such that $k^2 a^2 \to \infty$, the function $\tilde{I}_n(m_1)$ plays the principal role, and because of this, the function A_n tends to unity. This results from the fact that as a consequence of skin effect, the influence of conductivity vanishes, and as can be shown, the induced currents are concentrated mainly near that part of the surface of the cylinder which is close to the source of the primary field.

As follows from Faraday's law, the same behavior will be observed during the early stage of the transient response, provided that we use step function excitation. With respect to the late stage behavior, one can expect that it will commence at times that depend on the orientation and position of the dipole as well as of the observation point. As an example, we will consider the frequency- and time-domain responses for an angular dipole (see Fig. 3.66) with the assumption that the dipole is situated on the x-axis, that is, that ϕ_0 and z_0 are both 0.

In accord with eq. 3.465 for the potential of the secondary field, we have:

$$U_s^{an} = \frac{M}{r_0 \pi^2} \sum_{n=1}^{\infty} n \int_0^{\infty} A_n K_n(\lambda r_0) K_n(\lambda r) \cos \lambda z \, d\lambda \sin n\phi \tag{3.484}$$

As may be seen from Fig. 3.66, with increase in distance, r_0, the primary vortex electric field within the conductor becomes more nearly parallel

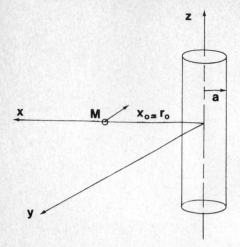

Fig. 3.66. Angular dipole in the presence of a conducting cylinder.

to the z-axis. As a consequence, we can expect that the induced currents which are parallel to the z-axis will also be dominant when the distance r_0 is markedly greater than z. Therefore, the field observed near the origin but outside the cylinder approaches that caused by a uniform magnetic field directed along the y-axis. To be specific, let us assume r_0 approaches infinity, and that $r/r_0 \ll 1$ while $z/r_0 \ll 1$. Introducing a new variable $x = \lambda r_0$, we have.

$$U_s^{an} = \frac{M}{\pi^2 r_0^2} \sum_{n=1}^{\infty} nL_n \sin n\phi \qquad (3.485)$$

where

$$L_n = \int_0^{\infty} A_n K_n(x) K_n \left(\frac{r}{r_0} x\right) \cos \frac{z}{r_0} x \, dx \qquad (3.486)$$

Because ratio r/r_0 tends to zero, one can replace function $K_n(rx/r_0)$ by its asymptotical expression:

$$K_n \left(\frac{r}{r_0} x\right) \approx \frac{(n-1)! \, 2^{n-1}}{r^n x^n} r_0^n$$

then we obtain:

$$L_n = \left(\frac{r_0}{r}\right)^n (n-1)! \, 2^{n-1} \int_0^{\infty} A_n \frac{K_n(x)}{x^n} \cos \frac{z}{r_0} x \, dx$$

Taking into account eq. 3.479 as well as the equality:

$$m = \lambda a = xa/r_0$$

we have:

$$L_n = -\left(\frac{a}{r}\right)^n \left(\frac{a}{r_0}\right)^n \frac{1}{n!} \frac{1}{2^n} T_n \int_0^\infty x^n K_n(x) \cos \frac{z}{r_0} x \, dx$$

These integrals are tabulated ones:

$$\int_0^\infty x^n K_n(x) \cos bx \, dx = \frac{\sqrt{\pi}}{2} 2^n \Gamma(n + \tfrac{1}{2}) \frac{1}{(1 + b^2)^n + \tfrac{1}{2}}$$

Inasmuch as

$$\Gamma(n + \tfrac{1}{2}) = \frac{(2n - 1)!!}{2^n} \sqrt{\pi}$$

we obtain:

$$L_n = -\left(\frac{a}{r}\right)^n \left(\frac{a}{r_0}\right)^n \frac{(2n - 1)!!}{n!} \frac{\pi}{2^{n+1}} T_n \tag{3.487}$$

Substituting 3.487 into 3.485, we have:

$$U_s^{an} = -\frac{M}{\pi r_0^2} \sum_{n=1}^\infty \frac{(2n - 1)!!}{(n - 1)!} \frac{1}{2^{n+1}} T_n \left(\frac{a}{r}\right)^n \left(\frac{a}{r_0}\right)^n \sin n\phi \tag{3.488}$$

Correspondingly, we obtain the following for the leading term:

$$U_s^{an} \approx -\frac{M}{4\pi r_0^3} \frac{a^2}{r} T_1 \sin \phi \tag{3.489}$$

From this, we find that the equations for the components of the magnetic field are:

$$H_r^s = -\frac{M}{4\pi r_0^3} \frac{a^2}{r^2} T_1 \sin \phi$$

$$\text{, when } r/r_0 \ll 1, \quad \frac{r}{r_0} \ll 1 \tag{3.490}$$

$$H_\phi^s = \frac{M}{4\pi r_0^3} \frac{a^2}{r^2} T_1 \cos \phi$$

where $M/4\pi r_0^3$ is the primary magnetic field near the cylinder, and

$$T_1 = I_2(ika)/I_0(ika)$$

It should be clear that eqs. 3.490 are the same as those for the case in which the primary magnetic field is uniform and directed perpendicularly to the z-axis, with the primary vortex electrical field being parallel to the z-axis. However, if one of the coordinates, r or z, for the observation point increases, regardless of how far away the magnetic dipole is situated, the influence of a change in the current density along the z-axis becomes more profound and the application of eq. 3.490 leads to significant error.

We will now consider the behavior of the field when the distance from an observation point to the cylinder increases, with the dipole being located relatively close to the conductor. We will again proceed from eq. 3.484.

Assuming that $r \to \infty$ and that $r_0/r \ll 1$, after defining a new variable $x = \lambda r$, we obtain:

$$U_s^{an} = \frac{M}{r_0 r \pi^2} \sum_{n=1}^{\infty} n M_n \sin n\phi \tag{3.491}$$

where

$$M_n = \int_0^{\infty} A_n K_n(x) K_n \left(\frac{r_0}{r} x\right) \cos \frac{z}{r} x \, dx \tag{3.492}$$

Considering that the integrals M_n are defined mainly by small values of $x r_0/r$ when r_0/r tends to 0, and replacing $K_n(r_0 x/r)$ by their asymptotic expressions, we obtain:

$$M_n = \left(\frac{r}{r_0}\right)^n (n-1)! \, 2^{n-1} \int_0^{\infty} A_n \frac{K_n(x)}{x^n} \cos \frac{z}{r} x \, dx \tag{3.493}$$

Substituting eq. 3.479 into this last equation, we finally have:

$$M_n = -\left(\frac{a}{r}\right)^n \left(\frac{a}{r_0}\right)^n \frac{(2n-1)!!}{n!} \frac{\pi}{2^{n+1}} T_n \tag{3.494}$$

Correspondingly,

$$M_n = L_n \tag{3.495}$$

In accord with eq. 3.494, we have:

$$U_s^{an} = -\frac{M}{r_0 r \pi} \sum_{n=1}^{\infty} \left(\frac{a}{r}\right)^n \left(\frac{a}{r_0}\right)^n \frac{(2n-1)!!}{(n-1)!} \frac{T_n}{2^{n+1}} \sin n\phi \tag{3.496}$$

Inasmuch as the number of alterations in sign for the induced currents along the coordinate ϕ increases with increase in n, the first harmonic plays the leading role in field behavior at large distances, and in accord with eq. 3.496, we have:

$$U_s^{an} \approx \frac{M a^2}{4\pi r_0^2} T_1 \frac{1}{r^2} \sin \phi$$

whence

$$H_r = -\frac{Ma^2}{4\pi r_0^2} T_1 \frac{1}{r^3} \sin \phi$$

$$\text{, if } \frac{r_0}{r} \ll 1, \quad \frac{z}{r} \ll 1 \qquad (3.497)$$

$$H_\phi = \frac{Ma^2}{4\pi r_0^2} T_1 \frac{1}{r^3} \cos \phi$$

Therefore, at relatively large distances, this field is equivalent to that of a magnetic dipole for which the spectrum would be described by the same function T_1. Correspondingly, during the late stage of time-domain response, as was true in the previous case, the behavior is described by a single exponential term. It is clear that the dipole behavior, observed only for certain conditions, results from the fact that the current density corresponding to the first harmonic decreases with increase in $|z|$.

Similar evaluations can be carried out for the other field components and for various orientations of the dipole source. Let us now describe the behavior of the frequency- and time-domain responses for various positions of the dipole and of an observation point. Numerical results from calculations based on eqs. 3.446–3.468 are shown in Figs. 3.67–3.82. From consideration of the inphase and quadrature components of the field H_ϕ, as well as of the transient responses of the electromotive force, we can make the following comments:

In contrast to the case of an infinitely long current filament as the source of excitation, a change in the parameters r/a and r_0/a manifests itself in a different manner. The progressive removal of the dipole away from the cylinder ($r_0/a \rightarrow \infty$), the secondary field which can be considered to be that corresponding to the first harmonic approaches the field for uniform excitation. Such behavior is observed with reasonable accuracy when the two distances r and r_0 satisfy the condition:

$$r/r_0 < \tfrac{1}{5}, \quad |z - z_0|/r_0 < \tfrac{1}{5} \qquad (3.498)$$

In this case, the secondary field is practically described by the first harmonic, and correspondingly, both in the frequency and the time domain, the field behaves as though there were a linear dipole source, and its frequency response is described by the function T_1, while the transient response can be written as the sum of exponential terms. It is obvious that during the late stage of transient response, the field decays as a simple exponential with the time constant τ_0, corresponding to uniform excitation, that is:

$$\tau_0 = \sigma\mu a^2 /5.783$$

In the contrasting case in which the distance between the cylinder and the observation point increases, the secondary field will not tend to the value it would have for a uniform primary field with the conductor, although the difference between the secondary field and that due to the first harmonic does decrease. In this second case, both in the frequency and time domains, the field behaves as a dipole magnetic field, and consequently it decreases as $1/r^3$. At the same time, in the frequency domain, its complex amplitude is again prescribed by the function T_1 while the transient responses are simple sums of exponential terms. Correspondingly, during the late state, the field decays exponentially with the same time constant:

$$\tau_0 = \sigma \mu a^2 / 5.783$$

In practical terms, the field behavior described here will be observed when:

$$r_0/r < \tfrac{1}{5} \tag{3.499}$$

Comparing both of these cases, we see that the complex amplitudes and the transient responses are the same, while the dependence on geometrical factors is different.

In the general case when conditions 3.498 and 3.499 do not apply, the frequency- and time-domain responses have the same features as in the previous two cases, although the dependence on geometric parameters are markedly different, inasmuch as in addition to the first azimuthal harmonic there is a significant contribution from the higher harmonics. However, it is important to stress that during the late stage, regardless of the position of the dipole or of the observation point, the field decays exponentially with the same time constant $\tau_0 = \sigma \mu a^2 / 5.783$. It is clear that this fact is of considerable importance from the practical point of view. As may be seen from Figs. 3.80 and 3.81, when either the dipole source or the observation point approaches the conductor, the late stage of transient response begins at progressively later times. This behavior can be observed for example in borehole logging measurements based on transient field behavior when the receiver approaches a conductor.

However, when the observation point is located relatively close to the conductor, determination of the time constant by measuring quadrature and inphase components at low frequencies can result in large errors.

The behavior of the field, as described above, in general can be explained as follows. First of all, with progressive removal of the dipole source away from the cylinder, the primary magnetic field within the conductor becomes more nearly uniform, and its associated vortex electric field begins to be more nearly parallel to the axis of the cylinder. However, in contrast to the case of uniform primary magnetic field everywhere, in this case, this uniformity occurs only within some limited range of changes in the coordinate z. Moreover, with dipole excitation, the induced currents are closed at some distance from the plane $z = z_0$, instead of being closed at infinity when the primary magnetic field is uniform and its associated vortex electric field is directed along the axis. For this reason, the influence of the non-uniformity of the primary magnetic field along the z-axis within the conductor is not particularly important, if the distance from the cylinder to the observation, r, is significantly less than the distance between the dipole source and the conductor, r_0 (condition 3.498). In other words, in this case, the secondary field is caused mainly by currents which are flowing parallel to the axis of the conductor, and in accord with this, the field approaches that for a linear magnetic dipole. With a further increase in the distance r, a further consequence of the fact that currents are closed at a finite distance from the plane $z = z_0$ (the non-uniformity of the primary magnetic field) begins to manifest itself more strongly. The system of closed currents is practically equivalent to a magnetic dipole so that condition 3.499 applies. Also it is clear that with an increase in the coordinate z of the observation point, the influence of the non-uniformity of the primary magnetic field becomes more pronounced. Up to this point, we have considered some general features of induced currents in the frequency domain. Now, suppose that the dipole moment is a step function. At the initial instant when the source current is turned off, induction currents are distributed along the surface of the cylinder in such a manner that within the conductor they generate the primary magnetic field. Their distribution on the surface at the initial instant of switching depends primarily on the position of the dipole (in our case, it is azimuthal with respect on the cylinder). For example, with an increase of separation r the area on the surface of the conductor over which the current density is practically constant along the z-axis becomes larger. Inasmuch as after the source has been turned off, the dipole moment is zero, the surface induced currents begin to defuse into the cylinder. Their distribution varies with time in all directions, including the direction along the z-axis. Currents which extend to greater distances along this cylinder decay more slowly and therefore they define the late stage of the transient field. Since their extent is much greater than the radius of this cylinder, the time constant for the decay of these currents tends towards the value for the time constant of the transient decay when the primary field is uniform. In conclusion, it should be stressed that conditions 3.498 and 3.499 also hold for time-domain behavior.

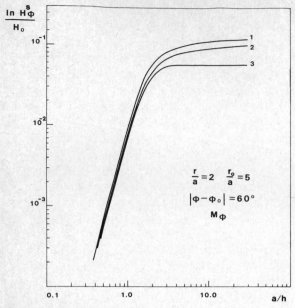

Fig. 3.67. Numerical results from calculations of the inphase component of the magnetic field. Legend for Figs. 3.67—3.78: *1* = uniform excitation; *2* = the first harmonic of the secondary field; *3* = secondary field.

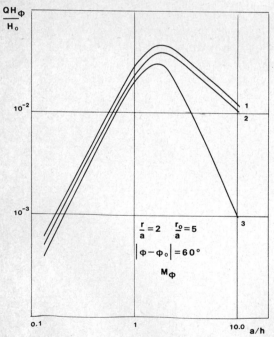

Fig. 3.68. Numerical results of calculations for the quadrature part of the magnetic field. See also Fig. 3.67.

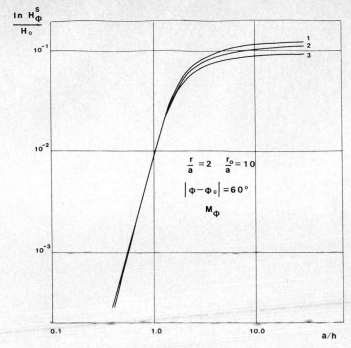

Fig. 3.69. Numerical results from calculations of the inphase part of the magnetic field. See also Fig. 3.67.

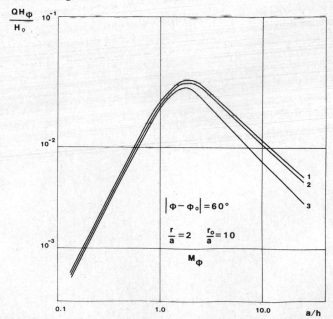

Fig. 3.70. Numerical results from calculations of the quadrature part of the tangential magnetic field. See also Fig. 3.67.

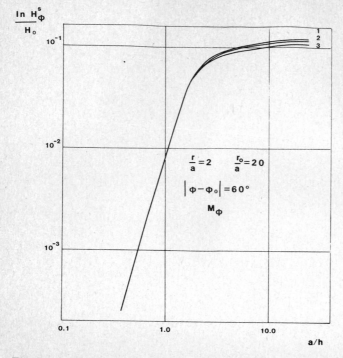

Fig. 3.71. Numerical results from calculations of the inphase part of the tangential magnetic field. See also Fig. 3.67.

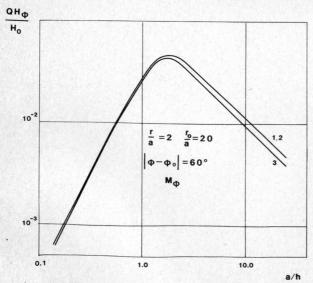

Fig. 3.72. Numerical results from calculations of the quadrature part of the magnetic field. See also Fig. 3.67.

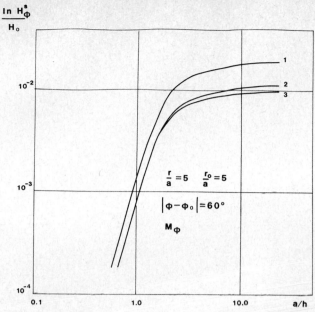

Fig. 3.73. Numerical results from calculations of the inphase part of the magnetic field. See also Fig. 3.67.

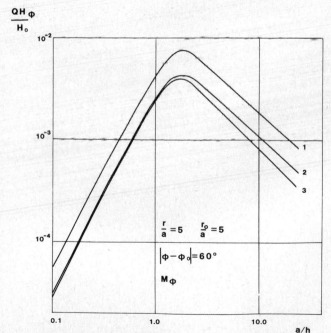

Fig. 3.74. Numerical results from calculations of the quadrature part of the magnetic field. See also Fig. 3.67.

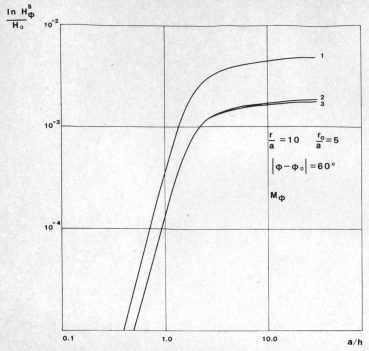

Fig. 3.75. Numerical results from calculations of the inphase part of the magnetic field. See also Fig. 3.67.

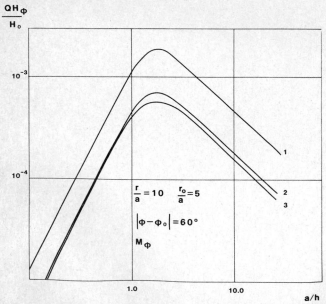

Fig. 3.76. Numerical results from calculations of the quadrature part of the magnetic field. See also Fig. 3.67.

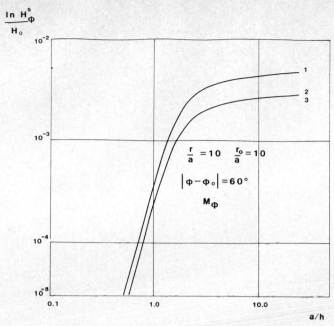

Fig. 3.77. Numerical results from calculations of the inphase part of the magnetic field. See also Fig. 3.67.

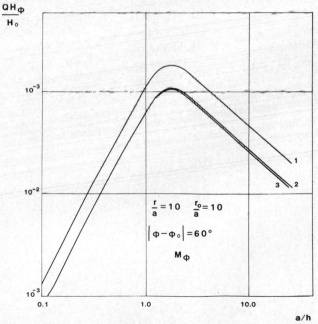

Fig. 3.78. Numerical results from calculations of the quadrature part of the magnetic field. See also Fig. 3.67.

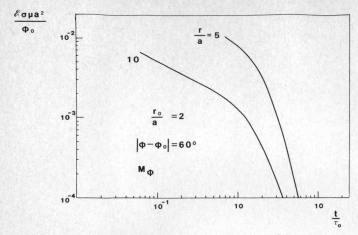

Fig. 3.79. Numerical results from calculations of the electromotive force.

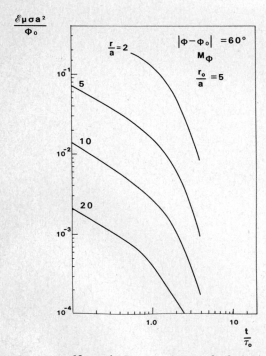

Fig. 3.80. Numerical results from calculations of the electromotive force.

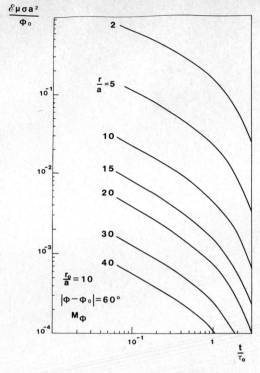

Fig. 3.81. Numerical results from calculations of the electromotive force.

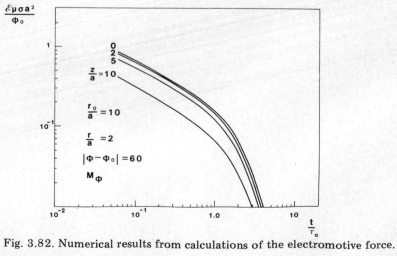

Fig. 3.82. Numerical results from calculations of the electromotive force.

Chapter 4

RESOLVING CAPABILITIES AND DEPTH OF INVESTIGATION OF
INDUCTIVE METHODS, WHEN GEOLOGIC NOISE IS A CONFINED
INHOMOGENEITY

INTRODUCTION

The principal use of inductive methods is in the search for ore bodies
made up of highly conductive minerals, so that the ore body as a whole has
a conductivity that is much greater than that of the surrounding medium.
However, it is well known, that anomalies in electrical and magnetic field
behavior can also be caused by accumulations of conductive but non-
economic minerals. In actuality, the sources for the quasi-stationary field
consist both of currents in the conductive medium, and electrical charges on
the surfaces between media with different resistivities. For this reason, the
electromagnetic field depends also on the electrical properties of the surrounding medium, which is the host for the ore body, and in particular, on
its degree of non-uniformity. The ratio between the fields caused by the
presence of economic conducting ore minerals and other conductors, commonly called geological noise, ultimately limits the resolution of induction
prospecting methods and the depth of investigation.

Clearly, the effectiveness of an induction prospecting method will also
depend on a number of other factors, including cultural noise, magneto-
variational noise, errors in the geometric parameters for the field system
(such as the separation and relative position of the transmitter and receiver),
the stability with which the primary field is cancelled, and so on. By decreasing the effect of these various factors to the level of geological noise, we can
increase the effectiveness of an induction prospecting method to the maximum amount possible. For this reason, an analysis of geological noise is
highly important because it permits us to determine the maximum effectiveness of any given prospecting method, and to determine which is the
optimum exploration method for a given set of geoelectric conditions.

One of the simplest methods used in the search for conductive rock
masses is the direct-current resistivity method. However, this method is
characterized by a relatively low resolution. This is caused by the fact that
sources for the stationary (or DC) field are electric charges that develop on
surfaces with the charges being only slightly dependent on the resistivity
within the conductive mass and in the surrounding host medium. For example,

the potential characterizing the anomalous field around a conductive sphere situated in a uniform medium is:

$$U_1^e = \frac{\sigma_i - \sigma_e}{\sigma_i + 2\sigma_e} a^3 E_0 \frac{\cos\theta}{R^2} \tag{4.1}$$

Here E_0 is the primary, uniform electric field, σ_i and σ_e are the conductivities of the material in the sphere and in the surrounding host medium, respectively, a is the radius of the sphere, and R and θ are the spherical coordinates of the point at which the potential is being observed. The influence of the electrical properties of the two media is characterized by the contrast coefficient:

$$K_{sph} = \frac{1 - \sigma_e/\sigma_i}{1 + 2\sigma_e/\sigma_i} \tag{4.2}$$

which ranges from -0.5 to $+1$ as the ratio of the conductivities goes from infinite to 0. In practice, when $\sigma_e/\sigma_i \leqslant 0.1$ the value for the coefficient K_{sph} is approximately unity. Under these conditions, it is impossible to determine the conductivity of an isometric body from measurements of the stationary field.

A similar result occurs when the primary electric field is directed perpendicularly to the axis of a right, circular cylinder. In this case, we have:

$$U_1^e = \frac{\sigma_i - \sigma_e}{\sigma_i + \sigma_e} \frac{a^2}{r} E_0 \cos\phi$$

where the value for the coefficient:

$$K_c = (\sigma_i - \sigma_e)/(\sigma_i + \sigma_e)$$

varies from -1 to $+1$.

In practical cases, the secondary electric field depends on many factors, including the geometry of the conductive mass, the ratio of conductivities, the coordinates at the observation point, and others. As an example, consider the field outside the spheroid contributed by electric charges on the surface of a conductor with major and minor axes a and b, when the primary electric field is directed along the major axis. As can readily be shown (Chapter 6), along the central plane perpendicular to the a-axis, we have:

$$E_{1x} = -\left(\frac{\sigma_i}{\sigma_e} - 1\right) E_{0x} \frac{\eta_0}{\eta_1} \frac{Q_1(\eta)}{\dfrac{1}{\eta_0^2 - 1} + \left(\dfrac{\sigma_i}{\sigma_e} - 1\right) Q_1(\eta_0)} \tag{4.3}$$

Here:

$$Q_1(\eta) = \frac{1}{2}\eta \ln \frac{\eta+1}{\eta-1} - 1, \quad \eta_0 = \frac{a}{(a^2-b^2)^{1/2}}, \quad \eta = \left(1+\frac{z^2}{a^2-b^2}\right)^{1/2}$$

where z is the distance from the axis a to the observation point, and the component E_{0x} is parallel to the axis a.

In accord with eq. 4.3, if the ratio $a/b \gg 1$, we have:

$$E_{1x} \approx \left(\frac{\sigma_i}{\sigma_e} - 1\right) \frac{E_{0x}}{2} \frac{b^2}{a^2} \ln \frac{b^2}{2a^2}, \quad \text{for } z/a < 1 \tag{4.4}$$

That is, the secondary field contributed by a conductive mass that is highly elongate parallel to the direction of the primary field is directly proportional to the ratio of conductivities, and at the same time $E_{1x} \ll E_{0x}$.

From a qualitative point of view, this behavior of the field might be explained as follows. Surface charges are concentrated mainly at the edges of the spheroid, so that the field on the z-axis is almost equal to that which would be caused by two charges having the same value and opposite signs located at distance a from the observation point. The value of the charge is readily defined from the expression for the surface divergence:

$$\Sigma = \frac{E_n^e - E_n^i}{4\pi} = \frac{\rho_e - \rho_i}{4\pi} J_n$$

From this we have:

$$e = \frac{\rho_e - \rho_i}{4\pi} I$$

where I is the total current passing through a cross-sectional area S (see Fig. 4.1).

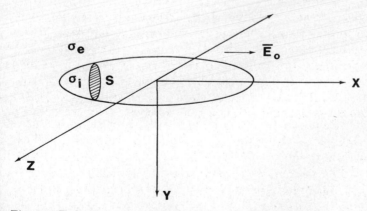

Fig. 4.1. Definition of a cross-sectional area S in a spheroid.

The field inside the elongate spheroid is practically the same as the primary field E_{0x} when the following condition is met:

$$\frac{a^2}{b^2} \gg \frac{1}{2} \frac{\sigma_i}{\sigma_e} \ln \frac{b^2}{2a^2} \tag{4.5}$$

Therefore, the current in the spheroid is $\pi b^2 \sigma_i E_{0x}$, and we have the following result for the electrical charge of e and the field along the z-axis:

$$e = \frac{\rho_e - \rho_i}{4\rho_i} b^2 E_{0x} \quad \text{and} \quad E_{1x} \approx \left(\frac{\sigma_i}{\sigma_e} - 1\right) \frac{b^2}{2a^2} E_{0x} \tag{4.6}$$

The disappearance of the multiplying term $\ln b/a$ in this last equation is a consequence of the fact that in deriving eq. 4.6, the surface distribution of the charge has not been taken into account, and the equivalent charges $\pm e$ are placed at the points $\pm a$.

Curves showing the behavior of secondary electric field normalized to the primary field strength and calculated from eq. 4.3, as shown in Fig. 4.2, demonstrate the influence that the geoelectric parameters exert. From a study of these curves, we can arrive at the following conclusions:

(1) If the dimensions of the conductive mass and the resistivity satisfy the condition:

$$\frac{\sigma_e}{\sigma_i} < \frac{1}{2} \frac{b^2}{a^2} \left| \ln \frac{b^2}{2a^2} \right| \tag{4.7}$$

the electric field is only slightly dependent on the conductivity of the conductive mass.

(2) With increasing length parallel to the direction of the primary field, the tightness of the relationship between the anomalous field strength and the conductivity increases, and in the limit, the strength of the secondary field becomes directly proportional to the conductivity ratio σ_i/σ_e. Also, the secondary field strength decreases rapidly and becomes markedly less than the strength of the primary field. This asymptotic behavior occurs when the following inequality is met:

$$\frac{a^2}{b^2} \gg \frac{1}{2} \frac{\sigma_i}{\sigma_e} \left| \ln \frac{b^2}{2a^2} \right| \tag{4.8}$$

Thus, the electric field strength within the conductor is practically the same as the primary field strength. With decreasing values for the conductivity ratio σ_i/σ_e, this asymptotic behavior is observed for smaller values of the ratio a/b. It should be noted that near the ends of a sufficiently long conductive mass, the secondary field strength, E_{1x}, may be comparable to the size to the primary field strength, E_{0x}.

(3) Independently of the shape or the dimensions of the conductive mass,

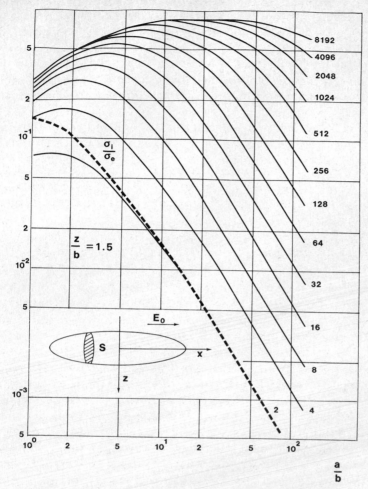

Fig. 4.2. Curves showing the behavior of the secondary electric field normalized with respect to the primary field strength for the case of a spheroid.

the secondary electric field is equally sensitive to changes in the resistivity of the body and of the surrounding medium. For this reason, increasing the accuracy with which measurements are made, does not result in an increase in the resolution of the direct current electrical methods of searching for relatively conductive ore bodies.

(4) Local relatively resistive inhomogeneities ($\sigma_i/\sigma_e \ll 1$) produce secondary electric fields which do not depend appreciably on the conductivity even though they may be of significant size.

These characteristics of the behavior of a stationary electric field illustrate why there is a high level of geological noise in such measurements, which will not permit one to apply direct current electrical methods effectively in the search for highly conductive ore bodies.

We should also note that when the condition 4.7 holds, the magnetic field is proportional to the conductivity of the surrounding medium, but not to that of the conductive rock mass. The external magnetic field becomes proportional to the conductivity of the conductive rock mass only when the field contributed by charges within the conductor is very small (that is, condition 4.8 holds).

Thus, resolution with direct current methods is low, and this is one of the main reasons that other methods based on the use of harmonic and non-stationary fields were first introduced.

We will now consider the resolution obtainable with inductive methods. The relationship of the behavior of an alternating electromagnetic field to the electrical properties of the medium is much more complicated than in the case of direct currents. Moreover, the relationship between the level of geological noise and meaningful signal depends on a great number of factors, including the type of field excitation (that is, whether the source is a closed loop, a long wire, a magnetic dipole, an electrical dipole, or something else), the location at which measurements are carried out (whether in the near zone of the source, in the wave zone of the source, or in an intermediate zone), whether the frequency domain or the time domain is used, which of the field components is measured, and so on. In this chapter, in considering resolution with inductive methods, we will make some significant simplifications to the theory. It will be assumed that the medium surrounding an ore body is insulating. In other words, induced current in the surrounding medium and the currents which are contributed by surface charges on interfaces will not be taken into consideration. In this approximation, the sources of geological noise are local inhomogeneities in which eddy currents can create anomalies of the same type as currents in the target ore body. As will be shown later, neglecting the conductivity of the host medium is a less stringent condition in cases in which axial symmetry is present. For example, if the source for the primary field is a circular loop located with its axis coincident with the center of the spheroid, and measurements are performed in the near-zone of the primary source, there will be only vortex currents in the local conductor. In this case, we can make use of the results obtained in the previous section, starting first with a frequency-domain method.

4.1. FREQUENCY-DOMAIN METHODS

We will first consider the high-frequency part of the spectrum where the inphase component of the field dominates. Also, the quadrature component will decrease slowly with increasing frequency, conductivity, in dimensions of the conductive body, and in the limit, it approaches zero. In accord with eq. 3.216, over the high-frequency range, the following relationships hold:

$$\text{In}\,H = H_0 \left(A - \frac{B}{(\sigma\mu\omega)^{1/2}b} \right)$$

$$Q H = H_0 \frac{C}{(\sigma\mu\omega)^{1/2}b}$$

$$(4.9)$$

where b is the characteristic dimension of the conductive mass, A, B, and C are constants which depend on the linear dimensions of the conductive mass and on the coordinates of the observation point and H_0 is the primary magnetic field strength. According to eq. 4.9, the influence exerted by the resistivity on the inphase component of the field is small. This is readily explainable, inasmuch as in the high-frequency limit, currents will concentrate near the surface of the conductive body and the density will approach that of currents induced in an ideal conductor. Therefore, if the magnetic field due to the currents in a relatively resistive body corresponds to the high-frequency part of the spectrum, the resolution of a method based on use of either direct or alternating current will be practically the same. Moreover, the relationship between the quadrature component and resistivity of the conductive mass for this part of the spectrum does not facilitate the discrimination of anomalies caused by excellent conductors. The principal reason is that the quadrature component is affected by currents in the surrounding medium in the same way as is the inphase component.

We will consider also the contrary case, that of the low-frequency part of the spectrum where the quadrature component dominates over the inphase component, and both increase monotonically with increasing frequency. In accord with eq. 3.183, we have:

$$Q H = H_0 \left\{ \frac{C_1}{\alpha}\,\omega - \frac{C_3}{\alpha^3}\,\omega^3 + \frac{C_5}{\alpha^5}\,\omega^5 - \ldots \right\}$$

$$\text{In}\,H = H_0 \left\{ -\frac{C_2}{\alpha^2}\,\omega^2 + \frac{C_4}{\alpha^4}\,\omega^4 - \frac{C_6}{\alpha^6}\,\omega^6 + \ldots \right\} \quad \text{for } \omega < q_1/\sigma\mu b^2$$

$$(4.10)$$

where $Q H$ and $\text{In}\,H$ represent an arbitrary component of the secondary field, and $\alpha = 1/\sigma\mu b^2$. At very low frequencies, the quadrature component is defined almost entirely by the first term in the series in eq. 4.10; that is:

$$Q H \approx \sigma\mu b^2 \omega C_1 H_0 = \frac{\omega}{\alpha} C_1 H_0$$

$$(4.11)$$

and so, is directly proportional to the conductivity. Thus, when eddy currents induced in the ore body and in an inhomogeneity of higher resistivity generate fields corresponding to the low-frequency part of the spectrum, the ratio of the quadrature components of the magnetic field, H_N and H_R, is directly proportional to the ratio of the conductivities:

$$\frac{H_N}{H_R} = \frac{C_{1N}}{C_{1R}}\frac{\alpha_R}{\alpha_N} = \frac{C_{1N}}{C_{1R}}\frac{\sigma_N}{\sigma_R}\frac{b_N^2}{b_R^2} \tag{4.12}$$

Here, H_N and H_R are, respectively, the quadrature components of the field generated by currents in an inhomogeneity of higher resistivity (N) and in the target ore body (R). This relationship shows that the inductive methods will be characterized by higher resolution than would be a method based on the use of direct current fields. However, to a considerable extent, the level of geological noise depends on the ratio between the parameters α_N and α_R which characterize the sources of geologic noise and of the meaningful signal. In particular, when $\alpha_N = \alpha_R$, the relationship between the two types of anomalies (H_R and H_N) is defined in the same way as in the case of direct current flow, with only geometric factors being involved. For example, when a high-resistivity inhomogeneity which has dimensions greater than those of a target ore body is situated closer to the observation point than is the ore body, the anomaly due to the noise source can be significantly greater than the signal contributed by the target ore body. As follows directly from eq. 4.12, at low frequencies the ratio of signals H_N/H_R is independent of frequency and with further decrease in frequency, there is no deduction of the contribution of geologic noise. On the other hand, with increasing frequency, as the response contributed by currents in the target ore body no longer reflects the behavior at low frequencies, the level of geological noise observed in measuring the quadrature component becomes greater. Inasmuch as the quadrature component is shifted in phase by $90°$ with respect to the primary field, the measurement of quite small signals is accomplished with relative ease.

We will now consider the inphase component of the secondary field at low frequencies where one can neglect all terms in a series given in eq. 4.10 but the first. Then we have:

$$\operatorname{In} H \approx -(\sigma\mu\omega b^2)^2 C_2 H_0 = -\frac{C_2}{\alpha^2}H_0\omega^2 \tag{4.13}$$

which shows that the inphase component of the field is directly proportional to the square of frequencies, and, of particular importance, to the square of the conductivity. For this reason, when the inphase component is measured, the inductive methods are characterized by high resolution, and the ratio between geological noise and meaningful signal is given by:

$$\frac{H_N}{H_R} \approx \frac{C_{2N}\alpha_R^2}{C_{2R}\alpha_N^2} = \frac{C_{2N}}{C_{2R}}\frac{\sigma_N^2 b_N^4}{\sigma_R^2 b_R^4} \tag{4.14}$$

Comparing eqs. 4.12 and 4.14, we can see that if α_R is less than α_N, we can markedly reduce the effect of geologic noise by measuring the inphase component instead of the quadrature component. However, the inphase component is more difficult to measure at low frequencies because it is much smaller than the quadrature component, and it is also in phase with the primary field.

It should be pointed out that in measuring the phase and amplitude of the total field at low frequencies, one can achieve the same resolution as in the case of the individual quadrature and inphase components. From eq. 4.10, we have the following expressions for the amplitude and the phase of the total field:

$$A = H_0 \left[\left(1 - \frac{C_2 \omega^2}{\alpha^2} \right)^2 + \frac{C_1^2 \omega^2}{\alpha^2} \right]^{1/2} \approx \left\{ 1 - \frac{\omega^2}{\alpha^2} (C_2 - \tfrac{1}{2} C_1^2) \right\} H_0 \qquad (4.15)$$

$$\phi = \tan^{-1} \frac{C_1 \omega}{\alpha \left(1 - \frac{C_2 \omega^2}{\alpha^2} \right)} \approx \tan^{-1} \frac{C_1 \omega}{\alpha} \approx \frac{C_1 \omega}{\alpha} \qquad (4.16)$$

The second term in eq. 4.15 is directly proportional to the square of the conductivity, which was the same behavior as seen in the inphase component. It is clear that the amplitude of the secondary field at low frequencies is proportional to the first power of conductivity, σ. Since the contribution of the meaningful signal is usually quite small, any method of measurement which reduces the contribution of the primary field is of distinct interest. Suppression of the primary field can be done in a variety of ways, including making measurements simultaneously at two frequencies.

As follows from this consideration, and from the series given in eq. 4.10, the use of the higher order terms for the expression at low frequencies (eq. 4.10) can markedly increase the resolution obtaining in induction methods. For example, the difference between two measurements of the quadrature components made at two frequencies can be used to eliminate the term proportional to frequency, which is characterized by a higher sensitivity to changes in conductivity:

$$\Delta H = QH(\omega_1) - \frac{\omega_1}{\omega_2} QH(\omega_2) \qquad (4.17)$$

$$\Delta \mathscr{E} = Q\mathscr{E}(\omega_1) - \left(\frac{\omega_1}{\omega_2} \right)^2 Q\mathscr{E}(\omega_2) \qquad (4.18)$$

here, $\mathscr{E}$ is the electromotive force, with its quadrature component, $Q\mathscr{E}$, being shifted in phase by $90°$ with respect to the electromotive force of the primary field.

Let us pose the following question: "Under what conditions is it possible to reduce the influence of geological noise of the type considered by measuring only the higher-order terms in a series in eq. 4.10?" The ratio of fields corresponding to the ith term in these series is:

$$\frac{H_{iN}}{H_{iR}} = \frac{C_{iN}}{C_{iR}} \frac{\alpha_R^i}{\alpha_N^i} \tag{4.19}$$

As has been demonstrated earlier (eq. 3.189), for relatively large values of the counter i, we have:

$$C_i \cong \frac{d_1}{q_1^i}$$

where d_1 is a coefficient depending on geometric parameters, and q_1 characterizes the first pole in the spectrum. Therefore, we obtain:

$$\frac{H_{iN}}{H_{iR}} = \frac{C_{iN}}{C_{iR}} \frac{\alpha_R^i}{\alpha_N^i} \approx \frac{d_{1N}}{d_{1R}} \frac{q_{1R}^i}{q_{1N}^i} \frac{\alpha_R^i}{\alpha_N^i} \tag{4.20}$$

or according to eq. 3.220:

$$\frac{H_{iN}}{H_{iR}} = \frac{d_{1N}}{d_{1R}} \left(\frac{\tau_{0N}}{\tau_{0R}}\right)^i \tag{4.21}$$

where

$$\tau_{0N} = \frac{\sigma_N \mu_N b_N^2}{q_{1N}}, \qquad \tau_{0R} = \frac{\sigma_R \mu_R b_R^2}{q_{1R}}$$

are time constants characterizing the geological noise and the target ore body, respectively.

From eq. 4.21, we can recognize three different cases as follows:

(1) The time constant for the ore body is longer than that for the geological noise:

$$\tau_{0R} > \tau_{0N} \tag{4.22}$$

In this case, in principle, one can decrease the contribution of geologic noise to any degree desired by making measurements only of later terms in the series 4.10. In fact, when the inequality of 4.22 is satisfied, as the number of terms, i, is increased, the ratio $(\tau_{0N}/\tau_{0R})^i$ decreases. It is important to note that condition 4.22 does not depend on the strength of the primary field in the vicinity of the source of geological noise or in the vicinity of the target ore body; that is, on the depth of burial of the ore body. It may occur that in measuring leading terms in the series 4.10, that is, the quadrature and inphase components of the field at low frequencies, the influence of geological noise will prevail. Then, in measuring later terms in these series, the influence of geologic noise becomes less and less, provided of course that the condition 4.22 is met. In other words, if the frequency at which measurements are being made actually corresponds to the low-frequency part of the spectrum, $\omega < 1/\tau_{0R}$, a further decrease in the frequency will not permit us to reduce the relative contribution of geological noise. Instead of a change in frequency, it is necessary to measure later terms in a series 4.10. This consideration is of great practical concern. In order to demonstrate this, let us imagine two models, as shown in Fig. 4.3a, b. First, suppose that the body contributing the geological noise and the target ore body are situated at practically the same depth from the surface of the earth (see Fig. 4.3a). If the quadrature component of the magnetic field is being measured, it may happen that the signal contributed by the geological noise source is greater than that contributed by induced currents in the target ore body. In measuring the inphase component, the signal observed over the body contributing the geologic noise can still be greater than that over the ore body, even though the ratio H_N/H_R is smaller than in the previous case. In both cases, the signal over the target ore body will be less than that over the source of geologic noise, and therefore, it will be hard to recognize the present of the ore body. However, if measurements are made of the later terms in the series 4.10, the relationship between these anomalies changes and signals measured over the ore body become dominant, $\tau_{0R} > \tau_{0N}$. This means that by measuring later terms in the series describing the low-frequency portion of the spectrum, a markedly increased resolution can be obtained with induction methods.

Let us now assume that the ore body is situated beneath some conductive rock mass, which is considered to be the source of geologic noise

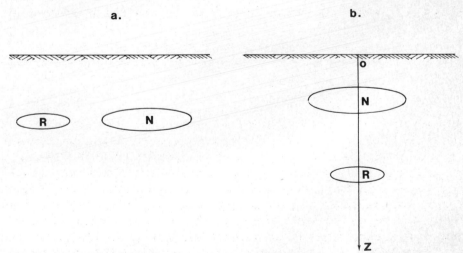

Fig. 4.3. Relative positions of spheroidal conductive bodies representing a target ore body (R) in one case and a noise source (N) in the other.

(see Fig. 4.3b) and that $\tau_{0R} > \tau_{0N}$. In this case, the signal will be the sum of contributions from currents in the ore body, and in the rock body contributing to geological noise:

$$H = H_R + H_N \tag{4.23}$$

where H is any of the components of the field being observed.

It might occur that in measuring the quadrature component of the field, the principal part of the signal is obtained from the source of geological noise, that is $QH_N \gg QH_R$. In other words, the inhomogeneity lying above the ore body serves as an electromagnetic screen, and therefore, the ore body located at greater depth cannot be detected. In measuring the inphase component, the ratio $\text{In}H_R/\text{In}H_N$ markedly increases, and one can consider two situations. The first one is the case in which the inphase component of the magnetic field contributed by currents within the target ore body is dominant, that is, $\text{In}H_R \gg \text{In}H_N$, then, it can be said that the body contributing the geologic noise "becomes transparent" and the ore body itself can be recognized. This means that the depth of investigation achieved by making measurements of the inphase component is greater than that achieved with the quadrature component. If it is impossible to remove the screening effect of the geological noise source even by measuring the inphase component, that is if $\text{In}H_N > \text{In}H_R$ then following terms of the series 4.10 must be measured. As follows from condition 4.22, the relative contribution of geological noise can be reduced to any degree desired, regardless of the position of the ore body relative to the source of geological noise, provided that τ_{0R} is greater than τ_{0N}. In principle at least, by measuring later terms in this series representing the low frequency part of the spectrum, one might achieve any depth of investigation for this type of geological noise (that is, a confined cross-section for the conductive rock mass) whenever condition 4.22 is met.

It is reasonable to expect that the nearer the observation point is to the source of the noise, and the deeper the ore body is, the more difficult it will be to separate the meaningful signal. In such cases, it is necessary to measure higher-order terms in the series 4.10. A similar requirement is imposed if there is a relatively small difference between the two time constants.

On the other hand, if the ore body lies at relatively shallow depths, and there is a large contrast in time constants between the ore body and the screening inhomogeneity, a significant reduction in the relative noise can be obtained by measuring the inphase component (the amplitude of the total field) or, for example, the difference between the quadrature components at two frequencies.

(2) So far, we have considered only the case in which the time constant for the ore body is greater than the time constant for the source of geological noise. Now let us assume the contrary case, that is, the one in which the time constant for the geological noise is longer:

$$\tau_{0N} > \tau_{0R} \qquad (4.24)$$

In this case, by measuring later terms in the series 4.10, the effect of geological noise is increased, causing a reduction in resolution and depth of investigation for the induction method. For example, when measurements are being made of the quadrature component, and the signal contributed by currents induced in the body contributing geologic noise is dominant, the effect of this body will be even greater when the inphase component is measured, and it becomes greater and greater as higher-order terms in the series 4.10 are measured. In this case, one can say that the use of induction methods will not permit one to distinguish between anomalies caused by "ore" and "non-ore" conductors.

(3) Finally, if the time constants of the ore body and the body contributing the geological noise are equal:

$$\tau_{0R} = \tau_{0N} \qquad (4.25)$$

the ratio of the signals contributed by the currents in the two types of conductors is determined by geometrical factors in just the way as in the case of direct current methods. It is obvious that by making measurements of later terms in the series 4.10, there is no change in the relative contribution caused by the current flow in either type of conductor. It can be said that the resolution and the depth of investigation achievable with an induction method in this case are essentially the same as those obtainable with the direct current method.

Next let us consider the capability of a time-domain or transient method for increasing the depth of investigation and for separating magnetic field anomalies caused by bodies with various conductivities.

During the early stage of the transient process, current flow within a conductive body is concentrated almost entirely in a thin surface layer, and as a consequence, the magnetic field contributed by the current depends only weakly on the conductivity, especially when the magnetic field is being detected (see eqs. 3.238, 3.239). Therefore, we will concentrate our attention on the late stage of the transient process during which the field contributed by currents in the confined conductive ore body decays exponentially. As has been shown earlier:

$$H = H_0 d_1 e^{-q_1 \alpha t} = H_0 d_1 e^{-t/\tau_0}$$

and $\qquad (4.26)$

$$\mathscr{E} = \Phi_0 d_1 \frac{q_1}{\sigma \mu b^2} e^{-q_1 \alpha t} = \frac{\Phi_0 d_1}{\tau_0} e^{-t/\tau_0}$$

where $\tau_0 = \sigma \mu b^2 / q_1$, is a time constant. Let us remember that Φ_0 is the flux representing the primary field B_0 within a closed path along which the EMF is to be determined, and d_1 is a function that depends only on geometry.

Inasmuch as the primary field is absent when the secondary field is being measured (this is an essential difference between time-domain and frequency-domain measurements), the background for signals generated by currents flowing in conductors is only that contributed by natural and man made fields. Inasmuch as equipment used in time-domain measurements is usually more sensitive to ambient noise than in the case of frequency-domain measurements, because measurements must be made over a wider spectrum, such noise sources can have a significant effect. For this reason, it is helpful to consider that the total signal which is measured, in either the frequency or time domain, is the sum of three parts:

$$H = H_R + H_g + H_{amb} \tag{4.27}$$

where H_R is signal contributed by the target ore body, H_g is the magnetic field contributed by currents flowing in bodies characterized as the geological noise, and H_{amb} is the ambient noise which includes natural and man-made but alien fields. By analogy, we can also write:

$$\mathscr{E} = \mathscr{E}_R + \mathscr{E}_g + \mathscr{E}_{amb} \tag{4.28}$$

where $\mathscr{E}$ represents the electromotive force.

For the moment, let us assume that the geological noise effect can be neglected. In order to define the maximum depth at which a conductive ore body can be detected, a natural approach is that of comparing the magnitude of the meaningful signal, $\mathscr{E}_{min}$, with the value of the ambient noise $\mathscr{E}_{amb}$ as follows:

$$\mathscr{E}_{min} > n\,\mathscr{E}_{amb} \qquad n \gg 1 \tag{4.29}$$

From this inequality, at any time, t, and for a specified set of properties for the conductive rock mass and the source, one can always find the minimum value of the primary magnetic field and therefore of the current in the source for which the meaningful signal will be several times (n) greater than the noise $\mathscr{E}_{amb}$. This question will be explored in more detail at the end of this chapter.

Now, as was the case with our consideration of harmonic field behavior, we will assume that the observation point is situated over two conductive rock masses, one which is the source of geologic noise, and the other which is the target ore body. The geological noise source is above the ore body and is characterized by a higher resistivity. The distance between the two conductive bodies is sufficiently great and in the first approximation we need not consider the interaction between the two. It should be clear that the exponential character of the late stage field defines the high resolution obtainable with time-domain methods, that is, the high sensitivity to a change in time constant when $t/\tau_0 > 1$.

In accord with eq. 4.26, the relationship between the fields and the EMF produced by currents in the ore body (R) and a noise source of higher resistivity (N) is of the form:

$$\frac{H_N}{H_R} = \frac{d_{1N}}{d_{1R}} e^{-(1/\tau_{0N} - 1/\tau_{0R})t}$$

and   (4.30)

$$\frac{\mathscr{E}_N}{\mathscr{E}_R} = \frac{d_{1N}\tau_{0N}}{d_{1N}\tau_{0N}} e^{-(1/\tau_{0N} - 1/\tau_{0R})t}$$

Just as in the case of our consideration of harmonic fields, we can recognize three cases, namely:

1. $\tau_{0R} > \tau_{0N}$ 2. $\tau_{0R} < \tau_{0N}$ 3. $\tau_{0R} = \tau_{0N}$ (4.31)

In the first case, when the time constant in the ore body is longer than that in the noise source, in accord with eq. 4.30, as time increases, the effect of the geological noise source becomes smaller, and there will always be some time after which the field is predominantly determined by currents flowing in the ore body.

In the second case, with increasing time, the effect of the geological noise becomes greater, and therefore, the depth of the investigation decreases.

Finally, when both time constants are the same, the ratio between meaningful signal and geological noise is independent of observation time, and is a function only of the geometry of the conductor and the coordinates of the observation point.

Conditions 4.31 are the same as conditions 4.22, 4.24, and 4.25 for harmonic fields. Moreover, they are also independent of the depth of burial of the ore body.

Although in principle, resolution and depth of investigation for the amplitude and phase method are the same as for the time-domain method, in the frequency domain the realization is often related to measuring new quantities not normally measured in exploration, these quantities being the magnitude of subsequent terms in the series describing the low-frequency spectrum given in eq. 4.10.

We will next compare resolution and depth of investigation in measuring the quadrature and the inphase components, the function $\Delta QH(\omega)$, and the transient field making use of calculations of the electromagnetic field contributed by induced currents in a spheroid. We will further assume that the spheroid is situated in free space, and that the primary magnetic field is uniform. However, first it will be useful to consider the effect of the depth of burial (z/b), the dimensions of the spheroid (a/b) and the range of parameters (b/h) where the maximum sensitivity to a change in the parameter τ_0 is observed, and then to evaluate signals normalized to the primary field strength.

First let us examine the effect of the distance from the observation point to the conductor, z/b, on the magnitude of the field. For this purpose, consider the ratio of coefficients:

$$C\left(\frac{a}{b},\frac{z}{b}\right)\bigg/C\left(\frac{a}{b},\frac{z_0}{b}\right)$$

which defines the nature of the change in the field strength with depth. Inasmuch as we have the following approximate equalities for the coefficients C_1, C_2, and C_3:

$$C_1 \approx d_1/q_1, \quad C_2 \approx d_1/q_1^2, \quad C_3 = d_1/q_1^3$$

the dependence of the depth of burial when measurements are made with the quadrature, and inphase components, the difference function ΔQH, and the transient field are all practically the same (see Fig. 4.4). Considering that for sufficiently large values of z/a, the field contributed by currents in the spheroid becomes equivalent to that for a magnetic dipole with a moment proportional to $a^4 b$, the ratio of fields caused by currents in two spheroids located at the same depth:

$$C\left(\frac{a_2}{b},\frac{z}{b}\right)\bigg/C\left(\frac{a_1}{b},\frac{z}{b}\right)$$

becomes greater with an increase in z/b, provided that a_2 is greater than a_1, and approaches an asymptotic value given by $(a_2/a_1)^4$, as shown in Fig. 4.5. In a comparison to resolution obtained with various methods, it is appropriate to assume that the target ore body and the more resistive homogeneity which is the source of geologic noise have different dimensions. Therefore, it is useful to investigate the effect of the geometry of a conductive rock mass on the measured quantities and with this in mind, let us consider some features of the distribution of induced currents. As has been shown earlier, at low frequencies, the currents induced in a confined conductor can be represented as:

$$\begin{aligned} Qj &\approx \alpha_1\omega - \alpha_3\omega^3 + \alpha_5\omega^5 + \\ \mathrm{In}\,j &\approx -\alpha_2\omega^2 + \alpha_4\omega^4 - \alpha_6\omega^6 + \ldots \end{aligned} \tag{4.35}$$

It is clear that each term in these series defines a corresponding term in a series for the magnetic field through the Biot-Savart law (eq. 3.183). In studying the electromagnetic field in a sphere, we have shown that the leading term of the quadrature component of the current ($\alpha_1\omega$) increases linearly to the surface of the sphere, while the corresponding term of the inphase component ($\alpha_2\omega^2$) increases more rapidly than linearly; that is, the greater part of the currents in comparison with the previous case are concentrated at the surface of the conductor. For this reason, the inphase

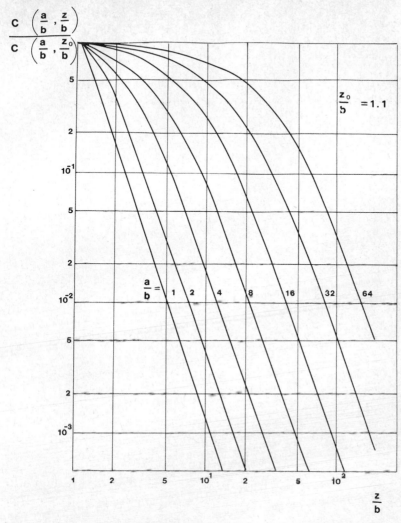

Fig. 4.4. Curves illustrating the effect of distance from the observation point to the conductor, z/b, on the magnitude of the field.

component of the magnetic field is more sensitive to a change in the dimensions of the conductor, than is the quadrature component. This phenomenon, reflecting the skin effect, is inherent also for the later terms of series 4.10 to an even greater extent. Curves for the functions:

$$C_i \left(\frac{a}{b}, \frac{z}{b} \right) \bigg/ \left(C_i \ 1, \frac{z}{b} \right)$$

as shown in Figs. 4.6—4.7 clearly illustrate the increase of the influence of the dimensions of the spheroid in going from measurements of the quadrature

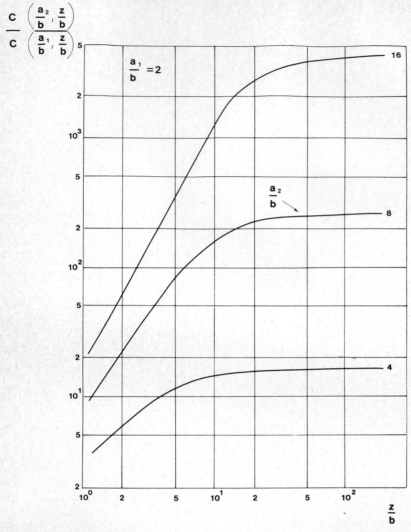

Fig. 4.5. The ratio of fields developed in two spheroids as a function of the ratio z/b.

component to the difference value ΔQH, measured with the dual frequency method. These relations between the field and the dimensions of the target have an affect on the resolution for various techniques for making measurements. In fact, if the more resistive inhomogeneity that contributes the noise has smaller dimensions than the ore body, measuring the function ΔQH in place of QH will result in a significant decrease in the effect of the noise. However, as has already been pointed out, in carrying out an analysis of the level of geologic noise, it is appropriate to assume a more complicated situation in which the source for geologic noise is situated close to the

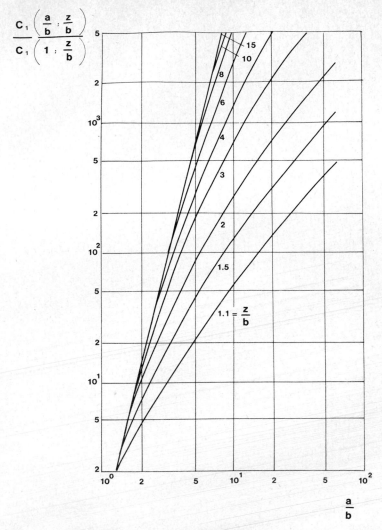

Fig. 4.6. Curves illustrating the influence of the dimensions of the spheroid in measuring the quadrature component of the field.

observation point and has larger dimensions than the target ore body. In this case, the higher sensitivity, for example, of the coefficient C_3 to a change in the dimensions of the spheroid causes a significant compensation of the positive effect inasmuch as the time constants of the ore body and of the source of geological noise are different and the time constant for the ore body is longer than the time constant for the noise. In other words, the level of geologic noise would be smaller if the effect of the dimensions of the bodies in the dual frequency method measuring $C_3 \omega^3 / \alpha^3$ were to be the same

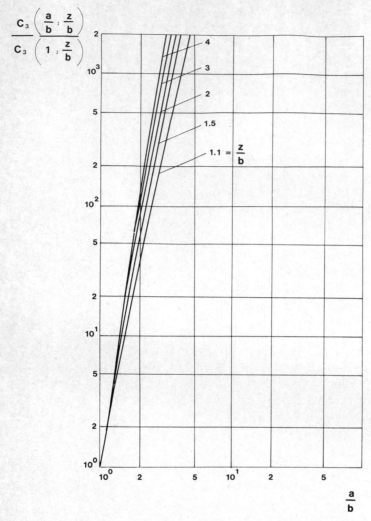

Fig. 4.7. Curves illustrating the influence of the dimensions of the spheroid in measuring the difference value ΔQH measured with a dual-frequency method.

as for the case in which the quadrature component were measured. When the time constants are equal, the effect of the geologic noise on the function ΔQH is found to be much greater than that in measuring the quadrature and the inphase components.

In our consideration of noise levels, it will be assumed that the primary field is uniform; that is, that currents flowing in the target ore body and in the homogeneity generating noise, which, as a rule, would be located closer to the observation point than the ore body, are contributed by the action of a field that has the same amplitude at both locations. This is the most favorable situation that can exist, and can be achieved for example using a circular ring source which a ratio r_1 which is greater than the distance z, or the axial dimension a of the spheroid (with a greater than b). We should note that the ratio of the amplitude of the vertical component of the magnitude field caused by the loop on the z-axis to the amplitude of the field at the center of the loop (the observation point in Fig. 4.8) is:

$$\frac{H_{0z}(z)}{H_{0z}(0)} = \frac{r_1{}^3}{(z^2 + r_1{}^2)^{3/2}}$$

and that it is a monotonic function of depth. Assuming that the strength of the current flowing in the loop source is constant, let us consider dependence of the magnetic field H_{0z} along the z-axis on radius r_1. When $r_1 \ll z$, the field from the source loop is equivalent to that from a magnetic dipole and increases in direct proportion to the square of the radius of the loop. However, when the loop radius is sufficiently large ($r_1 \gg z$), the primary field begins to decrease in inverse proportion to r_1. It can readily be seen that the maximum value for the vertical component of the field, H_{0z}, at a given depth z is observed when the loop's radius is:

$$r_1 = 2^{1/2}z \tag{4.33}$$

Moreover, the magnetic field strength at the depth $z = r_1/2^{1/2}$ is only slightly greater than half the strength of the primary field near the surface, as $z = 0$. For this reason, it is prudent to use loops with relatively large radii as sources, but also, it is wise to increase the current flowing in the loop as well in order to raise the signal level to meet the sensitivity and noise interference criteria

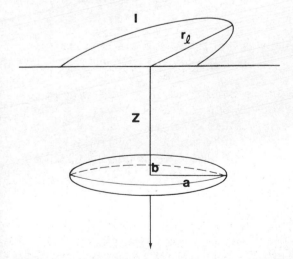

Fig. 4.8. A conducting spheroid in a field generated by a co-axial current-carrying loop.

at a receiver. For example, with the radius of the source loop $r_1 = 2z$, the magnetic field $H_{0z}(0)$ at the center of that loop is only 30% larger than the field at the depth $z = r_1/2^{1/2}$. Thus, in searching for relatively deep seated ore bodies with surface inhomogeneities being present, and with such inhomogeneities being the only source of geological noise in the model we have assumed, it is prudent to make use of loops with a large enough radius (r_1 being about $2z$) in order to provide a uniform primary field excitation over the depth range from 0 to z, and this will provide a favorable set of conditions for identifying anomalies with respect to their conductivity. However, in actual operating conditions in which the surrounding medium has a finite conductivity, it is often necessary to consider the effects of vortex currents in a host medium and, as will be shown in the next chapter, the radius of the source loop as well as the distance between the transmitter and receiver can play a vital role.

Let us now determine the relative sizes of anomalies generated by currents flowing in a highly conducting spheroid and define the range of parameters (b/h) over which the best resolution is obtained for each of the measured quantities we are considering. In measurements of the inphase component of the secondary field, or in measurements of the difference ΔQH, which is a combination of quadrature components, requires the determination of later terms in the series 4.10, used to describe the field at low frequencies. In principle, this can be achieved by making highly precise measurements. However, this approach is not practical, since the later terms in the series 4.10 are small in comparison with the primary field strength, and that part of the quadrature component which is directly proportional to the frequency ($C_1\omega/\alpha$). For this reason it is assumed that these quantities can be determined only through the use of some innovative measurement techniques. As for example, through the use of simultaneous measurements at two frequencies which are chosen in such as a way to cancel the primary field or to eliminate the leading term in the expression for the quadrature component, the term which is proportional to ω. In comparison with the established and widely used amplitude-phase measurement schemes, such a differential system based on simultaneous measurements at two frequencies will require a decrease in noise level in the receiver and an increase in the current in the transmitter, but it will not require any further increase in errors contributed by possible inaccuracies in the geometrical layout of a transmitter-receiver array. Inasmuch as the inphase component of the secondary field and particularly the difference of the quadrature components:

$$\Delta QH \ = \ QH(\omega_1) - \frac{\omega_1}{\omega_2} \, QH(\omega_2)$$

are quite small values at low frequencies, it will be very important to determine the upper-frequency boundary more precisely, $(b/h)_{\max}$, corresponding to the maximum signal for which high resolution is still obtained. Table 4.I contains lists of values for the parameter $(b/h)_{max}$ and values for the anomalous field normalized to the primary field and expressed as a percentage, for conductive spheriods situated at various distances from the observation point. These data permit us to draw the following conclusions:

(1) The upper limit for frequencies:

$$f_{max} = \frac{n^2}{4b^2} 10^6 \rho, \quad \text{where} \quad n = \left(\frac{b}{h}\right)_{max}$$

is shifted toward lower frequencies with an increase in the horizontal dimension of the spheriod (the primary magnetic field is assumed to be directed along a minor semi-axis b). However, the anomalies increase. For example, in going from a sphere to the spheroid with $a/b = 8$, the frequency limit, f_{max}, decreases by a factor of 15 regardless of which field parameter is measured.

(2) The magnitudes of the quadrature component of the magnetic field listed in Table 4.I are almost the same as the maximum values for the spectrum.

(3) Anomalies of the magnitude for the quadrature and the inphase components are practically the same when the frequency is at the limit $(b/h)_{max}$, but as might be expected, with lower frequencies, they begin to differ from each other, the greater the lower the frequency is.

(4) In considering the difference factor $\Delta Q H(\omega)$, the parameter $(b/h)_{max}$ becomes smaller and correspondingly the frequency decreases along with the meaningful signal strength.

(5) Under operating conditions in which the surrounding medium has a finite resistivity, a decrease in frequency (within the low-frequency portion of the spectrum) reducing the value of the quadrature component of the magnetic field, does not change the resolution capability. At the same time, in measuring other components, the choice of a low-frequency boundary of frequencies is important. This reflects the fact that regardless of the resistivity of the surrounding medium, one can always find some frequency ω_1, which marks the limit $(\omega < \omega_1)$ where the effect of currents induced in the host medium on the inphase component, on the amplitude of the total field, and on a difference function $\Delta Q H$, will be significantly greater than the influence of currents in the ore body. This behavior of the field will be illustrated in the next chapter.

When no geological noise is present and the source for the secondary field consists only of the conductive ore body, the depth of investigation is limited only by the sensitivity of the equipment and the array as a whole to detect relatively small signals in the presence of the primary field and natural and man-made electromagnetic noise fields. For this reason, assuming that the minimum value of the anomaly $(\eta = H/H_0\%)$ corresponding to the parameter $(b/h)_{max}$, let us define the maximum depth of an investigation

TABLE 4.I

	QH			ln H			Δ|H|			ΔQH		
a/b:	2	4	8	2	4	8	2	4	8	2	4	8
(b/h)$_{max}$:	1.0	0.6	0.4	1.1	0.65	0.45	1.0	0.6	0.4	0.7	0.45	0.3
z/b												
1.1	35	44	51	32	37	54	13	14	16	4.5	7.6	10
1.5	21	34	45	18	27	47	9.2	12	15	2.5	5.6	8.7
2.0	12	25	39	10	19	39	5.5	10	14	1.4	3.9	7.2
3.0	4.6	14	29	3.7	10	27	2.2	6.0	12	0.51	2.0	4.0
4.0	2.0	7.9	21	1.7	5.6	19	1.0	3.6	9.2	0.23	1.1	3.4
6.0	0.68	3.2	12	0.54	2.1	10	0.32	1.4	5.1	—	0.42	1.7
8.0	0.3	1.6	7.0	0.23	1.0	5.5	0.14	0.67	3.0	—	0.20	0.98
10.0	0.16	0.86	4.4	0.12	0.54	3.3	—	0.37	1.8	—	0.11	0.59
15.0	—	0.27	1.6	—	0.17	1.2	—	0.11	0.67	—	—	0.21
20.0	—	0.12	0.76	—	—	0.54	—	—	0.31	—	—	0.10
30.0	—	—	0.24	—	—	0.17	—	—	0.10	—	—	—
40.0	—	—	0.10	—	—	—	—	—	—	—	—	—

	QH			ln H			Δ|H|			ΔQH		
a/b:	16	32	64	16	32	64	16	32	64	16	32	64
(b/h)$_{max}$:	0.25	0.18	0.12	0.30	0.20	0.14	0.30	0.20	0.14	0.20	0.12	0.09
z/b												
1.1	51	54	53	46	51	53	21	19	19	11	5.4	8.0
1.5	48	52	52	52	50	52	20	19	19	9.7	5.2	7.9
2.0	44	51	51	47	47	51	20	19	19	8.8	4.9	7.7
3.0	38	47	49	39	43	49	19	18	18	7.2	4.4	7.6
4.0	33	44	47	33	39	46	17	18	18	5.9	4.0	6.9
6.0	24	38	44	23	33	42	13	17	17	4.1	3.3	6.3
8.0	18	32	41	16	27	38	10	15	17	2.8	2.7	5.6
10.0	13	28	38	12	23	35	7.7	13	15	2.0	2.2	5.1
15.0	6.7	19	31	5.7	15	28	3.9	9.6	13	0.95	1.4	4.0
20.0	3.6	13	26	3.0	9.8	22	2.1	6.8	9.3	0.50	0.91	3.1
30.0	1.4	6.7	18	1.1	4.7	14	0.79	3.5	6.5	0.18	0.43	2.0
40.0	0.63	3.7	12	0.50	2.5	9.7	0.36	1.9	3.3	—	0.22	1.3
60.0	0.20	1.4	6.2	0.16	0.91	4.6	0.11	0.70	1.8	—	—	0.61
80.0	—	0.64	3.4	—	0.42	2.4	—	0.32	1.1	—	—	0.32
100.0	—	0.34	2.0	—	0.22	1.4	—	0.17	0.37	—	—	0.18
150.0	—	0.11	0.70	—	—	0.49	—	—	0.17	—	—	—
200.0	—	—	0.31	—	—	0.22	—	—	—	—	—	—

z/b, that corresponds to various measured quantities (see Table 4.II). It can readily be seen from this table that in fact, in measuring the quadrature component, the depth of investigation is great and therefore under favorable geoelectric conditions, when the influence of geological noise is much less than that of ambient noise, it is not worthwhile to measure the signals of lesser strength such as the inphase component In H, the difference in amplitudes of the total field, ΔH, or the function $\Delta Q H(\omega)$. Under these conditions, the study of the frequency response of the quadrature component, $Q H(\omega)$ permits one by making use of the location of the maximum, to derive with sufficient accuracy a value for the time constant for the parameter $\alpha = 1/\sigma\mu b^2$. However, as was demonstrated earlier in this chapter, a meaningful model of the medium includes various types of sources which do not have any relationship to the target ore body and which represent geologic noise. Therefore, in general, when it is necessary to consider such noise, the conclusions about the maximum depth of an investigation that can be obtained in measurements of the quadrature component may very well be wrong. In order to illustrate this, we might assume that both an ore body and a more resistive inhomogeneity (the source of geological noise) are spheroids shown in Fig. 4.3. The two spheroids are situated at a significant distance (z) from each other, so that one would expect no significant interaction between the currents flowing in the two. We will further assume that the ore body (R) is situated beneath the source of geological noise (N). It is obvious that the measured field at the surface will be the sum of all the fields created by current flow in every conductive mass in the insulating host medium. The decrease in the obtainable depth of investigation with the local inhomogeneity present near the observation point is clearly demonstrated by the frequency responses for the quadrature component in the magnetic field, which are plotted on Fig. 4.9 for various values of the ratio of conductivity, σ_R/σ_N. As may be seen from this illustration, with a large ratio of conductivities ($\sigma_R/\sigma_N > 200$) the combined spectrum which is observed at the surface will be approximately the same as the computed spectrum for the quadrature component caused by currents flowing only in the ore body at low frequencies, but at high frequencies the combined spectrum will be essentially the same as that contributed by the noise source. For this reason, it is possible to separate the anomalies in the quadrature component with respect to the conductivity of their source, or more precisely, according to their time constants. However, with a relative increase in the conductivity σ_N, the frequency response of that part of the quadrature component contributed by the current flow in the noise source shifts toward lower frequencies and begins to distort the spectrum of the field generated by currents flowing in the ore body. At first, the character of the field changes over the range of frequencies in which the first descending branch of the spectrum is present, and then, the first maximum disappears, this being the character which permits one to determine the parameter α_R for the ore

TABLE 4.II

$\dfrac{a}{l}$ $\bigg\backslash$ η:	Q_H			$\ln H$			$\Delta\lvert H\rvert$			ΔQ_H		
	10^{-3}	5×10^{-2}	10^{-2}	10^{-3}	5×10^{-2}	10^{-2}	10^{-3}	5×10^{-2}	10^{-2}	10^{-3}	5×10^{-2}	10^{-2}
1	6	4	3	6	4	3	4	3	2	3	2	1.5
2	10	6	5	10	6	5	8	5	3	5	3	2.5
4	20	12	9	15	10	8	10	7	5	10	6	4
8	40	25	18	30	20	15	20	12	9	20	10	8
16	70	45	35	60	40	30	40	20	15	30	20	15
32	150	90	70	100	80	60	60	35	25	70	30	20
64	250	180	120	250	150	120	150	80	50	100	55	45

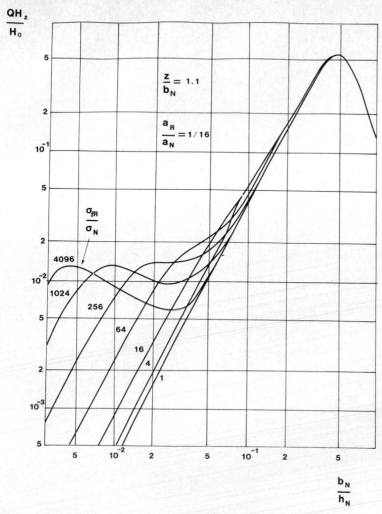

Fig. 4.9. Frequency response for the quadrature component of the magnetic field normalized to the strength of the primary field.

body. At this frequency, a relatively small part of the almost constant field is formed, and finally for some values of σ_R/σ_N, and for a given precision in measurement, the spectrum becomes identical with that caused only by current flow in the source of geologic noise. In this case, it is not possible to separate the meaningful signal QH_R. In a particular case when $\sigma_R/\sigma_N > 16$, the quadrature component of the magnetic field generated by currents in the ore body constitutes only 25% of the amplitude at low frequencies and even less at higher frequencies. For this reason, the main contribution to the spectral amplitude of the quadrature component of H when σ_R/σ_N is 16 is that of currents in the noise source, and therefore, we are only to determine

TABLE 4.III

$(a/b)_N = 16$, $(z/b)_N = 1.1$

$(z/b)_R$	$(a/b)_R = 1$			$(a/b)_R = 2$				$(a/b)_R = 4$			
	QH	ln H	Δ|H|	QH	ln H	Δ|H|	ΔQH	QH	ln H	Δ|H|	ΔQH
4	5×10^4	1.5×10^3	10^3	4×10^3	250	160	95	370	50	30	25
6	2×10^5	3×10^3	—	10^4	450	280	—	930	80	50	30
8	—	—	—	3×10^4	670	430	—	2×10^3	110	70	40
10	—	—	—	5×10^4	930	—	—	3.5×10^3	150	100	50
15	—	—	—	—	—	—	—	10^4	270	—	—
20	—	—	—	—	—	—	—	3×10^4	—	—	—
30	—	—	—	—	—	—	—	—	—	—	—
40	—	—	—	—	—	—	—	—	—	—	—
60	—	—	—	—	—	—	—	—	—	—	—
80	—	—	—	—	—	—	—	—	—	—	—
100	—	—	—	—	—	—	—	—	—	—	—
150	—	—	—	—	—	—	—	—	—	—	—
200	—	—	—	—	—	—	—	—	—	—	—

$(z/b)_R$	$(a/b)_R = 8$				$(a/b)_R = 16$				$(a/b)_R = 32$				$(a/b)_R = 64$			
	QH	ln H	Δ|H|	ΔQH	QH	ln H	Δ|H|	ΔQH	QH	ln H	Δ|H|	ΔQH	QH	ln H	Δ|H|	ΔQH
4	60	12	9	7	16	4	3	3	6	2	1	1	2	1	1	0.5
6	110	17	12	9	22	5	4	3	7	2	2	1	3	1	1	0.5
8	140	25	15	10	30	6	4	3	8	2	2	1	3	1	1	0.5
10	300	30	20	13	40	7	5	4	9	2	2	1	3	1	1	0.5
15	800	50	30	18	80	10	7	5	14	3	2	2	4	1	1	0.5
20	2×10^3	70	50	25	150	14	9	6	20	3	2	2	5	1	1	0.5
30	5×10^3	130	80	—	400	25	15	8	40	5	3	2	7	1	1	1
40	10^4	—	—	—	900	35	20	—	75	7	4	3	10	2	1	1
60	—	—	—	—	3×10^3	60	40	—	200	11	7	—	20	2	2	1
80	—	—	—	—	—	—	—	—	450	17	10	—	40	3	2	1
100	—	—	—	—	—	—	—	—	800	25	—	—	60	4	3	2
150	—	—	—	—	—	—	—	—	3×10^3	40	—	—	180	8	5	—
200	—	—	—	—	—	—	—	—	—	—	—	—	400	11	7	—

366

$(a/b)_N = 32$, $(z/b)_N = 1.1$

$(z/b)_R$	$(a/b)_R = 1$				$(a/b)_R = 2$				$(a/b)_R = 4$			
	QH	$\ln H$	$\Delta|H|$	ΔQH	QH	$\ln H$	$\Delta|H|$	ΔQH	QH	$\ln H$	$\Delta|H|$	ΔQH
4	10^5	3×10^3	2.10^3	—	8.10^3	530	340	205	800	100	70	50
6	4×10^5	6×10^3	—	—	2.10^4	950	600	—	2×10^3	170	110	70
8	—	—	—	—	6.10^4	1.5×10^3	900	—	4×10^3	240	150	90
10	—	—	—	—	10^5	2×10^3	—	—	8×10^3	220	210	110
15	—	—	—	—	—	—	—	—	2×10^4	600	370	—
20	—	—	—	—	—	—	—	—	5.5×10^4	—	—	—
30	—	—	—	—	—	—	—	—	—	—	—	—
40	—	—	—	—	—	—	—	—	—	—	—	—
60	—	—	—	—	—	—	—	—	—	—	—	—
80	—	—	—	—	—	—	—	—	—	—	—	—
100	—	—	—	—	—	—	—	—	—	—	—	—
150	—	—	—	—	—	—	—	—	—	—	—	—
200	—	—	—	—	—	—	—	—	—	—	—	—

$(z/b)_R$	$(a/b)_R = 8$				$(a/b)_R = 16$				$(a/b)_R = 32$				$(a/b)_R = 64$			
	QH	$\ln H$	$\Delta|H|$	ΔQH	QH	$\ln H$	$\Delta|H|$	ΔQH	QH	$\ln H$	$\Delta|H|$	ΔQH	QH	$\ln H$	$\Delta|H|$	ΔQH
4	130	25	20	15	35	9	7	5	12	4	3	2.5	5	1.5	1.5	1
6	240	35	25	20	50	11	8	6	15	4	3	2.5	6	2	1.5	1
8	400	50	32	23	65	13	9	7	17	4	3.5	3	6	2	1.5	1
10	650	65	40	27	90	15	10	8	20	5	4	3	7	2	1.5	1
15	2×10^3	100	70	40	180	22	14	10	30	6	4	3.5	8	2	2	1.5
20	4×10^3	150	100	50	330	30	19	13	45	7	5	4	10	3	2	1.5
30	3×10^4	280	180	—	900	50	32	18	90	10	7	5	15	4	2	2
40	—	—	—	—	2×10^3	75	45	—	160	15	9	6	20	5	2.5	2.5
60	—	—	—	—	6×10^3	130	80	—	450	25	15	—	45	7	3.5	3
80	—	—	—	—	—	—	—	—	950	35	23	—	80	9	5	4
100	—	—	—	—	—	—	—	—	2×10^3	50	30	—	140	16	6	—
150	—	—	—	—	—	—	—	—	6×10^3	90	—	—	400	25	10	—
200	—	—	—	—	—	—	—	—	—	—	—	—	900	—	15	—

the parameter α_N or the time constant τ_{0N} of the geological noise source from the frequency response curve. However, the maximum meaningful signal in this case is reasonably large and fully available for measurement.

From a comparison of the data given in Table 4.III for the model of the medium which is under consideration, it can be seen that the presence of a relatively resistive inhomogeneity (a spheroid) near the observation site decreases the obtainable depth of investigation with the quadrature component by a factor of nearly 5 when $\sigma_R/\sigma_N < 16$ and η is about 0.1%.

Thus, as the conductivity of the rock mass comprising the source of geological noise is even greater, that is, when $\sigma_N/\sigma_R > 1/16$, in this particular example, the quadrature component contributed by current flow in the ore body cannot be recognized in the total observed secondary field. However, it has been shown in this chapter that this problem can be resolved even for relatively highly conducting noise sources by making measurements of the inphase component of the secondary field or of the difference functions $\Delta|H|$ or ΔQH. The curves in Figs. 4.10–4.11 are frequency responses for $\text{In}H$ and ΔQH, measured over a model containing both a simulated ore body and a source of geological noise. As can be seen from these curves, the meaningful signal created by currents in the ore body dominates over the field generated by currents in the noise source when the noise source conductivity is even greater than in the previous case and over a wide range of frequency, exceeds the given level of ambient noise. As has been mentioned above, under actual conditions, the depth of investigation obtainable when these quantities are measured can be much greater than when the quadrature component is measured. In order to illustrate this more clearly, consider some numerical examples. As before, we will assume that one spheroid (N) is situated closer to the observation point than the ore body and has a relatively high resistivity in comparison with the ore body. The second spheroid is located at a greater depth directly beneath the noise source and represents the ore body in which induction currents generate a meaningful signal (H_R). The observation point (0) where the vertical component of the magnetic field is to be measured, is situated on the z-axis. Table 4.III lists minimum values of the ratio of conductivities, σ_R/σ_N, for which the influence of currents in the more resistive spheroid can be neglected (that is, by definition, when $H_N < 0.1 H_R$) for measurements of QH, $\text{In}H$, ΔH, and ΔQH. As may be seen from these calculations, the quadrature component is overly sensitive to the presence of the inhomogeneity with higher resistivity, and consideration of measuring quantities proportional to the second and third derivatives with respect to frequency is in principle a preferable means to reduce the contribution of currents flowing in the geological noise source, provided that $\tau_{0R} > \tau_{0N}$. However, there is always some limit to the value for σ_R/σ_N, depending on the geometry of the model (the linear dimensions of the two conductive bodies, the depth of burial, the position of the observation point, and the loop radius) below which the influence of the relatively

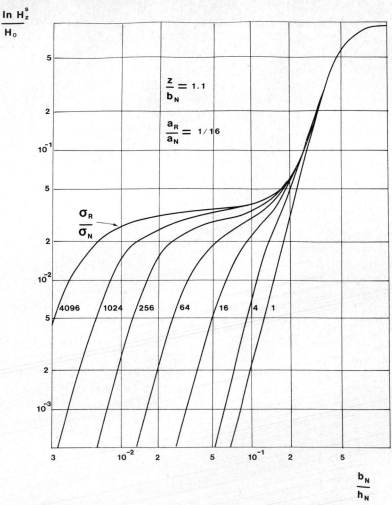

Fig. 4.10. Frequency response for the inphase component of the magnetic field normalized to the strength of the primary field.

resistive inhomogeneity becomes significant and it is impossible using harmonic fields or transient fields to distinguish between anomalies according to conductivity with any confidence.

We will now illustrate the capabilities of the transient method from the point of view of an increase in the achievable depth of investigation and possible separation of anomalies with respect to conductivity by making use of calculative fields contributed by currents flowing in spheroids. As has been mentioned previously at the early stage, the induced currents are concentrated principally near the surface of a spheroid and there is only a slight relationship with the conductivity. For this reason, we will concentrate on

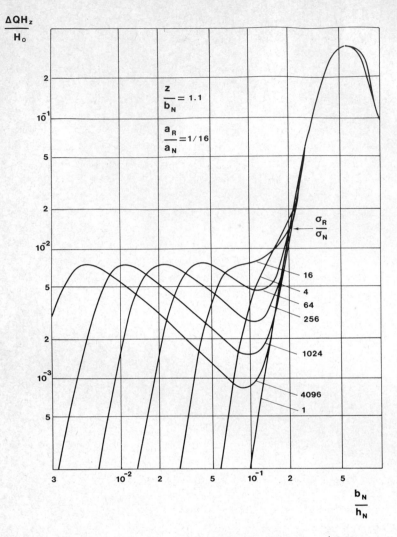

Fig. 4.11. Frequency response for the difference function ΔQH contributed by currents in an ore body and in a conductive body representing geologic noise.

the behavior of the late stage of the transient field, when the currents within a confined conductor decay exponentially, and we can write the following expression for the emf measured on the z-axis:

$$\mathscr{E} = \Phi_0 \frac{d_1 q_1}{\sigma \mu b^2} e^{-q_1 \alpha t} = \Phi_0 d_1 \frac{1}{\tau_0} e^{-t/\tau_0} \tag{4.34}$$

where Φ_0 is the flux of the primary uniform magnetic induction field B_0 penetrating the path along which the EMF is evaluated. The values for the parameters q_1 and d_1 are given in Table 3.VII. The primary magnetic field, as in the previous model with harmonic fields, is assumed to be directed along the minor axis of the spheroid.

As a first step, we will assume that no geological noise is present, and that the primary magnetic field is uniform in the vicinity of the ore body, which is a spheroid. The source of the primary field is a loop with a current I flowing in it and with a radius r_1.

Inasmuch as the source moment for the primary field is zero when measurements are to be made, in contrast to the situation with the amplitude-phase methods, the only background against which signals must be differentiated is the noise from natural and man-made fields, which can be more of a problem in the time domain than in the frequency domain because of the requirement for having a relatively broad-band response in time-domain equipment. For this reason, the maximum depth of investigation which can be obtained is defined by comparing the minimum meaningful $\mathscr{E}_{min}$ with the ambient noise background, $\mathscr{E}_0$, and in so doing, we will establish a signal-to-noise ratio, n, for which a signal can be discriminated with confidence as follows:

$$\mathscr{E}_{min} > n\mathscr{E}_0$$

where n is assumed to be greater than unity. Then, the expression for the emf normalized to the area of the receiver is:

$$\mathscr{E}_{min} = B_0 \alpha q_1 d_1 e^{-q_1 \alpha t} = n\mathscr{E}_0 \tag{4.35}$$

where

$$B_0 = \mu \frac{I r_1^2}{2(r_1^2 + z^2)^{3/2}} \tag{4.36}$$

Experience indicates that the most likely range of values for $\mathscr{E}_0$ will be:

$$10^{-11} < \mathscr{E}_0 < 10^{-8} \quad \text{(in V m}^{-2}\text{)}$$

For any given instant in time, t, and for any specified set of parameters for the conductive body, $(\alpha, a/b, z)$, one can always establish a minimum value for the magnetic induction vector representing the primary field for which the useful signal, $\mathscr{E}_{min}$, will be at least n times greater than the noise level, $\mathscr{E}_0$:

$$B_0 = \frac{n\mathscr{E}_0 e^{q_1 \alpha t}}{\alpha q_1 d_1} = n\mathscr{E}_0 \frac{e^{t/\tau_0}}{d_1} \tau_0 \tag{4.37}$$

Values of B_0 are listed in Table 4.IV corresponding to a noise level of $\mathscr{E}_0 = 10^{-9}$ V m^{-2}, $n = 5$, and a time for measurement of 4 milliseconds. The parameter α is assumed to have the value for the given conductor at which the late stage behavior begins. It is clear that using the information in this

TABLE 4.IV

z/b	a/b = 1, τ/b = 3, α = 28.5		a/b = 2, τ/b = 3, α = 28.5		a/b = 4, τ/b = 11, α = 383		a/b = 8, τ/b = 18, α = 1030	
	B_0	I/b	B_0	I/b	B_0	I/b	B_0	I/b
1.1	0.120×10^{-9}	0.55×10^{-3}	0.837×10^{-10}	0.38×10^{-3}	0.567×10^{-10}	0.26×10^{-3}	0.544×10^{-10}	0.25×10^{-3}
1.5	0.304×10^{-9}	0.19×10^{-2}	0.148×10^{-9}	0.92×10^{-3}	0.778×10^{-10}	0.48×10^{-3}	0.645×10^{-10}	0.40×10^{-3}
2	0.720×10^{-9}	0.60×10^{-2}	0.276×10^{-9}	0.23×10^{-2}	0.113×10^{-9}	0.94×10^{-3}	0.793×10^{-10}	0.66×10^{-3}
3	0.243×10^{-8}	0.30×10^{-1}	0.753×10^{-9}	0.94×10^{-2}	0.222×10^{-9}	0.28×10^{-2}	0.118×10^{-9}	0.15×10^{-2}
4	0.576×10^{-8}	0.95×10^{-1}	0.163×10^{-8}	0.27×10^{-1}	0.401×10^{-9}	0.66×10^{-2}	0.171×10^{-9}	0.28×10^{-2}
6	0.194×10^{-7}	0.48	0.516×10^{-8}	0.13	0.105×10^{-8}	0.26×10^{-1}	0.335×10^{-9}	0.83×10^{-2}
8	0.461×10^{-7}	0.15×10^{1}	0.119×10^{-7}	0.40	0.224×10^{-8}	0.74×10^{-1}	0.601×10^{-9}	0.20×10^{-1}
10	0.900×10^{-7}	0.37×10^{1}	0.230×10^{-7}	0.95	0.416×10^{-8}	0.17	0.100×10^{-8}	0.42×10^{-1}
15	0.304×10^{-6}	0.19×10^{2}	0.768×10^{-7}	0.48×10^{1}	0.133×10^{-7}	0.83	0.280×10^{-8}	0.17
20	0.720×10^{-6}	0.60×10^{2}	0.181×10^{-6}	0.15×10^{2}	0.308×10^{-7}	0.25×10^{1}	0.616×10^{-8}	0.51
30	0.243×10^{-5}	0.30×10^{3}	0.610×10^{-6}	0.76×10^{2}	0.103×10^{-6}	0.13×10^{2}	0.196×10^{-7}	0.24×10^{1}
40	0.576×10^{-5}	0.95×10^{3}	0.144×10^{-5}	0.24×10^{3}	0.242×10^{-6}	0.40×10^{2}	0.456×10^{-7}	0.75×10^{1}
60	0.194×10^{-4}	0.48×10^{4}	0.487×10^{-5}	0.12×10^{4}	0.813×10^{-6}	0.20×10^{3}	0.152×10^{-6}	0.38×10^{2}
80	0.461×10^{-4}	0.15×10^{5}	0.115×10^{-4}	0.38×10^{4}	0.193×10^{-5}	0.64×10^{3}	0.358×10^{-6}	0.12×10^{3}
100	0.900×10^{-4}	0.37×10^{5}	0.225×10^{-4}	0.93×10^{4}	0.376×10^{-5}	0.16×10^{4}	0.697×10^{-6}	0.29×10^{3}
150	0.304×10^{-3}	0.19×10^{6}	0.760×10^{-4}	0.47×10^{5}	0.127×10^{-4}	0.79×10^{4}	0.234×10^{-5}	0.14×10^{4}
200	0.720×10^{-3}	0.60×10^{6}	0.180×10^{-3}	0.15×10^{6}	0.300×10^{-4}	0.25×10^{5}	0.555×10^{-5}	0.46×10^{4}

z/b	$a/b = 16$, $\tau/b = 28$, $\alpha = 2480$		$a/b = 32$, $\tau/b = 40$, $\alpha = 5070$		$a/b = 64$, $\tau/b = 60$, $\alpha = 11400$	
	B_0	I/b	B_0	I/b	B_0	I/b
1.1	0.554×10^{-10}	0.25×10^{-3}	0.495×10^{-10}	0.23×10^{-3}	0.599×10^{-10}	0.27×10^{-3}
1.5	0.605×10^{-10}	0.37×10^{-3}	0.518×10^{-10}	0.32×10^{-3}	0.613×10^{-10}	0.38×10^{-3}
2	0.674×10^{-10}	0.56×10^{-3}	0.547×10^{-10}	0.45×10^{-3}	0.630×10^{-10}	0.52×10^{-3}
3	0.835×10^{-10}	0.10×10^{-2}	0.611×10^{-10}	0.76×10^{-3}	0.666×10^{-10}	0.83×10^{-3}
4	0.103×10^{-9}	0.17×10^{-2}	0.682×10^{-10}	0.11×10^{-2}	0.704×10^{-10}	0.12×10^{-2}
6	0.153×10^{-9}	0.38×10^{-2}	0.847×10^{-10}	0.21×10^{-2}	0.787×10^{-10}	0.19×10^{-2}
8	0.223×10^{-9}	0.74×10^{-2}	0.104×10^{-9}	0.35×10^{-2}	0.878×10^{-10}	0.29×10^{-2}
10	0.316×10^{-9}	0.13×10^{-1}	0.128×10^{-9}	0.53×10^{-2}	0.979×10^{-10}	0.40×10^{-2}
15	0.685×10^{-9}	0.43×10^{-1}	0.208×10^{-9}	0.13×10^{-1}	0.128×10^{-9}	0.80×10^{-2}
20	0.131×10^{-8}	0.11	0.324×10^{-9}	0.27×10^{-1}	0.165×10^{-9}	0.14×10^{-1}
30	0.369×10^{-8}	0.46	0.705×10^{-9}	0.88×10^{-1}	0.269×10^{-9}	0.33×10^{-1}
40	0.812×10^{-8}	0.14×10^{1}	0.136×10^{-8}	0.22	0.421×10^{-9}	0.70×10^{-1}
60	0.259×10^{-7}	0.64×10^{1}	0.382×10^{-8}	0.95	0.921×10^{-9}	0.23
80	0.601×10^{-7}	0.20×10^{2}	0.842×10^{-8}	0.28×10^{1}	0.178×10^{-8}	0.59
100	0.116×10^{-6}	0.48×10^{2}	0.159×10^{-7}	0.66×10^{1}	0.310×10^{-8}	0.13×10^{1}
150	0.389×10^{-6}	0.24×10^{3}	0.517×10^{-7}	0.32×10^{2}	0.924×10^{-8}	0.57×10^{1}
200	0.919×10^{-6}	0.76×10^{3}	0.121×10^{-6}	0.10×10^{3}	0.209×10^{-7}	0.17×10^{2}

table and eq. 4.37, the parameter B_0 can be established for any values of α and t corresponding to the late stage of transient behavior. Moreover, Table 4.IV contains values for the parameter I/b for which the magnetic induction vector in the area occupied by a spheroid is equal to B_0. This allows us to determine the maximum possible depth of investigation for the assumed model of the medium, taking into account the power in the primary source. These values for current have been obtained by assuming that there is an optimized relationship between the radius of the loop and the depth of burial, $r_1 = 2^{1/2} z$. However, calculations will show that a several fold change in the loop radius ($\frac{1}{4} < r_1/r_1^{\text{opt}} < 4$) has no significant effect on the maximum depth of investigation. This is a consequence of the fact that the value for $(z/b)_{\text{max}}$ is only weakly dependent on the strength of the primary field (z varies as $B_0^{1/3}$, when the depth of burial for the ore body is relatively great.

It should be stressed again that the nature of the behavior of the transient field during the late stage is almost the same as that for harmonic fields at low frequencies as the depth of burial for the conductive rock mass is increased. This follows directly from the previously written approximate relationships (eq. 3.188).

$$d_1 = C_1 q_1 \approx C_2 q_1^2 \approx C_3 q_1^3$$

We will next demonstrate a relationship between the emf induced in a receiver by a transient field during a late stage, and a harmonic field in which QH, $\text{In}H$, and ΔQH are measured at low frequencies. In accord with eqs. 4.10 and 4.26, we have:

$$\frac{\mathscr{E}(t)}{Q\mathscr{E}(\omega)} = \frac{q_1^2}{|(kb)^4|} e^{-q_1 \alpha t} \frac{H_0(t)}{H_0(\omega)}$$

$$\frac{\mathscr{E}(t)}{\text{In}\mathscr{E}(\omega)} = \frac{q_1^3}{|(kb)^6|} e^{-q_1 \alpha t} \frac{H_0(t)}{H_0(\omega)} \qquad (4.38)$$

$$\frac{\mathscr{E}(t)}{\Delta\mathscr{E}(\omega)} = \frac{q_1^4 e^{-q_1 \alpha t}}{|(kb)^8|(1 - \omega_2^2/\omega_1^2)} \frac{H_0(t)}{H_0(\omega)}$$

where $k = (i\sigma\mu\omega)^{1/2}$.

Values listed in Table 4.V, as $H_0(\omega) = H_0(t)$, demonstrate the relationship between the emf induced by the harmonic and the transient fields providing that the value for b/h corresponds to the upper frequency boundary for which the highest resolution for the methods still is obtained while the parameter τ/b characterizes the beginning of the late stage of the transient field. Note that the parameter τ is $(2\pi\rho t \times 10^7)^{1/2}$ and that it plays almost the same role as does skin depth, h, in the frequency domain. Up to this point it has been assumed that the nonconductive medium contains only a single spheroid, with that spheroid representing an ore body. Now, as before, we will additionally assume that the observation point is situated above two

TABLE 4.V

	a/b = 1			a/b = 2			a/b = 4			a/b = 8		
τ/b:	3.0	4.2	6.0	3.0	4.2	6.0	11	15	22	18	25	36
$\dfrac{\mathcal{E}(t)}{Q\mathcal{E}(\omega)}$	0.87	0.28	0.3×10^{-1}	2.0	1.3	0.61	0.48	0.61×10^{-1}	0.97×10^{-3}	0.36	0.34×10^{-1}	0.3×10^{-3}
$\dfrac{\mathcal{E}(t)}{\ln \mathcal{E}(\omega)}$	0.82	0.26	0.28×10^{-1}	1.9	1.3	0.60	0.53	0.67×10^{-1}	0.11×10^{-2}	0.29	0.27×10^{-1}	0.24×10^{-3}
$\dfrac{\mathcal{E}(t)}{\Delta \mathcal{E}(\omega)}$	15	4.8	0.51	34	23	11	5.3	0.67	0.11×10^{-1}	3.3	0.32	0.28×10^{-2}

	a/b = 16			a/b = 32			a/b = 64		
τ/b:	28	40	56	40	56	80	60	85	120
$\dfrac{\mathcal{E}(t)}{Q\mathcal{E}(\omega)}$	0.42	0.31×10^{-1}	0.17×10^{-1}	0.40	0.33×10^{-1}	0.23×10^{-3}	0.36	0.26×10^{-1}	0.13×10^{-3}
$\dfrac{\mathcal{E}(t)}{\ln \mathcal{E}(\omega)}$	0.26	0.19×10^{-1}	0.11×10^{-3}	0.35	0.29×10^{-1}	0.20×10^{-3}	0.25	0.18×10^{-1}	0.9×10^{-4}
$\dfrac{\mathcal{E}(t)}{\Delta \mathcal{E}(\omega)}$	3.0	0.22	0.12×10^{-2}	9.9	0.83	0.58×10^{-2}	4.1	0.29	0.15×10^{-2}

spheroids, one of which represents a source of geologic noise, and which is located closer to the surface and which has less conductivity than the target ore body. The distance between the two bodies is great enough so that the interaction between them can be neglected. As has been shown previously, the exponential character of the field behavior during late stage indicates the high resolution obtainable with the transient method and in correspondence with this, it permits us to increase the depth of investigation. In fact, re-writing an earlier expression (eq. 4.26) we have:

$$\frac{H_N}{H_R} = \frac{d_{1N}}{d_{1R}} e^{-(q_{1N}\alpha_N - q_{1R}\alpha_R)t}$$

$$\frac{\mathscr{E}_N}{\mathscr{E}_R} = \frac{d_{1N}}{d_{1R}} \frac{q_{1N}}{q_{1R}} \frac{\alpha_N}{\alpha_R} e^{-(q_{1N}\alpha_N - q_{1R}\alpha_R)t}$$

(4.39)

where

$$\alpha_N = \frac{1}{\sigma_N \mu_N b_N^2} \qquad \alpha_R = \frac{1}{\sigma_R \mu_R b_R^2}$$

As follows from eq. 4.39, if:

$$\tau_{0R} > \tau_{0N}$$

or

$$\frac{\sigma_R}{\sigma_N} > \frac{q_{1R}}{q_{1N}} \left(\frac{b_N}{b_R}\right)^2 \qquad \text{as } \mu_R = \mu_N$$

(4.40)

there is always some instant of time beginning at which the field is essentially determined only by the current flowing in the ore body (R). It is clear that condition 4.40 does not depend on the burial depth of the ore body, and that it defines the minimum ratio of conductivities, and the relative level of geological noise becomes smaller with increasing time for any given model. Curves for the function $(\sigma_R/\sigma_N)_{min}$ are given in Fig. 4.12 for various size spheroids. The index on each of the curves is the specific value for $(a/b)_R$. It is interesting to note that the ratio $(\sigma_R/\sigma_N)_{min}$ is more sensitive to a change in the length of the horizontal axis of the spheroid than the vertical, but if both axes increase simultaneously in such a way that the ratio a/b remains constant, the value for $(\sigma_R/\sigma_N)_{min}$ changes in proportion to the square of the semi-radius semi-axis b.

As has been well established, there are at least two factors which interfere to a greater or lesser extent with the resolution that can be obtained and the depth of investigation for the transient methods. First of all, there is the influence of induced currents in the host medium, which during the late stage decrease much more slowly than currents in a conductor of limited extent, and secondly with an increase in time, the problems of measuring the

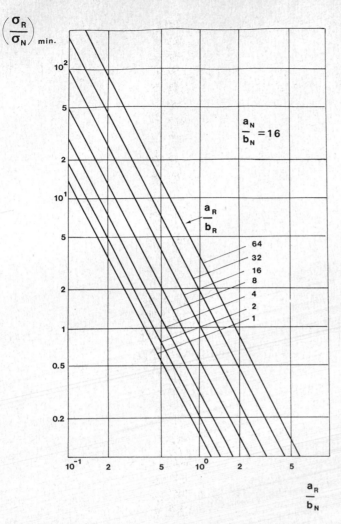

Fig. 4.12. Curves giving the minimum values for the ratio σ_R/σ_N for which the influence of geologic noise becomes insignificant.

signal accurately become significant. Therefore, it is of considerable practical interest to make measurements at earlier times, but corresponding to exponential behavior, and define values for σ_R/σ_N for which one can still consider the field caused by currents in the more resistive spheroid (the source of geological noise) to be sufficiently small, perhaps with a signal to noise ratio of 10. Table 4.VI lists values for $(\sigma_R/\sigma_N)_{min}$ for the same models which have studied for a comparative analysis of resolution in the frequency domain. Comparing the data listed in Tables 4.III and 4.VI, it may readily be seen that for the model of the medium under consideration, the level of the

TABLE 4.VI

$(a/b)_N = 16$, $(z/b)_N = 1.1$

| $(z/b)_R$ | $(a/b)_R = 1$ | | | $(a/b)_R = 2$ | | | $(a/b)_R = 4$ | | | $(a/b)_R = 8$ | | | $(a/b)_R = 16$ | | |
| | $(\tau/b)_R$ | | | $(\tau/b)_R$ | | | $(\tau/b)_R$ | | | $(\tau/b)_R$ | | | $(\tau/b)_R$ | | |
	3	5	7	3	5	7	11	15	20	18	25	35	28	40	55
4	370	140	90	320	110	60	20	13	9	7	4	3	3	2	2
6	410	160	98	360	125	67	23	14	10	7	5	3	3	2	2
8	440	170	103	400	140	73	25	15	11	8	5	4	3	2	2
10	470	180	108	420	148	77	27	16	11	9	5	4	3	2	2
15	510	200	115	460	165	85	30	18	12	10	6	4	4	2	2
20	540	210	125	490	175	91	32	19	13	10	6	4	4	2	2
30	590	220	130	540	190	99	35	21	14	12	7	5	4	2	2
40	620	230	135	570	200	105	38	22	14	13	7	5	5	3	2
60	660	250	145	610	215	115	41	24	15	14	8	5	5	3	2
80	690	260	148	640	225	120	43	25	16	15	9	5	6	3	2
100	720	270	153	670	235	125	45	26	17	15	9	6	6	3	2
150	760	280	160	710	250	130	48	28	18	16	10	6	6	3	2
200	790	290	165	740	265	135	50	29	18	17	10	6	7	3	2

$(a/b)_N = 32$, $(z/b)_N = 1.1$

$(z/b)_R$	$(a/b)_R = 1$			$(a/b)_R = 2$			$(a/b)_R = 4$			$(a/b)_R = 8$			$(a/b)_R = 16$		
	$(\tau/b)_R$			$(\tau/b)_R$			$(\tau/b)_R$			$(\tau/b)_R$			$(\tau/b)_R$		
	3	5	7	3	5	7	11	15	20	18	25	35	28	40	55
4	790	310	190	690	240	125	43	27	19	14	9	7	5	4	3
6	880	345	210	780	270	145	49	30	21	16	10	7	6	4	3
8	950	370	220	840	300	155	53	32	23	17	11	7	6	4	3
10	1000	390	230	900	315	165	57	34	24	18	11	8	7	4	3
15	1100	420	245	990	350	180	64	38	26	20	12	9	7	5	3
20	1150	445	260	1050	370	195	68	41	27	22	13	9	8	5	4
30	1250	480	275	1150	405	210	75	44	29	25	15	10	9	5	4
40	1350	500	290	1250	430	225	80	47	31	27	16	10	10	6	4
60	1400	535	305	1300	460	240	87	50	33	29	17	11	11	6	4
80	1450	560	320	1350	485	255	92	53	34	31	18	11	11	7	4
100	1550	575	330	1450	500	265	96	55	35	32	19	12	12	7	4
150	1650	610	345	1550	535	280	100	59	37	35	20	12	13	8	5
200	1700	630	355	1600	560	290	105	62	39	37	21	13	14	8	5

geological noise obtained measuring a transient EMF and the value for $\Delta QH(\omega)$ are practically the same for ore bodies at relatively shallow depths. On the other hand, with greater depth of burial, this relationship changes and the transient methods appear to offer the advantage. In addition, it should be noted that with sufficiently large values of $(z/b)_R$, it may be necessary to use extremely powerful current sources (see Table 4.IV) and therefore, the numerical values given in this table for such conditions probably do not have any practical interest.

In this chapter, we have compared the resolution and the depth of investigation obtainable with both frequency and transient methods, considering a single type of geological noise, that is, noise contributed by confined conductive masses surrounded by an insulating host rock. However, this is but one special type of noise and it will be necessary to investigate other types of geological noise in addition, including such things as:

(1) Noise contributed by a conducting uniform half-space.

(2) Noise contributed by a thin conductive layer representing the overburden.

(3) Noise contributed by a two-layer host medium.

(4) Noise contributed by electrical charges that develop on the interface between an ore body and the surrounding conductive medium.

All of these types of geological noise will be considered in the next two chapters.

In spite of the relative simplicity of the model that we have used here for geological noise, it is certainly useful to some extent to express some conclusions:

(1) In general, the total signal measured in the frequency domain or the time domain is the sum of three terms, namely:

$$H(\omega) = H_{us}(\omega) + H_{amb}(\omega) + H_g(\omega)$$
$$H(t) = H_{us}(t) + H_{amb}(t) + H_g(t)$$

(4.41)

Similar expressions can be written for electromotive force. In eq. 4.41, H_{us} is the signal caused by currents flowing in the ore body, H_{amb} is the ambient noise which consists of natural and cultural fields as well as measurement errors treated by the equipment and layout of the array, and H_g is the geological noise.

(2) For the case in which the two last terms in 4.41 are small:

$$H(\omega) = H_{us}(\omega), \quad H(t) = H_{us}(t)$$

(4.42)

With measurements in either the frequency or time domains, one can establish the characteristics of the ore body. In this case, the frequency and transient methods are exactly equivalent.

(3) This conclusion is still valid when it is possible to reduce the ambient noise for the frequency- and time-domain measurements to the same degree, as H_g is negligible.

(4) In the case in which geological noise plays a role, it is clear that with an increase in a depth of burial of the ore body or a decrease in its dimensions, the effect of the geological noise is greater. There is an essential difference between the ambient noise and the geological noise from the following point of view. With development of high-technology equipment, which can provide a significant decrease in both natural and cultural noise, the effect of the ambient noise will continue to become less and less important. In contrast to this situation, geological noise is a reality (contributed by overburden, by host medium, or by inhomogeneities in the host medium) which cannot be removed instrumentally, and which comprises part of the geoelectric section along with the ore body.

(5) It is clear that with an increase in the depth of burial of an ore body, the relative contribution of geologic noise consisting as an example of induced currents in the host medium increases and its contribution will ultimately establish the maximum depth of investigation which can be obtained. It is convenient to characterize this effect in terms of the ratio H_N/H_R or $\mathscr{E}_N/\mathscr{E}_R$ where H and $\mathscr{E}$ are any of the components of the magnetic field or the electromotive force which may be measured, respectively. It is obvious that when such a ratio becomes greater than unity, it cannot be expected that an ore body can be recognized.

(6) An analysis of the effect of geological noise contributed by a confined conductor permits us to recognize some fundamental features, such as:

(7) The quadrature and inphase components of the magnetic field are characterized by different degrees of resolutions and depth of investigation.

(8) When the time constant for the ore body is greater than that for the source of geological noise, the inphase component of the secondary field has a greater depth of investigation and is characterized by higher resolution in terms of conductivity than the quadrature component.

(9) A further increase in the depth of investigation and capability for resolution can be obtained by measuring later terms in a series that represents the low-frequency part of the spectrum under the condition that $\tau_{0R} > \tau_{0N}$.

(10) A decrease in frequency within the low-frequency part of the spectrum has no effect on the relative contribution of geological noise for the types of model considered here. It should be noted that when the host medium has a finite conductivity, a decrease in frequency merely increases the relative contribution of this type of noise and correspondingly, the depth of investigation which can be obtained becomes smaller if the inphase component of the magnetic field or function $\Delta Q H(\omega)$ is measured.

(11) The various terms in a series describing a low-frequency part of the spectrum, either the quadrature of the inphase components, are characterized by different resolving capabilities and depths of investigation. These various terms are sensitive to geological noise in different ways, and the higher order terms in the series are more and more transparent to geological noise, if $\tau_{0R} > \tau_{0N}$.

(12) It should not be surprising after these considerations that the resolution and depth of investigation obtainable with a transient method is different than the similar quantities for measurements of the quadrature or inphase component in the frequency domain.

(13) As the mathematical analysis has shown, there are in principle two ways to reduce geological noise contributed by currents in confined bodies, namely: (a) by measuring the transient response at the late stage; or (b) by measuring higher order terms in the series representing the low-frequency spectrum in the frequency domain. Both approaches permit one to achieve the same result, provided that the time constant, characterizing the ore body, τ_{0R}, is greater than the time constant, τ_{0N}, for the source of geological noise.

(14) Some of these conclusions will not apply for other types of geological noise.

Chapter 5

EFFECT OF INDUCED CURRENTS IN THE HOST MEDIUM ON FREQUENCY AND TRANSIENT RESPONSES CAUSED BY CURRENTS FLOWING IN A CONFINED CONDUCTOR

5.1. THE NORMAL FIELD FROM CURRENTS IN THE HOST MEDIUM

In this chapter, we will examine the influence of the surrounding medium when no electrical charges are present and the sole source for the electromagnetic field is induced currents. In general, such an analysis must include consideration of fields caused by the various types of sources which are used in exploration practise, such as vertical and horizontal magnetic dipoles, horizontal rectangular loops, grounded wires, and so on. Various types of electromagnetic methods will be described in a future monograph, and so, we will limit ourselves to consideration of a single type of source here, that being a circular loop, inasmuch as the principal purpose of the present analysis is a comparison of the effect of geological noise on measured fields in the frequency and time domain. It is reasonable to expect that the principal results which will be obtained for this case will also be useful in the examination of other types of equipment.

First, we will write the magnetic field or the electromotive force to be a sum of several parts:

$$H = H_0 + H_1, \quad \mathscr{E} = \mathscr{E}_0 + \mathscr{E}_1 \tag{5.1}$$

where H_0 and $\mathscr{E}_0$ are the normal magnetic field and the normal electromotive force, respectively, caused by the current flowing in the source and in the surrounding medium, when there is no confined conductor present. For example, it should be clear that:

$$H_0 = H_0^{(0)} + H_0^{(1)} \tag{5.2}$$

where $H_0^{(0)}$ is the primary field of the source, and $H_0^{(1)}$ is the field caused by current flowing in the surrounding medium. The second term in eq. 5.1 is the anomalous magnetic field contributed by currents flowing in the conductor and in the host medium, each of which is characterized by a different conductivity. Thus, the magnetic field observed in the host medium differs from the magnetic field observed in the case of a conductive body situated in a insulating host medium as a consequence of the appearance of currents in the host medium as well as surface charges, which affect both normal and secondary fields. The contribution of fields caused by electrical charges will be considered in the next chapter.

The analysis of the effect of the host medium will be done for four models, namely:

(1) A completely uniform medium.
(2) A uniform half-space.
(3) A conducting plate.
(4) A two-layer medium.

The behavior of the normal fields

Let us consider the behavior of the vertical component of both the magnetic field and the electromotive force as measured at the center of the source loop.

I. Observation point at the center of the source loop.
Let us start with the simplest case, a completely uniform medium as shown in Fig. 5.1a.

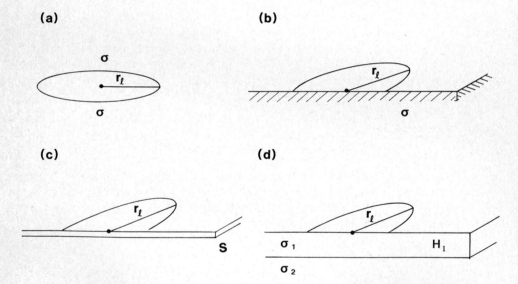

Fig. 5.1. Various types of host medium with nonzero conductivity. a. A conductive full-space. b. A conductive half-space. c. A conductive thin sheet. d. A layered half space.

a. Uniform medium. As is well known, the expression for the vertical component of the harmonic magnetic field is

$$H_0 = H_0^{(0)} e^{ikr_1}(1 - ikr_1) \tag{5.3}$$

where k is the complex wave number, and r_1 is the radius of the source loop. $H_0^{(0)}$ is the primary field observed at the center of the loop, and equal to $I/2r_1$. I is the current flowing in the loop. Expanding the exponential term as a power series in the small parameter $p = r_1/h$, where h is skin depth we obtain:

$$H_0 = H_0^{(0)} \left[1 - \sum_{m=2}^{\infty} \frac{(m-1)}{m!} 2^{m/2} e^{i3/4 \pi m} p^m \right] \tag{5.4}$$

Keeping only the first three terms in the series expansion, we have the following approximate expression for the low-frequency portion of the spectrum:

$$QH_0 \approx H_0^{(0)} [p^2 - \tfrac{2}{3}p^3 + \tfrac{2}{15}p^5 + \dots]$$

$$= \frac{I}{2r_1} \left[\frac{\sigma \mu \omega r_1^2}{2} - \frac{1}{3\sqrt{2}} (\sigma \mu \omega r_1^2)^{3/2} + \frac{1}{30\sqrt{2}} (\sigma \mu \omega r_1^2)^{5/2} + \dots \right]$$

$$\operatorname{In} H_0 \approx H_0^{(0)} [1 - \tfrac{2}{3}p^3 + \tfrac{1}{2}p^4] = \frac{I}{2r_1} \left[1 - \frac{1}{3\sqrt{2}} (\sigma \mu \omega r_1^2)^{3/2} \right.$$

$$\left. + \tfrac{1}{8}(\sigma \mu \omega r_1^2)^2 + \dots \right] \tag{5.5}$$

It should be noted that both components created by currents in the medium tend to zero as the radius of the loop decreases.

In the wave zone $(p > 1)$, vortex currents are concentrated near the source, while the field components oscillate, decaying exponentially

$$QH_0 = H_0^{(0)} p (\sin p - \cos p) e^{-p}$$
$$\operatorname{In} H_0 = H_0^{(0)} p (\sin p + \cos p) e^{-p} \tag{5.6}$$

The curves in Fig. 5.2. illustrate the frequency response for the field components:

$$\frac{QH_0}{H_0^{(0)}}, \quad \frac{\operatorname{In} H_0^{(1)}}{H_0^{(0)}} = (\operatorname{In} H_0 - H_0^{(0)})/H_0^{(0)}$$

and the difference function:

$$\frac{\Delta QH_0}{H_0^{(0)}} = \left[QH_0(\omega_1) - \frac{\omega_1}{\omega^2} QH_0(\omega_2) \right] \Big/ H_0^{(0)}$$

for various values of the parameter p. Nurmerical values are listed in Table 5.I. Applying the Fourier transform to eq. 5.3, we obtain the following expressions for the non-stationary magnetic field and electromotive force::

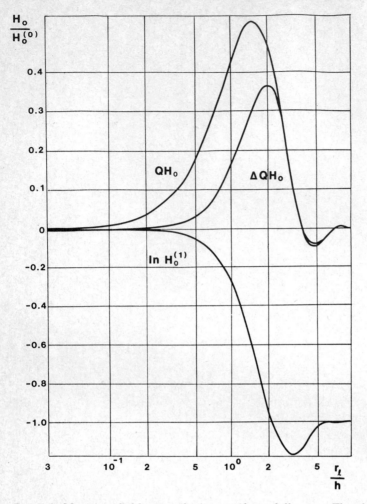

Fig. 5.2. Magnetic field strengths in a uniform full-space. The three curves represent the quadrature and inphase components of the magnetic field and the difference in quadrature strengths at two different frequencies.

$$H_0(t) = H_0^{(0)} \left[\Phi(u) - \frac{2}{\sqrt{\pi}} u e^{-u^2} \right]$$

$$\mathscr{E}_0(t) = -\mu S \frac{\partial H_0}{\partial t} = \frac{\Phi_0}{\sigma \mu r_1^2} \frac{8}{\sqrt{\pi}} u^5 e^{-u^2}$$

(5.7)

where S is the area of the received loop, Φ_0 is the flux in the primary magnetic field:

$$\Phi_0 = \mu H_0^{(0)} S = B_0^{(0)} S$$

TABLE 5.I

Uniform medium

$\dfrac{r_1}{h}$	$\dfrac{QH_0}{H_0^{(0)}}$	$\dfrac{\operatorname{In}H_0^{(1)}}{H_0^{(0)}}$	$\dfrac{\Delta QH_0}{H_0^{(0)}}$
0.125E + 00	0.143E − 01	−0.118E − 02	0.532E − 03
0.177E + 00	0.276E − 01	−0.322E − 02	0.149E − 02
0.250E + 00	0.522E − 01	−0.859E − 02	0.411E − 02
0.353E + 00	0.962E − 01	−0.224E − 01	0.111E − 01
0.500E + 00	0.170E + 00	−0.562E − 01	0.292E − 01
0.707E + 00	0.282E + 00	−0.134E + 00	0.716E − 01
0.100E + 01	0.420E + 00	−0.293E + 00	0.157E + 00
0.141E + 01	0.526E + 00	−0.569E + 00	0.285E + 00
0.200E + 01	0.482E + 00	−0.923E + 00	0.367E + 00
0.283E + 01	0.229E + 00	−0.116E + 01	0.239E + 00
0.400E + 01	−0.214E − 01	−0.111E + 01	−0.660E − 02
0.565E + 01	−0.296E − 01	−0.993E + 00	−0.313E − 01
0.800E + 01	0.338E − 02	− 0.998E + 00	0.347E − 02
0.113E + 02	−0.186E − 03	−0.100E + 01	−0.186E − 03
0.160E + 02	0.117E − 05	−0.100E + 01	0.117E − 05

and $\Phi(u)$ is the probability integral:

$$\Phi(u) = \frac{2}{\sqrt{\pi}} \int_0^u e^{-x^2} dx, \quad u = \frac{\pi\sqrt{2}r_1}{\tau}, \quad \tau = 2\pi(2t/\sigma\mu)^{1/2}$$

Equations 5.7 have been derived under the assumption that the current in the loop is turned on abruptly at the moment $t = 0$. During the early stage the transient EMF induced in the receiver is small; with increase in time, this EMF increases and passes through a maximum at τ/r_1 approximately equal to 3. With further increase in time, the EMF tends to zero as shown by the curve in Fig. 5.3. With an increase in the resistivity of the medium or a decrease in the loop radius, the maxium value for the electromotive force appears at earlier times. As follows from eq. 5.7, the magnetic field gradually and monotonically decreases, as shown by the curve in Fig. 5.4. Expanding the probability integral in a power series in the small parameter u (that is, for the condition of large time, small loop radius, or high resistivity in the medium) we can obtain approximate expressions for the late-stage behavior ($\tau/r_1 > 15$):

$$H_0(t) \approx \frac{Ir_1^2}{12\sqrt{\pi}}\left(\frac{\sigma\mu}{t}\right)^{3/2}, \quad \mathscr{E}_0(t) \approx \frac{\Phi_0}{4\sqrt{\pi}}r_1^3(\sigma\mu)^{3/2}\frac{1}{t^{5/2}} \tag{5.8}$$

Values for the magnetic field and the function $(\mathscr{E}_0/\Phi_0)\sigma\mu r_1^2$ are listed in Table 5.II.

388

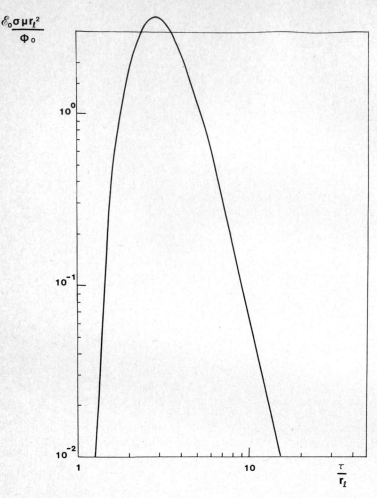

Fig. 5.3. Transient response of the electromotive force at the center of the source loop for a uniform full-space.

Now, let us consider a more complicated model.

b. Uniform half-space. Consider the case in which a source loop is situated on the surface of a uniform half-space, as shown in Fig. 5.1b. Then, as is known, the vertical component of the magnetic field at the center of the loop is:

$$H_0 = H_0^{(0)} \frac{6}{k^2 r_1^2} \left[1 - e^{ikr_1} \left(1 - ikr_1 - \frac{k^2 r_1^2}{3} \right) \right] \tag{5.9}$$

or

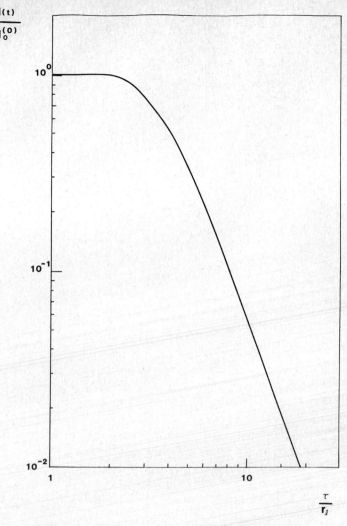

Fig. 5.4. Transient response of the magnetic field at the center of the source loop for a uniform full-space.

$$H_0 = H_0^{(0)} \left[1 - 2 \sum_{m=2}^{\infty} \frac{2^{m/2} e^{i 3\pi m/4} p^m}{(m-2)! \, m(m+2)} \right] \tag{5.10}$$

Then, for the quadrature and inphase components of the spectrum at low frequencies:

$$QH_0 = -H_0^{(0)} \left[\frac{\sigma\mu\omega r_1^2}{4} - \frac{\sqrt{2}}{15} (\sigma\mu\omega r_1^2)^{3/2} + \dots \right] \tag{5.11}$$

TABLE 5.II

Uniform full-space

$\dfrac{\tau}{r_1}$	$\dfrac{H_0(t)}{H_0^{(0)}}$	$\dfrac{\mathcal{E}_0(t)}{\Phi_0}\,\sigma\mu r_1^2$
0.100E + 01	0.100E + 01	0.209E − 04
0.141E + 01	0.100E + 01	0.714E − 01
0.200E + 01	0.980E + 00	0.176E + 01
0.283E + 01	0.823E + 00	0.366E + 01
0.400E + 01	0.519E + 00	0.222E + 01
0.566E + 01	0.255E + 00	0.728E + 00
0.800E + 01	0.107E + 00	0.175E + 00
0.113E + 02	0.416E − 01	0.361E − 01
0.160E + 02	0.154E − 01	0.690E − 02
0.226E + 02	0.556E − 02	0.127E − 02
0.320E + 02	0.199E − 02	0.228E − 03
0.452E + 02	0.708E − 03	0.408E − 04
0.640E + 02	0.251E − 03	0.724E − 05

$$\mathrm{In}H_0 \approx H_0^{(0)}\left[1 - \frac{\sqrt{2}}{15}(\sigma\mu\omega r_1^2)^{3/2} + \ldots\right] \tag{5.11}$$

Comparing these with eqs. 5.5, we can see that in the near zone ($p < 1$) field components created by current flowing in the uniform half-space are less by a factor of nearly two than the corresponding components of the field observed in a uniform full-space.

In the wave zone, the magnetic field at the surface of the earth decreases in inverse proportion to ω; that is, at a rate which is significantly slower than in the case of the uniform full-space:

$$H_0 \approx H_0^{(0)}\,\frac{6}{k^2 r_1^2} \tag{5.12}$$

Curves for the quadrature component and the inphase component, $-\mathrm{In}H_0^{(1)}$, caused by currents flowing in a half-space and the difference ΔQH_0 for various values of the parameters $p = r_1/h$ are shown in Fig. 5.5. The magnitudes of these same functions are given in Table 5.III. Applying the Fourier transform to eq. 5.9, we find that the non-stationary magnetic field and electromotive force at the center of the loop are:

$$H_0(t) = H_0^{(0)}\left[\frac{3}{\sqrt{\pi}}\frac{e^{-u^2}}{u} + \left(1 - \frac{3}{2u^2}\right)\Phi(u)\right] \tag{5.13}$$

and

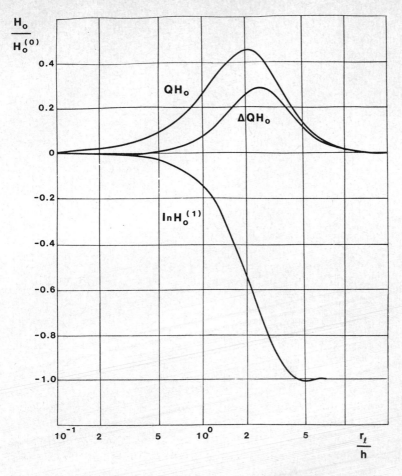

Fig. 5.5. Frequency responses for the quadrature and inphase components of the magnetic field observed at the center of the source loop and for the difference function ΔQH_0 for the case of a uniform half-space.

$$\mathscr{E}_0(t) = \frac{2\Phi_0}{\sigma\mu r_1^2}\left[3\Phi(u) - \frac{2}{\sqrt{\pi}}\,e^{-u^2}u(3 + 2u^2)\right] \tag{5.14}$$

In contrast to what was observed for a uniform full-space, the EMF observed at the surface of a half-space during the early stage tends to some constant value which is non-zero:

$$\mathscr{E}_0(t) = \frac{6\Phi_0}{\sigma\mu r_1^2} \tag{5.15}$$

However, beneath the surface of the half-space, the EMF is zero at the initial moment, $t = 0$. During the late stage, we have:

TABLE 5.III

Uniform half-space

$\dfrac{r_1}{h}$	$\dfrac{QH_0}{H_0^{(0)}}$	$\dfrac{\text{In}H_0^{(1)}}{H_0^{(0)}}$	$\dfrac{\Delta QH_0}{H_0^{(0)}}$
0.125E + 00	0.729E − 02	−0.480E − 03	0.214E − 03
0.177E + 00	0.142E − 01	−0.132E − 02	0.599E − 03
0.250E + 00	0.271E − 01	−0.355E − 02	0.167E − 02
0.353E + 00	0.509E − 01	−0.939E − 02	0.457E − 02
0.500E + 00	0.927E − 01	−0.241E − 01	0.122E − 01
0.707E + 00	0.161E + 00	−0.592E − 01	0.311E − 01
0.100E + 01	0.260E + 00	−0.136E + 00	0.731E − 01
0.141E + 01	0.373E + 00	−0.287E + 00	0.150E + 00
0.200E + 01	0.446E + 00	−0.526E + 00	0.246E + 00
0.283E + 01	0.400E + 00	−0.802E + 00	0.282E + 00
0.400E + 01	0.237E + 00	−0.980E + 00	0.188E + 00
0.566E + 01	0.972E − 01	−0.101E + 00	0.741E − 01
0.800E + 01	0.461E − 01	−0.100E + 00	0.344E − 01

$$H_0(t) \cong \frac{Ir_1^2}{30\sqrt{\pi}}\,(\sigma\mu)^{3/2}\,\frac{1}{t^{3/2}}$$

$$\mathscr{E}_0(t) \cong \frac{\Phi_0 r_1^3}{10\sqrt{\pi}}\,(\sigma\mu)^{3/2}\,\frac{1}{t^{5/2}}, \quad \text{if } \tau/r_1 > 16$$

(5.16)

Thus, the character of the decrease in magnetic field and electromotive force is the same as was observed for a uniform full-space. However, electromotive force in the later case is less by a factor of approximately 2.5. Corresponding values for the magnetic field and the electromotive force are listed in Table 5.IV, and curves describing the behavior of these functions are shown in Figs. 5.6 and 5.7.

c. *Thin conductive plate with a conductance S.* Consider a source loop situated on the surface of a plate with conductance S, with the plate representing the real case of a relatively thin but conductive overburden, as shown in Fig. 5.1c. As is well known, the expression for the vertical component of the magnetic field of an electric dipole source located on the plate S and oriented in the x-direction has the form:

$$H_0^e = \frac{Idx}{4\pi r^2}\,[1 - n^2 F'(n)]\,\sin\phi$$

(5.17)

where

$$n = \frac{i\omega\mu Sr}{2}, \quad F = \frac{\pi}{2}\,[H_0(n) - N_0(n)]$$

TABLE 5.IV

Uniform half-space

$\dfrac{\tau}{r_1}$	$\dfrac{H_0(t)}{H_0^{(0)}}$	$\dfrac{\mathscr{E}_0(t)}{\Phi_0}\,\sigma\mu r_1^2$
0.100E + 01	0.924E + 00	0.600E + 01
0.141E + 01	0.848E + 00	0.599E + 01
0.200E + 01	0.700E + 00	0.553E + 01
0.283E + 01	0.473E + 00	0.346E + 01
0.400E + 01	0.253E − 01	0.131E + 01
0.566E + 01	0.113E + 00	0.350E + 00
0.800E + 01	0.453E − 01	0.767E − 01
0.113E + 02	0.171E − 01	0.151E − 01
0.160E + 02	0.623E − 02	0.282E − 02
0.226E + 02	0.224E − 02	0.513E − 03
0.320E + 02	0.799E − 03	0.919E − 04
0.452E + 02	0.283E − 03	0.163E − 04
0.640E + 02	0.100E − 03	0.290E − 05

and where $H_0(n)$, $N_0(n)$ are the Struve and Neuman functions, respectively, and ϕ is the angle between the axis of the dipole source and the radius, r, connecting the center of the dipole to the observation point. Taking into account the axial symmetry characterizing the behavior of the vertical component of the magnetic field at the center of the loop, we obtain:

$$H_0 = \frac{I}{2r_1}\left[1 - n_1^2 F'(n_1)\right] \tag{5.18}$$

where $n_1 = i\omega\mu Sr_1/2$.

Making use of tabulated expansions for the Struve and Neuman functions, we obtain the following expressions for the low-frequency part of the spectrum:

$$QH_0 \approx -\frac{I}{2r_1}\left[\frac{\omega\mu Sr_1}{2} + \frac{(\omega\mu Sr_1)^3}{16}\ln\frac{\omega\mu Sr_1}{2} + \ldots\right]$$

$$\text{In}\,H_0 \approx \frac{I}{2r_1}\left[1 - \frac{(\omega\mu Sr_1)^2}{4} + \frac{\pi}{32}(\omega\mu Sr_1)^3 + \ldots\right] \quad \text{for } \omega\mu Sr_1 < 1 \tag{5.19}$$

At the opposite extreme, to derive the expression for the inphase component at high frequencies ($n_1 \gg 1$), it is more useful to make use of asymptotic expansions of the Struve function in terms of the Neuman functions and a sum of inverse powers of the argument n_1. Then:

$$\text{In}\,H_0 \approx -\frac{I}{2r_1}\left[\frac{12}{(\omega\mu Sr_1)^2} + \frac{720}{(\omega\mu Sr_1)^4} + \ldots\right] \tag{5.20}$$

$$\frac{H_0(t)}{H_o^{(0)}}$$

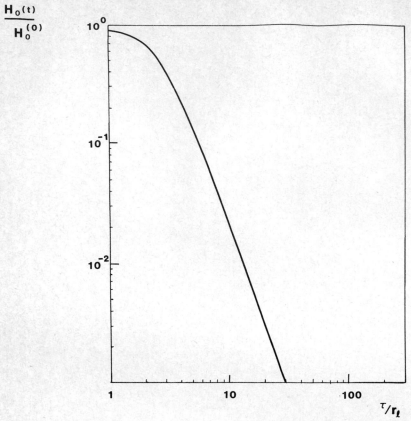

Fig. 5.6. Transient response of the magnetic field at the center of the source loop for a uniform half-space.

Making use of a tabulated representation for a Neuman function of an imaginary argument:

$$N_1(ix) = \frac{2}{\sqrt{\pi}} iK_1(x) - I_1(x)$$

we obtain the following expression for the quadrature component of the magnetic field:

$$QH_0 \approx \left(\frac{\pi}{\omega\mu Sr_1}\right)^{1/2} e^{-\omega\mu Sr_1/2} \left(\frac{\omega\mu Sr_1}{2}\right)^2 \tag{5.21}$$

Thus, in the wave zone, the inphase component is the strongest, and it decreases in inverse proportion to the square of the conductance S. Curves for the functions:

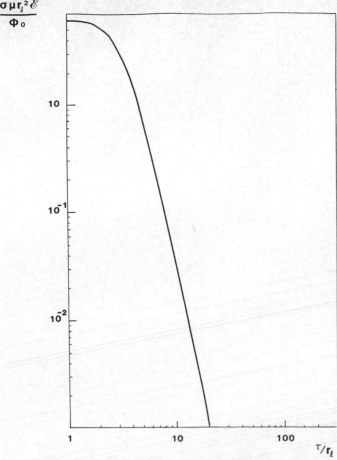

$$\frac{\sigma \mu r_\ell^2 \mathscr{E}}{\Phi_0}$$

Fig. 5.7. Transient response for the electromotive force at the center of the source loop for a uniform half-space.

$$\frac{QH_0}{H_0^{(0)}}, \quad \frac{\ln H_0^{(1)}}{H_0} = \frac{\ln H_0}{H_0} - 1, \quad \text{and} \quad \frac{\Delta Q H_0}{H_0}$$

are shown in Fig. 5.8, and numerical values are listed in Table 5.V.

At this point we would note a useful relationship between the parameters ω, μ, S, r_1 and r_1/h:

$$\omega \mu S r_1 = 2 \left(\frac{r_1}{h}\right)^2 \frac{H_1}{r_1} \tag{5.22}$$

where H_1 is the actual thickness of the layer which we have assumed to behave as an S plate, and h is the skin depth in this layer.

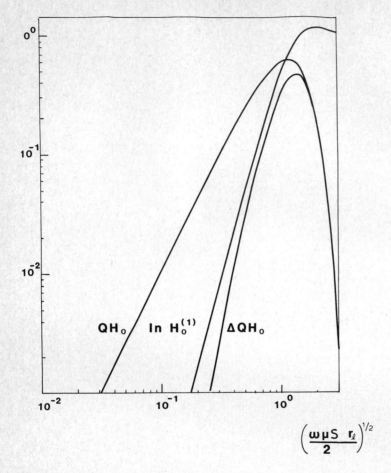

Fig. 5.8. Frequency response for the quadrature, inphase, and difference of the magnetic field at the center of the source loop for the case in which the earth can be represented as a thin conducting plane.

The non-stationary magnetic field at the center of the loop when the current in the source is abruptly turned on, can be presented as:

$$H_0(t) = \frac{I}{2r_1} \frac{1}{[1 + 4m^2]^{3/2}}$$ (5.23)

where

$$m = \frac{t}{\mu S r_1} = \frac{1}{8\pi^2} \left(\frac{\tau}{r_1}\right)^2 \frac{r_1}{H_1}$$

and where

TABLE 5.V

Plane S

$\left(\dfrac{\omega\mu Sr_1}{2}\right)^{1/2}$	$\dfrac{QH_0}{H_0^{(0)}}$	$\dfrac{\text{In}H_0^{(1)}}{H_0^{(0)}}$	$\dfrac{\Delta QH_0}{H_0^{(0)}}$
0.312E − 01	0.977E − 03	0.953E − 06	0.925E − 08
0.442E − 01	0.195E − 02	0.381E − 05	0.663E − 07
0.625E − 01	0.391E − 02	0.152E − 04	0.468E − 06
0.884E − 01	0.781E − 02	0.607E − 04	0.325E − 05
0.125E + 00	0.156E − 01	0.241E − 03	0.220E − 04
0.177E + 00	0.312E − 01	0.953E − 03	0.145E − 03
0.250E + 00	0.621E − 01	0.372E − 02	0.905E − 03
0.353E + 00	0.122E + 00	0.142E − 01	0.527E − 02
0.500E + 00	0.234E + 00	0.514E − 01	0.271E − 01
0.707E + 00	0.414E + 00	0.170E + 00	0.113E + 00
0.100E + 01	0.602E + 00	0.468E + 00	0.322E + 00
0.141E + 01	0.559E + 00	0.935E + 00	0.460E + 00
0.200E + 01	0.200E + 00	0.117E + 01	0.195E + 00
0.281E + 01	0.994E − 02	0.106E + 01	0.994E − 02
0.400E + 01	0.0	0.101E + 01	0.0

$$\tau = (2\pi\rho t \times 10^7)^{1/2}$$

We also have the following expression for the electromotive force:

$$\mathscr{E}_0(t) = \frac{12\Phi_0}{S\mu r_1} \frac{m}{[1 + 4m^2]^{5/2}} \tag{5.24}$$

From these two expressions, we see that during the early stage, the magnetic field tends to a value $H_0^{(0)}$, while the electromotive force in the time domain decreases as directly as time and inversely as the square of the longitudinal conductance S:

$$\mathscr{E}_0(t) \approx \frac{12\Phi_0}{\mu^2 S^2 r_1^2} t, \quad \text{if } m < 0.2 \tag{5.25}$$

Expanding the right-hand side of eqs. 5.23 and 5.24 in power series in terms of the small parameter $1/m$, we obtain well known asymptotic expressions for the magnetic field and the electromotive force during the late stage in the time domain:

$$H_0(t) \approx H_0^{(0)} \frac{(\mu Sr_1)^3}{8t^3} \left[1 - \frac{3}{8}\left(\frac{\mu Sr_1}{t}\right)^2 \right]$$

$$\mathscr{E}_0(t) \approx \frac{3}{8} \Phi_0 (S\mu r_1)^3 \frac{1}{t^4} \quad \text{when } m > 1 \tag{5.26}$$

398

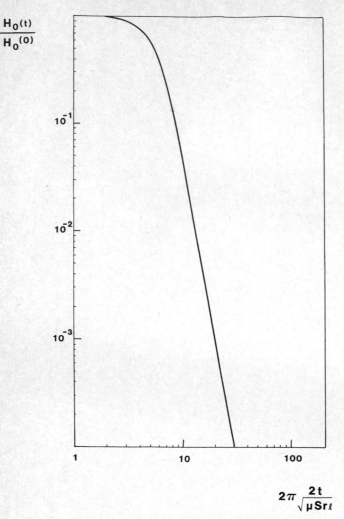

$$\dfrac{H_0(t)}{H_0(0)}$$

10^{-1}

10^{-2}

10^{-3}

1   10   100

$$2\pi\,\dfrac{2t}{\sqrt{\mu Sr\iota}}$$

Fig. 5.9. Transient response of the magnetic field at the center of the source loop when the earth can be represented as being a thin conducting sheet.

The transient magnetic field and the transient electromotive force are shown graphically in Figs. 5.9 and 5.10, and listed numerically in Table 5.VI.

d. Two-layer medium. According to Kaufman and Keller (1983), the magnetic field observed at the center of the loop situated on the surface of a two-layer sequence as shown in Fig. 5.1d can be written as:

$$H_0 = Ir_1 \int\limits_0^\infty \frac{\lambda^2}{\lambda + \lambda_1/R}\, J_1(\lambda r_1)\mathrm{d}\lambda \qquad (5.27)$$

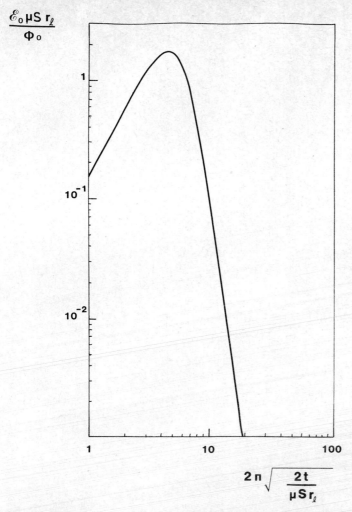

$$\frac{\mathscr{E}_0 \mu S\, r_\ell}{\Phi_0}$$

$$2\pi \sqrt{\frac{2t}{\mu S r_\ell}}$$

Fig. 5.10. Transient response for the electromotive force at the center of the source loop for the case in which the earth can be represented as being a thin conducting sheet.

where

$$R = \coth\,[\lambda_1 H_1 + \coth^{-1} \lambda_1/\lambda_2], \quad \lambda_i = (\lambda^2 + k_i^2)^{1/2}, \quad i = 1, 2$$

and where k_i^2 is the square of the wave number, and H_1 is the thickness of the surface layer.

Expanding the integral in eq. 5.27 in a power series in terms of the small parameter $k_1 r_1$, we have:

$$H_0 \approx \frac{I}{2r_1}\,[1 + a_1(k_1 r_1)^2 + a_2(k_1 r_1)^3 + a_3(k_1 r_1)^4 \ln k_1 r_1 + a_4(k_1 r_1)^4$$
$$+ a_5(k_1 r_1)^5 + \ldots]$$

(5.28)

TABLE 5.VI

Plane S

$2\pi\left(\dfrac{2t}{\mu S r_1}\right)^{1/2}$	$\dfrac{H_0(t)}{H_0^{(0)}}$	$\dfrac{\mathscr{E}_0(t)}{\Phi_0}\mu S r_1$
0.100E + 01	0.999E + 00	0.152E + 00
0.141E + 01	0.996E + 00	0.302E + 00
0.200E + 01	0.985E + 00	0.593E + 00
0.283E + 01	0.941E + 00	0.110E + 01
0.400E + 01	0.796E + 00	0.166E + 01
0.566E + 01	0.469E + 00	0.138E + 01
0.800E + 01	0.145E + 00	0.388E + 00
0.113E + 02	0.256E − 01	0.433E − 01
0.160E + 02	0.354E − 02	0.320E − 02
0.226E + 02	0.454E − 03	0.209E − 03
0.320E + 02	0.572E − 04	0.132E − 04
0.453E + 02	0.716E − 05	0.828E − 06
0.640E + 02	0.895E − 06	0.518E − 07
0.905E + 02	0.112E − 06	0.324E − 08
0.128E + 03	0.140E − 07	0.202E − 09

and therefore,

$$QH_0 \approx \frac{I}{2r_1}\left[a_1\,\sigma_1\,\mu\omega r_1^2 + \frac{a_2}{\sqrt{2}}(\sigma_1\,\mu\omega)^{3/2}\,r_1^3 - \frac{a_3}{2}\,\pi(\sigma_1\,\mu\omega r_1^2\,)^2\right] \tag{5.29}$$

$$\mathrm{In}H_0 \approx \frac{I}{2r_1}\left[1 - \frac{a_2}{\sqrt{2}}\,(\sigma_1\,\mu\omega)^{3/2}\,r_1^3 - \frac{a_3}{2}\,(\sigma_1\,\mu\omega)^2\,r_1^4\,\ln\,(\sigma_1\,\mu\omega)^2\,r_1^4 + \,.\,.\right] \tag{5.30}$$

where

$$a_1 = \tfrac{1}{4}\left[1 + (s-1)\,\frac{(4H_1{}^2 + r_1^2)^{1/2} - 2H_1}{r_1}\right]$$

$$a_2 = -\frac{2}{15}\,s^{3/2}, \quad a_3 = -\frac{H_1}{16r_1}\,s(s-1), \quad s = \frac{\sigma_2}{\sigma_1} \tag{5.31}$$

At low frequencies, the quadrature component of the magnetic field is principally defined by the first term in eq. 5.29:

$$\frac{QH_0}{H_0^{(0)}} = \frac{\sigma_1\,\mu\omega r_1^2}{4}\left[1 + (s-1)\,\frac{(4H_1{}^2 + r_1^2)^{1/2} - 2H_1}{r_1}\right] \tag{5.32}$$

When the radius of the loop is significantly less than the thickness of the surface layer, or more precisely, when

$$\frac{r_1}{2H_1}(s-1) \ll 1$$

the quadrature component of the magnetic field is practically independent of the conductivity of the lower region (the basement) and coincides with the quadrature component which would be observed on the surface of a uniform half-space characterized by the conductivity of the surface layer, σ_1. For a sufficiently large radius of the loop ($r_1/H_1 \gg 1$) and for a relatively low conductivity of the basement ($s \ll 1$), the quadrature component at low frequency is defined by the longitudinal conductance: $S = \sigma_1 H_1$, and it is no different than the quadrature component which would be observed with the loop situated on the surface of a thin conducting plate with the same conductance:

$$\frac{QH_0}{H_0^{(0)}} \approx \frac{\mu\omega Sr_1}{2}, \tag{5.33}$$

if $\sigma_1/\sigma_2 \gg 1$, $r_1/H_1 > 1$, $r_1/h_1 \ll 1$

With a further increase in the radius of the loop, with the provision that $r_1/h \ll 1$, the quadrature component approaches a value corresponding to the field that would be observed on the surface of a uniform half-space with a conductivity of the basement.

We should note also that for the range of parameters for which eq. 5.33 is valid, the quadrature component is constant:

$$QH_0 \approx \mu\omega SI/4$$

It should be clear that if the basement is more conductive than the surface layer, its influence begins to manifest itself for a relatively small radius of the loop, when $H_1/h \ll 1$. In particular, if σ_1 were to be zero, at low frequencies:

$$\frac{QH_0}{H_0^{(0)}} \approx \frac{\sigma_2\mu\omega r_1^2}{4} \frac{(4H_1^2 + r_1^2)^{1/2} - 2H_1}{r_1} \tag{5.34}$$

As an example, if the loop radius is less than H_1 we have:

$$QH_0 \approx -\frac{\sigma_2\mu\omega r_1^2 I}{32H_1} \qquad \frac{r_1}{H_1} \ll 1, \quad \frac{H_1}{h} \ll 1 \tag{5.35}$$

that is, the quadrature component in this particular case is defined by the product $\rho_2 H_1$.

At high frequencies, currents will concentrate near the surface of the earth and the quadrature component along with the other components will be essentially no different than the values which would be observed for a uniform half-space characterized by the conductivity of the surface layer.

Examples of the frequency responses for the quadrature component are given by the curves in Fig. 5.11.

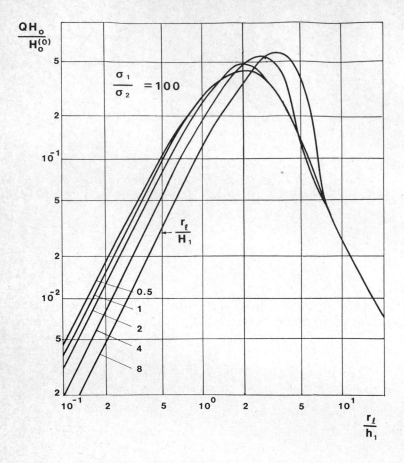

Fig. 5.11. Frequency response for the quadrature component of the magnetic field at the center of the source loop for a two-layered medium.

As follows from eq. 5.30, in measuring the inphase component, $\mathrm{In}H_0$, or the function ΔQH_0 at low frequencies, the currents flowing in the medium with conductivity σ_2 play a principal role, and this fact is of great practical importance from the point of view of detecting ore bodies hidden beneath a relatively conducting overburden. Frequency responses for the inphase component and for the function ΔQH_0 for various ratios of the parameter r_1/H_1 are shown by the curves in Figs. 5.12 and 5.13. As may be seen from these illustrations, the equivalence of the first layer to a conducting sheet with a conductance S will be observed in measuring the functions $\mathrm{In}H_0^{(1)}$ and ΔQH_0 for significantly smaller loop radii than when the quadrature component is being measured. At the same time, the approach of the curves, $\mathrm{In}H_0^{(1)}$ and ΔQH_0 to the range in which the S-zone behavior is observed at intermediate frequencies, inasmuch as at very low frequencies

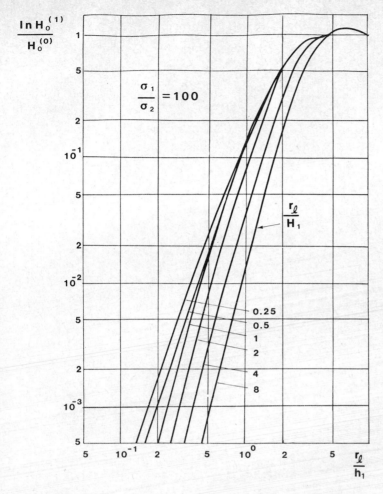

Fig. 5.12. Frequency response for the inphase component of the magnetic field at the center of the source loop for a two-layered medium.

these functions are controlled by the conductivity of the basement. It should be clear that with an increase in resistivity in the basement, the principal characteristics of the field behavior that identify the S-zone will be observed at lower frequencies.

Next let us examine the time-domain behavior in the case in which the current in the source loop is turned off abruptly. The transient responses of the magnetic field and the electromotive force induced at the center of the loop have been calculated from Fourier transforms of the spectrums given in eq. 5.27. Considering the relationship between the low-frequency parts of the spectrum and the late part of the transient field, we have:

404

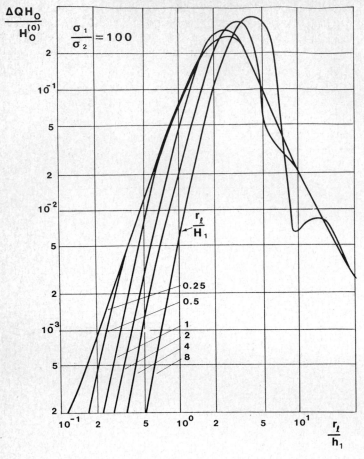

Fig. 5.13. Frequency response for the difference function observed at the center of the source loop for a two-layered medium.

$$H_0(t) \approx \frac{I}{2r_1}\left[\frac{r_1^3}{15\sqrt{\pi}}\left(\frac{\sigma_2\mu}{t}\right)^{3/2} - \frac{r_1^3 H_1}{16}s(s-1)\left(\frac{\sigma_1\mu}{t}\right)^2 - \right.$$

$$\left. \frac{r_1^3[r_1^2 s^{5/2} + 2H_1^2\sqrt{s}\,(1-s)(8s-9)]}{140\sqrt{\pi}}\left(\frac{\sigma_1\mu}{t}\right)^{5/2} + \dots\right] \qquad (5.36)$$

$$\mathcal{E}_0(t) \approx \Phi_0\left[\frac{r_1^3(\sigma_2\mu)^{3/2}}{10\sqrt{\pi}}\left(\frac{1}{t}\right)^{5/2} - \frac{r_1^3 H_1}{8}s(s-1)(\sigma_1\mu)^2\frac{1}{t^3}\right.$$

$$\left. - \frac{r_1^3[r_1^2 s^{5/2} - 2H_1^2\sqrt{s}\,(1-s)(8s-9)}{56\sqrt{\pi}}(\sigma_1\mu)^{5/2}\frac{1}{t^{7/2}}\right]$$

As may be seen from this expression during the late stage of the transient behavior of the magnetic field and of the electromotive force, there is no dependence on the parameters characterizing the upper layer for any dimensions of the current carrying source. This feature, in principle at least, demonstrates the feasibility for discovering an ore body covered by a relatively conductive overburden when time-domain measurements are employed, particularly if the loop radius is sufficiently small. Figure 5.14 shows curves for the electromotive force normalized by the parameter $\sigma_1 \mu H_1^2/\Phi_0$, as a function of the parameter τ_1/H_1 when $\sigma_2/\sigma_1 = 0$. During the early stage of the transient, the curve for a two-layer sequence nearly coincide with the

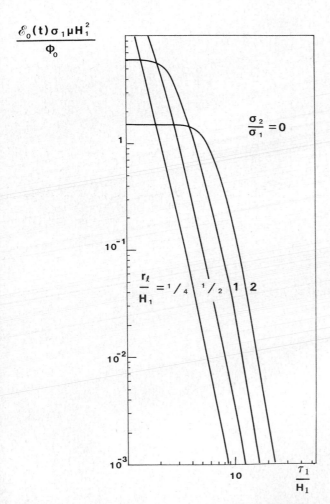

Fig. 5.14. Transient response for the scaled electromotive force observed at the center of the source loop.

curve for uniform half-space characterized by the conductivity σ_1. With an increase in time, the curves for the magnetic field and the electromotive force corresponding to the two-layer sequence with a more resistive basement coincide with the curves for $H(t)$ and $\mathscr{E}(t)$, respectively which would be caused by currents flowing in a vanishing thin plate, S. This range, which is also termed the S-zone, is one in which the currents in the upper layer are distributed almost uniformly along the vertical section, but have not penetrated to a significant degree into the basement. For larger values of the parameter τ_1/H_1, the transient magnetic field and EMF are described by the equations in 5.36. With an increase in the resistivity of the underlying

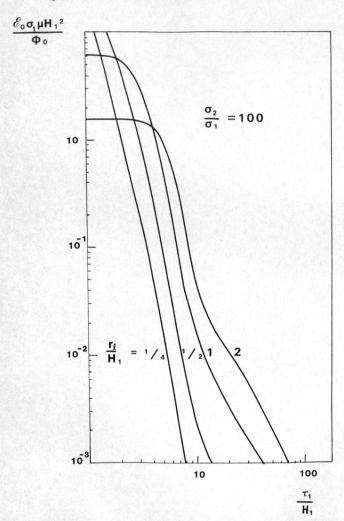

Fig. 5.15. Transient response for the scaled electromotive force observed at the center of the source loop.

medium, the duration of the S-zone in time increases, and when the basement in a two-layer sequence is insulating, the magnetic field and EMF during a late stage coincide with the values that would be observed over a vanishingly thin plate.

The time-domain responses for the electromotive force observed at the center of the loop are given graphically in Fig. 5.15 in the case in which the basement has a finite resistivity.

II. Electromotive force induced in combined loops

Let us now investigate the time-domain and frequency-domain behavior of the EMF observed in a receiving loop coincident with the current source loop.

a. Uniform medium. It has been shown previously that the vector potential for a current ring need have only a single component A_ϕ with that being related to the electric field by the expression:

$$E_{0\phi} = -\mu \frac{\partial A_{0\phi}}{\partial t}$$

In a uniform full-space, this single component of vector potential is:

$$A_{0\phi} = \frac{Ir_1}{2\pi} \int_0^\pi \frac{e^{ikR}}{R} \cos \phi d\phi \qquad (5.37)$$

where

$$R = (r_1^2 + r^2 + z^2 - 2r_1 r \cos \phi)^{1/2}$$

and where r, ϕ, z are the coordinates of the observation point in a cylindrical coordinate system. Thus, the electromotive force observed with combined source and receiver loops is:

$$\mathscr{E}_0 = 2\pi r_1 E_{0\phi} = -i\omega\mu I r_1^2 \int_0^\pi \frac{e^{ikR}}{R} \cos \phi d\phi \qquad (5.38)$$

where

$$R = \sqrt{2}r_1(1 - \cos \phi)^{1/2} = 2r_1 \sin \phi/2$$

We can write eq. 5.38 as:

$$\mathscr{E}_0 = -i\omega I(L_0 + Z_1) \qquad (5.39)$$

where L_0 is the self-inductance of a ring with the radius of the cross-section, r_0, and which is well known to be:

$$L_0 = r_1 \mu \left(\ln \frac{8r_1}{r_0} - 1.75 \right) \tag{5.39a}$$

and where Z_1 is the function, characterizing an influence of the conducting medium. In accord with eqs. 5.39 and 5.39a, we have:

$$Z_1 = \mu r_1^2 \int_0^\pi \frac{e^{ikR} - 1}{R} \cos \phi d\phi \tag{5.40}$$

Thus, the expression for the electromotive force induced by currents in the conducting medium is of the form:

$$\mathscr{E}_0^{(1)} = -i\omega\mu I r_1^2 \int_0^\pi \frac{e^{ikR} - 1}{R} \cos \phi d\phi = \alpha r_1 \int_0^\pi \frac{e^{ikR} - 1}{R} \cos \phi d\phi \tag{5.41}$$

where

$$\alpha = -i\omega\mu I r_1$$

Expanding the integrand in a power series in terms of the small parameter r_1/h, we can obtain the following asymptotic expansions for the quadrature and inphase components of the secondary electromotive force at low frequencies:

$$Q\mathscr{E}_0 \cong |\alpha| \left[\frac{2}{3} \sigma\mu\omega r_1^2 - \frac{\pi(\sigma\mu\omega)^{3/2} r_1^3}{6\sqrt{2}} \right]$$

$$\tag{5.42}$$

$$\text{In}\mathscr{E}_0^{(1)} \cong -|\alpha| \frac{\pi(\sigma\omega\mu)^{3/2} r_1^3}{6\sqrt{2}}$$

As may be seen from a comparison of eqs. 5.5 and 5.40, the value for the EMF normalized to the area of the receiver is practically the same for measurements made at the loop center or with a combined loop. At high frequencies, as follows from the asymptotic behavior of the integral on the right hand of eq. 5.40, the inphase component increases with an increase in frequency:

$$\text{In}\mathscr{E}_0^{(1)} \approx -|\alpha| \ln r_1/h, \quad \text{for } r_1/h \gg 1$$

while the quadrature component in the wave zone is practically independent of the conductivity of the medium, and is:

$$Q\mathscr{E}_0 \approx \frac{\pi}{4} |\alpha|$$

The frequency responses for the various observed functions:

$$\frac{Q\mathscr{E}_0}{|\alpha|}, \quad \frac{\text{In}\mathscr{E}_0^{(1)}}{|\alpha|} \quad \text{and} \quad \frac{\Delta Q\mathscr{E}_0}{|\alpha|} = \frac{1}{|\alpha|} \left[Q\mathscr{E}_0(\omega_1) - \left(\frac{\omega_1}{\omega_2}\right)^2 Q\mathscr{E}_0(\omega_2) \right],$$

where $\omega_1 = \omega_2/2^{1/2}$, are shown graphically in Fig. 5.16, and data are given in Table 5.VII.

Applying the Fourier transform to the expression for the vector potential $A_{0\phi}$, we obtain the time domain expression:

$$A_{0\phi}(t) = \frac{Ir_1}{2\pi} \int_0^\pi \frac{\Phi(u)}{R} \cos \phi \, d\phi \tag{5.43}$$

and

$$\mathscr{E}_0(t) = \frac{4}{\sqrt{\pi}} \frac{I}{\sigma r_1} u_1^3 \int_0^\pi e^{-u^2} \cos \phi \, d\phi$$

where

$$u = \frac{\pi\sqrt{2}R}{\tau}, \quad u_1 = \frac{\pi\sqrt{2}r_1}{\tau} \tag{5.44}$$

During the early stage of the transient response, the EMF observed in a combined loop increases without limit in inverse proportion to time. In fact, at the limit when t approaches zero, the integral of eq. 5.44 can be written as:

TABLE 5.VII

Full-space

$\dfrac{r_1}{h}$	$\dfrac{QH_0}{H_0^{(0)}}$	$\mathrm{In}H_0^{(1)}$	$\dfrac{\Delta QH_0}{H_0^{(0)}}$
0.312E − 01	0.127E − 02	−0.309E − 04	0.132E − 04
0.442E − 01	0.251E − 02	−0.864E − 04	0.373E − 04
0.625E − 01	0.495E − 02	−0.240E − 03	0.105E − 03
0.884E − 01	0.969E − 02	−0.660E − 03	0.296E − 03
0.125E + 00	0.188E − 01	−0.180E − 02	0.827E − 03
0.177E + 00	0.359E − 01	−0.481E − 02	0.229E − 02
0.250E + 00	0.673E − 01	−0.126E − 01	0.621E − 02
0.345E + 00	0.122E + 00	−0.319E − 01	0.163E − 01
0.550E + 00	0.212E + 00	−0.769E − 01	0.404E − 01
0.707E + 00	0.343E + 00	−0.173E + 00	0.911E − 01
0.100E + 01	0.504E + 00	−0.353E + 00	0.178E + 00
0.141E + 01	0.651E + 00	−0.636E + 00	0.283E + 00
0.200E + 01	0.737E + 00	−0.989E + 00	0.355E + 00
0.283E + 01	0.764E + 00	−0.135E + 01	0.377E + 00
0.400E + 01	0.774E + 00	−0.170E + 01	0.384E + 00
0.566E + 01	0.780E + 00	−0.204E + 01	0.388E + 00
0.800E + 01	0.782E + 00	−0.239E + 01	0.390E + 00
0.113E + 02	0.784E + 00	−0.274E + 01	0.392E + 00
0.160E + 02	0.785E + 00	−0.308E + 01	0.392E + 00
0.226E + 02	0.785E + 00	−0.343E + 01	0.392E + 00
0.320E + 02	0.785E + 00	−0.378E + 01	0.393E + 00
0.452E + 02	0.785E + 00	−0.412E + 01	0.393E + 00
0.640E + 02	0.785E + 00	−0.447E + 01	0.393E + 00

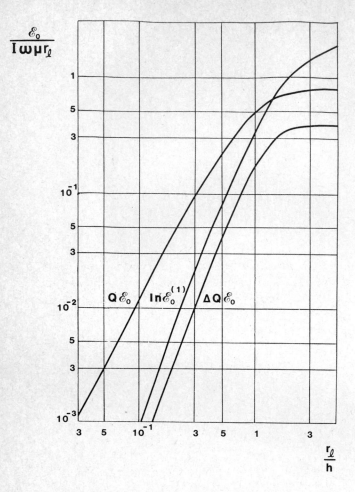

Fig. 5.16. Frequency responses for the electromotive force and the electromotive difference function in coincidence loops on a uniform space.

$$I_1 \approx \int\limits_0^\pi e^{-u_1^2\phi^2}\, \mathrm{d}\phi = \frac{1}{u_1} \int\limits_0^\infty e^{-x^2}\, \mathrm{d}x = -\frac{\sqrt{\pi}}{2u_1}$$

Therefore,

$$\mathscr{E}_0(t) \approx \frac{2I}{\sigma r_1} u_1^2 \approx \frac{Ir_1}{2}\frac{\mu}{t} \tag{5.45}$$

With increase in time, the EMF decreases gradually, as shown graphically in Fig. 5.17 and numerically in Table 5.VIII, and in particular, during the late stage:

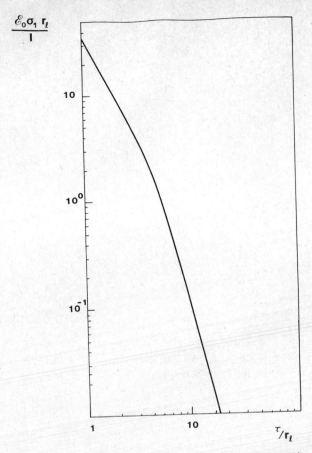

Fig. 5.17. Transient response for the electromotive force observed with coincidence source and receiver loops on a uniform space.

$$\mathscr{E}_0(t) \approx 16\pi^6 \left(\frac{2}{\pi}\right)^{1/2} \frac{I r_1^4}{\sigma \tau^5} \tag{5.46}$$

It should not be difficult to recognize that eq. 5.46 coincides with the expression for the EMF that would be observed at the center of the loop if the areas around which the EMF were measured were to be the same.

b. *Uniform half-space*. Let us apply the expression developed for the horizontal component of vector potential from an electric dipole source with its axis oriented along the x-axis:

$$A_{0x} = \frac{I\,dx}{2\pi k^2 r^3} \left[e^{ikr}(1 - ikr) - 1\right] \tag{5.47}$$

TABLE 5.VIII

Full-space

$\dfrac{\tau}{r_1}$	$\dfrac{0}{I}\,\sigma r_1$
0.100E + 01	0.391E + 02
0.141E + 01	0.194E + 02
0.200E + 01	0.948E + 01
0.283E + 01	0.453E + 01
0.400E + 01	0.201E + 01
0.566E + 01	0.742E + 00
0.800E + 01	0.212E + 00
0.113E + 02	0.492E − 01
0.160E + 02	0.101E − 01
0.226E + 02	0.192E − 02
0.320E + 02	0.352E − 02
0.453E + 02	0.634E − 04
0.640E + 02	0.113E − 04
0.905E + 02	0.201E − 05
0.128E + 03	0.356E − 06

Considering the axial symmetry characterizing this model, we have the following expressions for the vector potential component $A_{0\phi}$ for a current ring and for the electromotive force along this same periphery:

$$A_{0\phi} = \frac{Ir_1}{\pi k^2}\int_0^\pi \frac{e^{ikR}(1-ikR)-1}{R^3}\cos\phi\,d\phi \tag{5.48}$$

and

$$\mathscr{E}_0 = -i\omega\mu\,\frac{2r_1^2 I}{k^2}\int_0^\pi \frac{e^{ikR}(1-ikR)-1}{R^3}\cos\phi\,d\phi \tag{5.49}$$

By analogy with eq. 5.41, we can write:

$$\mathscr{E}_0^{(1)} = \alpha r_1\int_0^\pi\left[\frac{2[e^{ikR}(1-ikR)-1]}{k^2R^2}-1\right]\frac{\cos\phi}{R}\,d\phi \tag{5.50}$$

At low frequencies, the quadrature and inphase components of the secondary EMF can be expressed as assymptotic expansions:

$$Q\mathscr{E}_0 \cong |\alpha|\left\{\frac{\sigma\mu\omega r_1^2}{3}-\frac{\pi(\sigma\mu\omega)^{3/2}r_1^3}{15\sqrt{2}}\right\} \tag{5.51}$$

and

$$\ln \mathscr{E}_0^{(1)} \approx - |\alpha| \pi \frac{(\sigma\mu\omega)^{3/2} r_1^3}{15\sqrt{2}} \tag{5.52}$$

As was the case for a uniform full-space, the values for the electromotive force observed at the center of the loop and with combined loops, when normalized to the area enclosed within the measurement periphery, are practically the same.

At high frequencies, the inphase and quadrature components of the electromotive force observed on the surface of a uniform half-space does not differ in any signficant degree from the corresponding values that would be observed in a uniform full-space (see Fig. 5.18). Table 5.IX lists values for the EMF for different values of the parameter r_1/h.

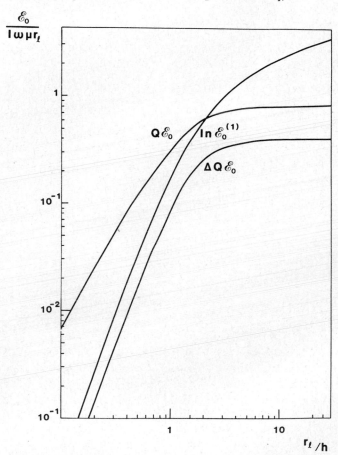

Fig. 5.18. Frequency responses for the quadrature and inphase components of the electromotive force and the difference electromotive force function when coincident source and receiver loops are situated on the surface of a uniform half-space.

414

TABLE 5.IX

Uniform half-space

$\dfrac{r_1}{h}$	$\dfrac{QH_0}{H_0^{(0)}}$	$\dfrac{\mathrm{In}\,H_0^{(1)}}{H_0^{(0)}}$	$\dfrac{\Delta QH_0}{H_0^{(0)}}$
0.312E − 01	0.638E − 03	−0.124E − 04	0.529E − 05
0.442E − 01	0.127E − 02	−0.348E − 04	0.149E − 04
0.625E − 01	0.250E − 02	−0.969E − 04	0.422E − 04
0.884E − 01	0.492E − 02	−0.268E − 02	0.119E − 02
0.125E + 00	0.960E − 02	−0.735E − 03	0.333E − 03
0.177E + 00	0.185E − 01	−0.199E − 02	0.926E − 03
0.250E + 00	0.352E − 01	−0.527E − 02	0.254E − 02
0.354E + 00	0.654E − 01	−0.136E − 01	0.680E − 02
0.500E + 00	0.117E + 00	−0.338E − 01	0.174E − 01
0.707E + 00	0.199E + 00	−0.793E − 01	0.417E − 01
0.100E + 01	0.315E + 00	−0.172E + 00	0.891E − 01
0.141E + 01	0.452E + 00	−0.338E + 00	0.163E + 00
0.200E + 01	0.579E + 00	−0.584E + 00	0.245E + 00
0.283E + 01	0.667E + 00	−0.887E + 00	0.308E + 00
0.400E + 01	0.718E + 00	−0.121E + 01	0.344E + 00
0.566E + 01	0.748E + 00	−0.155E + 01	0.366E + 00
0.800E + 01	0.765E + 00	−0.189E + 01	0.378E + 00
0.113E + 02	0.774E + 00	−0.224E + 01	0.384E + 00
0.160E + 02	0.779E + 00	−0.258E + 01	0.388E + 00
0.226E + 02	0.782E + 00	−0.293E + 01	0.390E + 00
0.320E + 02	0.783E + 00	−0.328E + 01	0.391E + 00
0.452E + 02	0.784E + 00	−0.362E + 01	0.392E + 00
0.640E + 02	0.785E + 00	−0.397E + 01	0.392E + 00

Finally the EMF induced in the ring after the current is abruptly terminated is:

$$\mathscr{E}_0(t) = \frac{2Ir_1^2}{\sigma} \int_0^\pi \frac{\Phi(u) - \dfrac{2}{\sqrt{\pi}}\,u\,e^{-u^2}}{R^3} \cos\phi\,d\phi \tag{5.53}$$

During the early stage of the transient response, the EMF increases without limit in inverse proportion to time, and is almost the same as the EMF that would be observed in a uniform full-space. With increase in time, the EMF decreases gradually, and for sufficiently large values of the parameter τ/r_1, it decays as $t^{-5/2}$:

$$\mathscr{E}_0(t) \approx \frac{32}{5}\sqrt{\frac{2}{\pi}}\,\frac{\pi^6 Ir_1^4}{\sigma\tau^5} = \frac{Ir_1^4\,\sigma^{3/2}\mu^{5/2}}{t^{5/2}}\,\frac{\sqrt{\pi}}{20} \tag{5.54}$$

The transient electromotive force is shown graphically in Fig. 5.19; numerical values are listed in Table 5.X.

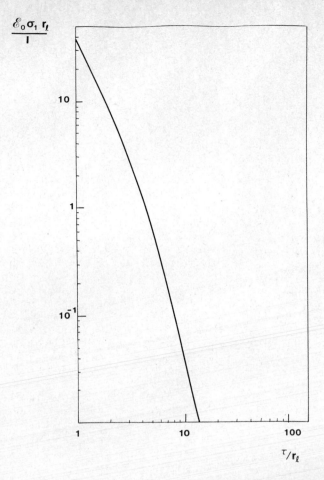

Fig. 5.19. Transient response of the electromotive force observed with coincident source and receiver loops situated on a uniform half-space.

c. *Thin conductive plane S and a two-layer medium.* As is well known, the expression for the horizontal component of the vector potential from an electric dipole situated on a surface of a thin conducting sheet with conductance S is:

$$A_{0x}^e = \frac{I\,dx}{4\pi r}[1 + nF(n)] \tag{5.55}$$

where

$$F(n) = \frac{\pi}{2}[H_0(n) - N_0(n)], \quad n = i\omega\mu Sr/2$$

TABLE 5.X

Half-space

$\dfrac{\tau}{r_1}$	$\dfrac{\mathscr{E}_0}{I} \sigma r_1$
0.100E + 01	0.372E + 02
0.141E + 01	0.177E + 02
0.200E + 01	0.813E + 01
0.283E + 01	0.347E + 01
0.400E + 01	0.130E + 01
0.566E + 01	0.399E + 00
0.800E + 01	0.997E − 01
0.113E + 02	0.214E − 01
0.160E + 02	0.420E − 02
0.226E + 02	0.784E − 03
0.320E + 02	0.142E − 03
0.453E + 02	0.255E − 04
0.640E + 02	0.454E − 05
0.905E + 02	0.805E − 06
0.128E + 03	0.143E − 06

Thus, we have the following expressions for the vector potential $A_{0\phi}$ and the EMF caused by currents flowing in the conducting plane:

$$A_{0\phi} = \frac{Ir_1}{2\pi} \int_0^\pi \frac{1 + n_1 F(n_1)}{R} \cos \phi d\phi \tag{5.56}$$

and

$$\mathscr{E}_0^{(1)} = \frac{(\omega \mu r_1)^2 SI}{2} \int_0^\pi F(n_1) \cos \phi d\phi = \frac{i\omega \mu S r_1}{2} \alpha \int_0^\pi F(n_1) \cos \phi d\phi \tag{5.57}$$

where

$$n_1 = i\omega \mu SR/2, \quad \alpha = -i\omega \mu Ir_1$$

Expanding the Neuman and Struve functions, we obtain expressions for the low-frequency part of the spectrum:

$$Q \mathscr{E}_0 \approx |\alpha| \frac{\pi}{4} [\omega \mu S r_1 + \tfrac{1}{8}(\omega \mu S r_1)^3 \ln (\omega \mu S r_1)] \tag{5.58}$$

and

$$\text{In} \, \mathscr{E}_0^{(1)} \approx |\alpha| \frac{(\omega \mu S r_1)^2}{3} \tag{5.59}$$

These two equations are almost the same as the corresponding formulas for the EMF observed at the center of the loop when the moments for the receivers in the two cases are the same. Curves for the functions $Q\mathscr{E}_0/|\alpha|$, $\mathrm{In}\mathscr{E}_0^{(1)}/|\alpha|$ and $\Delta Q\,\mathscr{E}_0/|\alpha|$, for various values of the parameter $p_s = (\omega\mu Sr_1/2)^{1/2}$ are shown graphically in Fig. 5.20 and given numerically in Table 5.XI. One can demonstrate the vector potential for a current ring and the electromotive force in combined loops have the following expressions:

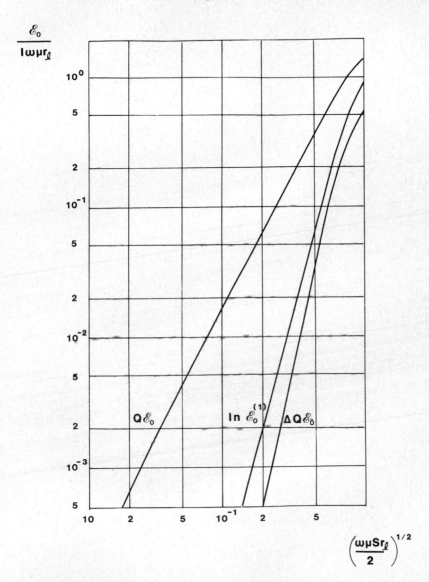

Fig. 5.20. Frequency responses for the quadrature and inphase components of the electromotive force and the difference function for electromotive force observed with coincident loops lying in an earth which can be considered to be a thin conducting sheet.

TABLE 5.XI

Plane S

$(\omega\mu Sr_1/2)^{1/2}$	$\dfrac{QH_0}{H_0^{(0)}}$	$\dfrac{\text{In}H_0^{(1)}}{H_0^{(0)}}$	$\dfrac{\Delta QH_0}{H_0^{(0)}}$
0.312E − 01	0.153E − 02	0.127E − 05	0.140E − 07
0.442E − 01	0.307E − 02	0.508E − 05	0.997E − 07
0.625E − 01	0.613E − 02	0.203E − 04	0.700E − 06
0.884E − 01	0.123E − 01	0.808E − 04	0.482E − 05
0.125E + 00	0.245E − 01	0.321E − 03	0.324E − 04
0.177E + 00	0.490E − 01	0.126E − 02	0.209E − 03
0.250E + 00	0.976E − 01	0.492E − 02	0.128E − 02
0.354E + 00	0.193E + 00	0.186E − 01	0.721E − 02
0.500E + 00	0.371E + 00	0.666E − 01	0.352E − 01
0.707E + 00	0.671E + 00	0.215E + 00	0.137E + 00
0.100E + 01	0.107E + 01	0.585E + 00	0.370E + 00
0.141E + 01	0.140E + 01	0.123E + 01	0.632E + 00
0.200E + 01	0.153E + 01	0.199E + 01	0.750E + 00
0.283E + 01	0.156E + 01	0.272E + 01	0.777E + 00
0.400E + 01	0.156E + 01	0.342E + 01	0.783E + 00
0.566E + 01	0.156E + 01	0.412E + 01	0.785E + 00
0.800E + 01	0.156E + 01	0.481E + 01	0.785E + 00

$$A_{0\phi}(t) = \frac{Ir_1}{2\pi}\int_0^\pi \frac{\cos\phi}{\left[R^2 + \left(\dfrac{2t}{\mu S}\right)^2\right]^{1/2}}\,d\phi \tag{5.60}$$

and

$$\mathscr{E}_0(t) = \frac{4Ir_1^2 t}{\mu S^2}\int_0^\pi \frac{\cos\phi}{\left[R^2 + \dfrac{4t^2}{\mu^2 S^2}\right]^{3/2}}\,d\phi = \frac{I}{S}\tau_s k\left[\frac{2-k^2}{2(1-k^2)}\,E(k) - K(k)\right] \tag{5.61}$$

where

$$\tau_s = 2t/S\mu r_1, \quad k = 2/(4 + \tau_s^2)^{1/2}$$

and where $K(k)$ and $E(k)$ are elliptical integrals of the first and second kind, respectively.

During the early stage of transient coupling, the EMF increases without limit in inverse proportion to time, and is independent of the longitudinal conductance S. With increase in time, the electromotive force gradually decreases, and for sufficiently large values of the parameter τ_s, begins to behave in inverse proportion to the fourth power of time:

$$\mathcal{E}_0(t) \approx \frac{3\pi r_1^4}{16} IS^3 \left(\frac{\mu}{t}\right)^4 \tag{5.62}$$

This expression is the same as that for the EMF observed on the surface of a two-layer sequence when the basement is insulating, and with $S = \sigma_1 H_1$. If the basement is characterized by finite resistivity, the electromotive force during the late stage of transient response can be written as:

$$\mathcal{E}_0(t) \approx \frac{32 I \pi^6 r_1^4}{\sigma_1 \tau_1^5} s \left[\left(\frac{2}{\pi}\right)^{1/2} \frac{\sqrt{s}}{5} - \frac{\pi(s-1)H_1}{\tau_1} \right.$$
$$\left. - \frac{4}{7} \left(\frac{2}{\pi}\right)^{1/2} \frac{\pi^2 H_1^2}{\tau_1^2} \left\{ \frac{s^{3/2} r_1^2}{H_1^2} - \frac{(1-s)(8s-9)}{\sqrt{s}} \right\} \right] \tag{5.63}$$

where $s = \sigma_2/\sigma_1$ and $\tau_1 = 2\pi (2t/\mu\sigma_1)^{1/2}$.

Thus, the EMF induced in a receiver that is combined with the source, and the EMF induced in a small loop situated at the center of the source loop do not differ from one another if the receiver moments are taken to be the same.

Curves for the function $(\sigma r_1/I) \mathcal{E}_0$ in the case of the plate S and for a two-layer sequence are shown in Figs. 5.21—5.23, and numerical values are given in Table 5.XII for the behavior for a thin plate.

In conclusion, we should note the following:

(1) In a horizontally layered medium at low frequencies, the magnetic field can always be written as

$$\Sigma a_{1n}\omega^n + \Sigma a_{2n}\omega^{n+1/2} + \Sigma a_{3n}\omega^n \ln \omega \tag{5.64}$$

In other words, it is fundamentally different than the behavior of the low-frequency part of the spectrum characterizing currents flowing in a confined conductive material.

(2) With inductive excitation, the leading term in the expression for the magnetic field at low frequencies is directly proportional to ω. This corresponds to the case where electromagnetic interaction between current filaments is negligible.

(3) During the late stage of transient response, the transient field is the sum of terms each of which is inversely proportional to integer and fractional powers of t:

$$\Sigma \frac{b_{1n}}{t^{n+1/2}} + \Sigma \frac{b_{2n}}{t^n} \tag{5.65}$$

The coefficients a_n and b_n are related to each other in a simple manner. As follows from eq. 5.65 the late stage time domain behavior of the field caused by currents flowing in a layered sequence differs drastically from the

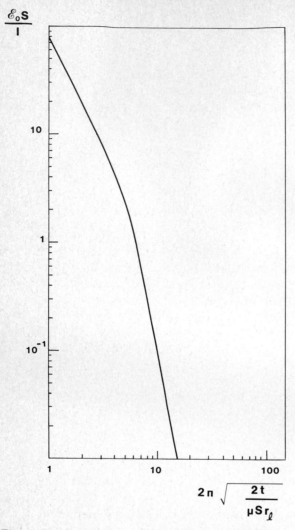

Fig. 5.21. Transient response of the electromotive force for coincident source and receiver loops when the earth can be represented by a thin conducting sheet.

exponential decay that will be observed for the field during a late stage generated by currents flowing in a confined conductor. This fundamental difference plays a vital role in the choice the optimum time range to minimize the effect of currents flowing in the host medium.

(4) The difference between the frequency responses caused by currents flowing in confined conductor and those flowing in a layered medium is also important for determining the range of frequencies over which the effect of the layered medium (which can be thought of as contributed geological noise) is minimal.

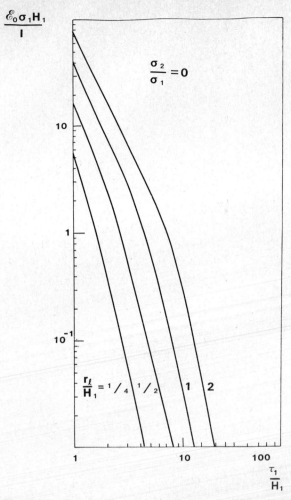

Fig. 5.22. Transient response of the electromotive force observed in coincident source and receiver loops on the surface of a two-layer medium in which the lowermost medium is an insulator.

(5) The first term in the expression for the low-frequency part of the spectrum for the magnetic field, being proportional to $\omega(\omega \to 0)$ does not contribute to the late stage of the transient field in the time domain.

(6) The nature of the dependence of the quadrature component of the field on the electrical properties of the layers and on the transmitter-receiver separation differs fundamentally at low frequencies from corresponding dependencies for the inphase component, $\ln H_0^{(1)}$ and the function $\Delta Q H(\omega)$.

(7) The conditions under which the uppermost layer in a two-layer sequence $(\rho_2 \gg \rho_1)$ can be replaced by a thin conducting sheet with a

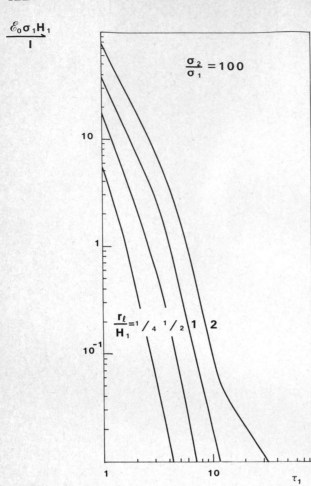

Fig. 5.23. Transient response of the electromotive force observed in coincident source and receiver loops on the surface of a two-layer medium in which the conductivity at the lower half-space is greater than that of the surface layer.

conductance S depends on which component is being measured: QH_0, $\text{In} H_0^{(1)}$, ΔQH_0, $H(t)$.

5.2. THE INFLUENCE OF CURRENTS INDUCED IN THE HOST MEDIUM ON THE SECONDARY ELECTROMAGNETIC FIELD

In this section, we will examine the behavior of frequency-domain and time-domain fields when the host medium surrounding a spheroidal conductor is characterized by a finite resistivity. The source for the primary

TABLE 5.XII

Plane S

$2\pi\left(\dfrac{2t}{\mu Sr_1}\right)^{1/2}$	$\dfrac{\mathcal{E}_0}{I}S$
0.100E + 01	0.789E + 02
0.141E + 01	0.393E + 02
0.200E + 01	0.195E + 02
0.283E + 01	0.944E + 01
0.400E + 01	0.429E + 01
0.566E + 01	0.164E + 01
0.800E + 01	0.415E + 00
0.113E + 02	0.568E − 01
0.160E + 02	0.476E − 02
0.226E + 02	0.323E − 03
0.320E + 02	0.207E − 04
0.453E + 02	0.130E − 05
0.640E + 02	0.813E − 07
0.905E + 02	0.508E − 08
0.128E + 03	0.318E − 09

field will be a current-carrying loop lying in the horizontal plane with its center situated on an axis of symmetry. With this geometry, the normal electrical field will not intersect the surface of the conductor, and so, electrical charges will not appear on that surface. Models consisting of a uniform full-space, a uniform half-space, a conductive plate, and a two-layer sequence will be considered. At the end of this section, some numerical information about the relative sizes of anomalies in the frequency and time domains will be given. These data will provide some idea about the extent of the influence of finite conductivity in the host medium on the depth of investigation that can be achieved using various inductive methods.

A spheroid in a full-space

First, let us examine the effect of the host medium in the simplest case, that in which a sphere is situated in a uniform conducting full-space. In this case, the electromagnetic field can be represented with a closed form expression. The components H_r and E_ϕ of the secondary field can be written as follows in spherical coordinates (see Fig. 5.24 for the geometry of the system):

$$H_{1r} = -\frac{ir_1 I}{2rr_1}\sum_{n=1}^{\infty}(2n+1)\xi_n(k_e r_1)T_n\frac{\xi_n(k_e r)}{k_e r}P_n(\cos\theta)P_n^{(1)}(\cos\theta_1)$$

$$(5.66)$$

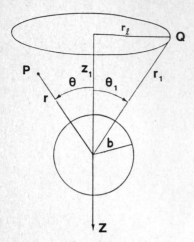

Fig. 5.24. A spherical conductor in the field generated by a circular current-carrying loop.

and

$$E_{1\phi} = \frac{\omega \mu r_1 I}{2r_1} \sum_{n=1}^{\infty} \frac{2n+1}{n(n+1)} \xi_n(k_e r_1) T_n \frac{\xi_n(k_e r)}{k_e r} P_n^{(1)}(\cos\theta) P_n^{(1)}(\cos\theta_1)$$

$$(5.67)$$

where

$$T_n = \frac{m\psi_n(\rho)\psi_n'(m\rho) - \psi_n'(\rho)\psi_n(m\rho)}{\xi_n'(\rho)\psi_n(m\rho) - m\xi_n(\rho)\psi_n'(m\rho)}$$

$$P_n^{(1)}(u) = (1-u^2)^{1/2} P_n'(u) \qquad m = (\sigma_i/\sigma_e)^{1/2}$$

$$(5.68)$$

and where σ_i and σ_e are the conductivities for the sphere and for the host medium respectively, $\rho = k_e b$, with b the radius of the sphere, k_e and k_i are wave numbers for the host medium and for the sphere, respectively,

$$k_e = (i\sigma_e \mu\omega)^{1/2}, \quad k_i = (i\sigma_i \mu\omega)^{1/2}, \quad \mu = 4\pi \times 10^{-7}\,\text{H/m}$$

$\omega = 2\pi f$ is the frequency in radians per second, $P_n(u)$ and $P_n'(u)$ are the Legendre function and its derivative, respectively:

$$\xi_n(x) = (\pi x/2)^{1/2} H_{n+1/2}^{(1)}(x)$$

and

$$\psi_n(x) = (\pi x/2)^{1/2} J_{n+1/2}(x)$$

$J_\nu(x)$ and $H_\nu^{(1)}(x)$ are Bessel functions of the first and third kinds of orders ν, r is the distance from the center of the sphere to the observation site (see Fig. 5.24), r_1 is the radius of the loop, I is the current flowing in the loop, and

$$r_1 = (z^2 + r_1^2)^{1/2}, \quad \cos\theta = z/r, \quad \cos\theta_1 = z_1/r_1$$

It can readily be shown from eqs. 5.66 and 5.67 that for high frequencies, in which case the skin depth in the host medium is smaller than the distance from the loop to the sphere, the secondary fields will be exponentially small and are virtually independent of the conductivity of the sphere. Therefore, let us concentrate our attention on fields observed at relatively low frequencies. First, assume that a skin depth in the host medium, h_e, is significantly greater than the radius of the sphere, b:

$$|\rho| = |k_e b| \ll 1 \quad \text{and} \quad z/b > 2 \tag{5.69}$$

As is well known, if $|x| \ll 1$:

$$\psi_n(x) \approx \frac{x^{n+1}}{(2n+1)!!}, \quad \psi_n'(x) \approx \frac{n+1}{(2n+1)!!} x^n$$

$$\xi_n(x) \approx \frac{i(2n-1)!!}{x^n}, \quad \xi_n'(x) \approx -in \frac{(2n-1)!!}{x^{n+1}} \tag{5.70}$$

where $n!!$ is a double factorial, defined as:

$$(2n-1)!! = 1 \cdot 3 \cdot 5 \cdot 7 \cdot 9 \ldots (2n-1)$$

Substituting eq. 5.70 into eq. 5.68, we obtain:

$$T_n = -i \frac{\rho^{2n+1}}{(2n+1)[(2n-1)!!]^2} D_n \tag{5.71}$$

where

$$D_n = \frac{\psi_{n+1}(m\rho)}{\psi_{n-1}(m\rho)} \tag{5.72}$$

In this last expression, D_n is a function which depends on the parameters characterizing the sphere, inasmuch as $m\rho = k_i b$. In particular:

$$T_1 = -i \frac{\rho^3}{3} D_1, \quad T_2 = -i \frac{\rho^5}{45} D_2$$

Thus, the field will be almost completely described by the first term in either eq. 5.66 or 5.67 when condition 5.69 holds. For example, for the vertical component of the secondary magnetic field at the center of the source loop, we have:

$$H_{1z} = \frac{M}{2\pi z^3} e^{ik_e z}(1 - ik_e z) \tag{5.73}$$

where

$$M = 2\pi D_1 b^3 \frac{Ir_1^2}{2r_1^3} e^{ik_e r_1} (1 - ik_e r_1)$$

(5.74)

$$D_1 = \frac{3}{k_i b} \operatorname{ctn} k_i b - \frac{3}{k_i^2 b^2} + 1$$

Whence

$$H_{1z} = H_0^{(0)} \frac{b^3}{z^3} \frac{r_1^3}{r_1^3} D_1 e^{ik_e R} (1 - ik_e R - k_e^2 zr_1)$$

(5.75)

where $R = z + r_1$ and $H_0^{(0)} = I/2r_1$ is the vertical component of the primary field at the center of the loop.

Equation 5.73 describes the electromagnetic field at frequencies for which the currents in the host medium do not affect the interaction of currents flowing within the sphere. We might note at this point that an approximate solution which can be used to find fields for bodies of more complicated shape is also based on this assumption, and it will be described in detail in the following section.

The quadrature component QH_1 for a spheroid in a full-space

Letting $|k_e R|$ be less than one in eq. 5.75 we obtain the following approximate formula for the quadrature component of the magnetic field QH_1, at the center of the loop:

$$QH_1 = H_0^{(0)} \left(\frac{br_1}{zr_1}\right)^3 \{B[1 + 2(c - \tfrac{1}{3})p_e^3] + 2p_e^2 A[(c - \tfrac{1}{2}) - p_e(c - \tfrac{1}{3})]\}$$

(5.76)

This expression remains valid even when the parameter $p_i = b/h_i$ which controls the behavior of the function D_1 is significantly greater than unity. In this expression:

$$p_e = \frac{R}{h_e}, \quad p_i = \frac{b}{h_i}, \quad c = \frac{zr_1}{R^2}, \quad D_1 = -A(p_i) + iB(p_i)$$

and

$$h_i = (2/\sigma_i \mu\omega)^{1/2}, \quad h_e = (2/\sigma_e \mu\omega)^{1/2}$$

where $-A(p_i)$ and $B(p_i)$ are the real and imaginary parts of the function D_1.

The parameter c takes on a maximum value of 0.25 when $z = r_1$ (that is, with the source of the primary field being a magnetic dipole). Suppose for a moment that the host medium has a sufficiently high resistivity that eq. 5.76 will give the correct representation for the field over a significant range of frequencies in the high-frequency part of the spectrum of the function D_1. This assumption will allow us to investigate rather simply the principal features of the secondary magnetic field. From eq. 5.76, as the parameter p_e decreases, the quadrature component of the secondary field (or more precisely, the first term of the sum of spatial harmonics in eq. 5.66) will approach the appropriate value for a nonconducting medium:

$$QH_{1z} = H_0^{(0)} \left(\frac{br_1}{zr_1} \right)^3 B \quad \text{for } p_e \to 0 \tag{5.77}$$

If the skin distance in the sphere is significantly larger than the radius of the sphere, b:

$$QH_{1z} \approx \frac{\sigma_i \mu \omega b^2}{15} \left(\frac{br_1}{zr_1} \right)^3 H_0^{(0)} \tag{5.78}$$

As follows from eq. 5.76, for large values of the function D_1 for the condition B being much greater than the $p_e^2 A$ applying, the rate of decrease of the quadrature component with decrease in frequency in contrast to the behavior in free space is proportional to $\omega^{5/2}$. This results from the fact that the inphase component of the "normal" currents in the host medium causes a field which is in opposition to the primary magnetic field. From the practical point of view, over this part of the spectrum eq. 5.77 defines the quadrature component of the secondary field. With increasing frequency, the second factor becomes significant and leads to a decrease in the quadrature component as well. Under the action of the magnetic field contributed by the inphase current flow in the sphere, secondary currents will arise in the host medium, with these being shifted by $90°$ in phase with respect to the primary field. Therefore, over the high-frequency part of the spectrum D_1, there is always a zero value for the quadrature component of the secondary field when:

$$B \approx p_e^2 A (1 - 2c) \tag{5.79}$$

With a further increase in frequency, the quadrature component, which is now negative, will increase in absolute value in proportion to p_e^2 (still with the condition that p_e is much less than 1). The remaining part of the spectrum cannot be described accurately using eq. 5.76. However, from physical considerations which are in agreement with calculations based on the exact equation (eq. 5.66), as a consequence of conversion of energy to heat in the conducting host medium, the modulus of the quadrature component passes through a maximum with increase in frequency, after which the modulus of the quadrature component decreases rapidly, oscillating in sign and values near zero (see Fig. 5.25). At the same time, the maximum negative value for the secondary field is almost twice the amplitude of the maximum positive value.

428

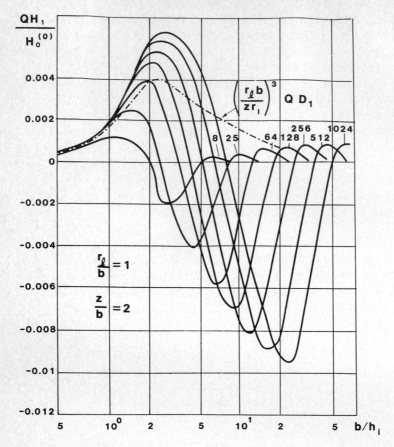

Fig. 5.25. Frequency response of the quadrature component of the secondary field normalized to the strength of the primary field observed at the center of the source loop (index of curves is ratio σ_i/σ_e).

Calculations have shown that with increase in conductivity in the host medium, increase in radius of the source loop, or increase in deep of burial of the conductive body, the extreme values for the quadrature component are observed at lower frequencies. This same behavior is also observed when the horizontal dimension of the spheroid is increased.

Consider the spectrum of the quadrature component of the total field, QH, normalized with respect to the field value in free space.

$$\frac{QH_1 + QH_0}{H_0^{(0)}} = \frac{QH}{H_0^{(0)}}$$

where QH_0 is the quadrature component in the host medium when the confined conductor is absent, that is, it is the normal field of the source. At some frequencies, including low frequencies, the presence of the conductive body will cause an increase in the field in comparison with that which would be observed in the surrounding host medium. There will always be some frequency, determined by the geoelectric parameters and by the source loop radius, for which the total field is exactly equal to the normal field. At higher frequencies, as the quadrature component of the secondary field becomes negative, the total field will be less than the normal field. The frequency characteristic for the quadrature component of the total field with the spherical conductor present is shown by the curves in Fig. 5.26. The spectrums are practically no different in their general form from the spectrum for the quadrature component of the normal field. This is because, at frequencies above the frequency for which the quadrature component of the secondary field reaches its maximum value, the normal field dominates. At very high frequencies, the behavior is exceptional, inasmuch as QH_0 and QH_1, both decaying exponentially, and oscillate such that they equal zero at various frequencies.

With an increase in the strength of the secondary field such as might result from increase in the horizontal dimension of the conducting spheroid, two maximums appear in the spectrum. The first corresponds to the maximum of the frequency characteristic for the secondary field, while the second

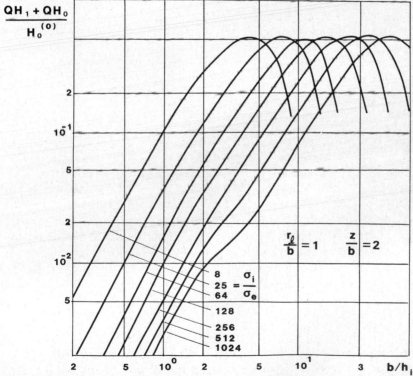

Fig. 5.26. Frequency reponse for the total quadrature component of the magnetic field observed at the center of the source loop normalized to the strength of the primary field.

430

corresponds to the maximum in the spectrum for the normal field (see Fig. 5.27). With a further increase in horizontal dimension of the spheroid, the second maximum becomes smaller in amplitude because the secondary field over this part of the spectrum is defined by secondary currents in the surrounding medium, and this leads to a decrease in the normal field. We should take note that numerical calculations of fields in the presence of a spheroid have been carried out using the solution of a system of integral equations.

Thus, in the general case, one can recognize the following characteristic ranges in the spectrum for the quadrature component of the total field:

(1) A low-frequency part of the spectrum where the field is directly proportional to frequency and the secondary field is practically independent of the conductivity of the surrounding medium, provided that σ_i/σ_e is much more than unity,

(2) A range of frequencies over which a maximum in the secondary field strength occurs. In this part of the spectrum, when the parameter R/h_e is sufficiently small, the field component QH_1 is only weakly related to the conductivity in the host medium as well.

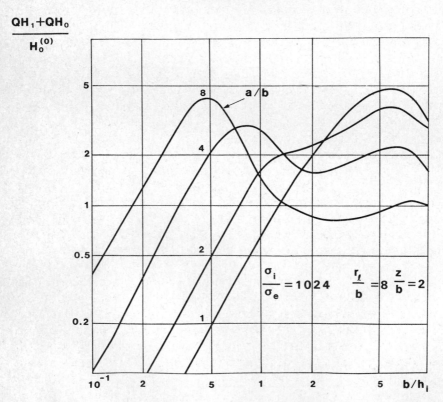

Fig. 5.27. Frequency response curves for the total quadrature component of the magnetic field observed at the center of the source loop for spheroids with various axial ratios.

(3) A range of frequencies within which the normal field is characterized by a maximum.

(4) A range of frequencies where the normal field is dominant.

(5) The high-frequency part of the spectrum where the normal and secondary fields are oscillatory and exponentially damped.

The relative importance of the various parts of the spectrum differs, depending on the geometrical parameters describing a medium, and geometrical characteristics of the measurement system.

When measurements are made on the earth's surface, the secondary field at high frequencies is only oscillatory, and therefore, one might combine frequency ranges 4 and 5.

At this point let us consider the results of calculations that have been carried out for the quadrature component expressed in terms of the ratio:

$$Qh = \frac{QH_0 + QH_1}{QH_0} = 1 + \frac{QH_1}{QH_0}$$

a quantity that reflects the influence of the host medium and characterizes the depth of investigation when the quadrature component is measured.

The first example consists of a set of curves for a sphere enclosed by a uniform conducting medium (Figs. 5.28 and 5.29). The left-hand asymptote for the curves corresponds to the low-frequency part of the spectrum over which the normal and secondary fields are directly proportional to frequency. It is clear that a decrease in this frequency range for which the left-hand asymptote is valid does not reduce the influence of the surrounding medium. With an increase in the value of the parameter $p_i = b/h_i$, the secondary fields begin to increase more slowly than does the normal field, and therefore, the relative anomaly becomes smaller. The higher the conductivity of the host medium, the sooner this aspect of behavior appears. As was noted earlier, for certain values of the parameter p_i, the secondary field is zero, and the function Qh is unity. At higher frequencies, the relative anomaly Qh decreases until the effect of exponential damping is no longer apparent, and the curves approach an asymptotic value of unity.

It should also be noted that for the sphere, the minimum value for the relative anomaly is observed closed to unity because the negative values of the secondary field are relatively small compared to the normal field. As may be seen from an analysis of the set of curves Qh, the low-frequency asymptote corresponding to the maximum relative anomalies observed with the quadrature component extends over a significant range of frequencies, which for a sphere will be:

$$b/h_i < 1.5 \quad \text{or} \quad f < 5 \times 10^5 \frac{\rho_i}{b^2} \text{ (in Hz)} \tag{5.80}$$

432

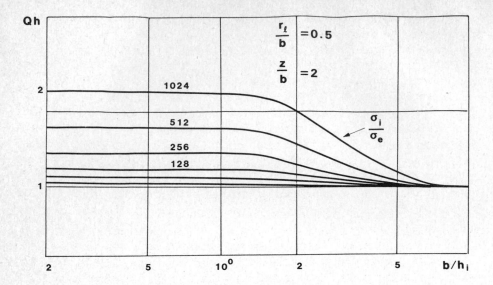

Fig. 5.28. Frequency response for the relative anomaly in the quadrature component of the magnetic field observed at the center of the source loop.

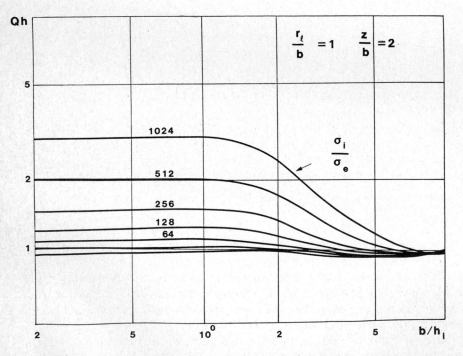

Fig. 5.29. Frequency response for the relative anomaly in the quadrature component of the magnetic field observed at the center of the source loop.

Making use of the expressions for the secondary field given in eq. 5.78 and for the normal field given in eq. 5.5, we have the following result for this part of the spectrum:

$$Qh = 1 + \frac{\sigma_i}{\sigma_e} \frac{2}{15} \frac{\overline{r}_1}{\overline{z}^3 (\overline{r}_1^2 + \overline{z}^2)^{3/2}} \tag{5.81}$$

where

$$\overline{r}_1 = r_1/b, \quad \overline{z} = z/b$$

It can be shown that the relative anomaly Qh takes on a maximum value with variation of loop size when $r_1 = z/2^{1/2}$. The maximum is defined by the value of the primary field at the center of the sphere and by the quadrature component of the normal field at the observation point (located at the center of the source loop). Actually, when r_1 is much less than z, the primary field is the same as that for a magnetic dipole source, and with increase in loop radius, it increases in proportion to r_1^2 merely because $H_0^{(0)} = Ir_1^2/2z^3$. However, for sufficiently large values of r_1 in comparison with depth of burial for the conductor, the primary field decreases in direct proportion to r_1. Thus, at the optimum loop radius:

$$Qh = 1 + \frac{s}{20\overline{z}^5} \tag{5.82}$$

where $s = \sigma_i/\sigma_e$.

The frequency characteristics for the function $Qh(b/h_i)$ are shown in Fig. 5.30 for various ratios between the semi-axial lengths, a/b. Here, a and b are the major (horizontal) and minor (vertical) semi-axes for the spheroid, respectively.

When the skin depth in the conductive body is significantly greater than the length of the minor axis, as the horizontal dimensions of the spheroid increases, the relative anomaly will increase and the curve for h_a approaches an asymptotic value at low frequencies defined by the condition:

$$\frac{b}{h_i} \leqslant \frac{(q_1)^{1/2}}{2} \tag{5.83}$$

where q_1 is the characteristic for the first pole in the spectrum (see Table 3.VII).

With increase in frequency, the relative anomaly becomes smaller, and for highly elongate conductors, this behavior is accentuated. Therefore, over some range of frequencies, the relative anomaly will become larger with a decrease in the horizontal dimension of the spheroid.

A change in sign of the secondary field related to the elongation of the spheroid occurs at progressively lower frequencies. In this case, there can be significantly large negative anomalies, the values of which are nearly constant

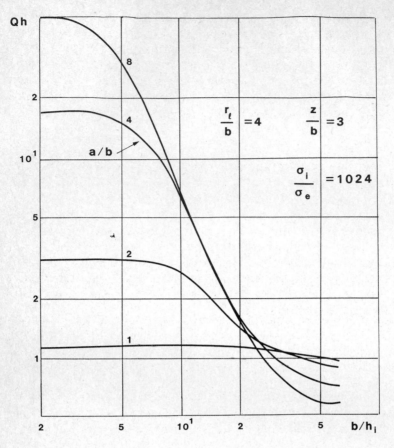

Fig. 5.30. Frequency response curves for the relative anomaly in the strength of the quadrature component of the magnetic field observed at the center of the source loop for spheroids having various values for their ratio of axial lengths.

over a relatively wide range corresponding to a minimum in Qh and to a strong skin effect in the spheroid. This means that when the field is measured at several fixed frequencies, the observed smoothed minimum over the relatively high-frequency part of the spectrum for the function Qh can be interpreted as being the effect of a body with a relatively high resistivity. This mistake will be made only if one does not take into account the sign of the value for the secondary field, QH_1.

The quadrature component of a spheroid in a two-layer half-space

We will now consider the effect of currents induced in an inhomogenous host medium on secondary fields. First, consider the case in which a sphere is surrounded by an insulating host medium and is located beneath a thin conducting sheet, characterized by a conductance value, S. As has been shown in a previous paragraph this can be equivalent to a conducting layer with a finite thickness under certain conditions. It will be shown in the next section that over a wide range of frequencies, the secondary currents induced in the conducting sheet do not affect the interaction of currents within the sphere, and the field of these currents is equivalent to that of a magnetic dipole. Therefore, for the magnetic field along the z-axis, we have:

$$H_{1z} = \frac{M_z}{4\pi} \int_0^\infty \frac{m^3 e^{-mz}}{m - ix} \, dm \tag{5.84}$$

where

$$M_z = 2\pi D_1 H_{0z}$$

and

$$H_{0z} = \frac{Ir_1}{2} \int_0^\infty \frac{m^2 e^{-mz}}{m - ix} J_1(mr_1) dm, \quad x = \mu\omega S/2 \tag{5.85}$$

where H_{0z} is the vertical component of the magnetic field at the center of the sphere and S is the conductance of the sheet.

Using conventional methods for expanding the integrals in eqs. 5.84 and 5.85, we can represent the secondary magnetic field at low frequencies in the form:

$$H_{1z} \approx H_0^{(0)} \left(\frac{r_1 b}{z r_1} \right)^3 D_1 [1 + ixr_1 a_1 - x^2 r_1^2 a_2 + ix^3 r_1^3 a_3 \ln(ixz) + \ldots]$$

where

$$a_1 = \frac{2r_1^3(r_1 - z) + r_1^2 z r_1}{2 r_1 z^3}$$

$$a_2 = \frac{z^2}{2r_1^2} + \frac{r_1^3}{2r_1^2(r_1 + z)} + \frac{z r_1^2}{2r_1^4}(r_1 - z)$$

$$a_3 = \frac{r_1^3}{r_1^3}, \quad r_1 = (r_1^2 + z^2)^{1/2}, \quad a_1 > 0, \quad a_2 > 0, \quad a_3 > 0$$

or

$$QH_1 = H_0^{(0)} \left(\frac{r_1 b}{z r_1} \right)^3 \left\{ B\left[1 - p_s^2 a_2 - p_s^3 a_3 \frac{\pi}{2}\right] - A\left[p_s a_1 + p_s^3 a_3 \ln p_s \frac{z}{r_1}\right] \right\} \tag{5.87}$$

where

$$p_s = xr_1 = \omega\mu Sr_1/2, \quad D_1 = -A + iB$$

It should be clear that with decrease in conductance, S, eq. 5.87 will hold at higher frequencies. From eq. 5.87, we see that at low frequencies as the parameter p_s approaches zero, the quadrature component of the secondary field is nearly the same as the field in an insulator:

$$QH_1 \approx H_0^{(0)}\left(\frac{r_1 b}{zr_1}\right)^3 B$$

As the exact solution shows, with increase in frequency, the function QH_1 increases, passes through a maximum, and then decreases to become negative. This is a consequence of the fact that the magnetic field of the time-varying inphase component of the current flow in the sphere induces currents in the conducting sheet which are shifted in phase by $-90°$ with respect to the primary field. Under the condition that $B \approx Ap_s q_1$ the function QH_1 is zero and with further increase in frequency, it becomes large in absolute value, passes through a maximum, and then approaches zero. The amplitudes of both extremes decrease and are shifted toward lower frequencies with increasing longitudinal conductance in the overburden.

Curves for the relative anomaly in the quadrature component, Qh, have the same form as those for uniform full-space (see Fig. 5.31). With an increase in the ratio:

$$\frac{\sigma_i}{\sigma_1}\frac{b}{H_1} = \frac{\sigma_i b}{S}$$

the relative anomalies become larger and the curves approach their left-hand asymptotes at higher frequencies. Over this portion of the spectrum for the function Qh, we have:

$$Qh \approx 1 + \frac{2}{15}\frac{r_1^2 b^4}{z^3 r_1^3}\frac{\sigma_i b}{S} \tag{5.88}$$

The effect of the dimensions of the conductive body as well as of the finite thickness and conductivity of the first layer, based on the exact solution, is illustrated by curves for the function Qh when this spheroid is located beneath the upper layer as shown in Fig. 5.32.

Inphase component for a spheroid in full-space

We will now examine the inphase component of the magnetic field, starting with the case in which a conducting sphere is situated in a uniform full-space, with calculations being carried out using eq. 5.66.

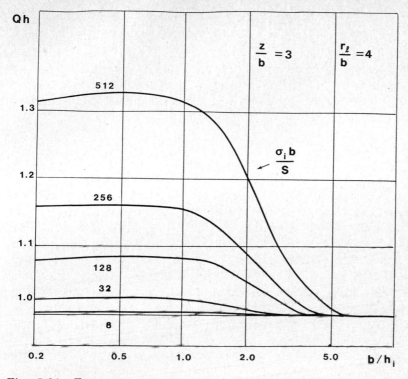

Fig. 5.31. Frequency response curves for the relative anomaly in the quadrature component observed at the center of the source loop for the case in which a conducting sphere is located beneath a conducting surface sheet.

According to eq. 5.75, we have the following for the inphase component, $\text{In}H_1$:

$$\text{In}H_1 = H_0^{(0)}\left(\frac{br_1}{zr_1}\right)^3\{-A[1 + 2(c - \tfrac{1}{3})p_e^3] + 2p_e^2B[(c - \tfrac{1}{2}) - p_e(c - \tfrac{1}{3})]\} \tag{5.89}$$

where

$$p_e = R/h_e, \quad p_i = b/h_i, \quad c = zr_1/R^2, \quad D_1 = -A(p_i) + iB(p_i)$$

It can readily be shown that in a nonconducting medium, the inphase component will be:

$$\text{In}H_1 = -A\left(\frac{br_1}{zr_1}\right)^3 H_0^{(0)} \tag{5.90}$$

and in particular, when the skin depth in the sphere is significantly greater than the radius, b:

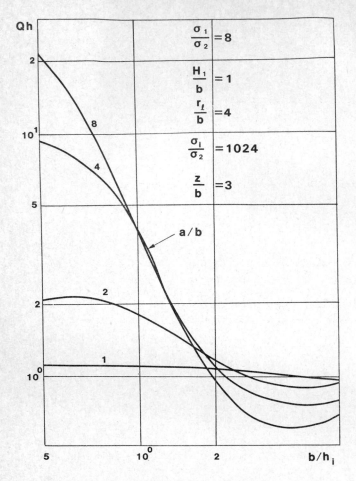

Fig. 5.32. Frequency response curves for the relative anomaly of the quadrature component observed at the center of the source loop for spheroids with various axial ratios located in a two-layer medium.

$$\mathrm{In}H_1 \;=\; -\frac{2}{315}\,(\sigma_i \mu \omega b^2)^2 \left(\frac{br_1}{zr_1}\right)^3 H_0^{(0)} \tag{5.91}$$

Curves for the inphase component of the secondary field normalized to the strength of the primary field are given in Fig. 5.33.

According to eq. 5.79, the effect of the surrounding medium on the inphase component, $\mathrm{In}H_1$, manifests itself in two ways. In the first place, the inphase component of the magnetic field caused by normal currents in the host medium, which is proportional to $\omega^{7/2}$ (as ω approaches zero) opposes the primary field causing a decrease in the total field. Secondly, as the magnetic field created by the quadrature component of current flow in the

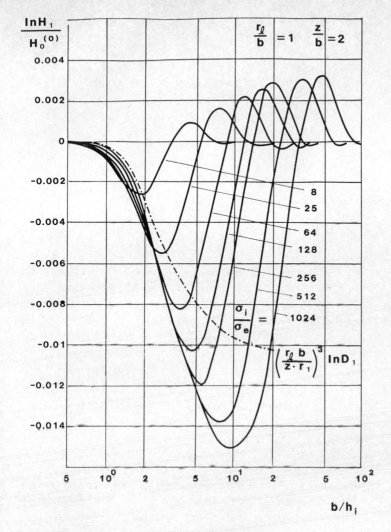

Fig. 5.33. Frequency response curves for the inphase component of the secondary field referred to the strength of the primary field.

sphere changes in time, secondary currents arise in the host medium. They coincide in phase with the inphase component of currents in the sphere. Therefore, the second factor, which is proportional to ω^2 at low frequencies, is the important one and causes an increase in the inphase component. As the frequency becomes larger, the component $\text{In}H_1$ increases in absolute value, passes through a maximum, and then decreases rapidly as a consequence of absorption of energy as heat in the host medium. At high frequencies, the spectrum decays exponentially, oscillating about zero. With increasing resistivity in the host medium, the absolute value at the maximum increases and shifts towards higher frequencies.

Thus, in contrast to the behavior of the quadrature component, the effect of the surrounding medium on the inphase component of the secondary field at low frequencies does not decrease with decrease in frequency. However, under certain conditions, the secondary field $\text{In} H_1$, is practically the same as the field when the surrounding medium is an insulator. According to eq. 5.89, for a sphere these conditions can be written as:

$$p_e \ll 1, \quad \frac{\sigma_e}{\sigma_i} p_i^2 B \ll A$$

and generalizing for the case of a spheroidal conductor, we have:

$$p_e \ll 1, \quad \frac{\sigma_e}{\sigma_i} < \frac{1}{q_1} \quad (q_1 > 1) \tag{5.92}$$

Curves for the inphase component of the total field are shown in Fig. 5.34. The curves have the same general character as those for the normal field. With an increase in resistivity in the host medium, the presence of the confined conductor manifests itself more strongly, especially over the part of the spectrum where the absolute value of the inphase component of the secondary field increases.

Now, consider the relative anomalies of the inphase component assuming that the primary field $H_0^{(0)}$ can be cancelled instrumentally:

$$\text{In} h = \frac{\text{In} H_0^{(1)} + \text{In} H_1}{\text{In} H_0^{(1)}} = 1 + \frac{\text{In} H_1}{\text{In} H_0^{(1)}}$$

Here, $\text{In} H_0^{(1)}$ is the inphase component of the magnetic field caused by the inphase component of the induced currents in the host medium only. At low frequencies, the relative anomaly $\text{In} h$ tends towards unity, in contrast to the behavior of the function Qh, inasmuch as over this part of the spectrum, the secondary field $\text{In} H_1$ decreases more rapidly with increase in frequency (ω^2) than does the normal field $H_0^{(1)}$, which varies as $\omega^{3/2}$.

With increase in frequency, the function $\text{In} h$ increases, passes through a maximum, and (with further increase in frequency) tends to unity again. As may be seen from the curves in Fig. 5.35, the frequency at which the maximum occurs is virtually independent of the conductivity of the host medium and the parameter, b/h_i, has a value of approximately 1.7 (for a sphere). This behavior is a consequence of the fact that for the properties of the medium being considered here, at the loop radii used (see condition 5.92) the inphase component of the secondary field differs only slightly from the field in the non-conducting medium. If the host medium has a relatively high resistivity for this range of values for the parameters p_i, the inphase anomaly $\text{In} H_1 / \text{In} H_0^{(1)}$ varies in inverse proportion to $\sigma_e^{3/2}$.

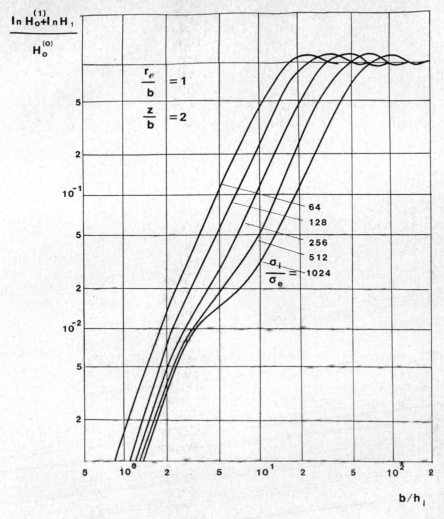

Fig. 5.34. Frequency response curves for the inphase component of the total magnetic field observed at the center of the source loop.

From results obtained from an analysis of secondary fields in Chapter 3, we realize that the maximum for the function $\ln h$ occurs at frequencies for which the inphase component is the more sensitive to the conductivity of confined conductors. In other words, under these conditions we have the highest resolving capability for separating anomalies into "ore" and "non-ore" anomalies. Correspondingly, a maximum depth of investigation will be obtained.

Figure 5.36 shows the behavior of the function $\ln h$ for spheroids having various ratios for their semi-axial lengths, a/b. It can be seen that as the

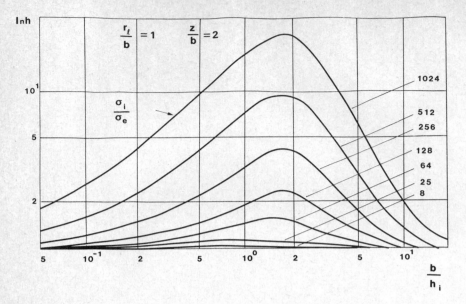

Fig. 5.35. Frequency response curves for the relative anomaly of the inphase component of the magnetic field observed at the center of the source loop.

horizontal dimension of the spheroid is increased, the value for the relative anomaly will also increase and the abscissa at which the maximum is observed shifts towards lower frequencies. It is related to the minimum pole for the spectrum by the equation:

$$b/h_i \approx q_1^{1/2}/2$$

where b is the minor semi-axis of the spheroid, along which the primary field is directed. As in the case for a sphere, the parameter $(q_1^{1/2})$ represents the frequencies at which the resolution obtainable with the inphase component is high (under the condition σ_i/σ_e is much greater than unity).

With decrease in loop radius, the relative anomaly $\mathrm{In}h$ first increases (see Fig. 5.37) and then approaches the limit for which the primary field in the area occupied by the conductor is equivalent to a dipole field while the normal field at the center of the source loop is defined by the main term in an expansion of $\mathrm{In}H_0^{(1)}$ at low frequencies. Therefore, both the secondary and normal fields are proportional to r_1^2.

At least two factors complicate measurements of the inphase components made with loops having small radii. First, with decrease in radius, the primary field becomes less at depths z greater than r_1, and in the second place, the requirements for cancellation of the primary field against which the small anomalies must be recognized will be more demanding.

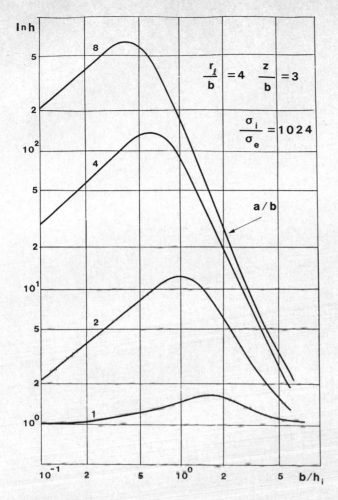

Fig. 5.36. Frequency response curves for the inphase component of the magnetic field observed at the center of the source loop for spheroids with various axial ratios.

Inphase component for a spheroid in a two-layer half space

Let us now consider the effect of the currents induced in the sheet on the inphase component. From eq. 5.86, we will have the following expression for a sphere.

$$\mathrm{In}H_1 = H_0^{(0)} \left(\frac{r_1 b}{z r_1}\right)^3 \left[-A\left(1 - p_s^2 a_2 - p_s^3 a_3 \frac{\pi}{2}\right) - B\left(p_s a_1 - p_s^3 a_3 \ln p_s \frac{z}{r_1}\right) \right]$$

$$(5.93)$$

The appropriate values for A, a_1, a_2, a_3, and p_s have been listed previously. As may be seen from this equation, at low frequencies, the inphase component

444

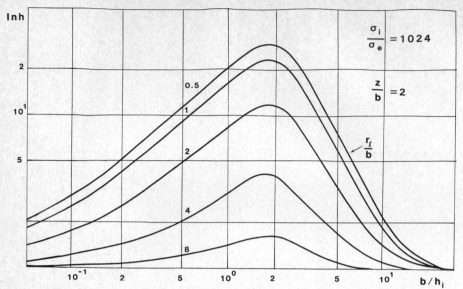

Fig. 5.37. Frequency response curves for the inphase part of the magnetic field observed at the center of the source loop for various radii, r_1.

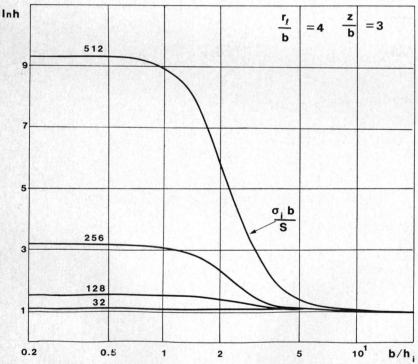

Fig. 5.38. Frequency response curves for the inphase part of the magnetic field observed at the center of the source loop.

of the secondary field decreases as ω^2 and in a similar manner to that of the quadrature component, and the effect of currents in the conducting plane remains constant over this range of frequencies (see Fig. 5.38). Because the inphase component of currents in this sphere and in this sheet are the same in direction, the inphase component of the magnetic field at low frequency becomes larger with increase in conductivity in the overburden (the S-plane).

With increase in frequency the absolute value of the inphase component of the secondary field passes through a maximum, and then decreases, changes sign, and exhibits a positive extremum, and decays rapidly in the wave zone due to skin effect. The frequency response for the relative anomaly of $\text{In}h$ in the case of a conducting sheet differs from the function $\text{In}h$ for other models, principally due to the behavior of the normal field at

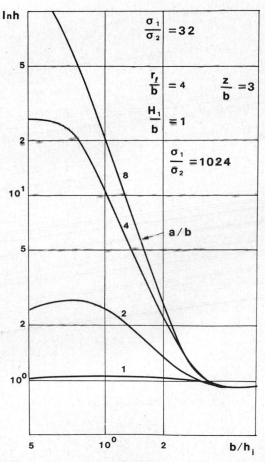

Fig. 5.39. Frequency response curves for the inphase part of the magnetic field for a two-layer medium.

446

low frequencies. As in the case for relative anomalies of the quadrature component, the function Inh tends to a horizontal asymptotic value as shown in Fig. 5.38 which results from a similar dependence of the secondary and normal fields on frequency.

If the lower layer in the two-layer sequence is characterized by a finite resistivity, the function Inh for spheroids having different ratios for the semi-axes a/b tends to unity at low frequencies as shown by the curves in Figs. 5.39 and 5.40. This corresponds to the behavior of the normal field which decreases in proportion to the factor $\omega^{3/2}$. At higher frequencies, Inh passes through a maximum, the amplitude of which becomes larger with

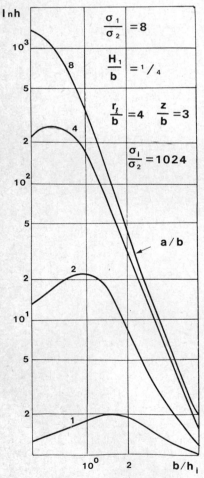

Fig. 5.40. Frequency response curves for the inphase part of the magnetic field for a two-layer sequence.

increase in horizontal axial dimensions of the spheroid. At the same time, the abscissa will shift to lower frequencies.

It is interesting to note the calculations for the inphase component in the case in which the confined body is a sphere show that the layer in the two-layer sequence can be replaced by a sheet having a conductance S when:

$$H_1/h_i < 1, \quad z/b > 2$$

The parameter h_i is the skin depth within this sphere.

Difference of the quadrature components ΔQH for a spheroid in a full-space and a sphere beneath a conducting sheet

Consider the effect of the surrounding medium when the difference between quadrature components at two frequencies is measured:

$$\Delta QH_1 = QH_1(\omega_1) - \frac{\omega_1}{\omega_2} QH_1(\omega_2)$$

The general features of the spectrum of ΔQH_1 are the same as those for the quadrature component QH_1, and which are shown in Figs. 5.41 and 5.42.

From eq. 5.76, the expression for the parameter ΔQH_1 can be written as follows for a sphere:

$$\Delta QH_1 = H_0^{(0)} \left(\frac{br_1}{zr_1}\right)^3 \left\{ -\frac{4}{15}(1-3c)\left(\frac{\sigma_e}{\sigma_i}\right)^{3/2}\left(\frac{R}{b}\right)^2 p_i^5\left[1-\left(\frac{\omega_2}{\omega_1}\right)^{3/2}\right] \right.$$

$$\left. + \left[B_1(\omega_1) - \frac{\omega_1}{\omega_2}B_1(\omega_2)\right] + p_e^2(1-2c)[A(\omega_2)-A(\omega_1)] + \ldots \right\}$$

$$(5.94)$$

where

$$B_1(\omega) = B(\omega) - \frac{p_i^2}{15}, \quad p_e \ll 1$$

At low frequencies, the first term in eq. 5.94 dominates. It is proportional to $\omega^{5/2}$ and to the conductivity of the body and it is sensitive to changes in the resistivity in the surrounding medium as well. This term describes the field which is generated by the time variations of the magnetic field accompanying the normal currents when the spectrum of the secondary field of currents in the sphere is defined by the function $p_i^2/15$. The second term coincides with the parameter ΔQH_1 provided that the surrounding medium is insulating, and as a consequence it would depend only on the parameters of the conductor. The third term is related to the magnetic field of secondary currents in the host medium. They in turn are generated by the time rate of change of the magnetic field accompanying the inphase component of currents in the

448

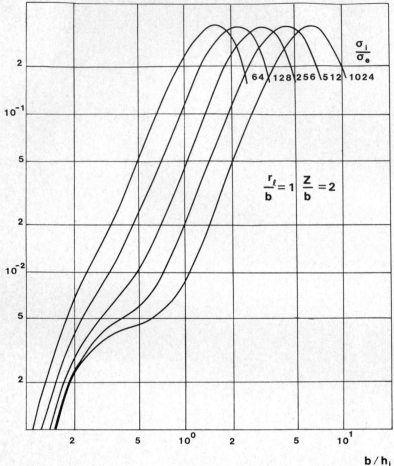

b / h$_i$

Fig. 5.41. Frequency response curves for the difference in quadrature values of the magnetic field.

confined body. The first three terms have the same sign so that at low frequencies, values for the parameter ΔQH_1 in the host medium are larger than in the case for an insulator.

With increase in frequency, the fourth term (which is proportional to $\omega^{7/2}$) begins to play a role. Because this term has a negative sign, there is a point at which the parameter ΔQH_1 is 0. Then, over a broad range of frequencies that takes on negative values.

Now consider the relative anomaly in the difference of the quadrature components measured at two frequencies, a quantity which clearly reflects the influence of the surrounding medium. Curves for the functions:

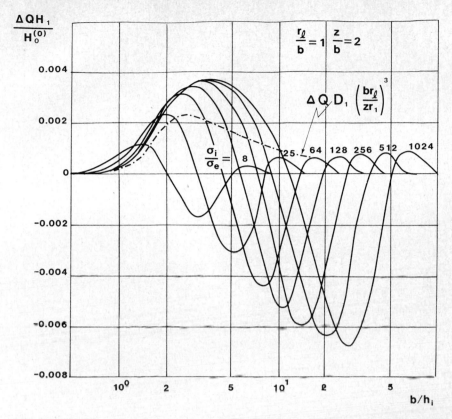

Fig. 5.42. Frequency response curves for the difference function observed at the center of the source loop.

$$\Delta Q h \;=\; 1 + \frac{\Delta Q H_1}{\Delta Q H_0}$$

are shown in Fig. 5.43. For small values of the parameter $p_i = b/h_i$, the values for the relative anomaly tend to unity, and they approach this horizontal asymptote at higher frequencies than in the case of the inphase component. With increase in frequency, the value $\Delta Q h$ increases, and there is a maximum which is almost twice the amplitude as the maximum in the function $In\,h$. The parameter $\Delta Q h$ then decreases approaching unity at high frequencies.

A comparison with results obtained for an insulating host medium shows that the maximum value for the parameter $\Delta Q h$ is observed at frequencies when the two frequency method has a high resolving capability (see Chapter 3). The abscissa of the maximum is almost the same as for the inphase component, $In\,h$, for the range of properties of the medium which has been considered, and does not depend strongly on the conductivity of the host

450

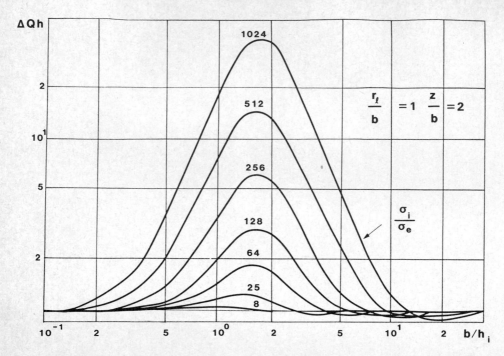

Fig. 5.43. Frequency response curves for the function ΔQh.

medium or on the loop radius. Curves for the relative anomaly ΔQh are given in Fig. 5.44 for the case of spheroids with various ratios of axial dimensions. Unlike the relative anomaly of the quadrature component, the effect of the surrounding medium on the function ΔQh can be small only over the mid part of the spectrum. This range of frequencies is wider with larger dimensions for the conductor, higher resistivities in the host medium, and lesser depth of burial for the conductor. Figures 5.45 through 5.47 show the responses of the relative anomaly ΔQh for cases in which the sphere is situated beneath the conducting sheet with a conductance S or beneath a layer with finite thickness, H_1. As may be seen from these various curves, when the difference of the quadrature components, that is the function ΔQh, is measured at two frequencies, the effect of the overburden can be reduced markedly.

One can demonstrate that by measuring parameters that correspond to higher-order terms in the low-frequency part of the spectrum, we can obtain a still further reduction in the influence of the layered medium over some middle portion of the spectrum.

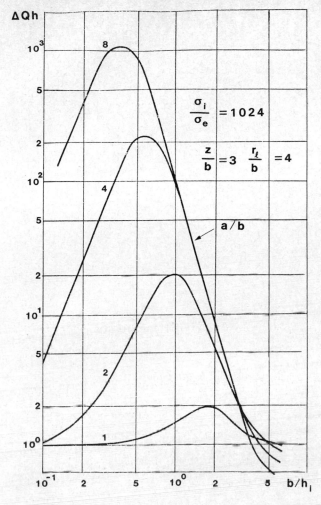

Fig. 5.44. Frequency response curves for the function $\Delta Q h$ observed at the center of the source loop for spheroids with various axial ratios.

A transient field for a spheroid in a full-space and for a sphere beneath a thin conducting sheet

Consider the behavior of a transient field arising from currents in a confined body embedded in a host rock. Assume that the current in a loop is switched off at the moment $t = 0$:

$$I = \begin{cases} I & t < 0 \\ 0 & t > 0 \end{cases}$$

452

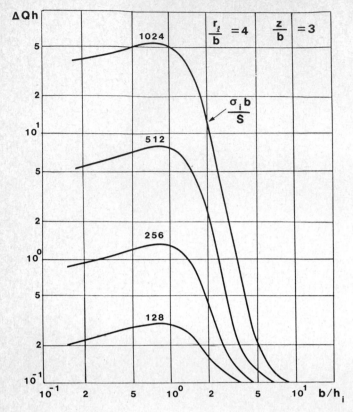

Fig. 5.45. Frequency response curves for the function ΔQh observed at the center of the source loop.

The transient response for the total and secondary fields shown in Figs. 5.48 and 5.49 have been obtained by using the Fourier transform. We should pay main attention to the relative anomaly, which is defined as:

$$\epsilon(t) \; = \; \frac{E_0(t) + E_1(t)}{E_0(t)} \; = \; 1 + \frac{E_1(t)}{E_0(t)}$$

Examples of the behavior of this anomaly function are shown by the curves in Fig. 5.50. Here, $E_0(t)$ is the EMF of the normal field when the surrounding medium contains no conductor, while $E_1(t)$ is the electromotive force of the secondary field. The confined body is a spheroid. At times zero, because of the skin effect, the EMF of the secondary field will be zero. The rate of decrease for the secondary field is greater than that for the normal field. For this reason, the asymptote for the curves representing $\epsilon(t)$ is unity when t approaches zero.

During the early part of the transient response, the absolute value of the secondary field increases in a manner different from that of the normal field.

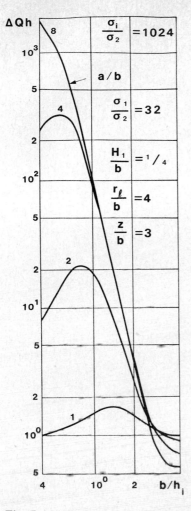

Fig. 5.46. Frequency response curves for the function ΔQh observed at the center of the source loop for a two-layer earth.

Because $E_0(t)$ and $E_1(t)$ are of opposite polarity for this geometry, the relative anomaly $\epsilon(t)$ is less than unity over this time interval. When the secondary magnetic field approaches its extreme value, the function $\epsilon(t)$ is unity. With increase in time $\epsilon(t)$ increases, passes through a maximum, and tends to unity again during the late part of the transient process because the EMF in the secondary field decreases more rapidly than that in a normal field (this will be demonstrated in the next section):

$$E_1(t) \approx k_1 \frac{1}{t^{7/2}}, \quad E_0(t) \approx k_2 \frac{1}{t^{5/2}} \tag{5.95}$$

454

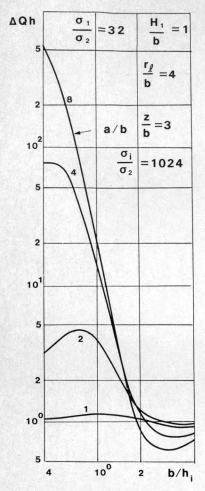

Fig. 5.47. Frequency response curves for the function ΔQh observed at the center of the source loop for a two-layer earth.

Thus, in the general case, there is an intermediate range of time over which it is most desirable to measure the EMF (this is comparable to measuring values for ΔQh and Inh over an optimum frequency range). As will be shown in the next section, over most of this interval we can use an approximate expression for the EMF which is valid when the confined body is a sphere:

$$E_1(t) = \frac{6\Phi_0}{\sigma_i \mu b^2}\left(\frac{r_1 b}{r_1 z}\right)^3 \left\{ e^{-\pi^2 \alpha t}\left[-1 - \frac{\sigma_e \pi^2 (z^2 + r_1^2)}{2\sigma_i b^2} + \left(\frac{\sigma_e}{\sigma_i}\pi^2\right)^2 \frac{R^2(z-r_1)^2}{8b^4}\right] \right.$$

$$\left. - \frac{5}{8\sqrt{\pi}\pi^4}\left(\frac{\sigma_i}{\sigma_e}\right)^2 b^4(z^3 + r_1^3)\left(\frac{\sigma_e \mu}{t}\right)^{7/2}\right\} \tag{5.96}$$

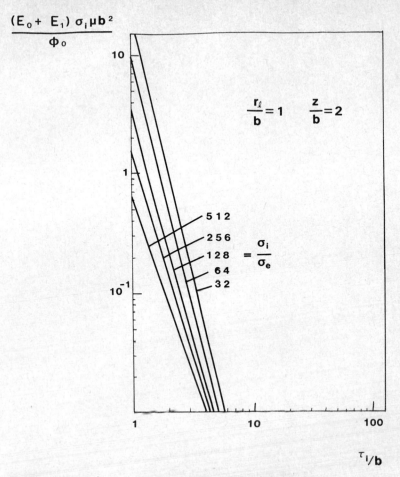

Fig. 5.48. Transient response curves for the total electromotive force.

where

$$r_1 = (z^2 + r_1^2)^{1/2}, \quad R = z + r_1, \quad \alpha = 1/\sigma_i \mu b^2$$

and Φ_0 is the flux of the primary field, B_0, through an enclosed path along which the EMF is evaluated.

A similar expression can be derived when the combined body is a spheroid.

The relationship between $\epsilon(t)$ and the radius of the source loop is analogous to the behavior of the function ΔQh in the frequency domain. With a decrease in the radius of the loop, the anomaly $\epsilon(t)$ first $(r_1 > z)$ increases approximately as r_1^{-3}, and then, at sufficiently small values of r_1 in comparison with the depth of burial, it becomes constant (see Fig. 5.51). From the practical point of view, the interval over which the relative anomaly reaches its maximum is of principal interest. Here, the influence of the host medium

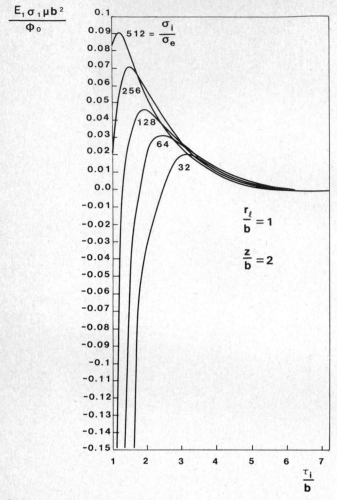

Fig. 5.49. Transient response curves for the secondary electromotive force.

is at a minimum, and under certain conditions, the secondary field is almost the same as that for currents in a conductor surrounded by an insulating medium. In this case, the transient method has its highest resolving capability and therefore the maximum depth of investigation. With increase in horizontal dimensions of the spheroid, the maximum for the relative anomaly $\epsilon(t)$ is shifted toward later times, as shown in Fig. 5.52. As an example of this, limiting values for the parameter τ_i/b and time are given in Table 5.XIII. For these values, the EMF measured in the conducting medium differs by less than 25% from the EMF observed when the sphere is located in an insulating host rock $(z/b < 3)$.

As may be seen from Table 5.XIII, with a decrease in the ratio of the

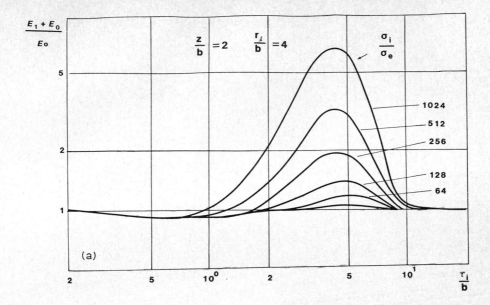

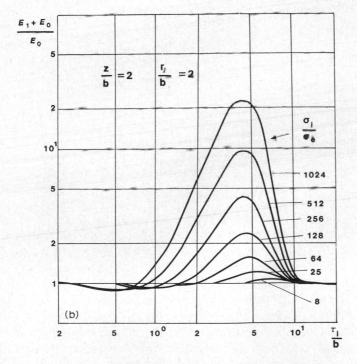

Fig. 5.50a and b. Transient response curves for the relative anomaly and electromotive force observed at the center of the source loop.

458

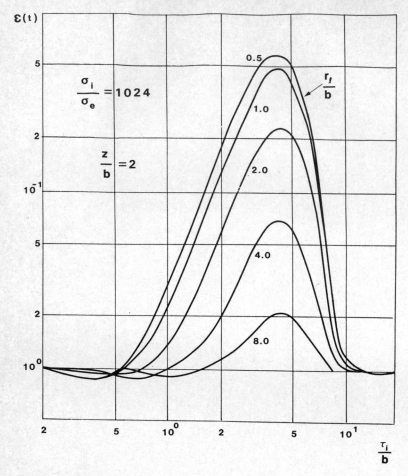

Fig. 5.51. Transient response of the electromotive force in the time domain as a function of the radius of the source loop.

conductivity σ_i/σ_e, the time interval over which the currents induced in the surrounding medium have no significant effect on the secondary field is reduced. As also is the case with the uniform medium, for a two-layer medium with a transient field measured in place of the frequency responses such as QH, InH, and ΔQH, the relative anomaly increases, and the maximum time-domain response occurs at later times than is the case in the uniform medium (see Fig. 5.53). As calculations of field strength show, for a wide range of properties of the medium, when the relative anomaly in the transient response is at a maximum significantly greater than unity ($\epsilon(t) > 1$) the following conditions hold: (1) the secondary field is independent of the conductivity and of the host medium, and decays exponentially; and (2) the normal field is described by asymptotic expressions representing a late stage of the transient response. For this reason, it is not difficult to

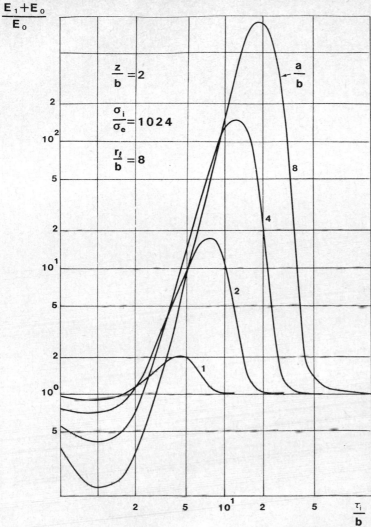

Fig. 5.52. Transient response curves for the electromotive force measured in the time domain at the center of the source loop.

obtain an expression which defines the maximum depth of investigation. In view of the assumptions we have made about the EMF for the secondary and normal fields in a uniform half space, we have:

$$E_1 \approx \frac{\Phi_0 r_1^3}{r_1^3} \frac{d_1 q_1}{\sigma_i \mu b^2} e^{-q_1 \alpha t}$$

$$E_0 \approx \frac{\Phi_0 r_1^3}{10\sqrt{\pi}\sigma_e \mu} \left(\frac{\sigma_e \mu}{t}\right)^{5/2}$$

(5.97)

460

TABLE 5.XIII

	σ_i/σ_e									
	4096		1024		512		256		128	
	min	max	min	max	min	max	min	max	min	max
τ_i/b	1.5	8	2.5	5.5	2.8	5.0	3.5	5	4	4
$t(m)$	1.5	40	4	20	4	10	7.6	10	10	10

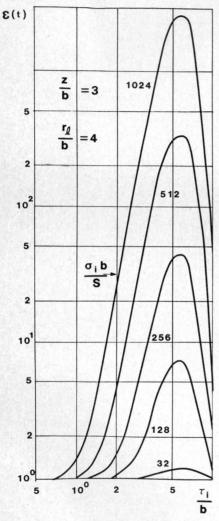

Fig. 5.53. Transient response curves for the electromotive force observed at the center of the source loop.

and

$$\epsilon(t) = 1 + \frac{E_1}{E_0} = 1 + \frac{10\sqrt{\pi}\, q_1 d_1 e^{-q_1 \alpha t}}{r_1^3 b^2} \left(\frac{\sigma_e}{\sigma_i}\right)\left(\frac{t}{\sigma_e \mu}\right)^{5/2} \tag{5.98}$$

where Φ_0 is the flux of the primary field:

$$\Phi_0 = \mu S_0 H_0^{(0)}$$

q_1 is one of the parameters characterizing the first pole of the spectrum of currents in a conducting spheroid, and d_1 is a coefficient that depends on geometric factors, as shown earlier in Chapter 3:

$$d_1 \approx \frac{q_1}{15} \frac{a^4}{bz^3}$$

when z is greater than $2a$.

It follows from this that the maximum relative anomaly is observed for spheroids situated in a uniform magnetic field when the power of the exponential term is:

$$q_1 \alpha t^* = t^*/\tau_0 = 2.5 \tag{5.99}$$

The greater the conductivity of the spheroid and the greater the linear dimensions, the larger will be the time at which the maximum in $\epsilon(t)$ occurs.

Considering that the secondary field decreases in the same manner as the field of a dipole source (with $z > 2a$) and $t = t^*$, we find that the maximum depth at which a conductor can be buried and yet cause a relative anomaly of ten is related to the properties of the medium and the conductivity of the body as follows:

$$z = 0.46 \frac{(a^4 b^2)^{1/3}}{r_1} \left(\frac{1}{q_1}\right)^{1/6} \left(\frac{\sigma_i}{\sigma_e}\right)^{1/2} \tag{5.100}$$

And in particular, when $r_1 = z$, we have:

$$z = 0.57 \frac{(a^2 b)^{1/3}}{q_1^{1/12}} \left(\frac{\sigma_i}{\sigma_e}\right)^{1/4} \tag{5.101}$$

If the normal field corresponds to "S-zone" behavior, that is, if

$$E_0 \approx \frac{3}{8} \Phi_0 \frac{(S\mu r_1)^3}{t^4} \tag{5.102}$$

the time at which the function $\epsilon(t)$ reaches its maximum, satisfies the condition:

$$q_1 \alpha t^* = t^*/\tau_0 = 4 \tag{5.103}$$

Therefore, the appropriate expression for the limiting depth is:

$$z = 0.4 \left(\frac{a^4 b^5}{q_1^2}\right)^{1/3} \frac{\sigma_i}{r_1 S} \tag{5.104}$$

If $r_1 = z$, we have:

$$z = 0.55 \left(\frac{a^2 b}{q_1}\right)^{1/3} \left(\frac{\sigma_i b}{S}\right)^{1/2} \tag{5.105}$$

where S is the longitudinal conductance of the overburden.

When the spheroid is quite flat (that is, $b \ll a$):

$$q_1 = 7.7b/2a$$

In this case we will have:

$$z = 0.25a(S_d/S)^{1/2} \tag{5.106}$$

where S_d is the conductance of the disk in its central plane. Note that if the ratio of the useful signal E to noise E_0 decreases, for example, by a factor of two, the maximum depth of investigation does not change significantly because the secondary field decreases very rapidly.

Computation of frequency-domain and time-domain fields using exact and approximate expressions permits us to obtain a general idea about the minimum values for the ratio σ_i/σ_e for which the effect of the host medium is no greater than 10% of the total field, and thus, to estimate the depth of investigation when such quantities as QH, $\mathrm{In}H$, and the parameter ΔQH, as well as the transient field $E(t)$ are measured (see Table 5.XIV). The values for σ_i/σ_e listed in this table apply for a uniform conducting medium. When considering the model of a uniform half-space, the relative anomalies increase but at the same time the maximum values for z/b and σ_i/σ_e do not change signficantly, particularly when ΔQH and $E(t)$ are measured.

Comparison of the values given in Table 5.XIV shows that as in the case for confined conductors (covered in Chapter 3), the influence of the surrounding medium becomes less respectively, when measurements are made of QH, $\mathrm{In}H$, ΔQH, and $E(t)$.

Over an intermediate range of times, when the secondary field decreases exponentially (under the condition that $t/\tau_0 > 1$) and the normal field corresponds to either the S-zone or the late stage of the transient process in a conducting medium, one can reduce the effect of this surrounding medium significantly by measuring the EMF at two times. In fact, for the total EMF in a two-layer medium, when $\sigma_2 = 0$, we will have

$$E(t) = \frac{M_1(S)}{t^4} + N_1 e^{-q_1 \alpha t} \quad (\text{zone } S)$$

TABLE 5.XIV

QH

z/b \ a/b:	1	2	4	8
2	8×10^3		2.5×10^2	2×10^2
3	6×10^4	10^3	7×10^2	3×10^2
4	2.5×10^5	6×10^3	2×10^3	5×10^2
6	—	2×10^4	10^4	1.5×10^3
8	—	8×10^4	5×10^4	4×10^3
10	—	—	—	10^4
15	—	—	—	—
20	—	—	—	5×10^4

lnH

z/b \ a/b:	1	2	4	8
2	10^3	1.5×10^2	80	70
3	5.5×10^3	7×10^2	2×10^2	10^2
4	1.5×10^4	2×10^3	3.5×10^2	2×10^2
6	9×10^4	9×10^3	1.5×10^3	3.5×10^2
8	2.5×10^5	3×10^4	4×10^3	6×10^2
10	7×10^5	8×10^4	9×10^3	1.5×10^3
15	—	—	6×10^4	9×10^3
20	—	—	—	2×10^4

ΔQH

z/b \ a/b:	1	2	4	8
2	7×10^2	10^2	60	50
3	3.5×10^3	5×10^2	10^2	80
4	10^4	1.5×10^3	2.5×10^2	10^2
6	6×10^4	7×10^3	10^3	2.5×10^2
8	1.8×10^5	2×10^4	3×10^3	4.5×10^2
10	5×10^5	6×10^4	6.5×10^3	10^3
15	—	—	4×10^4	6×10^3
20	—	—	—	1.5×10^4

E(t)

z/b \ a/b:	1	2	4	8
2	6×10^2	10^2	50	45
3	3×10^3	5×10^2	10^2	70
4	10^4	1.5×10^3	2×10^2	10^2
6	5×10^4	6×10^3	9×10^2	2×10^2
8	1.5×10^5	2×10^4	2.5×10^3	3.5×10^2
10	4×10^5	5×10^4	6×10^3	8×10^2
15	—	2.5×10^5	3×10^4	4×10^3
20	—	8×10^5	10^5	10^5

or

$$E(t) = \frac{M_2(\sigma_2)}{t^{5/2}} + N_2 e^{-q_1 \alpha t} \quad \text{(the late stage in a conducting medium)}$$

In both cases, the coefficients M and N do not depend on time. For this reason the differences for the S-zone:

$$E(t_1) - \left(\frac{t_2}{t_1}\right)^4 E(t_2)$$

and

$$E(t_1) - \left(\frac{t_2}{t_1}\right)^{5/2} E(t_2)$$

in a conducting medium depend only weakly on the conductivity of the medium surrounding the confined conductor over a certain range of times.

The analysis of the behavior of electromagnetic fields presented in this section permit us to arrive at the following conclusions.

In the frequency domain with decrease in frequency, the effect of induced currents in the medium surrounding a conductor on the quadrature component of the secondary field tends to zero, while the effect of the surrounding medium on the inphase component of the secondary field does not become zero at low frequencies. It should be noted that the maximum in relative anomalies for the quadrature component occurs at low frequencies where it is independent of frequency, while the maximum in the relative anomaly in the case of the inphase component is observed at intermediate frequencies. If the difference in the quadrature component at two frequencies is considered as the measured quantity in the field, the maximum relative anomaly is also observed at intermediate frequencies, and the magnitude of this anomaly is significantly greater than that of the relative anomaly of the inphase component. Thus the effect of the surrounding medium (geological noise) is less on measurements of the difference in the quadrature component and consequently, the depth of investigation can be made larger.

In the application of time domain measurements for the same model, it has been shown that the secondary field is subject to a strong influence from the surrounding medium during both the early and late stages of the transient process, and the maximum relative anomaly of the transient field will be observed at intermediate times. In many cases, the geoelectric characteristics of the target body can be determined from the behavior of the transient field over this intermediate range in time. The maximum relative anomaly in the transient is even larger than the maximum relative anomaly in measurements of the difference of the quadrature component at two frequencies. Thus, the transient method permits elimination of the effect of the host medium to a greater extent than does the observation of the inphase, the quadrature, or the difference in the quadrature components of the magnetic field in the frequency domain.

5.3. AN APPROXIMATE METHOD FOR CALCULATING THE FIELDS CAUSED BY VORTEX CURRENTS

Determination of the most favorable conditions for the use of induction prospecting methods is directly related to an evaluation of the influence of currents induced in the conducting host medium surrounding an ore body, on time-domain and frequency-domain electromagnetic field behavior. The well recognized difficulties met in the solution of direct problems of electrodynamic theory in non-uniform conducting media force us to resort to approximation theory valid for specific ranges of frequencies or time intervals over which measurements are made in the field using modern exploration technology. Relatively simple mathematical formulas derived by this approximation approach permit us in general to investigate the relationship between fields and geoelectric parameters characterizing the medium and with types of array, as well as to understand better the distribution of currents in a confined conductor and in the surrounding host medium. Therefore, we will describe in detail the physical principles for this approximation theory and derive formulas for components of the time-domain and frequency-domain field behavior as well as define the range of frequencies and time intervals and parameters for the medium for which the results of calculations by approximate and exact expressions will be essentially the same.

Let us assume that a conductor such as a spheroid is placed in a horizontally layered medium, and that the source for a primary field is a current loop with its center on the z-axis as shown in Fig. 5.54. As a consequence of the axial symmetry of this model, there will be no surface charges at the interfaces between the parts of the medium with different resistivities, and the only source for the field will be vortex currents, which in a cylindrical system of coordinates will have only the component j_ϕ.

In developing approximate methods for evaluating quasi-stationary fields, we will proceed from an integral equation representation for the electric field. If the whole of the conducting space including the ore body and the host medium is conceptually represented as being a system of elementary coaxial current tubes, by applying Ohm's law we can obtain the following integral equation for the current density:

$$j_\phi(p) = j_\phi^0(p) + i\omega\mu \int\limits_S G_0(p, q) j_\phi(q) \mathrm{d}S_q \qquad (5.107)$$

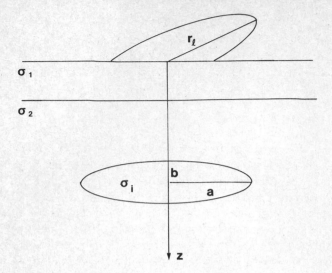

Fig. 5.54.

where $G_0(p, q)$ is a function that depends on the geometric parameters and which describes the interaction between any two coaxial current loops. The integral equation 5.107 has been developed using the same approach as was described earlier in Chapter 3.

In contrast to the integral equation describing current density in a conductor surrounded by an insulator, such as was developed in Chapter 3, in this case, the integration is carried out over all of space, and as a consequence, the actual calculations are not practical, and it does not allow us to apply a method of successive approximations. Therefore, we will make use of the Green's formula and obtain an integral equation for a scalar component of the electric field, E_ϕ, over which the integration will be carried out only over the cross-section of the conductor. In order to simplify the algebraic operations, let us assume that the surrounding medium is a uniform conducting full-space with conductivity σ_e. Later, this assumption will be removed. For the models of the medium and for the arrays that we will consider, the electric field is characterized by a single component:

$$E = E_\phi I_\phi$$

which satisfies the Helmholtz equation:

$$\nabla^2 E + k^2 E = 0 \tag{5.108}$$

where k^2 is the square of the wave number for the host medium. We can write the electric field as a sum:

$$E = E_0 + E_1 \tag{5.109}$$

where E_0 is the field in the uniform surrounding medium, consisting of the field from the current loop in free space (a primary field term) and a field caused by vortex currents induced in this medium for the case in which the conductor would be absent. It is clear that E_0 is a normal field satisfying the following equation over all of space:

$$\nabla^2 E_0 + k_e^2 E_0 = 0 \tag{5.110}$$

where k_e^2 is the square of a wave number in the medium with conductivity σ_e. The second term in eq. 5.109 is the secondary field, E_1, contributed by the presence of the ore body.

In accord with eq. 5.108, the field E_1 will be the solution of the following equation:

$$\nabla^2 E_1 = -k^2 E_1 - k^2 E_0 - \nabla^2 E_0$$

Therefore, within the host medium and in the volume occupied by the local conductive ore body, we will have, respectively:

$$\nabla^2 E_1 = -k_e^2 E_1 \tag{5.111}$$

$$\nabla^2 E_1 = -k_i^2 E_1 + (k_e^2 - k_i^2) E_0 \tag{5.112}$$

where k_i^2 is the square of the wave number in the ore body with a conductivity σ_i.

We will now introduce a function $G = GI_\phi$ which is continuous along with its first derivative everywhere except at the point p, where the field E_1 is being defined. In addition, the function G is independent of the angle ϕ and at the point p it has a logarithmic singularity, and it satisfies the equation:

$$\nabla^2 G + k_e^2 G = 0 \tag{5.113}$$

Taking the axial symmetry into account, we will use a two-dimensional analogy to Green's formula and first place the point p inside the confined conductor. Then, for the internal and external areas, we will have respectively (see Fig. 5.54):

$$\int_{S_i} (G \nabla^2 E_{1\phi} - E_{1\phi} \nabla^2 G) \, dS = \int_L \left(G \frac{\partial E_{1\phi}}{dn_+} - E_{1\phi} \frac{\partial G}{\partial n_+} \right) dl$$

$$+ \int_L \left(G \frac{\partial E_{1\phi}}{\partial n_+} - E_{1\phi} \frac{\partial G}{\partial n_+} \right) dl \tag{5.114}$$

and

$$\int_{S_e} (G \nabla^2 E_{1\phi} - E_{1\phi} \nabla^2 G) \, dS = \int_L \left(G \frac{\partial E_{1\phi}}{\partial n_-} - E_{1\phi} \frac{\partial G}{\partial n_-} \right) dl \tag{5.115}$$

468

inasmuch as:

$$I_\phi \nabla^2 E_{1\phi} = \nabla^2 E_1 - E_{1\phi} \nabla^2 I_\phi, \quad I_\phi \nabla^2 G = \nabla^2 G - G \nabla^2 I_\phi$$

In accord with eqs. 5.111 through 5.113, we must have the following in the host medium:

$$G \nabla^2 E_{1\phi} - E_{1\phi} \nabla^2 G = 0$$

and within the spheroid:

$$G \nabla^2 E_{1\phi} - E_{1\phi} \nabla^2 G = (k_e^2 - k_i^2) E_{0\phi} G + (k_e^2 - k_i^2) E_{1\phi} G$$

Near the point p, the field $E_{1\phi}$ will be finite, while the function G increases without limit as $\ln R$, where R is the radius of the circumference L_0. For this reason, the value of the integral along the contour L_0 tends towards the value $-2\pi E_{1\phi}(p)$ as R approaches 0.

After combining expressions 5.114 and 5.115 as a consequence of the continuity of tangential components of the field, the integrals along the contour L will disappear and we obtain an integral equation which contains only a surface integral over a limited area corresponding to the cross-section of the confined conductor:

$$E_{1\phi}(p) = \frac{k_i^2 - k_e^2}{2\pi} \int\limits_{S_i} E_{0\phi}(q) G(k_e, p, q) \mathrm{d}S_q$$

$$+ \frac{k_i^2 - k_e^2}{2\pi} \int\limits_{S_i} E_{1\phi}(q) G(k_e, p, q) \mathrm{d}S_q \tag{5.116}$$

When the point p is located in the host medium, eq. 5.116 is a relationship that permits us to define the electric field in the external medium when the field inside the conductor is known. The function $G(k_e, p, q)$ describes the electric field of the current ring in a uniform conducting medium with conductivity σ_e with an uncertainty of a constant multiplier.

If the host medium is not uniform but is characterized by axial symmetry in the same respect as the source, as would be the case for a horizontally layered medium, by choosing a Green's function which is proportional to the electric field in a current ring in a horizontally layered medium, we can again obtain the integral equation in 5.116. Thus, the derivation carried out above for uniform host medium is still valid when the medium is a horizontally layered medium as Green's function is chosen properly. As can readily be shown, in a horizontally layered medium, the function G can be expressed in terms of an indefinite integral.

Let us write the integral equation with respect to the current density and the expression for determining the field outside the spheroid in the forms:

$$j_{1\phi}(p_i) = \frac{k_i^2 - k_e^2}{2\pi} \int\limits_{S_i} j_{0\phi}(q) G(k_e, p_i, q) dS$$

$$+ \frac{k_i^2 - k_e^2}{2\pi} \int\limits_{S_i} j_{1\phi}(q) G(k_e, p_i, q) dS \tag{5.117}$$

and

$$E_{1\phi}(p_e) = \frac{i\omega\mu}{2\pi} \int\limits_{S_i} j_{0\phi}^* G dS + \frac{i\omega\mu}{2\pi} \int\limits_{S_i} j_{1\phi}^* G dS \tag{5.118}$$

where $j_{0\phi} = \sigma_i E_{0\phi}$ is the current density in the spheroid caused by the normal field, $E_{0\phi}$, $j_{1\phi} = \sigma_i E_{1\phi}$ is the current density in the spheroid caused by the secondary field $E_{1\phi}$, and

$$j_{0\phi}^* = (\sigma_i - \sigma_e) E_{0\phi}, \quad j_{1\phi}^* = (\sigma_i - \sigma_e) E_{1\phi} \tag{5.119}$$

With the help of eq. 5.117 the actual magnitude of the current density inside the spheroid:

$$j_\phi = \sigma_i (E_{0\phi} + E_{1\phi})$$

is defined, but the secondary currents in the host medium remain unknown. Therefore, by using the Biot-Savart law for the currents $j_\phi(p_i)$, we can find only a portion of the anomalous magnetic field, H, which must be supplemented by the magnetic field contributed by secondary currents in the host medium. The solution of this problem is accomplished by the application of eq. 5.118, where the values $j_{0\phi}^*$ and $j_{1\phi}^*$ play the role of the special currents for the secondary field. In fact, in accord with eq. 5.118, the anomalous field in the host medium is the same as if it were in the part of the space where the spheroid is situated and the current rings with intensity $j_\phi^* = j_{0\phi}^* + j_{1\phi}^*$ are placed. Because of the change of the magnetic field of these currents with time, the secondary currents will appear in the surrounding medium and their influence is taken into account by the proper choice of Green's function. Let us note that the current density j_ϕ^* differs from the actual current density by the amount:

$$\sigma_e (E_{0\phi} + E_{1\phi})$$

It should be emphasized here that the integral equation in 5.116 also describes the field for two-dimensional problems (with E-polarization) and Green's function is the electric field from linear current filaments oriented parallel to the primary source. Finally, eq. 5.116 proves to be a convenient relationship for developing approximate methods for calculating fields that can be applied in many aspects of induction prospecting, including borehole measurements.

First of all, let us examine the case for which the skin depth inside the spheroid and in this host medium is significantly greater than the linear dimension of this spheroid. Then, writing the expressions for the field $E_{1\phi}$ and Green's function as series expansions

$$E_{0\phi} = \sum_{n=2}^{\infty} a_n k_e^n, \quad E_{1\phi} = \sum_{n=4} b_n k_e^n, \quad G = G_0 + \sum_{n=2}^{\infty} c_n k_e^n \qquad (5.120)$$

and substituting these into eq. 5.117, we find a relationship between the unknown coefficients b_n defining the low-frequency part of the spectrum for the field $E_{1\phi}$. In particular, by neglecting the influence of the secondary field within the spheroid in comparison with the normal field (that is, making the assumption that $E_{0\phi}$ is much greater than $E_{1\phi}$ for points in the host medium, we have:

$$E_{1\phi}(p) = \frac{k_i^2 - k_e^2}{2\pi} \int_{S_i} E_{0\phi} G \, dS \qquad (5.121)$$

and for the total field:

$$E_{\phi} = E_{0\phi} + E_{1\phi} = E_{0\phi} + \frac{k_i^2 - k_e^2}{2\pi} \int_{S_i} E_{0\phi} G(k_e, p, q) \, dS \qquad (5.122)$$

It is readily seen that this approximation will be valid if we can neglect the interaction between the currents in the confined conductor and do not consider the effect of the magnetic field caused by anomalous currents in the surrounding medium on the strength of the currents flowing within the spheroid. According to eq. 5.121, the determination of the secondary field merely consists of calculating the surface integral over the cross-section of the spheroid in the case in which the integrand is the product of the normal field $E_{0\phi}$ and Green's function. As is well known, in a horizontally layered medium and in the medium with coaxial cylindrical interfaces, both of these functions are expressed in terms of indefnite integrals. This approximate method of calculation has proved to be highly effective in problems of induction well-logging for media with cylindrical interfaces. A comparison with results of calculations by exact formulations has shown that for the most interesting range of frequencies, for which the minimum skin depth is greater than the radius of the intermediate zone, an even simpler representation for the field is valid, namely:

$$E_{\phi} = E_{0\phi} + \frac{k_i^2 - k_e^2}{2\pi} \int_{S_i} E_{0\phi}^{(0)} G_0(p, q) \, dS \qquad (5.123)$$

where $E_{0\phi}^{(0)}$ is the electric field of the source in free space (the primary field), $G_0(p, q)$ is a function which is proportional to the coefficient of mutual induction between elementary current rings. In this approximation, the anomalous magnetic field is directly proportional to k^2, and therefore, the quadrature component only, QH_1, is defined. If the observation point is situated at a distance L from the source of the primary field that is significantly less than the skin depth in the surrounding medium, the first term in eq. 5.123 can also be replaced by $E_{0\phi}^{(0)}$. By so doing, we obtain expressions valid for small values of the parameter $|k_e L|$, in which the magnetic field is directly proportional to the frequency, conductivity and geometry of those parts of the medium which have a constant resistivity.

For induction logging problems in which measurements are made in a borehole, the source of the electromagnetic field is situated directly within the borehole; that is, in the range over which the integration described by eq. 5.122 must be performed. Therefore, over a wide range of frequencies the current density in the borehole and in the intermediate zone surrounding the borehole is controlled principally by a primary field $E_{0\phi}^{(0)}$ and one can introduce geometric factors for these parts of the medium. The range of application of the approximate method for calculating the fields using geometric factors is not limited to problems in induction logging where the radial characteristics for probes carrying many coils are investigated but can be used as well in interpretation in shallow surface-based induction prospecting applications. The approximate method of calculations based on eq. 5.122 has been shown to be useful in studying electromagnetic fields in the magnetotelluric sounding method and in minerals prospecting where the mapping approach is used. For example, the theoretical basis for magnetotelluric soundings in media containing small structures and with the field polarized in the E-direction can be readily derived starting from eq. 5.122, with the result that the secondary field is expressed in terms of a single integration. This approach is also useful for investigating the effect of moderately conductive cylindrical inhomogeneities in the interpretation of data obtained with such methods as TURAM, VLF and AFMAG. It is also useful in support of magneto-variational profiling over deep structures of higher conductivity, for analysis of the effect of surface inhomogeneities which comprise significant noise in the interpretation of deep magnetotelluric soundings.

A comparison of calculations done using the integral equation and using the approximate formulation in eq. 5.122 shows that this latter method results in only relatively small errors when:

$$\frac{a}{h_i} < 0.25 \quad \text{as} \quad \frac{\sigma_i}{\sigma_e} > 1, \quad \frac{a}{h_e} \leqslant 0.50 \quad \text{as} \quad \frac{\sigma_i}{\sigma_e} < 1 \qquad (5.124)$$

where $2a$ is the maximum linear dimension of a cross-section of a cylindrical conductor, and h is the skin depth. However, in solving problems in mining prospecting when the target is a highly conducting ore body, this method of calculation is not of any practical interest since the condition on a/h_i is valid only at relatively low frequencies. It is obvious that in order to construct an approximate theory permitting us to obtain information about electromagnetic fields over a much wider range of frequencies, in contrast to the asymptotic methods considered earlier, it will be necessary to take into account the skin effect in the confined conductor.

In order to study this problem, it will be convenient to make use of the fact that in practice in inductive minerals prospecting, for the case of conductive bodies which are not particularly elongate, usually two conditions are valid simultaneously, these being:

(1) the conductivity of the ore body is significantly greater than the conductivity of the surrounding medium; and

(2) the skin depth in the surrounding medium is significantly greater than the linear dimension of the cross-section of the conductive body.

Thus, we can assume that

$$\frac{\sigma_i}{\sigma_e} \gg 1 \quad \text{and} \quad |k_e a| \ll 1 \tag{5.125}$$

but the value for $|k_i a|$ might be arbitrary.

Now, letting the parameter $|k_e a|$ be small and making use of the method of successive approximations to the integral equation for current density given in eq. 5.117, but limiting ourselves to the first term, we obtain:

$$j_{1\phi}(p) = \frac{k_i^2}{2\pi} \int_{S_i} j_{0\phi} G_0(p, q) \mathrm{d}S + \frac{k_i^2}{2\pi} \int_{S_i} j_{1\phi} G_0(p, q) \mathrm{d}S \tag{5.126}$$

This is the integral equation for the current density in a conductor surrounded by an insulator. Thus, if the conditions expressed in 5.125 are met, the current density in the spheroid to a first approximation will not depend on the conductivity of the surrounding medium, which for this particular case can be assumed to be zero. In other words, the current density in an ore body surrounded either by a conducting medium or by an insulator will be the same if in both cases the normal fields $E_{0\phi}$ at each point in the spheroid are the same. This means that the secondary currents in the surrounding medium do not have any effect on the interaction of currents in the spheroid, where the skin effect manifests itself in exactly the same manner as if the conductor were to be implaced in free space. In accord with eq. 5.118, we have the following expression for the anomalous electric field outside the spheroid:

$$E_{1\phi}(p_e) = \frac{i\omega\mu}{2\pi} \int_{S_i} j_\phi^{**}(q) G(k_e, p_e, q) \mathrm{d}S \tag{5.127}$$

where

$$j_\phi^{**} = \sigma_i(E_{0\phi} + E_{1\phi})$$

or, that is, $E_{1\phi}(p_e)$ is the electric field elementary current rings with a current density j_ϕ^{**} when the resistivity of the spheroid and the medium immediately outside the conductor are the same.

Thus, determination of the anomalous field consists of two steps, namely:

(1) making use of the integral eq. 5.126, the current density, j_ϕ^{**} in the conductor as located in free space is defined; and (2) by making use of eq. 5.127 the field caused by these currents is defined for observation points in the surrounding medium.

Because an increase in frequency leads to the secondary currents in the surrounding medium having an effect on the current strength in the conductor, the high-frequency portion of the spectrum cannot be determined using this method of calculation. Therefore, in determining the current density j_ϕ^{**}, it is convenient to make use of the fact that over a relatively wide range of frequencies, the intensity of the induced currents depends mainly on the normal electric field and on the distribution of the first poles in the spectrum which are characterized by the values $q_i(q_1 < q_2 < q_3$, etc.). In particular, as has been shown in Chapter 3 in order to describe the currents using the parameter q_1, it is sufficient to have a representation of the low-frequency part of spectrum in the form of a series of integer powers of $k_i^2 a^2$. This can be accomplished by solving the integral equation using the method of successive approximations which simplifies the calculations of the field.

Let us note that the condition $|k_e a| < 1$ indicates the absence of a phase shift in the normal field within the area occupied by the spheroid.

In order to illustrate the use of this method for calculating the field, we must make some assumptions which have no fundamental importance, but they will allow us to obtain relatively simpler expressions. These assumptions are:

(1) The source of the primary field is a current loop.

(2) The host medium is uniform.

(3) The conductor consists of a spheroid, coaxially situated with the source of the primary field.

(4) The normal magnetic field is uniform within the area occupied by the spheroid.

(5) The observation point at which the magnetic field is measured is situated on the z-axis as shown in Fig. 5.54.

As has already been shown in our consideration of the electromagnetic field for a conducting spheroid embedded in an insulating medium and illuminated by a uniform magnetic field H_{0z}, the equation for induced current density can be written as (see eq. 3.164):

$$j_\phi(p) = k_i^2 b^2 H_{0z} \sum_{n=1}^{\infty} \frac{\beta_n(p)}{q_n - k_i^2 b^2} \tag{5.128}$$

For a sufficiently wide range of frequencies that includes the low frequency part of the spectrum, the first term in this series is the important one. For this reason, considering the physical principles of the methods, we can assume that the current density at any point in the spheroid is given by:

$$j_\phi^{**}(p_i) = k_i^2 b^2 H_{0z} \frac{\beta_1(p_i)}{q_1 - k_i^2 b^2} \qquad (5.129)$$

where b is the length of the vertical semi-axis of the spheroid, q_1 is the value characterizing the position of the first pole in the spectrum and which is a function of the ratio of axial lengths in the spheroid, and $\beta_1(p_i)$ is a function defined by the behavior of the spectrum near the first pole and which depends on the coordinates of the point p_i within this spheroid.

In accord with eq. 5.127, the anomalous magnetic field along the z-axis will be:

$$H_{1z}(p_e) = \frac{k_i^2 b^2}{q_1 - k_i^2 b^2} H_{0z} \int\limits_{S_i} \beta_1(p_i) G_{II}^z(k_e, p_i, p_e) dS \qquad (5.130)$$

where H_{0z} is the normal magnetic field from the current loop within the spheroid and G_{II}^z is the vertical component of the magnetic field on the z-axis contributed by a circular current lying in a horizontal plane, with unit intensity.

As has been demonstrated in the first paragraph of this chapter, in a uniform medium, these functions can be expressed in turn through the elementary functions as follows:

$$H_{0z} = \frac{I r_1^2}{2 r_1^3} e^{ik_e r_1}(1 - ik_e r_1)$$

$$G_{II}^z = \frac{r_p^2}{2 r^3} e^{ik_e r_1}(1 - ik_e r_1) \qquad (5.131)$$

where r_p is the radius of the current ring passing through the point p_i, r is the distance between the points p_e and p_i, r_1 is the radius of the source for the primary field, and r_1 is the distance from the loop to the center of the spheroid, as shown in Fig. 5.54.

An examination of numerical results obtained using the integral equation method has shown that the error in determining the field using eq. 5.130 will not be greater than that involved with eq. 5.116 when the following conditions apply:

$$\frac{b}{h_i} < \frac{q_1^{1/2}}{2} \quad \text{or} \quad \frac{fb^2}{\rho_i} < 0.6 q_1 \times 10^5 \qquad (5.132)$$

and

$$\frac{a}{h_e} < 0.4 \quad \text{or} \quad \frac{fa^2}{\rho_e} < 0.4 \times 10^5 \quad \text{as } a/b > 1 \qquad (5.133)$$

where a and b are the semi-axial lengths of the spheroid, and ρ_e and ρ_i are the resistivities of the host medium and the ore body respectively.

In deriving the expression for the field, it will be assumed that the frequency response for currents induced in the spheroid will be determined almost entirely by the position of the least pole, that is by the parameter q_1, and the phase of the normal magnetic field within the area occupied by the constant will be constant. It is obvious that with an increase in frequency, first of all the role of higher-order poles will increase, and secondly, the phase of the normal field within the spheroid will begin to change. As follows from the analysis of the normal field and the spectrum of the currents flowing in a spheroid surrounded by an insulating host medium, constancy of the phase of the normal field within the spheroid will be observed for most reasonable models of the medium (that is for $\sigma_i/\sigma_e \gg 1$) for higher frequencies than will be the equivalence of the fraction:

$$\frac{k_i^2 b^2}{q_1 - k_i^2 b^2} \beta_1$$

to the spectrum of currents flowing in the conductor. For this reason, a determination of the values q_n and β_n characterizing the higher-order poles of the spectrum will permit us to significantly extend the range of application of this approximate method for cases in which the host medium has a relatively high resistivity. In particular, if one makes use of only the first two terms of the sum in eq. 5.128 instead of eq. 5.130, we obtain the following expression for the anomalous magnetic field

$$H_{1z}(p_e) = k_i^2 b^2 H_{0z} \left\{ \frac{1}{q_1 - k_i^2 b^2} \int\limits_{S_i} \beta_1(p_i) G_H^z(k_e, p_i, p_e) \mathrm{d}S \right.$$

$$\left. + \frac{1}{q_2 - k_i^2 b^2} \int\limits_{S_i} \beta_2(p_i) G_H^z(k_e, p_1, p_e) \mathrm{d}S \right\} \tag{5.134}$$

We should note that in view of the exponential decay of currents in a confined conductor embedded in an insulating medium, the transient response is to a large extent determined by the position of the first pole in the spectrum, and therefore, the approximate expressions for the time domain field (eq. 5.96) will have a wider range of applicability than those describing the frequency domain field. Therefore, the complicated problem of determining the electromagnetic field in the presence of a spheroid surrounded by a conducting medium, for example, is reduced to two essentially simpler problems for a relatively wide range of frequencies, these problems being:

(1) the determination of the numbers q_1 and $\beta_1(p_i)$ using the method of successive approximations; and (2) the integration over a cross-section of the conductor.

If the spheroid is located in a horizontally layered medium, the functions

H_{0z} as well as G_H^z or G_H^r are themselves the magnetic field of current rings in this medium and they are expressed in terms of indefinite integrals, and finally, if the normal magnetic field is not uniform within the area occupied by the conductor, the expression for the current density j_ϕ^{**} and the secondary magnetic field will have the form:

$$j_\phi^{**}(p_i) = \frac{k_i^2 b^2 \beta_1^*(p_i)}{q_1 - k_i^2 b^2} H_{0z}(p_i)$$

$$H_{1z}(p_e) = \frac{k_i^2 b^2}{q_1^2 - k_i^2 b^2} \int_{S_i} \beta_1^*(p_i) H_{0z}(p_i) G_H^z(k_e, p_e, p_i) dS$$

(5.135)

Returning to the case in which the normal field is uniform in the vicinity of the spheroid, we should note that the expression for the secondary field given by eq. 5.130 is considerably simplified when the confined conductor is a sphere. As can readily be shown, the current field induced in this sphere is equivalent to that from a magnetic dipole source with the moment (see Chapter 3):

$$M = 2\pi D a^3 H_0$$

(5.136)

where

$$D = \frac{3}{x} \coth x - \frac{3}{x^2} - 1, \quad x = ik_i a$$

Therefore, in place of eq. 5.130, we have

$$H_{1z} = \frac{M}{2\pi z^3} e^{ik_e z} (1 - ik_e z) \quad \text{for } \frac{a}{h_e} < 0.25$$

(5.137)

With a similar generalization of eq. 5.137, it can be used to describe the secondary field of currents in a sphere located in a horizontally layered medium and in this case, there is only one difficulty, this being related to the calculation of single indefinite integrals defining the field of the current loop and a magnetic dipole in this medium.

At low frequencies and when the spheroid is located in a conducting space, the normal magnetic field and the secondary field can be represented by a power series in k_e in accord with eqs. 5.10 and 5.75:

$$H_{0z} \approx a_0 + a_1 k_e^2 + a_2 k_e^3 + \ldots$$

$$H_{1z} \approx b_1 k_e^2 + b_2 k_e^4 + b_3 k_e^5 + b_4 k_e^6 + \ldots$$

(5.138)

where b_n are a set of coefficients that depend on the conductivity of the spheroid and the geoelectric parameters describing the surrounding medium as well as on the type of primary source.

As may be seen from an analysis of successive approximations in the solution of eq. 5.116, the assumption about the absence of any effect from the surrounding medium on the interaction of currents within the spheroid (that is, $G = G_0$) and making use of only the normal field H_{0z}, the coefficients in the series 5.138 will change beginning with b_4. It is quite important to observe that the third term in this series, $b_3 k_e^5$ does not depend on the current density within the spheroid, that is, on its conductivity, but that it is defined by the electric field $E_{0\phi}$ and as is well known (see eq. 5.5) it characterizes the asymptotic behavior in the difference of quadrature components, ΔQH, in the dual-frequency method, and controls the late stage of the transient field behavior.

At this point, we will consider the transient response, again assuming that the magnetic field from secondary currents in the surrounding medium does not have any effect on the interaction of currents within the confined conductor, which in our case, is a spheroid. With this approximation, the development of the secondary field can be considered to occur as follows. When the current in the source changes instantaneously, a normal electromagnetic field, H_0 and E_0, arises instantly (provided displacement currents are neglected) at each point in the medium, and the character of the time rate of change will depend on the distance from the loop and on the conductivity of the medium. In particular, for relatively high resistivities in the host medium, and with the observation point being relatively close to the source, the magnetic field H_0 will decrease relatively more rapidly, while a maximum will manifest itself at earlier times in the electric field.

Under the action of the electric field $E_{0\phi}$, induced currents will develop in the spheroid with an intensity as if the conductor were located in free space. It is clear that the magnetic field H_0 at any instant t can be written as the sum of elementary step functions of various amplitudes, situated within the interval 0 to t (representation by the Duhamel integral). As was previously demonstrated in Chapter 3, for an instantaneous change in the primary field obeying the relationship:

$$H_0(t) = \begin{cases} H_0 & t \leqslant T \\ 0 & t > T \end{cases}$$

currents flowing in a conductor embedded in an insulating host will be described by a sum of exponential terms and at sufficiently large times, a single one of these exponentials will play the main role, this being the exponential term defined by the minimum pole in the spectrum for the secondary field (that is, by q_1). Of course, this behavior also holds for an arbitrary type of field excitation when the current in the source vanishes.

In the approximate method for calculating transient fields, it is assumed that the expression for current density in the conductor will contain only this single exponential term. This character in the behavior of the current is reasonable when the influence of the normal electric field on the current intensity at the measurement instant is reasonably small.

The anomalous magnetic field from the vortex current flowing in the spheroid $j_\phi^{**}(p_i)$ is defined as being the time domain field of the current rings located in the surrounding medium. Inasmuch as at relative early times, the time rate of change of the magnetic field is reasonably large, and therefore the secondary currents in the host medium have a strong influence on the interaction between the currents within the spheroid, the approximate method being proposed is not applicable for the early stage. During the late stage of transient field behavior, the normal electric field, $E_{0\phi}$, decreases as $t^{-5/2}$ (for a conducting half space) and because of the lower rate of change of the field with time, we can neglect the interaction between these currents within the conductor. Therefore, during the very late stage of transient behavior, their density will change with time as $t^{-5/2}$. Therefore, the anomalous magnetic field will depend in time in exactly the same manner as does the normal electric field, $E_{0\phi}$, and moreover, it will be significantly less than the normal magnetic field:

$$H_0(t) \sim \frac{1}{t^{3/2}}, \quad H_1 \sim \frac{1}{t^{5/2}} \tag{5.139}$$

It is clear that this conclusion will hold for any layered medium, and for an arbitrary conductor, provided that the only source of the field is a vortex current. However, there is one case that should be mentioned specifically. If the underlying medium is highly resistive, there is a range of times corresponding to the "S-zone". In the case, in accord with eq. 5.26, we will have the following behavior for the normal and secondary fields, respectively;

$$H_0(t) \sim \frac{1}{t^3}, \quad H_1(t) \sim \frac{1}{t^4} \tag{5.140}$$

That is, the normal field will prevail during the late stage.

The approximate method of calculation of time-domain fields is valid for times at which one can neglect the effect of secondary currents in the host medium on the interaction between induced currents in the confined conductor, and we can assume that these currents essentially decay exponentially. At the same time, the minimum pole in the spectrum characterizes the power in this exponential term.

It should be clear that any additional information concerning higher-order poles in the spectrum will permit us to improve the accuracy of calculations and to enlarge to some extent the range of applicability to earlier times.

In determining the time-domain field, we can use either the Duhamel or the Fourier integral approach. Making use of the Fourier transform applied to eq. 5.130, we have:

$$H_{1z}(t) = \frac{1}{2\pi} \int_{S_i} dS \int_{-\infty}^{\infty} \frac{j_\phi^{**}(p_i)}{i\omega} G_{II}^z e^{-i\omega t} d\omega \qquad (5.141)$$

Similar equations can be written for the other components of the magnetic field. Using eqs. 5.130 and 5.131, and accepting the same limitations on the model of the medium and on the normal field, which applied for our examination of the frequency domain case, we obtain the following expressions for the vertical component of the secondary field:

$$H_{1z}(t) = \frac{Ir_1^2}{4r_1^3} \int_{S_i} \frac{\beta_1 r_p^2}{r^3} \left[\frac{1}{2\pi} \int_{-\infty}^{\infty} \frac{e^{ik_e R} e^{-i\omega t} (1 - ik_e R - k_e^2 r r_1)}{a_1 - i\omega} d\omega \right] dS \qquad (5.142)$$

where

$$R = r + r_1, \quad a_1 = q_1/\sigma_i \mu b^2$$

or

$$H_{1z}(t) = \frac{Ir_1^2}{4r_1^3} \int_{S_i} \frac{\beta_1 r_p^2}{r^3} \left[V - R \frac{\partial V}{\partial R} + r r_1 \frac{\partial^2 V}{\partial R^2} \right] dS \qquad (5.143)$$

where

$$V = \frac{1}{2\pi} \int_{-\infty}^{\infty} \frac{e^{-(i\omega\alpha)^{1/2}} e^{-i\omega t}}{a_1 - i\omega} d\omega = e^{-a_1 t} \frac{1}{2\pi} \int_{-\infty}^{\infty} \frac{e^{-(i\omega\alpha)^{1/2}} e^{(a_1 - i\omega)t}}{a_1 - i\omega} d\omega$$

$$= e^{-a_1 t} N \qquad (5.144)$$

and where $\alpha = \sigma_e \mu R^2$.

The derivative of the integral N with respect to time is:

$$\frac{\partial N}{\partial t} = \frac{1}{2\pi} \int_{-\infty}^{\infty} e^{-(i\omega\alpha)^{1/2}} e^{(a_1 - i\omega)t} d\omega = e^{a_1 t} \frac{1}{2\pi} \int_{-\infty}^{\infty} e^{-(i\omega\alpha)^{1/2}} e^{-i\omega t} d\omega$$

$$= -\frac{1}{2} \left(\frac{\alpha}{\pi} \right)^{1/2} \frac{e^{a_1 t - \alpha/4t}}{t^{3/2}}$$

Whence

$$N = -\frac{\alpha^{1/2}}{2\pi^{1/2}} \int \frac{e^{a_1 t - \alpha/4t}}{t^{3/2}} dt$$

$$= \tfrac{1}{2} [e^{-i(a_1\alpha)^{1/2}} \{\Phi(z_-) - 1\} + e^{i(a_1\alpha)^{1/2}} \{\Phi(z_+) - 1\}] + C \qquad (5.145)$$

where

$$z_+ = u + iv, \quad z_- = u - iv, \quad u = \frac{R}{2}(\sigma_e \mu/t)^{1/2}$$

$$v = (q_1 t/\sigma_i \mu b^2)^{1/2}, \quad \Phi(x) = \int_0^x e^{-t^2} dt$$

and where $\Phi(x)$ is the probability integral. On integration, the unknown constant C is assumed to be zero in eq. 5.145, which can be demonstrated from the physical behavior of the field as t approaches infinity. Thus:

$$V = \frac{e^{-a_1 t}}{2} \left[e^{-i(a_1 \alpha)^{1/2}} \{ \Phi(z_-) - 1 \} + e^{i(a_1 \alpha)^{1/2}} \{ \Phi(z_+) - 1 \} \right] \tag{5.146}$$

and the secondary magnetic field of currents flowing in the spheroid is expressed in terms of probability integrals:

$$H_{1z}(t) = \frac{I r_1^2}{4 r_1^3} \int_{S_i} \frac{\beta_1(p_i) r_p^2}{r^3} W dS \tag{5.147}$$

where

$$W = V - R \frac{\partial V}{\partial R} + r r_1 \frac{\partial^2 V}{\partial R^2}$$

In the particular case of a sphere, integration over the cross-section of eq. 5.147 can be accomplished in explicit form, such that:

$$H_{1z}(t) = \frac{3 I r_1^2}{\pi^2 r_1^3} \left(\frac{a}{r} \right)^3 W_{sph} \quad \text{if } \frac{\tau_i}{b} > 3.5, \quad \frac{\sigma_i}{\sigma_e} > 30 \tag{5.148}$$

where $W_{sph} = V_{sph} - z \partial V_{sph}/\partial z$ and V_{sph} is the same as that with 5.146, as $R = z$, and:

$$\tau_i = 2\pi(2t/\sigma_i \mu)^{1/2}$$

Table 5.XV lists minimum values for the parameter τ_i/b marking the lower value for which eq. 5.147 is valid.

TABLE 5.XV

a/b	1	2	4	8
τ_i/b	3.5	4.0	8	16

Considering the transient total EMF related to the normal field, we can see that the approximate method of calculation permits us to obtain a reasonably accurate representation of the field over the range of time which is most important from the practical point of view.

Expanding the function W in a power series for small values of the parameter u (this means large times, relatively small radius of the loop, or relatively short distance from the center of the spheroid to the source, or reasonably high resistivity near the host medium), we have

$$W = e^{-t/\tau_0}\left[-1 - \frac{\sigma_e}{\sigma_i}\,q_1\,\frac{r^2+r_1^2}{2b^2} + \left(\frac{\sigma_e}{\sigma_i}\,q_1\right)^2 R^2\,\frac{(r-r_1)^2}{8b^4}\right]$$

$$+ \frac{2}{\sqrt{\pi}}\,\frac{u^3}{v^2}\,\frac{3rr_1 - R^2}{R^2} \qquad \text{if } \tau_i/b > 4, \quad u \ll 1 \tag{5.149}$$

where τ_0 is a time constant given by:

$$\sigma_i\mu b^2/q_1$$

The first term in this expression describes a field which decays in the same manner as the field from currents induced in a conductor surrounded by an insulator during the late stage. If the ratio σ_e/σ_i is relatively small, these fields are practically the same.

The second term in eq. 5.149 characterizes the principal part of the secondary field during the late stage when one can neglect the interaction of currents within the spheroid. In this case the current density is:

$$j_\phi^{**} = \sigma_i E_{0\phi}$$

and the secondary magnetic field decreases as $t^{-5/2}$. When the conductive body is spherical, the late stage of the transient behavior begins when:

$$\frac{\tau_i}{b} \approx 10 \quad \left(\frac{r_1}{b} < 8, \quad \frac{\sigma_i}{\sigma_e} > 30\right)$$

Returning to the expressions for the transient field, we should note that when the second pole in the spectrum is known, we will have as an example for the vertical component of the magnetic field:

$$H_{1z}(t) = \frac{Ir_1^2}{4r_1^3}\int_{s_i}\frac{r_B^2}{r^3}\,[\beta_1(p_i)W_1 + \beta_2(p_i)W_2]\,dS \tag{5.150}$$

where

$$W_1 = W(q_1), \quad W_2 = W(q_2)$$

Consideration of the transient response for a confined conductor situated in the horizontally layered medium shows that determination of the field at relatively late times can be accomplished using numerical integration in eq. 5.141.

In conclusion it should be noted that some of the ideas presented here can be applied for approximate calculations of time domain and frequency domain fields even when the model is not characterized by axial symmetry and in which case electrical charges will appear on the surface of the conductor.

Chapter 6

THE EFFECT OF SURFACE ELECTRICAL CHARGE ON THE BEHAVIOR OF SECONDARY ELECTROMAGNETIC FIELDS

INTRODUCTION

In the general case, the sources for a quasi-stationary electromagnetic field consist both of currents and of electrical charges which develop principally on interfaces between portions of the medium characterized by different resistivity values. In many of the cases which have been treated theoretically in induction mining prospecting, axial symmetry is exhibited, and the primary electrical field does not intersect surfaces between media characterized by different values of resistivity. As a consequence, in such special cases, the only source for a field will be currents that are closed on themselves either within the confirmed conductor or in the surrounding medium. A change with time of the strength of the magnetic field associated with these currents results in the appearance of a vortex, or inductive electric field, E^v in accord with the law of electromagnetic induction. The most simple example of such an electric field with a source consisting only of vortex currents is the field about a magnetic dipole sited in a uniform conducting medium.

It is obvious that the assumption of axial symmetry of the field and of the medium which is often used is a very strong limitation of a real earth. However, it provides favorable conditions for the application of inductive methods to classify anomalies with respect to conductivity. In this relation, the analysis of the effect of charges is of very practical interest. It is important in providing a better understanding of the behavior of electromagnetic fields used in induction prospecting as well as for developing more appropriate interpretations of field data. The present examination of this effect will assist us in appreciating the role played by field excitation, by the particular zone in which measurements are carried out, and also the influence of the part of the field which is caused by the existence of electrical charges when measuring either the electric or magnetic fields. Moreover, it will be demonstrated that this effect manifests itself in different ways when measuring the quadrature and inphase components of the field or the time-domain response. In our examination of this problem, we will make use of previously developed solutions for the field in the presence of a conducting sphere or a circular cylinder situated in an otherwise uniform medium. In spite of the fact that these models obviously cannot be considered as being particularly general, the analyses of the effects on the electromagnetic field will permit us to arrive at many useful conclusions which are almost universally valid. Although we fully comprehend that there are limitations to such an analysis in view of the simplicity of the models which we will use, we must still note that this approach will provide a very fruitful supplement to the exact solutions which might be developed for much more complicated and at the same time realistic earth models.

The development of electrical charges is usually a consequence of the fact that a primary electric field intersects surfaces between two portions of the medium that have different conductivities. Usually for our considerations in mining exploration, such a surface would separate an inhomogeneity such as a conductive ore body from the surrounding or host medium. In view of the development of charges, there will be a current distribution arising under the action of the Coulomb (E^{g}) and the inductive (E^{v}) electric fields, and their combined behavior will usually be very complicated. However, in a few cases, the secondary field can be represented as being the sum of independent parts which are often called oscillations of the magnetic and electric types. The sources of the oscillations of the magnetic type consist only of induction currents which do not intersect any of the surfaces between various parts in the medium (only the inductive part of the field). Oscillations of the electric type arise as the consequence of the presence of charges and of currents which are closed through a conductor and the surrounding medium (the inductive-galvanic portion). It might be appropriate to note at this time that inasmuch as the skin depth in the host medium is usually much greater than the distance from the observation point to the conductor, oscillations of electric type almost always coincide with the direct current field. Based on experience with electrical methods, particularly the application of direct current methods, it follows that the presence of an electric charge can lead to a significant decrease in the resolution obtainable with inductive methods. In view of this, an examination of the relationship between electromagnetic fields caused by currents and charges, and the resistivity of a conductor is of considerable practical interest.

First, we will derive expressions for the behavior of an electromagnetic field in the presence of a sphere considering various types of excitation for the primary field. Then, the secondary fields caused by a plane wave and by a magnetic dipole will be analyzed. Following this, the secondary fields due to the presence of a cylindrical conductor will be investigated considering various types of excitation in both the frequency and time domains.

6.1. DIFFRACTION OF ELECTROMAGNETIC FIELDS ON A CONDUCTING SPHERE

In this section, we will derive equations for the secondary field, caused by the presence of a conducting sphere for various types of sources which will be used in following sections of this book. Regardless of the type of field excitation, the main approach used for the solution of a boundary value problem is based on the fact that the field can be written as the sum of two terms, each independent of the other and with each term being described in using a single scalar potential function.

Potentials of the electric and magnetic types

Maxwell's equations in the frequency domain take the form:

$$\text{curl } E = -i\omega\mu H \qquad \text{div } E = 0$$

$$\text{curl } H = (\sigma + i\omega\epsilon)E \qquad \text{div } H = 0 \tag{6.1}$$

or

$$\text{curl curl } E = \text{grad div } E - \nabla^2 E = -i\omega\mu \text{ curl } H = -i\omega\mu(\sigma + i\omega\epsilon)E$$

Inasmuch as div $E = 0$, we have

$$\nabla^2 E - i\omega\mu(\sigma + i\omega\epsilon)E = 0$$

or

$$\nabla^2 E + k^2 E = 0, \qquad \nabla^2 H + k^2 H = 0 \tag{6.2}$$

where

$$k^2 = -i\omega\mu(\sigma + i\omega\epsilon)$$

is the square of the wave number characterizing the medium. In a non-conducting medium, we have:

$$k^2 = \omega^2\epsilon\mu, \quad \text{and} \quad k = (\omega/c)(\epsilon^*\mu^*)^{1/2}$$

where ϵ^* and μ^* are the relative dielectric and magnetic permeabilities for the medium respectively with c being the velocity of light 3×10^8 m s^{-1}.

Assuming that $k_0 = \omega(\epsilon_0\mu_0)^{1/2}$ (here $\epsilon = \epsilon^*\epsilon_0$, $\mu = \mu^*\mu_0$), we can write the square of wave number as:

$$k^2 = \omega^2\epsilon\mu\left(1 - i\frac{\sigma}{\omega\epsilon}\right) = \epsilon^*\mu^*k_0^2\left(1 - i\frac{\sigma}{\omega\epsilon}\right)$$

Let us denote:

$$m^2 = \epsilon^*\mu^*\left(1 - i\frac{\sigma}{\omega\epsilon}\right) \tag{6.3}$$

so that we have:

$$k = mk_0 \tag{6.4}$$

where

$$m = (\epsilon^* \mu^*)^{1/2} \left(1 - i \frac{\sigma}{\omega\epsilon}\right)^{1/2} \tag{6.5}$$

In future development, where we use the parameter:

$(\sigma + i\omega\epsilon)/ik$

we can represent it in the various forms:

$$\frac{\sigma + i\omega\epsilon}{ik} = \frac{k^2}{-i\omega\mu i k} = \frac{k}{\omega\mu} = \frac{mk_0}{\omega\mu} = m \left(\frac{\epsilon_0}{\mu_0}\right)^{1/2} \frac{1}{\mu^*}$$

In a non-magnetic medium we have:

$$\frac{\sigma + i\omega\epsilon}{ik} = m \left(\frac{\epsilon_0}{\mu_0}\right)^{1/2} \tag{6.6}$$

Let us represent curl M in a spherical coordinate system:

$$\text{curl } M = \frac{1}{r^2 \sin\theta} \begin{vmatrix} I_r & r I_\theta & r \sin\theta\, I_\phi \\ \dfrac{\partial}{\partial r} & \dfrac{\partial}{\partial \theta} & \dfrac{\partial}{\partial \phi} \\ M_r & r M_\theta & r \sin\theta\, M_\phi \end{vmatrix}$$

Then, in accord with eq. 6.6, the second equation of the field can be written as:

$$i \left(\frac{\epsilon_0}{\mu_0}\right)^{1/2} m^2 k_0 E_r = \frac{1}{r \sin\theta} \left\{ \frac{\partial}{\partial \theta} \sin\theta\, H_\phi - \frac{\partial}{\partial \phi} H_\theta \right\} \tag{6.7}$$

$$i \left(\frac{\epsilon_0}{\mu_0}\right)^{1/2} m^2 k_0 E_\theta = \frac{1}{r \sin\theta} \left\{ \frac{\partial}{\partial \phi} H_r - \sin\theta \frac{\partial}{\partial r} r H_\phi \right\} \tag{6.8}$$

and

$$i \left(\frac{\epsilon_0}{\mu_0}\right)^{1/2} m^2 k_0 E_\phi = \frac{1}{r} \left\{ \frac{\partial}{\partial r} r H_\theta - \frac{\partial}{\partial \theta} H_r \right\} \tag{6.9}$$

By analogy from the first equation:

$\text{curl } E = -i\omega\mu\, H$

we obtain:

$$-ik_0 \left(\frac{\mu_0}{\epsilon_0}\right)^{1/2} H_r = \frac{1}{r \sin\theta} \left\{ \frac{\partial}{\partial \theta} \sin\theta\, E_\phi - \frac{\partial E_\theta}{\partial \phi} \right\} \tag{6.10}$$

$$-ik_0 \left(\frac{\mu_0}{\epsilon_0}\right)^{1/2} H_\theta = \frac{1}{r \sin\theta} \left\{ \frac{\partial}{\partial \phi} E_r - \frac{\partial}{\partial r} r \sin\theta\, E_\phi \right\} \tag{6.11}$$

and

$$-ik_0 \left(\frac{\mu_0}{\epsilon_0}\right)^{1/2} H_\phi = \frac{1}{r} \left\{ \frac{\partial}{\partial r} r E_\theta - \frac{\partial}{\partial \theta} E_r \right\} \tag{6.12}$$

inasmuch as:

$$i\omega\mu = i\omega (\epsilon_0 \mu_0)^{1/2} \left(\frac{\mu_0}{\epsilon_0}\right)^{1/2} = ik_0 \left(\frac{\mu_0}{\epsilon_0}\right)^{1/2}, \quad \text{when } \mu^* = 1$$

The unknown field must satisfy eqs. 6.7—6.12 as well as boundary conditions at the surface of the sphere which require continuity of the tangential components of the field:

$$E_\theta^e = E_\theta^i \qquad H_\theta^e = H_\theta^i$$

$$E_\phi^e = E_\phi^i \qquad H_\phi^e = H_\phi^i \tag{6.13}$$

With increase in distance from the sphere, secondary field tends to zero, and this behavior defines a condition which must be met at infinity.

We will represent the field caused by surface charges and currents as the superposition of two types of oscillations, namely electric and magnetic. The radial component of the magnetic field, H_r, of the electric type of oscillation is zero, while the radial component of the electric field E_r of the magnetic type oscillation is zero.

Equations and potential for oscillations of the electric type ($H_r = 0$)

In accord with eq. 6.10, we have:

$$\frac{\partial}{\partial \theta} \sin\theta \, E_\phi = \frac{\partial}{\partial \phi} E_\theta \tag{6.14}$$

This equation can be satisfied if we assume that:

$$\sin\theta \, E_\phi = \partial\Phi_1/\partial\phi, \quad \text{and} \quad E_\theta = \partial\Phi_1/\partial\theta \tag{6.15}$$

Substituting this equation into eqs. 6.8 and 6.9, we have:

$$i\left(\frac{\epsilon_0}{\mu_0}\right)^{1/2} m^2 k_0 \frac{\partial\Phi_1}{\partial\theta} = -\frac{1}{r} \frac{\partial}{\partial r} r H_\phi$$

$$i\left(\frac{\epsilon_0}{\mu_0}\right)^{1/2} m^2 k_0 \frac{\partial\Phi_1}{\partial\phi} = \frac{\sin\theta}{r} \frac{\partial}{\partial r} r H_\theta \tag{6.16}$$

Letting:

$$\Phi_1 = \frac{1}{r} \frac{\partial f_1}{\partial r} \tag{6.17}$$

we obtain:

$$H_\phi = -i \left(\frac{\epsilon_0}{\mu_0} \right)^{1/2} m^2 k_0 \frac{1}{r} \frac{\partial f_1}{\partial \theta} \tag{6.18}$$

and

$$H_\theta = i \left(\frac{\epsilon_0}{\mu_0} \right)^{1/2} m^2 k_0 \frac{1}{r \sin\theta} \frac{\partial f_1}{\partial \phi} \tag{6.19}$$

From eq. 6.7, we have:

$$E_r = \frac{1}{r \sin\theta} \left\{ \frac{\partial}{\partial \theta} \left(-\frac{\sin\theta}{r} \frac{\partial f_1}{\partial \theta} - \frac{\partial}{\partial \phi} \frac{1}{r \sin\theta} \frac{\partial f_1}{\partial \phi} \right) \right\}$$

Introducing the notation:

$$U_1 = f_1/r \tag{6.20}$$

we obtain:

$$E_r = \frac{1}{r \sin\theta} \left\{ \frac{\partial}{\partial \theta} \left(-\sin\theta \frac{\partial U_1}{\partial \theta} \right) - \frac{\partial}{\partial \phi} \frac{1}{\sin\theta} \frac{\partial U_1}{\partial \phi} \right\} \tag{6.21}$$

In determining the potential U_1, we will make use of eq. 6.12. It is clear that after integration with respect to θ, we have:

$$-k^2 U_1 = \frac{1}{r} \frac{\partial^2}{\partial r^2} (r U_1) - \frac{1}{r^2 \sin\theta} \left\{ \frac{\partial}{\partial \theta} \left(-\sin\theta \frac{\partial U_1}{\partial \theta} \right) - \frac{1}{\sin\theta} \frac{\partial^2 U_1}{\partial \phi^2} \right\}$$

and

$$\frac{1}{r} \frac{\partial^2}{\partial r^2} (r U_1) + \frac{1}{r^2 \sin\theta} \frac{\partial}{\partial \theta} \sin\theta \frac{\partial U_1}{\partial \theta} + \frac{1}{r^2 \sin^2\theta} \frac{\partial^2 U_1}{\partial \phi^2} + k^2 U_1 = 0 \tag{6.22}$$

or

$$\nabla^2 U_1 + k^2 U_1 = 0 \tag{6.23}$$

Therefore, the scalar potential of the electric type U_1 satisfies the Helmholtz equation in eq. 6.13, and it is related to the various field components through the expressions:

$$E_r = \frac{\partial^2}{\partial r^2} (r U_1) + k^2 r U_1 \qquad H_r = 0$$

$$E_\theta = \frac{1}{r} \frac{\partial^2 (r U_1)}{\partial \theta\, \partial r} \qquad\qquad H_\theta = i \left(\frac{\epsilon_0}{\mu_0} \right)^{1/2} m^2 k_0 \frac{1}{r \sin\theta} \frac{\partial r U_1}{\partial \phi}$$

$$E_\phi = \frac{1}{r \sin\theta} \frac{\partial^2 (r U_1)}{\partial r\, \partial \phi} \qquad H_\phi = -i \left(\frac{\epsilon_0}{\mu_0} \right)^{1/2} m^2 k_0 \frac{1}{r} \frac{\partial r U}{\partial \theta} \tag{6.24}$$

Equations and potential for oscillations of the magnetic type

As follows from eq. 6.7 and the condition that $E_r = 0$:

$$\frac{\partial}{\partial\theta}\sin\theta\,H_\phi = \frac{\partial}{\partial\phi}H_\theta \tag{6.25}$$

This relationship will be satisfied if we assume that:

$$\sin\theta\,H_\phi = \frac{\partial\Phi_2}{\partial\phi}, \quad\text{and } H_\theta = \frac{\partial\Phi_2}{\partial\theta} \tag{6.26}$$

Substituting this last result into eqs. 6.11 and 6.12, we have:

$$-ik_0\left(\frac{\mu_0}{\epsilon_0}\right)^{1/2}\frac{\partial\Phi_2}{\partial\theta} = -\frac{1}{r\sin\theta}\frac{\partial}{\partial r}(r\sin\theta\,E_\phi)$$

$$-ik_0\left(\frac{\mu_0}{\epsilon_0}\right)^{1/2}\frac{1}{\sin\theta}\frac{\partial\Phi_2}{\partial\phi} = \frac{1}{r}\frac{\partial}{\partial r}(rE_\theta) \tag{6.27}$$

Letting:

$$\Phi_2 = \frac{1}{r}\frac{\partial f_2}{\partial r}$$

we have:

$$-ik_0\left(\frac{\mu_0}{\epsilon_0}\right)^{1/2}\frac{\partial}{\partial\theta}\frac{1}{r}\frac{\partial f_2}{\partial r} = -\frac{1}{r}\frac{\partial}{\partial r}(rE_\phi)$$

$$-ik_0\left(\frac{\mu_0}{\epsilon_0}\right)^{1/2}\frac{1}{\sin\theta}\frac{\partial}{\partial\phi}\left(\frac{1}{r}\frac{\partial f_2}{\partial r}\right) = \frac{1}{r}\frac{\partial}{\partial r}(rE_\theta)$$

Integrating with respect to r, we obtain:

$$E_\phi = ik_0\left(\frac{\mu_0}{\epsilon_0}\right)^{1/2}\frac{1}{r}\frac{\partial f_2}{\partial\theta}$$

$$E_\theta = -ik_0\left(\frac{\mu_0}{\epsilon_0}\right)^{1/2}\frac{1}{r\sin\theta}\frac{\partial f_2}{\partial\phi} \tag{6.28}$$

Assume that:

$$-U_2 = \frac{1}{r}f_2 \tag{6.29}$$

Then:

$$\Phi_2 = \frac{1}{r}\frac{\partial}{\partial r}(rU_2)$$

and

$$E_r = 0$$

$$E_\theta = -ik_0 \left(\frac{\mu_0}{\epsilon_0}\right)^{1/2} \frac{1}{r\sin\theta} \frac{\partial}{\partial\phi}(r\,U_2) \qquad H_\theta = \frac{1}{r} \frac{\partial^2(r\,U_2)}{\partial r\,\partial\theta}$$

$$E_\phi = ik_0 \left(\frac{\mu_0}{\epsilon_0}\right)^{1/2} \frac{1}{r} \frac{\partial}{\partial\theta}(rU_2) \qquad H_\phi = \frac{1}{r\sin\theta} \frac{\partial^2(r\,U_2)}{\partial r\,\partial\phi} \qquad (6.30)$$

In order to find H_r, we will make use of eq. 6.10:

$$-ik_0 \left(\frac{\mu_0}{\epsilon_0}\right)^{1/2} H_r = \frac{1}{r\sin\theta} ik_0 \left(\frac{\mu_0}{\epsilon_0}\right)^{1/2} \left\{ \frac{1}{r} \frac{\partial}{\partial\theta}\sin\theta \frac{\partial}{\partial\theta}(r\,U_2) + \frac{1}{r\sin\theta} \frac{\partial^2(r\,U_2)}{\partial\phi^2} \right\}$$

and

$$H_r = -\frac{1}{r^2\sin\theta} \left\{ \frac{\partial}{\partial\theta}\sin\theta \frac{\partial}{\partial\theta}(r\,U_2) + \frac{1}{\sin\theta} \frac{\partial^2(r\,U_2)}{\partial\phi^2} \right\} \qquad (6.31)$$

or

$$H_r = \frac{\partial^2(r\,U_2)}{\partial r^2} + k^2 r\,U_2$$

inasmuch as in accord with eq. 6.9, the potential of the magnetic type, U_2 is a solution of the Helmholtz equation:

$$\nabla^2 U_2 + k^2 U_2 = 0 \qquad (6.32)$$

Boundary conditions for the potentials U_1 and U_2 at the surface of the sphere

In order to assure continuity of the tangential components of the fields, it will be sufficient to demonstrate continuity of the following quantities:

$$\frac{\partial}{\partial r}(rU_1) \qquad k^2 r U_1$$

$$r U_2 \qquad \frac{\partial}{\partial r}(rU_2) \qquad (6.33)$$

It can be seen that the boundary conditions are not mixed, and that they permit us to define independently fields corresponding to oscillations of the electric and magnetic types.

Solutions to Helmholtz's equation for potentials in a spherical coordinate system

Using the method of separation of variables, we can write the solution to the equation for potentials as the product of two functions:

$$U = f(r)Y(\theta, \phi) \tag{6.34}$$

Then, in place of Helmholtz's equation, we obtain two ordinary differential equations:

$$\frac{1}{r}\frac{d^2(rf)}{dr^2} + \left(k^2 - \frac{\lambda}{r^2}\right)f = 0 \tag{6.35}$$

and

$$\frac{1}{\sin\theta}\left\{\frac{\partial}{\partial\theta}\sin\theta\frac{\partial Y(\theta, \phi)}{\partial\theta} + \frac{1}{\sin\theta}\frac{\partial^2 Y(\theta, \phi)}{\partial\phi^2}\right\} + \lambda Y(\theta, \phi) = 0 \tag{6.36}$$

Equation 6.36 has a unique and continuous solution for $\lambda = n(n + 1)$:

$$Y_n(\theta, \phi) = a_0 P_n(\cos\theta) + \sum_{m=1}^{\infty}(a_m\cos m\phi + b_m\sin m\phi)P_n^{(m)}(\cos\theta) \tag{6.37}$$

The associated Legendre polynomials $P_n^{(m)}$ are related to the Legendre polynomials P_n through the identities:

$$P_n^{(m)}(\mu) = (1 - \mu^2)^{m/2}\frac{d^m P_n(\mu)}{d\mu^m} \tag{6.38}$$

where $\mu = \cos\theta$. In particular:

$$P_n^{(1)}(\mu) = (1 - \mu^2)^{1/2}\frac{d}{d\mu}P_n(\mu) = -\frac{d}{d\theta}P_n(\cos\theta) \tag{6.39}$$

The Legendre polynomials satisfy the differential equation:

$$\frac{d}{d\mu}\left[(1 - \mu^2)\frac{dP_n}{d\mu}\right] + n(n + 1)P_n = 0 \tag{6.40}$$

and they are characterized by the following identities:

$$nP_{n-1} + (n + 1)P_{n+1} = (2n + 1)\mu P_n$$

$$P'_{n+1} - P'_{n-1} = (2n + 1)P_n$$

$$P'_{n+1} = \mu P'_n + (n + 1)P_n \tag{6.41}$$

or

$$\mu P'_n = P'_{n-1} + nP_{n-1}$$

A solution of eq. 6.35 is:

$$f_n(kr) = \frac{1}{(kr)^{1/2}} Z_{n+1/2}(kr) \tag{6.42}$$

where $Z_{n+1/2}(kr)$ is the Bessel function of order $n + 1/2$.

Let us define the following notation:

$$\psi_n(x) = \left(\frac{\pi x}{2}\right)^{1/2} J_{n+1/2}(x)$$

$$\xi_n(x) = \left(\frac{\pi x}{2}\right)^{1/2} H^{(2)}_{n+1/2}(x) \tag{6.43}$$

The functions $\psi_n(x)$ and $\xi_n(x)$ satisfy recurrence relationships of the following type:

$$\epsilon_{l+1}(x) = \frac{2l+1}{x} \epsilon_l(x) - \epsilon_{l-1}(x)$$

$$\epsilon_l'(x) = \epsilon_{l-1}(x) - \frac{l}{x} \epsilon_l(x) \tag{6.44}$$

Thus, particular solutions of Helmholtz's equation will take the form:

$$\frac{\psi_n(k_i r)}{k_i r} Y_n(\theta, \phi), \quad \text{if } r < a \tag{6.45}$$

and

$$\frac{\xi_n(k_e r)}{k_e r} Y_n(\theta, \phi), \quad \text{if } r > a \tag{6.46}$$

where k_i and k_e are the wave numbers characterizing the media inside and outside the sphere, respectively.

Potentials of the electric and magnetic types for the primary field

In determining the secondary field, it is necessary to find potentials that describe the primary field and express these in terms of spherical functions. In order to carry out these operations, we will make use of well-established representations for the functions e^{ikR} and e^{ikR}/R. First, let us assume that the primary field is a plane wave propagating downwards along the z-axis as shown in Fig. 6.1a. We will consider the case in which the electric field component E of the plane wave is directed along the x-axis, and the vector $-H$ has a single component, H_y. Under these conditions, eq. 6.2 has the form:

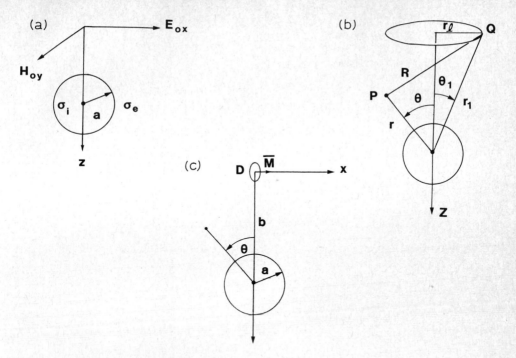

Fig. 6.1. Various types of field excitation about a spherical conductor.

$$\frac{\mathrm{d}^2 E_{0x}}{\mathrm{d}z^2} + k_e^2 E_{0x} = 0, \quad \frac{\mathrm{d}^2 H_{0y}}{\mathrm{d}z^2} + k_e^2 H_{0y} = 0 \tag{6.47}$$

It is obvious that:

$$E_{0x} = E_{0x}^{(0)} e^{ik_e z}, \quad H_{0y} = H_{0y}^{(0)} e^{ik_e z} \tag{6.48}$$

From the second of Maxwell's equations:

$$\mathrm{curl}\, H = (\sigma + i\omega\epsilon)\, E$$

we have:

$$-\frac{\mathrm{d}H_{0y}}{\mathrm{d}z} = (\sigma + i\omega\epsilon)\, E_{0x}$$

or

$$H_{0y} = -\frac{\sigma_e + i\omega\epsilon}{ik_e}\, E_{0x} \tag{6.49}$$

In accord with eq. 6.6:

$$H_{0y} = -m_e \left(\frac{\epsilon_0}{\mu_0}\right)^{1/2} E_{0x} \frac{1}{\mu^*} \tag{6.50}$$

In a non-conducting medium, we obtain:

$$\frac{H_{0y}}{E_{0x}} = -\frac{1}{\mu^*} \left(\frac{\epsilon_0}{\mu_0}\right)^{1/2} (\epsilon^*\mu^*)^{1/2} = -\left(\frac{\epsilon}{\mu}\right)^{1/2} \tag{6.51}$$

but for the quasi-stationary field in a conducting medium, we have:

$$\frac{H_{0y}}{E_{0x}} = -\frac{1}{\mu^*} m_e \left(\frac{\epsilon_0}{\mu_0}\right)^{1/2} = -\frac{1}{\mu^*} \left(\frac{\epsilon_0}{\mu_0}\right)^{1/2} \sqrt{-i}\,(\epsilon^*\mu^*)^{1/2} \left(\frac{\sigma}{\omega\epsilon}\right)^{1/2}$$

$$= -\left(-i\frac{\sigma}{\omega\mu}\right)^{1/2} \tag{6.52}$$

Thus, the impedance characterizing a plane wave propagating through the uniform medium is:

$$Z = \frac{E_{0x}}{H_{0y}} = \left(\frac{\omega\mu}{\sigma}\right)^{1/2} e^{i(\pi/4)} \tag{6.53}$$

We can make use of a well-known expression:

$$e^{ik_e z} = e^{ik_e r \cos\theta} = \sum_{l=1}^{\infty} i^l (2l+1) \frac{\psi_l(k_e r)}{k_e r} P_l(\cos\theta) \tag{6.54}$$

As can be shown:

$$(r_0\,i) = \sin\theta\,\cos\phi \qquad (r_0\,j) = \sin\theta\,\sin\phi$$

$$(\theta_0\,i) = \cos\theta\,\cos\phi \qquad (\theta_0\,j) = \cos\theta\,\sin\phi$$

Therefore, we have the following expressions for the components of a plane wave in spherical coordinates:

$$E_{0r}^{(0)} = E_{0x}^{(0)} \sin\theta\,\cos\phi \qquad H_{0r}^{(0)} = -m_e \left(\frac{\epsilon_0}{\mu_0}\right)^{1/2} E_{0x}^{(0)} \sin\theta\,\sin\phi$$

$$E_{0\theta}^{(0)} = E_{0x}^{(0)} \cos\theta\,\cos\phi \qquad H_{0\theta}^{(0)} = -m_e \left(\frac{\epsilon_0}{\mu_0}\right)^{1/2} E_{0x}^{(0)} \cos\theta\,\sin\phi \tag{6.55}$$

$$E_{0\phi}^{(0)} = -E_{0x}^{(0)} \sin\phi \qquad H_{0\phi}^{(0)} = -m_e \left(\frac{\epsilon_0}{\mu_0}\right)^{1/2} E_{0x}^{(0)} \cos\phi \qquad \text{if } \mu^* = 1$$

Inasmuch as the radial components of the electric and magnetic fields are defined in terms of potentials of a single type, it is appropriate to obtain the expressions for the potentials characterizing plane wave by making use of these already known components. Differentiating eq. 6.54 with respect to θ and considering eq. 6.39, we have:

$$E_{0r} = E_{0x}^{(0)} \sum_{l=1}^{\infty} i^{l-1} (2l + 1) \frac{\psi_l(k_e r)}{(k_e r)^2} P_l^{(1)}(\cos\theta) \cos\phi$$

$$H_{0r} = H_{0y}^{(0)} \sum_{l=1}^{\infty} i^{l-1} (2l + 1) \frac{\psi_l(k_e r)}{(k_e r)^2} P_l^{(1)}(\cos\theta) \sin\phi \qquad (6.56)$$

These expressions permit us to find expressions for the potentials $U_1^{(0)}$ and $U_2^{(0)}$. Actually, the radial components are related to the potentials as follows:

$$E_{0r} = \frac{\partial^2}{\partial r^2}(rU_1^{(0)}) + k^2 rU_1^{(0)} \text{ and } H_{0r} = \frac{\partial^2}{\partial r^2}(rU_2^{(0)}) + k^2 rU_2^{(0)}$$

Therefore, we will seek expressions for U_1 and U_2 in the form:

$$\sum M(r) P_l^{(1)}(\cos\theta) \begin{matrix} \cos\phi \\ \sin\phi \end{matrix}$$

Then, we will have the following expression for the component E_r:

$$E_{0r} = \sum \left[\frac{d^2}{dr^2}(rM(r)) + k^2 rM(r) \right] P_l^{(1)}(\cos\theta) \cos\phi \qquad (6.57)$$

In accord with eq. 6.35,

$$\frac{d^2}{dr^2}(rM(r)) + k^2 rM(r) = \frac{l(l + 1) M(r)}{r}$$

so that we obtain the following in place of eq. 6.57:

$$E_{0r} = \frac{1}{r} \sum_{l=1}^{\infty} l(l + 1) M(r) P_l^{(1)}(\cos\theta) \cos\phi \qquad (6.58)$$

Comparing eqs. 6.56 and 6.58, we see that:

$$E_{0x} i^{l-1}(2l + 1) \frac{\psi_l(k_e r)}{k_e^2 r^2} = \frac{1}{r} l(l + 1) M(r)$$

Hence, for the unknown function $M(r)$, we have:

$$M(r) = \frac{E_{0x}}{k_e^2 r} \frac{2l + 1}{l(l + 1)} \psi_l(k_e r) i^{l-1} \qquad (6.59)$$

Therefore,

$$U_1^{(0)} = \sum_{l=1}^{\infty} d_{1l} \frac{\psi_l(k_e r)}{k_e r} P_l^{(1)}(\cos\theta) \cos\phi \qquad (6.60)$$

where

$$d_{1l} = E_{0x} \frac{i^{l-1}}{k_e} \frac{2l+1}{l(l+1)} \tag{6.61}$$

By analogy, we obtain the following result for the potential of the magnetic type:

$$U_2^{(0)} = \sum_{l=1}^{\infty} e_{1l} \frac{\psi_l(k_e r)}{k_e r} P_l^{(1)}(\cos\theta) \sin\phi \tag{6.62}$$

where

$$e_{1l} = H_{0y} \frac{i^{l-1}}{k_e} \frac{2l+1}{l(l+1)} \tag{6.63}$$

Next, we can find the potential of the primary field when the source is a current ring located co-axially with respect to the sphere as shown in Fig. 6.1b. The analysis of this particular field behavior has already been done in the previous chapter. The alternating magnetic field which is caused by the current in the ring creates a vortex electric field which has but a single component, E_ϕ, and induced currents that arise in the conducting medium, and which obviously have only the component j_ϕ. For this reason, the primary field, and therefore, the secondary field are defined using only a potential of the magnetic type $U_2^{(0)}$.

In determining $U_2^{(0)}$, we obtain expressions for the electric field of a loop source in a conducting medium, by representing the current ring as the sum of elementary electric dipoles in series. The vector potential of the electric type, A, for the electric dipole source is directed along the dipole axis, and is equal to:

$$A = \frac{I \, dl}{4\pi} \frac{e^{ikR}}{R} \tag{6.64}$$

The field components are derivable from the potential as follows:

$$H = \text{curl } A \quad \text{and} \quad E = -\frac{\partial A}{\partial t} - \text{grad } U \tag{6.65}$$

Consider the configuration in Fig. 6.1b. Assume that $P(r, \theta, \phi)$ is an observation point, and that $Q(r_1, \theta_1, \phi_1)$ is a point characterizing one of the elements of the current ring. The origin for the coordinate system is located at the center of the sphere. The distance between the points P and Q is:

$$R = (r^2 + r_1^2 - 2rr_1 \cos\gamma)^{1/2}$$

where

$$\cos\gamma = \cos\theta \cos\theta_1 + \sin\theta \sin\theta_1 \cos(\phi - \phi_1)$$

We can make use of the identities:

$$\frac{e^{ik_eR}}{ik_eR} = \begin{cases} \dfrac{1}{k_e^2 r\, r_1} \sum\limits_{l=0}^{\infty} (2l+1)P_l(\cos\gamma)\,\psi_l(k_e r)\,\xi_l(k_e r_1), & \text{if } r < r_1 \\[3mm] \dfrac{1}{k_e^2 r\, r_1} \sum\limits_{l=0}^{\infty} (2l+1)P_l(\cos\gamma)\,\psi_l(k_e r_1)\,\xi_l(k_e r), & \text{if } r > r_1 \end{cases} \tag{6.66}$$

and

$$P_l(\cos\gamma) = P_l(\cos\theta)P_l(\cos\theta_1)$$
$$+ 2\sum_{m=1}^{l} \frac{(l-m)!}{(l+m)!}\,P_l^{(m)}(\cos\theta)P_l^{(m)}(\cos\theta_1)\cos m\,(\phi-\phi_1) \tag{6.67}$$

In solving the boundary value problem, it is necessary to represent the primary field in terms of the spherical functions only on the surface of the sphere. For this reason, we will restrict our consideration to the case $r < r_1$ and rewrite eq. 6.66 in the form:

$$\frac{e^{ik_e r}}{ik_e r} = \sum_{l=0}^{\infty}\sum_{m=0}^{l} (2l+1)\delta_m \frac{(l-m)!}{(l+m)!} \frac{\xi_l(k_e r_1)}{k_e r_1}\frac{\psi_l(k_e r)}{k_e r}$$
$$\times\, P_l^{(m)}(\cos\theta_1)P_l^{(m)}(\cos\theta)\cos m\,(\phi-\phi_1) \tag{6.68}$$

where

$$\delta_m = 2, \quad \text{if } m \neq 0 \text{ and } \delta_0 = 1$$

In view of the axial symmetry, the vector potential for the current ring does not depend on the coordinate ϕ, and it has but a single component A_ϕ. It is obvious that

$$A_\phi = \frac{I}{4\pi}\int_L \frac{e^{ikR}}{R}\,dl_\phi \tag{6.69}$$

and considering that:

$$dl_\phi = r_l\cos(\phi-\phi_1)\,d\phi$$

we have:

$$A_\phi = \frac{I\,r_l}{4\pi}\int_0^{2\pi} \frac{e^{ikR}}{R}\cos(\phi-\phi_1)\,d\phi \tag{6.70}$$

Substituting eq. 6.68 into this last expression and making use of the orthogonality properties of trigonometric functions, we obtain

$$A_\phi = I r_l i \sum_{l=1}^{\infty} \frac{2l+1}{l(l+1)} \frac{\xi_l(k_e r_1)\psi_l(k_e r)}{r_1}\,P_l^{(1)}(\cos\theta_1)P_l^{(1)}(\cos\theta) \tag{6.71}$$

In accord with eq. 6.65 for the vortex electric field, we have:

$$E_{0\phi} = I\, r_l\, \omega\mu \sum_{l=1}^{\infty} \frac{2l+1}{l(l+1)} \frac{\xi_l(k_e r_1)}{r_1} \frac{\psi_l(k_e r)}{k_e r} P_l^{(1)}(\cos\theta_1) P_l^{(1)}(\cos\theta) \qquad (6.72)$$

inasmuch as the integral of the second term of eq. 6.65 along a closed path is zero.

In determining the potential of the magnetic type, $U_2^{(0)}$ for the current ring, we will make use of eqs. 6.28 and 6.29:

$$E_{0\phi} = i\omega\mu \frac{1}{r} \frac{\partial}{\partial\theta} r\, U_2^{(0)}$$

along with the identity in eq. 6.39:

$$P_l^{(1)}(\cos\theta) = \sin\theta \frac{\mathrm{d}\, P_l(\cos\theta)}{\mathrm{d}\cos\theta}$$

On integration, we obtain:

$$U_2^{(0)} = \sum_{l=1}^{\infty} e_{2l} \frac{\psi_l(k_e r)}{k_e r} P_l(\cos\theta) \qquad (6.73)$$

where

$$e_{2l} = i \frac{I\, r_l}{r_1} \frac{2l+1}{l(l+1)} \xi_l(k_e r_1) P_l^{(1)}(\cos\theta_1) \qquad (6.74)$$

Now, let us consider the third type of excitation, that corresponding to a magnetic dipole source. If a magnetic dipole with its axis arbitrarily oriented with respect to this sphere is situated at a point D as shown in Fig. 6.1c, it is convenient to consider the normal field as being the sum of fields caused by two magnetic dipoles, these being a transversely oriented one with its axis perpendicular to the line OD, and a longitudinally oriented one with its axis along the line OD. The field for the longitudinal dipole source is the particular case of the field of the current ring when the radius of the ring tends to zero. In this case:

$$\cos\theta_1 \to 1 \ \text{ and } \ P_l^{(1)}(1) \to \tfrac{1}{2}l(l+1)\sin\theta_1 = \tfrac{1}{2}l(l+1)\frac{r_l}{r_1}$$

and in place of eqs. 6.73 and 6.74, we obtain an expression for the potential of the magnetic type for a longitudinal (radial) dipole source in the form:

$$U_2^{(0)} = \sum_{l=1}^{\infty} e_{3l} \frac{\psi_l(k_e r)}{k_e r} P_l(\cos\theta) \qquad (6.75)$$

where

$$e_{3l} = \frac{Mi}{2\pi b^2} (2l+1)\xi_l(k_e b) \tag{6.76}$$

and where M is the dipole moment, given by $\pi r_l^2 I$, and with b being the distance from the center of the sphere to the dipole.

Let us assume that the moment of a transverse (azimuthal) dipole source is directed along the x-axis and that we will first find the radial component of the normal electric field. As can readily be shown, the field of a magnetic dipole in a uniform medium can be described using one component of a vector potential of the magnetic type, A_x^*:

$$E = \operatorname{curl} A^*, \quad H = (\sigma + i\omega\epsilon) A^* - \frac{1}{i\omega\mu} \operatorname{grad} \operatorname{div} A^*$$

and

$$A_x^* = \frac{i\omega\mu M}{4\pi} \frac{e^{ik_e R}}{R} \tag{6.77}$$

In a spherical coordinate system with its origin at the center of the sphere, the components of the potential as can be seen from Fig. 6.1c are:

$$A_r^* = A_x^* \sin\theta \cos\phi, \quad A_\theta^* = A_x^* \cos\theta \cos\phi, \quad A_\phi^* = -A_x^* \sin\phi \tag{6.78}$$

where, in accord with eq. 6.66 we have:

$$\frac{e^{ik_e R}}{R} = \frac{i}{b} \sum_{l=0}^{\infty} (2l+1)\xi_l(k_e b) \frac{\psi_l(k_e r)}{k_e r} P_l(\cos\theta), \quad r < b \tag{6.79}$$

and

$$\frac{e^{ik_e R}}{R} = \frac{i}{b} \sum_{l=0}^{\infty} (2l+1)\psi_l(k_e b) \frac{\xi_l(k_e r)}{k_e r} P_l(\cos\theta), \quad r > b \tag{6.80}$$

Making use of eq. 6.77, we can define the radial component of the electric field:

$$E_{0r} = \frac{1}{r \sin\theta} \left\{ \frac{\partial \sin\theta\, A_\phi^*}{\partial\theta} - \frac{\partial A_\theta^*}{\partial\phi} \right\} \tag{6.81}$$

Substituting eq. 6.78 in this last expression and carrying out some elementary algebraic operations, we have:

$$E_{0r} = -\frac{\omega\mu M \sin\phi}{4\pi\, br} \sum_{l=1}^{\infty} (2l+1)\xi_l(k_e b) \frac{\psi_l(k_e r)}{k_e r} P_l^{(1)}(\cos\theta) \tag{6.82}$$

By analogy with the plane wave field behavior, we have the following expressions for the potential of the electric type for a transversal dipole source:

$$U_1^{(0)} = \sin\phi \sum_{l=1}^{\infty} d_{4l} \frac{\psi_l(k_e r)}{k_e r} P_l^{(1)}(\cos\theta) \tag{6.83}$$

where

$$d_{4l} = -\frac{\omega\mu M}{4\pi b} \frac{2l+1}{l(l+1)} \xi_l(k_e b) \tag{6.84}$$

In finding a potential $U_2^{(0)}$ of the magnetic type, we will seek an expression for the radial component of the magnetic field, H_{0r}. In accord with eqs. 6.1 and 6.77 we have:

$$-i\omega\mu H_r = \frac{1}{r\sin\theta} \left\{ \frac{\partial}{\partial\theta}(\sin\theta\, E_\phi) - \frac{\partial E_\theta}{\partial\phi} \right\}$$

and

$$E_\phi = \frac{1}{r} \left\{ \frac{\partial}{\partial r}(rA_\theta^*) - \frac{\partial A_r^*}{\partial\theta} \right\}$$

$$E_\theta = -\frac{1}{r\sin\theta} \left\{ \sin\theta\, \frac{\partial}{\partial r}(rA_\phi^*) - \frac{\partial A_r^*}{\partial\phi} \right\}$$

It should be obvious that:

$$\frac{\partial}{\partial r}(rA_\theta^*) = -\frac{\omega\mu M}{4\pi b}\cos\phi\,\cos\theta \sum_{l=0}^{\infty}(2l+1)\xi_l(k_e b)\,\psi_l'(k_e r)P_l(\cos\theta)$$

$$\frac{\partial A_r^*}{\partial\theta} = -\frac{\omega\mu M}{4\pi b}\cos\phi \sum_{l=0}^{\infty}(2l+1)\xi_l(k_e b)\frac{\psi_l(k_e r)}{k_e r}\frac{\partial}{\partial\theta}(\sin\theta\, P_l(\cos\theta))$$

$$E_\phi = -\frac{\mu\omega M}{4\pi b r}\cos\phi \sum_{l=0}^{\infty}(2l+1)\xi_l(k_e b)\left[\left\{\psi_l' - \frac{\psi_l}{k_e r}\right\}\cos\theta P_l(\cos\theta)\right.$$

$$\left. + \frac{\psi_l(k_e r)}{k_e r}\sin\theta\, P_l^{(1)}(\cos\theta)\right] \tag{6.85}$$

$$\frac{\partial}{\partial r}(rA_\phi^*) = \frac{\omega\mu M}{4\pi b}\sin\phi \sum_{l=0}^{\infty}(2l+1)\xi_l(k_e b)\psi_l'(k_e r)P_l(\cos\theta)$$

$$\frac{\partial A_r^*}{\partial\phi} = \frac{\omega\mu M}{4\pi b}\sin\theta\,\sin\phi \sum_{l=0}^{\infty}(2l+1)\xi_l(k_e b)\frac{\psi_l(k_e r)}{k_e r}P_l(\cos\theta)$$

and

$$E_\theta = -\frac{\mu\omega M}{4\pi b r}\sin\phi \sum_{l=0}^{\infty}(2l+1)\xi_l(k_e b)\left\{\psi_l' - \frac{\psi_l}{k_e r}\right\}P_l(\cos\theta) \tag{6.86}$$

Whence

$$H_{0r} = -i \frac{M}{4\pi b r^2} \frac{\cos\phi}{\sin\theta} \left[\sum_{l=0}^{\infty} (2l+1) \xi_l(k_e b) \left\{ N_1 \frac{\partial}{\partial\theta} (\sin\theta \, \cos\theta \, P_l(\cos\theta)) \right.\right.$$

$$\left.\left. + N_2 \frac{\partial}{\partial\theta} (\sin^2\theta \, P_l^{(1)}(\cos\theta)) - N_1 P_l(\cos\theta) \right\} \right] \qquad (6.87)$$

where

$$N_1 = \psi_l'(k_e r) - \frac{\psi_l(k_e r)}{k_e r}, \quad N_2 = \frac{\psi_l(k_e r)}{k_e r}, \quad P_0^{(1)}(x) = 0 \qquad (6.88)$$

After some rather tedious manipulations, making use of Legendre's equation, and the recurrence relationships for the functions $P_l(\cos\theta)$ and $\psi_l(k_e r)$ we obtain the following expression for the potential:

$$U_2^{(0)} = \cos\phi \sum_{l=1}^{\infty} e_{4l} \frac{\psi_l(k_e r)}{k_e r} P_l^{(1)}(\cos\theta) \qquad (6.89)$$

where

$$e_{4l} = -\frac{i k_e M}{4\pi b} \frac{2l+1}{l(l+1)} \xi_l'(k_e b) \qquad (6.90)$$

Potentials for the secondary field

Taking into account the behavior of the functions $\psi_l(k_e r)$ and $\xi_l(k_e r)$ in the vicinity of the origin and at infinity, we can represent the electric and magnetic potentials for the secondary field as follows:

$$U_1^{(\bullet)} = \sum_{l=1}^{\infty} d_l \, C_l \frac{\xi_l(k_e r)}{k_e r} Y(\theta, \phi)$$

$$\qquad\qquad\qquad \text{if } r \geqslant a \qquad (6.91)$$

$$U_2^{(e)} = \sum_{l=1}^{\infty} e_l \, B_l \frac{\xi_l(k_e r)}{k_e r} Y(\theta, \phi)$$

and inside the sphere:

$$U_1^{(i)} = \sum_{l=1}^{\infty} d_l \, C_l^* \frac{\psi_l(k_i r)}{k_i r} Y(\theta, \phi)$$

$$\qquad\qquad\qquad \text{if } r \leqslant a \qquad (6.92)$$

$$U_2^{(i)} = \sum_{l=1}^{\infty} e_l \, B_l^* \frac{\psi_l(k_i r)}{k_i r} Y(\theta, \phi)$$

Let us define the following notation:

$$\frac{2\pi a}{\lambda_0} = k_0 a = \rho_0, \quad \frac{2\pi a}{\lambda_0} m_e = k_e a = \rho \quad \text{and} \quad \frac{m_i}{m_e} = \frac{k_i}{k_e} = m \qquad (6.93)$$

In accord with eq. 6.33, on the surface of the sphere we have:

$$k_e^2(U_1^{(0)} + U_1^{(e)}) = k_i^2 U_1^{(i)}$$

$$\frac{\partial}{\partial r}(r(U_1^{(0)} + U_1^{(e)})) = \frac{\partial}{\partial r}(r U_1^{(i)}) \qquad \text{if } r = a \qquad (6.94)$$

and

$$U_2^{(0)} + U_2^{(e)} = U_2^{(i)}$$

$$\frac{\partial}{\partial r}(r(U_2^{(0)} + U_2^{(e)})) = \frac{\partial}{\partial r}(r U_2^{(i)}) \qquad \text{if } r = a \qquad (6.95)$$

Considering the orthogonality property of Legendre polynomials, we obtain two systems for determining the unknown coefficients:

$$m\, C_l^*\, \psi_l(m\rho) - C_l \xi_l(\rho) = \psi_l(\rho)$$

$$C_l^*\, \psi_l'(m\rho) - C_l \xi_l(\rho) = \psi_l'(\rho) \qquad (6.96)$$

and

$$B_l^*\, \psi_l(m\rho) - m\, B_l \xi_l(\rho) = m\, \psi_l(\rho)$$

$$B_l^*\, \psi_l'(m\rho) - B_l \xi_l'(\rho) = \psi_l'(\rho) \qquad (6.97)$$

Solving this system, we have:

$$C_l = \frac{\psi_l(\rho)\psi_l'(m\rho) - m\,\psi_l'(\rho)\psi_l(m\rho)}{m\,\xi_l'(\rho)\psi_l(m\rho) - \xi_l(\rho)\psi_l'(m\rho)} \qquad (6.98)$$

$$B_l = \frac{m\,\psi_l(\rho)\psi_l'(m\rho) - \psi_l'(\rho)\psi_l(m\rho)}{\psi_l(m\rho)\xi_l'(\rho) - m\,\xi_l(\rho)\psi_l'(m\rho)} \qquad (6.99)$$

$$C_l^* = \frac{\psi_l(\rho)\xi_l'(\rho) - \psi_l'(\rho)\xi_l(\rho)}{m\,\xi_l'(\rho)\psi_l(m\rho) - \xi_l(\rho)\psi_l'(m\rho)} \qquad (6.100)$$

$$B_l^* = \frac{m\,\psi_l(\rho)\xi_l'(\rho) - \psi_l'(\rho)\xi_l(\rho)}{\psi_l(m\rho)\xi_l'(\rho) - m\,\xi_l(\rho)\psi_l'(m\rho)} \qquad (6.101)$$

It has been established in the theory for spherical Bessel functions that:

$$\psi_l(\rho)\,\xi_l'(\rho) - \psi_l'(\rho)\,\xi_l(\rho) = -i$$

For this reason, we have the following expressions for the coefficients C_l^* and B_l^*:

$$C_l^* = i\,\frac{1}{\xi_l(\rho)\psi_l'(m\rho) - m\,\xi_l'(\rho)\psi_l(m\rho)}$$

$$B_l^* = i \frac{1}{m\,\xi_l(\rho)\,\psi_l'(m\rho) - \psi_l(m\rho)\,\xi_l'(\rho)} \tag{6.102}$$

The coefficients given in eqs. 6.98—6.102 represent the amplitudes of the corresponding spatial harmonics in the secondary field.

The equations which have been developed in this section are sufficient as the basis for an investigation of the behavior of electromagnetic fields caused by a plane wave or by a transversal magnetic dipole. This analysis will be carried out in detail in the next two sections.

6.2. A CONDUCTING SPHERE IN A PLANE WAVE FIELD

We will place the origin of a coordinate system at the center of the sphere with the cartesian and spherical coordinate axes oriented as shown in Fig. 6.1a. The radius of the sphere is taken to be a. Assume that a linearly polarized electromagnetic wave propagates in the downward direction on the z-axis. The x-axis characterizes the polarization of the electric field, E^0, while the y-axis characterizes the polarization of the magnetic field, H^0. Both of these vectors, E^0 and H^0 comprise the primary field, that is, the plane wave. The electric and magnetic fields comprising the plane wave can be described using the expressions:

$$E^0 = E_0\,e^{ik_e z}, \qquad E_{0x} = E_0, \qquad E_{0y} = E_{0z} = 0$$

$$H^0 = H_0\,e^{ik_e z}, \qquad H_{0y} = -m_e E_{0x}, \qquad H_{0x} = H_{0z} = 0 \tag{6.103}$$

where

$$k_e = (-i\,\sigma_e\mu\omega)^{1/2} = m_e k_0$$

is the wave number in the host medium and

$$m_e = \left(-i\,\frac{\sigma_e}{\mu\omega}\right)^{1/2}, \qquad k_0 = \omega\mu$$

where σ_e is the conductivity in the host medium, taken to be uniform, ω is the frequency in radiants per second, and μ is the magnetic permeability.

As follows from the preceding section, the expressions for various components of the secondary electromagnetic field caused by the presence of the sphere take the form:

$$E_r^s = \frac{E_0}{k_e^2 r^2} \sum_{l=1}^{\infty} C_l l(l+1)\,\xi_l(k_e r)\,P_l^{(1)}(\cos\theta)\,\cos\phi \tag{6.104}$$

$$E_\theta^s = \frac{-E_0}{k_e r}\cos\phi \sum_{l=1}^{\infty}\left\{ C_l\,\xi_l'(k_e r)P_l^{(1)\prime}(\cos\theta)\sin\theta - iB_l\xi_l(k_e r)\frac{P_l^{(1)}(\cos\theta)}{\sin\theta}\right\} \tag{6.105}$$

$$E_\phi^s = -\frac{\sin\phi\, E_0}{k_e r} \sum_{l=1}^{\infty} \left\{ C_l \xi_l'(k_e r) \frac{P_l^{(1)}(\cos\theta)}{\sin\theta} - i\, B_l\, \xi_l(k_e r)\, P_l^{(1)}(\cos\theta)\, \sin\theta \right\}$$

$$\tag{6.106}$$

$$H_r^s = -\frac{\sin\phi\, E_0}{k_0 k_e r^2} \sum_{l=1}^{\infty} B_l\, l(l+1)\, \xi_l(k_e r)\, P_l^{(1)}(\cos\theta) \tag{6.107}$$

$$H_\theta^s = -\frac{i\sin\phi\, E_0}{k_0 r} \sum_{l=1}^{\infty} \left\{ C_l \xi_l(k_e r) \frac{P_l^{(1)}(\cos\theta)}{\sin\theta} + i B_l \xi_l'(k_e r)\, P_l^{(1)\prime}(\cos\theta)\, \sin\theta \right\}$$

$$\tag{6.108}$$

$$H_\phi^s = \frac{i\cos\phi\, E_0}{k_0 r} \sum_{l=1}^{\infty} \left\{ C_l\, \xi_l(k_e r)\, P_l^{(1)\prime}(\cos\theta)\, \sin\theta + i B_l \xi_l'(k_e r) \frac{P_l^{(1)}(\cos\theta)}{\sin\theta} \right\}$$

$$\tag{6.109}$$

where

$$C_l = i^{l+1} \frac{2l+1}{l(l+1)} \frac{\psi_l(\rho)\psi_l'(m\rho) - m\,\psi_l'(\rho)\psi_l(m\rho)}{\xi_l(\rho)\psi_l'(m\rho) - m\,\xi_l'(\rho)\psi_l(m\rho)} \tag{6.110}$$

$$B_l = i^{l+1} \frac{2l+1}{l(l+1)} \frac{\psi_l'(\rho)\psi_l(m\rho) - m\,\psi_l(\rho)\psi_l'(m\rho)}{\xi_l'(\rho)\psi_l(m\rho) - m\,\xi_l(\rho)\psi_l'(m\rho)} \tag{6.111}$$

where $\psi_l(\rho)$ and $\xi_l(\rho)$ are spherical functions related to the cylindrical functions of fractional order by the following identities:

$$\psi_l(\rho) = \left(\frac{\pi\rho}{2}\right)^{1/2} J_{l+1/2}(\rho), \quad \xi_l(\rho) = \left(\frac{\pi\rho}{2}\right)^{1/2} H_{l+1/2}^{(2)}(\rho) \tag{6.112}$$

k_i is the wave number of the material within the sphere, $P_l(\cos\theta)$ is a Legendre polynomial, and $P_l^{(1)}(\cos\theta)$ is an associated Legendre function.

In accord with eqs. 6.103 through 6.109, the total electromagnetic field is actually the sum of a primary and a secondary field, that is:

$$E = E_0 + E^s, \quad H = H_0 + H^s$$

At the same time, as indicated by eqs. 6.104–6.109, the secondary field can be interpreted as being the sum of various spherical waves with magnitudes defined by the numbers C_l and B_l, which depend primarily on the parameters $k_e a = \rho$ and $k_i a = m\rho$.

Let us consider the secondary field for the case in which $|\rho| \ll 1$. This is equivalent to assuming that the skin depth in the surrounding medium is much greater than the radius of the sphere. Actually:

$$|\rho| = |k_e a| = (\sigma_e \mu\omega)^{1/2} a = a\sqrt{2}/h_e$$

where h_e is the skin depth in a medium with the resistivity ρ_e. The condition:

$$|\rho| \ll 1 \tag{6.113}$$

usually applies in inductive mineral prospecting. For example, if the radius of the sphere is 30 m, the resistivity of this host medium is 200 Ωm, and the frequency of the field is 200 Hz, the parameter $|\rho|$ is no greater than 0.1. It is clear that if the parameter:

$$|\rho| = a\sqrt{2}/h_e$$

is not sufficiently small in value, the distance between the observation point and the center of the sphere, r, becomes greater than a skin distance, h_e. As a consequence, due to the absorption of electromagnetic energy in the conducting medium, the secondary field will be small in comparison with the primary field. For this reason, the range of frequencies which correspond to condition 6.113 can be considered as being the most favorable for practical applications.

For small values $|\rho|$ the functions ψ_l and ξ_l and their derivatives can be approximated as:

$$\psi_l(\rho) = \frac{\rho^{l+1}}{(2l+1)!!} \qquad \xi_l(\rho) = (2l-1)!! \frac{i}{\rho^l} e^{-i\rho}$$

$$\psi_l'(\rho) = \frac{(l+1)}{(2l+1)!!} \rho^l \qquad \xi_l'(\rho) = -(2l-1)!! li \frac{e^{-i\rho}}{\rho^{l+1}} \tag{6.114}$$

Substituting these expressions into the general formulas for the coefficients C_l and B_l, after some simple algebraic operations, we arrive at the following results:

$$C_l = \frac{(-i)^l \rho^{2l+1}}{l^2 [(2l-1!!]^2} \frac{m^2-1}{m^2+(l+1)/l} (-1)^{l+1}$$

$$B_l = \frac{(-i)^l \rho^{2l+1}}{l(l+1)[(2l-1)!!]^2} \frac{\psi_{l+1}(m\rho)}{\psi_{l-1}(m\rho)} (-1)^{l+1} \tag{6.115}$$

For example, we have:

$$C_1 = -i\rho^3 \frac{m^2-1}{m^2+2}, \quad C_2 = \frac{\rho^5}{36} \frac{m^2-1}{m^2+1.5}, \quad C_3 = \frac{i\rho^7 (m^2-1)}{9 \cdot 225 (m^2+4/3)} \tag{6.116}$$

$$B_1 = -i\frac{\rho^3}{2} \frac{\psi_2(m\rho)}{\psi_0(m\rho)}, \quad B_2 = \frac{\rho^5}{54} \frac{\psi_3(m\rho)}{\psi_1(m\rho)}, \quad B_3 = \frac{i\rho^7}{12 \cdot 225} \frac{\psi_4(m\rho)}{\psi_2(m\rho)}$$

If field measurements are made at distances which are at least two times greater than the radius of the sphere, in accord with eq. 6.116, only the first terms C_1 and B_1 have any significant influence. Therefore, in the expressions for the field given in eqs. 6.104–6.109, we can discard all terms except the first, so that these expressions take on the following approximate forms:

$$E_r^s = \frac{2 E_0}{k_e^2 r^2} C_1 \xi_1(k_e r) P_1^{(1)}(\cos\theta) \cos\phi \tag{6.117}$$

$$E_\theta^s = -\frac{\cos\phi E_0}{k_e r} \left\{ C_1 \xi_1'(k_e r) P_1^{(1)'}(\cos\theta) \sin\theta - i B_1 \xi_1(k_e r) \frac{P_1^{(1)}(\cos\theta)}{\sin\theta} \right\} \tag{6.118}$$

$$E_\phi^s = -\frac{\sin\phi}{k_e r} E_0 \left\{ C_1 \xi_1'(k_e r) \frac{P_1^{(1)}(\cos\theta)}{\sin\theta} - i B_1 \xi_1(k_e r) P_1^{(1)'}(\cos\theta) \sin\theta \right\} \tag{6.119}$$

$$H_r^s = -\frac{2 E_0}{k_0 k_e r^2} B_1 \xi_1(k_e r) P_1^{(1)}(\cos\theta) \sin\phi \tag{6.120}$$

$$H_\theta^s = -\frac{i \sin\phi E_0}{k_0 r} \left\{ C_1 \xi_1(k_e r) \frac{P_1^{(1)}(\cos\theta)}{\sin\theta} + i B_1 \xi_1'(k_e r) P_1^{(1)'}(\cos\theta) \sin\theta \right\} \tag{6.121}$$

$$H_\phi^s = \frac{i \cos\phi E_0}{k_0 r} \left\{ C_1 \xi_1(k_e r) P_1^{(1)'}(\cos\theta) \sin\theta + i B_1 \xi_1'(k_e r) \frac{P_1^{(1)}(\cos\theta)}{\sin\theta} \right\} \tag{6.122}$$

By rewriting the functions: $\xi_1, \xi_1', P_1^{(1)}$ and $P_1^{(1)'}$ in terms of elementary functions, we have:

$$E_r^s = 2 \frac{\sigma_i - \sigma_e}{\sigma_i + 2\sigma_e} \frac{a^3}{r^3} E_0 \, e^{-ik_e r}(1 + ik_e r) \cos\phi \sin\theta$$

$$E_\theta^s = \frac{\sigma_i - \sigma_e}{\sigma_i + 2\sigma_e} \frac{a^3}{r^3} E_0 \, e^{-ik_e r}(k_e^2 r^2 - ik_e r - 1) \cos\phi \cos\theta$$

$$\qquad - \frac{i\omega\mu}{2} D_1 \frac{a^3}{r^2} H_0 \, e^{-ik_e r}(1 + ik_e r) \cos\phi$$

$$E_\phi^s = -\frac{\sigma_i - \sigma_e}{\sigma_i + 2\sigma_e} E_0 \frac{a^3}{r^3} e^{-ik_e r}(k_e^2 r^2 - ik_e r - 1) \sin\phi$$

$$\qquad + \frac{i\omega\mu}{2} D_1 \frac{a^3}{r^2} H_0 \, e^{-ik_e r}(1 + ik_e r) \sin\phi \cos\theta \tag{6.123}$$

$$H_r^s = D_1 H_0 \frac{a^3}{r^3} e^{-ik_e r}(1 + ik_e r) \sin\theta \sin\phi$$

$$H_\theta^s = -\sigma_e E_0 \frac{\sigma_i - \sigma_e}{\sigma_i + 2\sigma_e} \frac{a^3}{r^2} e^{-ik_e r}(1 + ik_e r)\sin\phi$$

$$+ \frac{D_1}{2} H_0 \frac{a^3}{r^3} e^{-ik_e r}(k_e^2 r^2 - ik_e r - 1)\sin\phi \cos\theta$$

and

$$H_\phi^s = -\sigma_e E_0 \frac{\sigma_i - \sigma_e}{\sigma_i + 2\sigma_e} \frac{a^3}{r^2} e^{-ik_e r}(1 + ik_e r)\cos\phi \cos\theta$$

$$+ \frac{D_1}{2} H_0 \frac{a^3}{r^3} e^{-ik_e r}(k_e^2 r^2 - ik_e r - 1)\cos\phi$$

As has already been pointed out, the set of equations in 6.123 describes the secondary field with reasonable accuracy when

$$r/a > 2 \quad \text{and} \quad a/h_e \ll 1 \tag{6.124}$$

The set of equations in 6.123 permit us to arrive at some useful conclusions concerning the behavior of the electromagnetic field. First of all, the electromagnetic field is equivalent to the field of two dipoles situated in a uniform conducting medium at the origin, that is at the center of the sphere. The first of these dipoles is an electric dipole with the moment P_e directed along the x-axis (direction of the primary electric field E_0) and its strength is:

$$P_e = \frac{\sigma_i - \sigma_e}{\sigma_i + 2\sigma_e} a^3 E_0 \tag{6.125}$$

The second of the dipoles is a magnetic type with a moment given by:

$$M_1 = 2\pi D_1 a^3 H_0 \tag{6.126}$$

directed along the y-axis (the direction of the primary field H_0) and as has been demonstrated in Chapter 3, the function D can be written in the form:

$$D_1 = \frac{3 \coth ik_i a}{ik_i a} + \frac{3}{k_i^2 a^2} - 1 \tag{6.127}$$

The nature of its behavior has been described in considerable detail in Chapter 3, but it is important to note that this function depends only on the parameter a/h_i defined by the conductivity and by the radius of the sphere.

Let us assume that the skin depth in the surrounding medium, h_e, is considerably greater than the distance from this sphere, that is:

$$r/h_e \ll 1 \tag{6.128}$$

In this case, the set of expressions in 6.123 will be simplified considerably, so that we will have:

508

$$E_r^s = 2E_0 \frac{a^3}{r^3} \frac{\sigma_i - \sigma_e}{\sigma_i + 2\sigma_e} \cos\theta \sin\phi \tag{6.129}$$

$$E_\theta^s = -E_0 \frac{a^3}{r^3} \frac{\sigma_i - \sigma_e}{\sigma_i + 2\sigma_e} \cos\theta \cos\phi - \frac{i\omega\mu}{2} D_1 H_0 \frac{a^3}{r^2} \cos\phi \tag{6.130}$$

$$E_\phi^s = E_0 \frac{a^3}{r^3} \frac{\sigma_i - \sigma_e}{\sigma_i + 2\sigma_e} \sin\phi + \frac{i\omega\mu}{2} H_0 D_1 \frac{a^3}{r^2} \cos\theta \sin\phi \tag{6.131}$$

$$H_r^s = D_1 H_0 \frac{a^3}{r^3} \sin\theta \sin\phi \tag{6.132}$$

$$H_\theta^s = -\sigma_e E_0 \frac{a^3}{r^2} \frac{\sigma_i - \sigma_e}{\sigma_i + 2\sigma_e} \sin\phi - \frac{D_1}{2} H_0 \frac{a^3}{r^3} \sin\phi \cos\theta \tag{6.133}$$

$$H_\phi^s = -\sigma_e E_0 \frac{a^3}{r^2} \frac{\sigma_i - \sigma_e}{\sigma_i + 2\sigma_e} \cos\phi \cos\theta - \frac{D_1}{2} H_0 \frac{a^3}{r^3} \cos\phi \tag{6.134}$$

It is clear that the simplicity in this last set of equations is based on the assumption that measurements are performed in the near zone of the secondary field, that is, when $r/h_e \ll 1$.

The set of equations 6.129 through 6.134 permit us to interpret the behavior of the secondary field in the following way: Because of the change in the primary magnetic field with time, induced currents will appear in this sphere. In our approximation, the electromagnetic field from these currents is equivalent to that from a magnetic dipole located in a uniform conducting medium, and these induced currents are tangential to the surface of the sphere. Inasmuch as the expression for the moment of the magnetic dipole (eq. 6.126) contains the function D_1, this part of the field is usually called the vortex or inductive part, and depends only on the conductivity of the sphere (for the condition $r/h_e \ll 1$). For example, when the resistivity of the sphere is infinitely large (that is, it is an insulator), the electromagnetic field from this magnetic dipole is zero. In contrast, when the sphere has a very high conductivity, the function D_1 approaches its asymptote of -1, and the magnitiude of the secondary field is maximum.

At this point, let us examine the second part of the field, often called the galvanic part. Under the action of the primary electric field, E_0, electrical charges of both signs appear on the surface of the sphere as shown in Fig. 6.2. The electromagnetic field of these alternating charges is equivalent to that from an electric dipole source in a uniform conducting medium. In the approximation we are treating, these charges cause an electric field in accord with Coulomb's law. This electric field, in correspondence with Ohm's law, causes secondary currents to arise and their flow lines intersect the surface of the sphere. In turn, these currents create the magnetic field. It is obvious that if the conductivity in the host medium is zero, the galvanic part of

Fig. 6.2. Development of electric charge on the surface of a sphere under the action of a primary electric field.

the magnetic field disappears. In fact, in accord with eqs. 6.133 and 6.134, this part of the field is directly proportional to σ_e ($\sigma_i/\sigma_e \gg 1$ or $\sigma_i/\sigma_e \ll 1$), while the galvanic part of the electric field does not vanish when σ_e is 0, because the surface electric charges still remain.

In the case of an insulating sphere, the galvanic part of the magnetic and electric fields will not vanish because the surface charges still appear and there are secondary currents in the surrounding medium flowing around the insulator. It is extremely important to note that the galvanic part of the field is only weakly related to the conductivity of the sphere, as follows from eq. 6.125. In fact, the coefficient:

$$(\sigma_i - \sigma_e)/(\sigma_i + 2\sigma_e)$$

can be written as:

$$\left(\frac{\sigma_i}{\sigma_e} - 1\right)\Big/\left(\frac{\sigma_i}{\sigma_e} + 2\right)$$

and it is obvious that it varies from values $-1/2$ to $+1$ as the ratios of conductivities varies from zero to infinity. We should note that even when the ratio σ_i/σ_e amounts to a factor only as large as 10, the coefficient:

$$\left(\frac{\sigma_i}{\sigma_e} - 1\right)\Big/\left(\frac{\sigma_i}{\sigma_e} + 2\right)$$

is very nearly equal to unity and the influence of the conductivity in the sphere practically vanishes.

Sometimes the galvanic part of the field is considered as a channeling or gathering effect. Inasmuch as the term "galvanic part" emphasizes the main cause of this field (the existence of charges), the presence of which quite clearly explains the behavior in the electric and magnetic fields along with the distribution of currents, we will use this term for the electrical type of oscillations, and they turn out to be directly proportional to the primary electric field strength.

In accord with the equation set 6.129 through 6.134, the galvanic part of the field coincides with the direct current field behavior, that is, in the near zone the skin effect has no significant effect on the galvanic part of the field. Similarly, the interaction between currents flowing in the host medium is also negligible when the vortex part of the field is considered in the near zero. However, this portion of the field can be subjected to the skin effect inside the sphere, which in turn is controlled by the behavior of the function $D_1(a/h_i)$.

Up until this point, we have made use of the equations set 6.129 to 6.134 which describes the secondary field in the near zone. If the observation point is located beyond this zone, but still meets the conditions that $r/a > 2$ and $a/h_e \ll 1$, one can make use of the equations in the set 6.123. In this case, the skin effect in the host medium can be quite significant; that is, the influence of the induced currents in this medium can be detected. Thus, both parts of the secondary field turn out to be functions of the conductivity of the host medium. However, the effect of the conductivity, σ_e, is exerted on the vortex and the galvanic parts in different ways.

First of all, in the near zone, the vortex part of the secondary field is nearly totally insensitive to influence from induced currents in the host medium. With increase in values for the parameter r/h_e, this influence becomes significant. In contrast, the galvanic part is subject to an influence from the conductivity σ_e as a consequence of two factors, namely:

(1) The magnetic field associated with the galvanic part, at least in the near zone, turns out to be directly proportional to the conductivity, σ_e, when the contrast coefficient $(\sigma_i - \sigma_e)/(\sigma_i + 2\sigma_e)$ approaches its maximum value. However, the electrical charge density and the corresponding electric field E^g are almost indpendent of the conductivities when $\sigma_i/\sigma_e \gg 1$ or $\sigma_i/\sigma_e \ll 1$ in the near zone.

(2) The second factor is the same as that for the vortex part, that is, the appearance of induced currents due to the secondary field in the host medium.

It is appropriate at this point to emphasize again that in accord with eqs. 6.123, in contrast to the behavior of the vortex part of the field, the galvanic part is practically independent of the conductivity in the sphere. This means that classification of anomalies as "ore" or "non-ore" based on measuring the galvanic part of the field is not likely to be feasible, just as in the case for direct current methods. Therefore, it is of considerable practical interest to investigate the relationship between the galvanic and vortex parts of the field. First, we will examine the ratio between the electric fields caused by magnetic and electric dipoles. As an example, in accord with eqs. 6.131 we have:

$$\eta_\phi^e = \left| \frac{E_\phi^M}{E_\phi^e} \right| = \frac{|i\omega\mu r| \, |D_1| \, |H_0|}{2 \, |E_0|} |\cos\theta| = \frac{|k_e r| \, |D_1| |\cos\theta|}{2} \tag{6.135}$$

since

$$H_0 = \left(- i \, \frac{\sigma_e}{\mu\omega} \right)^{1/2} E_0, \quad \frac{\sigma_i - \sigma_e}{\sigma_i + 2\sigma_e} \approx 1$$

Considering that the parameter $k_e r$ is usually quite small, the electric field of the galvanic part is significantly greater than the electric field of vortex origin. This means that in measuring the electric field in the case in which the primary field is a plane wave, significant anomalies can be caused by both conductive and resistive inhomogeneities. In other words, the resolution obtainable with an electromagnetic technique in which the electric field is measured as the relationship between the primary fields E_0 and H_0 is that for plane wave behavior, is practically nonexistence. This conclusion applies particularly to such methods as VLF, AFMAG, audio-frequency soundings, and TURAM, where measurements are mainly performed in the wave zone.

At this point, let us examine the relationship between the magnetic fields caused by both parts of the secondary field. As an example, the magnitude of the ratio ϕ-components of the magnetic fields generated by electric and magnetic dipoles is

$$\eta_\phi^M = \left| \frac{H_\phi^e}{H_\phi^M} \right| = \frac{|k_e r|}{|D_1|} \, 2 \, |\cos\theta| \tag{6.136}$$

Let $2\cos\theta = 1$ and with the definitions:

$$\frac{\sigma_i}{\sigma_e} = s, \quad r = an$$

we have:

$$\eta_\phi^M = \frac{n}{\sqrt{s}} \frac{|k_1 a|}{|D_1|} \tag{6.137}$$

We will now consider how this function varies with frequency.

At high frequencies, the function D_1 takes on the value unity, and we obtain:

$$\mathrm{n}_\phi^M = \frac{n}{\sqrt{s}} \, |k_i a| \tag{6.138}$$

This is, with an increase in the frequency, the magnetic field of the galvanic part becomes the greater. At sufficiently low frequencies, the function D_1 is nearly equal to $|k_i^2 a^2|/15$. Substituting this approximation into eq. 6.137, we have:

$$\eta_\phi^M = \frac{15 \, n}{\sqrt{s} \, |k_i a|} \tag{6.139}$$

Therefore, with decrease in frequency, the relative importance of the galvanic part of the magnetic field grows; that is, the magnetic field from currents that close through the host medium begins to dominate over that of the induced currents within the sphere. This behavior can be easily explained. In fact, with a decrease in frequency, the magnetic field from the eddy currents flowing in the sphere decreases in direct proportion to ω, while the magnetic field of the galvanic part is proportional to the square root of ω, because the primary electric field is related to the primary magnetic field as:

$$E_0 = \left(\frac{\omega\mu}{-i\sigma_e}\right)^{1/2} H_0$$

As may be seen from eq. 6.137, as well as being obvious from the physical point of view, with an increase in the resistivity of the host medium, the relative contribution of the vortex part of the magnetic field increases.

Let us examine two numerical examples. Suppose that $n = 2$, $|k_i a| = 5$, and $|D_1| = 1$. Then:

$$n_\phi^M \cong \frac{10}{\sqrt{s}}, \quad \text{and} \quad \text{if} \begin{cases} s = 1000 & \eta_\phi^M = 0.3 \\ s = 10000 & \eta_\phi^M \approx 0.1 \end{cases}$$

Whence it is possible to say that only for very large ratios of the resistivity between the host medium and the sphere, the magnetic field of the galvanic part of the secondary field can be neglected.

We arrive at the conclusion that electromagnetic methods can be applied in the wave zone to classify anomalies on the basis of conductivity only in the rare cases when the host medium is very resistive and only the magnetic field is measured.

Before we investigate the secondary field for the case in which the source is a magnetic dipole, let us note from an examination of eqs. 6.129 through 6.139, that in principle the possibility exists of making measurements in a way in which the galvanic part of the field would be largely suppressed. For example, in accord with eq. 6.134, the difference:

$$\frac{H_\phi^s(\omega_1)}{\sqrt{\omega_1}} - \frac{H_\phi^s(\omega_2)}{\sqrt{\omega_2}}$$

is defined almost entirely by the vortex part of the field.

6.3. A CONDUCTING SPHERE IN THE FIELD OF A MAGNETIC DIPOLE

In this section, we will examine the frequency- and time-domain behavior of an electromagnetic field in the case in which a conducting sphere is situated in the field from an arbitrarily oriented dipole and in which the host medium is conductive as well. Particular attention will be paid to the relationship between vortex and galvanic parts of the secondary electromagnetic field for various ranges in the frequency and time domains. As has been described in detail in the first paragraph of this chapter, the solution for a dipole source arbitrarily oriented with respect to the position of the sphere can be reduced to two problems, which are: (1) that of a radial dipole in the presence of a sphere; and (2) that of a transversal dipole in the presence of the sphere.

Before we start the analysis of the electromagnetic field, we will make one comment. In contrast to the first paragraph, we will use somewhat different notations for the potentials of the electric and magnetic types. In comparing the corresponding formulas in the two sections, relating components of the magnetic and electric fields with potentials, it can readily be seen that the functions U_1 and U_2 and Π_1 and Π_2 are related with each other in a very simple manner. Our investigation will start from the simplest case, that of a primary vortex electric field which does not intersect the surface of the sphere.

Radial dipole

The field of the radial dipole can be expressed in terms of a single vector potential of the magnetic type, Π_2:

$$\Pi_2 = \frac{M_{0r}}{4\pi\, b} \sum_{n=1}^{\infty} e_n \left[\frac{\xi_n(k_e r)}{r} \right] P_n(\cos\theta) \tag{6.140}$$

where M_{0r} is the dipole moment, b is the distance from the dipole to the center of the sphere as shown in Fig. 6.3, and:

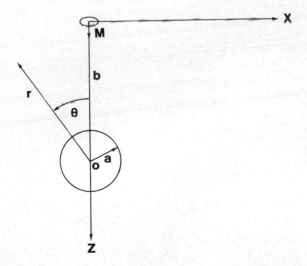

Fig. 6.3. Magnetic dipole source illuminating a conducting sphere.

$$e_n = \frac{k_i \, \psi_n \, (k_e a) \, \psi'_n \, (k_i a) - k_e \, \psi'_n \, (k_e a) \, \psi_n \, (k_i a)}{k_i \, \xi_n \, (k_e a) \, \psi'_n \, (k_i a) - k_e \, \xi'_n \, (k_e a) \, \psi_n \, (k_i a)} \, d_n \tag{6.141}$$

where

$$d_n = -i \, (2n + 1) \frac{\xi_n \, (k_e b)}{k_e b} \tag{6.142}$$

From this point on, we will consider the case in which the skin depth in the host medium is greater than the radius of the sphere, that is:

$$|k_e a| \ll 1$$

Using asymptotic expressions for the functions $\psi_n \, (k_e a)$ and $\xi_n \, (k_e a)$ and their derivatives as listed in eq. 6.44, in making use of the recurrence relationships for the functions $\psi_n \, (z)$:

$$\psi_{n+1}(z) = \frac{2n + 1}{z} \, \psi_n \, (z) - \psi_{n-1}(z)$$

we obtain the following expression for the function e_n:

$$e_n = \frac{\rho^{2n+1}}{k_e b \, [(2n - 1)!!]^2} D_n \, \xi_n \, (k_e b) \tag{6.143}$$

where

$$\rho = k_e a, \quad D_n = \frac{\psi_{n+1} \, (k_i a)}{\psi_{n-1} \, (k_i a)}$$

In accord with eq. 6.143, we have:

$$\Pi_2 = \frac{M_{0r}}{4\pi b} \sum_{n=1}^{\infty} \frac{\rho^{2n+1}}{[(2n - 1)!!]^2} D_n \, \frac{\xi_n \, (k_e b)}{k_e b} \, \frac{\xi_n \, (k_e r)}{r} \, P_n \, (\cos\theta)$$

The components of the electromagnetic field are derivable from the potential Π_2 as follows:

$$E_r = 0 \qquad\qquad H_r = \frac{\partial^2 (r \, \Pi_2)}{\partial r^2} + k^2 r \, \Pi_2$$

$$E_\theta = 0 \qquad\qquad H_\theta = \frac{1}{r} \, \frac{\partial^2 (r \, \Pi_2)}{\partial r \, \partial\theta}$$

$$E_\phi = \frac{i\omega\mu}{r} \, \frac{\partial (r \, \Pi_2)}{\partial\theta} \qquad H_\phi = 0$$

Therefore, we obtain the following expression for the field components for a radial dipole source:

$$E_\phi^s = -\frac{i\omega\mu\sin\theta}{4\pi b} M_{0r} \sum_{n=1}^{\infty} \frac{\rho^{2n+1}}{[(2n-1)!!]^2} D_n \frac{\xi_n(k_e b)}{k_e b} \frac{\xi_n(k_e r)}{r} P_n'(\cos\theta)$$

$$H_r^s = \frac{k_e^2}{4\pi b} M_{0r} \sum_{n=1}^{\infty} \frac{\rho^{2n+1}}{[(2n-1)!!]^2} D_n \frac{\xi_n(k_e b)}{k_e b} [\xi_n''(k_e r) + \xi_n(k_e r)] P_n(\cos\theta)$$

$$H_\theta^s = -\frac{k_e\sin\theta}{4\pi b} M_{0r} \sum_{n=1}^{\infty} \frac{\rho^{2n+1}}{[(2n-1)!!]^2} D_n \frac{\xi_n(k_e b)}{k_e b} \frac{1}{r} \xi_n'(k_e r) P_n'(\cos\theta)$$

$$\text{if } a/h_e \ll 1 \qquad (6.144)$$

If the distance from the center of the sphere to the observation point is less than a skin depth in this host medium, and if $|k_e b| < 1$, the expressions for the field given in eqs. 6.144 are markedly simplified. Substituting expressions for $\xi_n(k_e r)$ and $\xi_n(k_e b)$ and their derivatives valid for small values of the argument, we have:

$$E_\phi^s = \frac{i\omega\mu a^3}{4\pi b^3 r^2} M_{0r} \sin\theta \left[D_1 + \frac{3a^2}{br} D_2 \cos\theta + \frac{3a^4}{2b^2 r^2} D_3 (5\cos^2\theta - 1) + \ldots \right]$$

$$H_r^s = -\frac{2a^3 M_{0r}}{4\pi b^3 r^3} \left[D_1 \cos\theta + D_2 \frac{3}{2}\frac{a^2}{br} (3\cos^2\theta - 1) \right.$$

$$\left. + \frac{3a^4}{b^2 r^2} D_3 (5\cos^2\theta - 3\cos\theta) + \ldots \right] \qquad (6.145)$$

$$H_\theta^s = -\frac{a^3 M_{0r}}{4\pi b^3} \frac{\sin\theta}{r^3} \left[D_1 + \frac{6a^2}{br} D_2 \cos\theta + \frac{9a^4}{2b^2 r^2} D_3 (5\cos^2\theta - 1) + \ldots \right]$$

Equations 6.145 describe the secondary field caused by currents induced in the sphere for the condition that $r/h_e \ll 1$.

An elementary numerical analysis shows that for those cases in which the distances b and r are at least two times greater than the radius of the sphere, a, we can restrict ourselves to considering only the first term in each of the series in eqs. 6.145, and yet have reasonable accuracy. By so doing, we have:

$$E_\phi^s = \frac{i\omega\mu a^3}{4\pi b^3 r^2} M_{0r} D_1 \sin\theta$$

$$H_r^s = -\frac{a^3 D_1 M_{0r}}{2\pi b^3 r^3} \cos\theta \qquad (6.146)$$

$$H_\theta^s = -\frac{a^3 M_{0r}}{4\pi b^3 r^3} D_1 \sin\theta$$

These last equations correspond to the case in which a sphere is located in a uniform magnetic field with the strength equal to that from a radial magnetic dipole at the center of the sphere, that is, $M_{0r}/2\pi b^3$. At the same time, in place of eqs. 6.144 we can write

$$E^s_\phi = i\omega\mu \frac{M_{0r}}{4\pi} \frac{a^3}{b^3} D_1 e^{-ik_e b} (1 + ik_e b) \frac{e^{-ik_e r}}{r^2} (1 + ik_e r) \sin\theta$$

$$H^s_r = -\frac{2D_1}{4\pi} \frac{M_{0r}}{b^3} a^3 e^{-ik_e b} (1 + ik_e b) \frac{e^{-ik_e r}}{r^3} (1 + ik_e r) \cos\theta \qquad (6.147)$$

$$H^s_\theta = \frac{M_{0r}}{4\pi} D_1 \frac{a^3}{b^3} e^{-ik_e b} (1 + ik_e b) \frac{e^{-ik_e r}}{r^3} (k_e^2 r^2 - ik_e r - 1) \sin\theta$$

if $r/a > 2$, $b/a > 2$, $a/h_e \ll 1$

Let us now introduce the notation:

$$M_1 = 2\pi D_1 H_0 a^3 \qquad (6.148)$$

where

$$H_0 = \frac{M_{0r}}{2\pi b^3} e^{-ik_e b} (1 + ik_e b)$$

is the magnetic field of the radial magnetic dipole observed at the center of the sphere. Then, we obtain:

$$E^s_\phi = \frac{i\omega\mu}{4\pi} M_1 \frac{e^{-ik_e r}}{r^2} (1 + ik_e r) \sin\theta$$

$$H^s_r = -\frac{M_1}{2\pi r^3} e^{-ik_e r} (1 + ik_e r) \cos\theta \qquad (6.149)$$

$$H^s_\theta = \frac{M_1}{4\pi r^3} e^{-ik_e r} (k_e^2 r^2 - ik_e r - 1) \sin\theta$$

Therefore, the field caused by currents induced in the sphere in the presence of the radial dipole source is equivalent to the field of a magnetic dipole if the skin depth in the host medium is significantly greater than the radius of the sphere and if the distance from the radial dipole to the center of the sphere as well as the distance from the observation site, that is both b and r, are at least twice as great as the radius of the sphere. The moment of this magnetic dipole M_1 describing the secondary field is oriented along the same line as the moment M_{0r}.

The conditions which define the equivalence of the secondary field to that of the magnetic dipole, supplementing each other, provide uniformity of the primary magnetic field at all points on the sphere, and a stronger influence for the first term in eq. 6.144. For this reason, with a decrease in frequency, or in the case of a very resistive host medium, the expressions for the components of the secondary electromagnetic field given in eq. 6.149 correspond with those for a sphere placed in a uniform magnetic field in free space. Thus, at very low frequencies, the effect of the host medium on the magnitude of the secondary field is almost negligible. However, the magnitude of the magnetic field caused by currents induced both in the sphere and in the host medium (that is, the total field) is subject to the influence of the host medium regardless of how low the frequency is. This question has been investigated in detail in an earlier chapter.

The main characteristics of the field when the primary source is a radial dipole are based on the fact that the flow lines for the primary electric field do not intersect on the surface of the sphere and therefore, electrical charges do not arise. In all cases, when the primary electric field is tangential to the surface of conductor, the main features of the behavior of the field which have been described previously will be observed. For example, the secondary field due to vortex currents in a conductor that is characterized by axial symmetry and with an axis that coincides with the vertical axis of a circular loop transmitter tends to the value for the field in free space as the frequency decreases.

It is relevant to note here that the absence of electric charges is important for classifying anomalies according to conductivity.

The transversal dipole

The case in which a transverse dipole source eliminates a spherical body is shown in Fig. 6.4. It is obvious that in this case the flow lines for the primary electric field cross the surface of the sphere, and that therefore, electric charges must arise. From the mathematical point of view, this means that a secondary field can be expressed in terms of potentials of both the electric and magnetic types, Π_1 and Π_2, which in accord with the basic material developed in Paragraph 6.1 can be written as:

$$\Pi_1 = \frac{\sin\phi\, M_{0t}}{4\pi} \sum_{n=1}^{\infty} a_{1n} \frac{\xi_n(k_e r)}{r} P_n^{(1)}(\cos\theta) \tag{6.150}$$

$$\Pi_2 = \frac{\cos\phi\, M_{0t}}{4\pi} \sum_{n=1}^{\infty} a_{2n} \frac{\xi_n(k_e r)}{r} P_n^{(1)}(\cos\theta) \tag{6.151}$$

where

$$a_{1n} = - \frac{\psi_n(k_e a)\psi_n'(k_i a) - \dfrac{k_i}{k_e}\psi_n'(k_e a)\psi_n(k_i a)}{\xi_n(k_e a)\psi_n'(k_i a) - \dfrac{k_i}{k_e}\xi_n'(k_e a)\psi_n(k_i a)} \alpha_n$$

$$\tag{6.152}$$

518

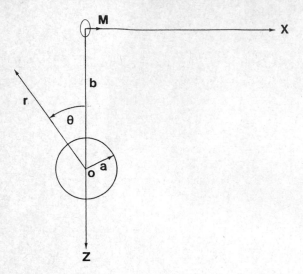

Fig. 6.4. Excitation of a conducting sphere by a transversal dipole source.

$$\alpha_n = -\frac{\omega\mu\,(2n+1)}{n(n+1)k_e b}\,\xi_n(k_e b)$$

and

$$a_{2n} = \frac{\psi_n'(k_e a)\,\psi_n(k_i a) - (k_i/k_e)\,\psi_n(k_e a)\,\psi_n'(k_i a)}{\xi_n'(k_e a)\,\psi_n(k_i a) - (k_i/k_e)\,\xi_n(k_e a)\,\psi_n'(k_i a)}\,\beta_n$$

$$\beta_n = -\frac{i\,(2n+1)}{n\,(n+1)}\,\frac{\xi_n'(k_e b)}{b}$$

(6.153)

For small values of the parameter $k_e a$, the expressions for the coefficients a_{1n} and a_{2n} are much simplified, and we have:

$$a_{1n} = \frac{i\omega\mu\,\rho^{2n+1}}{[(2n-1)!!]^2 n^2}\,\frac{m^2-1}{m^2+(n+1)/n}\,\frac{\xi_n(k_e b)}{k_e b}$$

$$\text{if } |k_i a| \ll 1 \qquad (6.154)$$

$$a_{2n} = \frac{\rho^{2n+1}}{b\,[(2n-1)!!]^2}\,\frac{D_n}{n(n+1)}\,\xi_n'(k_e b)$$

where $m = k_i/k_e$. In accord with eqs. 6.150—6.151 we have:

$$\Pi_1 = \frac{i\omega\mu}{k_e b}\,\frac{\sin\phi}{4\pi}\,M_{0t}\sum_{n=1}^{\infty}\frac{\rho^{2n+1}}{[(2n-1)!!]^2 n^2}\,\frac{m^2-1}{m^2+\dfrac{n+1}{n}}\,\xi_n(k_e b)$$

$$\times\,\frac{\xi_n(k_e r)}{r}\,P_n^{(1)}(\cos\theta)$$

(6.155)

$$\Pi_2 = \frac{\cos\phi}{4\pi b} M_{0t} \sum_{n=1}^{\infty} \frac{\rho^{2n+1} D_n \, \xi_n'(k_e b)}{[(2n-1)!!]^2 n(n+1)} \frac{\xi_n(k_e r)}{r} P_n^{(1)}(\cos\theta) \qquad (6.156)$$

Using the various relationships that exist between the magnetic potential, Π_2, and the various field components, we obtain:

$$E_r^s = 0$$

$$E_\theta^s = \frac{i\omega\mu \, \sin\phi}{4\pi b \, \sin\theta} M_{0t} \sum_{n=1}^{\infty} \frac{\rho^{2n+1} D_n \, \xi_n'(k_e b)}{[(2n-1)!!]^2 n(n+1)} \frac{\xi_n(k_e r)}{r} P_n^{(1)}(\cos\theta)$$

$$E_\phi^s = \frac{i\omega\mu \cos\phi}{4\pi \, b} M_{0t} \sum_{n=1}^{\infty} \frac{\rho^{2n+1} D_n \, \xi_n'(k_e b)}{[(2n-1)!!]^2 n(n+1)} \frac{\xi_n(k_e r)}{r} \frac{\partial}{\partial\theta} P_n^{(1)}(\cos\theta)$$

$$H_r^s = \frac{k_e^2 \cos\phi}{4\pi b} M_{0t} \sum_{n=1}^{\infty} \frac{\rho^{2n+1} D_n \, \xi_n'(k_e b)}{[(2n-1)!!]^2 n(n+1)} [\xi_n''(k_e r) + \xi_n'(k_e r)] P_n^{(1)}(\cos\theta)$$

$$\qquad\qquad (6.157)$$

$$H_\theta^s = \frac{\cos\phi}{4\pi b} M_{0t} \sum_{n=1}^{\infty} \frac{\rho^{2n+1} D_n \, \xi_n'(k_e b)}{[(2n-1)!!]^2 n(n+1)} \frac{\xi_n'(k_e r)}{r} \frac{\partial}{\partial\theta} P_n^{(1)}(\cos\theta)$$

$$H_\phi^s = -\frac{\sin\phi \, M_{0t}}{4\pi b \, \sin\theta} \sum_{n=1}^{\infty} \frac{\rho^{2n+1} D_n \xi_n'(k_e b)}{[(2n-1)!!]^2 n(n+1)} \frac{\xi_n'(k_e r)}{r} P_n^{(1)}(\cos\theta)$$

Numerical analysis of the exact solution shows that for ratios r/a and b/a that are greater than two, we need only retain the first terms in each of the equations of 6.157. So doing, we have:

$$E_r^s = 0$$

$$E_\theta^s = -\frac{i\omega\mu \, M_2}{4\pi} \frac{e^{-ik_e r}}{r^2} (1 + ik_e r) \sin\phi$$

$$E_\phi^s = -\frac{i\omega\mu}{4\pi} M_2 \frac{e^{-ik_e r}}{r^2} (1 + ik_e r) \cos\phi \, \cos\theta$$

$$\qquad\qquad (6.158)$$

$$H_r^s = -\frac{2 M_2}{4\pi r^3} e^{-ik_e r} (1 + ik_e r) \cos\phi \, \sin\theta$$

$$H_\theta^s = -\frac{M_2}{4\pi r^3} e^{-ik_e r} (k_e^2 r^2 - ik_e r - 1) \cos\phi \, \cos\theta$$

$$H_\phi^s = \frac{M_2}{4\pi r^3} e^{-ik_e r} (k_e^2 r^2 - ik_e r - 1) \sin\phi$$

where

$$M_2 = \frac{M_{0t}}{b^3} e^{-ik_e b} (k_e^2 b^2 - ik_e b - 1) \frac{D_1 a^3}{2}$$

Therefore, this part of the secondary field, described by the vector potential Π_2, is the same as that from a magnetic dipole source with a moment:

$$M_2 = 2\pi D_1 a^3 H$$

where H is the magnetic field from the transverse dipole in the uniform medium as observed at the origin.

The expressions in equation sets 6.158 are particularly simple when the parameters $|k_e r|$ and $|k_e b|$ are small. Expanding the right-hand side of eq. 6.158 in a series of powers of the small parameter $k_e r$ and $k_e b$, we obtain:

$$E_r^s = 0 \qquad\qquad H_r^s = \frac{D_1 a^3 M_{0t}}{4\pi b^3 r^3} \cos\phi \, \sin\theta$$

$$E_\theta^s = \frac{i\omega\mu}{4\pi} \frac{D_1 a^3}{2b^3 r^2} M_{0t} \sin\phi \qquad\qquad H_\theta^s = -\frac{D_1 a^3 M_{0t}}{8\pi b^3 r^3} \cos\phi \, \cos\theta \qquad (6.159)$$

$$E_\phi^s = \frac{i\omega\mu}{4\pi} \frac{D_1 a^3}{2b^3 r^2} M_{0t} \cos\phi \, \cos\theta \qquad H_\phi^s = \frac{D_1 a^3}{8\pi b^3 r^3} M_{0t} \sin\phi$$

If the values for $|k_e r|$ and $|k_e b|$ are no larger than 0.2, the error involved in discarding higher order terms amounts to only about 10 to 15% of the value of the secondary field.

Now, we will consider the field related to the potential of the electric type: Π_1. In spherical coordinates, the electromagnetic field can be expressed in terms of the potential Π_1 as follows:

$$E_r = \frac{\partial^2(r\Pi_1)}{\partial r^2} + k_e^2 r \Pi_1 \qquad H_r = 0$$

$$E_\theta = \frac{1}{r} \frac{\partial^2(r\Pi_1)}{\partial r \partial \theta} \qquad\qquad H_\theta = \frac{\sigma_e}{r \sin\theta} \frac{\partial(r\Pi_1)}{\partial \phi}$$

$$E_\phi = \frac{1}{r \sin\theta} \frac{\partial^2(r\Pi_1)}{\partial r \partial \phi} \qquad H_\phi = -\frac{\sigma_e}{r} \frac{\partial(r\Pi_1)}{\partial \theta}$$

Substituting these expressions of Π_1 into eq. 6.155, we obtain:

$$E_r^s = \frac{i\omega\mu \, k_e}{4\pi b} \sin\phi \, M_{0t} \sum_{n=1}^{\infty} \frac{\rho^{2n+1}}{[(2n-1)!!]^2 n^2} \frac{m^2 - 1}{m^2 + \dfrac{n+1}{n}}$$

$$\times \, \xi_n(k_e b)[\xi_n''(k_e r) + \xi_n(k_e r)] P_n^{(1)}(\cos\theta)$$

$$E_\theta^s = \frac{i\omega\mu}{4\pi b}\ \sin\phi\ M_{0t}\ \sum_{n=1}^{\infty} \frac{\rho^{2n+1}}{[(2n-1)!!]^2 n^2}\ \frac{m^2-1}{m^2+\dfrac{n+1}{n}}\ \xi_n(k_e b)$$

$$\times\ \frac{\xi_n(k_e r)}{r}\ \frac{\partial}{\partial\theta}\ (P_n^{(1)}(\cos\theta))$$

$$E_\phi^s = \frac{i\omega\mu}{4\pi b}\ \frac{\cos\phi}{\sin\theta}\ M_{0t}\ \sum_{n=1}^{\infty} \frac{\rho^{2n+1}\,\xi_n(k_e b)}{[(2n-1)!!]^2 n^2}\ \frac{\xi_n'(k_e r)}{r}\ \frac{m^2-1}{m^2+\dfrac{n+1}{n}}\ P_n^{(1)}(\cos\theta)$$

$$H_r^s = 0 \tag{6.160}$$

$$H_\theta^s = -\frac{k_e}{4\pi b}\ \frac{\cos\phi}{\sin\theta}\ M_{0t}\ \sum_{n=1}^{\infty} \frac{\rho^{2n+1}\,\xi_n(k_e b)}{[(2n-1)!!]^2 n^2}\ \frac{\xi_n(k_e r)}{r}\ \frac{m^2-1}{m^2+\dfrac{n+1}{n}}\ P_n^{(1)}(\cos\theta)$$

$$H_\phi^s = \frac{k_e}{4\pi b}\ \sin\phi\ M_{0t}\ \sum_{n=1}^{\infty} \frac{\rho^{2n+1}\,\xi_n(k_e b)}{[(2n-1)!!]^2 n^2}\ \frac{\xi_n(k_e r)}{r}\ \frac{m^2-1}{m^2+\dfrac{n+1}{n}}\ \frac{\partial}{\partial\theta}\ (P_n^{(1)}(\cos\theta))$$

As before, we will consider the case: $r/a > 2$ and $b/a > 2$. Neglecting all terms except those of first order, we obtain:

$$E_r^s = -\frac{2M_e}{r^3}\ e^{-ik_e r}\,(1+ik_e r)\ \sin\phi\ \sin\theta$$

$$E_\theta^s = -\frac{M_e}{r^3}\ e^{-ik_e r}\,(k_e^2 r^2 - ik_e r - 1)\ \sin\phi\ \cos\theta$$

$$E_\phi^s = -\frac{M_e}{r^3}\ e^{-ik_e r}\,(k_e^2 r^2 - ik_e r - 1)\ \cos\phi \tag{6.161}$$

$$H_r^s = 0$$

$$H_\theta^s = -\sigma_e\ M_e\ \frac{e^{-ik_e r}}{r^2}\,(1+ik_e r)\ \cos\phi$$

$$H_\phi^s = \sigma_e\ M_e\ \frac{e^{-ik_e r}}{r^2}\,(1+ik_e r)\ \sin\phi\ \cos\theta$$

where

$$M_e = \frac{i\omega\mu M_{0t}}{4\pi}\ \frac{\sigma_i - \sigma_e}{\sigma_i + 2\sigma_e}\ \frac{a^3}{b^2}\ e^{-ik_e b}\,(1+ik_e b)$$

Thus, this part of the secondary field is equivalent to that which would be observed from alternating electric dipole situated in a uniform conducting medium and having a moment equal to the product of the electric field from the transverse dipole at the center of the sphere and the multiplier:

$$\frac{\sigma_i/\sigma_e - 1}{\sigma_i/\sigma_e + 2} a^3$$

If the values for $k_e r$ and $k_e b$ are small, the equations in set 6.161 are much simplified:

$$E_r^s = -2i\omega\mu \frac{M_{0t}}{4\pi} \frac{\sigma_i/\sigma_e - 1}{\sigma_i/\sigma_e + 2} \frac{a^3}{b^2 r^3} \sin\phi \sin\theta$$

$$E_\theta^s = i\omega\mu \frac{M_{0t}}{4\pi} \frac{\sigma_i/\sigma_e - 1}{\sigma_i/\sigma_e + 2} \frac{a^3}{b^2 r^3} \sin\phi \cos\theta$$

$$E_\phi^s = i\omega\mu \frac{M_{0t}}{4\pi} \frac{\sigma_i/\sigma_e - 1}{\sigma_i/\sigma_e + 2} \frac{a^3}{b^2 r^3} \cos\phi \qquad\qquad (6.162)$$

$$H_r^s = 0$$

$$H_\theta^s = -i\sigma_e\mu\omega \frac{M_{0t}}{4\pi} \frac{\sigma_i/\sigma_e - 1}{\sigma_i/\sigma_e + 2} \frac{a^3}{b^2 r^2} \cos\phi$$

$$H_\phi^s = i\sigma_e\mu\omega \frac{M_{0t}}{4\pi} \frac{\sigma_i/\sigma_e - 1}{\sigma_i/\sigma_e + 2} \frac{a^3}{b^2 r^2} \sin\phi \cos\theta$$

Using eqs. 6.146, 6.159, and 6.162, we can arrive at the following expressions for the secondary field in the near zone of an arbitrarily oriented magnetic dipole source:

$$E_r^s = -2i\omega\mu \frac{M_{0t}}{4\pi} \frac{\sigma_i/\sigma_e - 1}{\sigma_i/\sigma_e + 2} \frac{a^3}{b^2 r^3} \sin\phi \sin\theta$$

$$E_\theta^s = \frac{i\omega\mu}{2} \frac{D_1 a^3}{b^3 r^2} \frac{M_{0r}}{4\pi} \sin\phi + i\omega\mu \frac{M_{0t}}{4\pi} \frac{\sigma_i/\sigma_e - 1}{\sigma_i/\sigma_e + 2} \frac{a^3}{b^2 r^3} \sin\phi \cos\theta$$

$$E_\phi^s = \frac{i\omega\mu}{4\pi} \frac{a^3 D_1}{b^3 r^2} \frac{M_{0r}}{4\pi} \sin\theta + \frac{i\omega\mu}{2} \frac{M_{0t}}{4\pi} D_1 \frac{a^3}{b^3 r^2} \cos\phi \cos\theta$$

$$\qquad + i\omega\mu \frac{M_{0t}}{4\pi} \frac{\sigma_i/\sigma_e - 1}{\sigma_i/\sigma_e + 2} \frac{a^3}{b^2 r^3} \cos\phi$$

$$H_r^s = -\frac{2a^3}{b^3 r^3} D_1 \frac{M_{0r}}{4\pi} \cos\theta + \frac{D_1 a^3}{b^3 r^3} \frac{M_{0t}}{4\pi} \cos\phi \sin\theta \qquad (6.163)$$

$$H_\theta^s = -\frac{D_1 a^3}{b^3 r^3} \frac{M_{0r}}{4\pi} \sin\theta - \frac{D_1 a^3}{2b^3 r^3} \frac{M_{0t}}{4\pi} \cos\theta \cos\phi$$

$$- i\sigma_e \mu\omega \frac{M_{0t}}{4\pi} \frac{\sigma_i/\sigma_e - 1}{\sigma_i/\sigma_e + 2} \frac{a^3}{b^2 r^2} \cos\phi$$

$$H_\phi^s = \frac{D_1 a^3}{2b^3 r^3} \frac{M_{0t}}{4\pi} \sin\phi + i\sigma_e \mu\omega \frac{\sigma_i/\sigma_e - 1}{\sigma_i/\sigma_e + 2} \frac{M_{0t}}{4\pi} \frac{a^3}{b^2 r^2} \sin\phi \cos\theta$$

Somewhat similar expressions can be written for the cases in which the parameters $k_e r$ and $k_e b$ are arbitrary in size. Both sets of expressions will be valid when the skin depth in the host medium is significantly greater than the radius of the sphere and the distances from the center of the sphere to the observation site and the dipole source, by a factor of at least twice the radius, a. With these conditions, the secondary field due to the presence of the sphere is equivalent to that which would be observed for magnetic and electric dipoles in a uniform medium. The main features of these fields have been described in an earlier section, including appearance of surface electric charges when the source of the primary field is a plane wave. In other words, regardless of the type of excitation (whether it be plane wave or magnetic dipole), the secondary field due to the presence of a sphere is equivalent to that for two dipoles when r/a and b/a are greater than 2, and a/h_e is less than one. However, there is one very significant difference in that in low and intermediate frequencies, namely the relationship between a primary electric and magnetic field for magnetic dipole excitation differs significantly from the case for plane wave excitation. Correspondingly, the relation between the galvanic and vortex parts of the secondary field also is essentially different.

At this point, let us consider the modulus of the ratio of electric fields of vortex and galvanic origin. In accord with eq. 6.161, we have the following result for the behavior of the component E_θ^s in the near zone:

$$\eta_\theta^e = \left|\frac{E_\theta^v}{E_\theta^g}\right| = \frac{|D_1|r}{2b\,|\cos\theta|}, \quad \text{if } \frac{\sigma_i - \sigma_e}{\sigma_i + 2\sigma_e} = 1 \tag{6.164}$$

For example, when $r = b$ and $2\cos\theta = 1$, we have:

$$\eta_\theta^e \leqslant |D_1| < 1$$

Because of this, anomalies in electric field strength can be caused either by conducting or resistive inhomogeneities.

Examining the equations in set 6.163, it can readily be demonstrated that with an increase in frequency, for measurements made beyond the near zone in the secondary field, the influence of the galvanic part of the electric field progressively increases. Therefore, one can predict that the resolution obtainable with methods based on the measurement of the electric field will be

524

very poor, regardless of the type of source used to generate the primary field.

The ratio of the quadrature components of the magnetic field

Let us now consider in detail the behavior of the ratio of the vortex and galvanic parts of the magnetic field based on calculations using exact expressions for the magnetic field. First of all, we will consider the behavior of the quadrature component of the magnetic field.

The curves shown in Figs. 6.5—6.8 demonstrate the behavior of the ratio of the vortex and galvanic parts of the quadrature components as a function of the parameter a/h_i. As may be seen from these various curves, the maximum value for the ratio QH_θ^v/QH_θ^g occurs at low frequencies, such that a/h_i is less than 1, and with increase in frequency, the influence of the galvanic part of the current contributed by current flow along lines closed

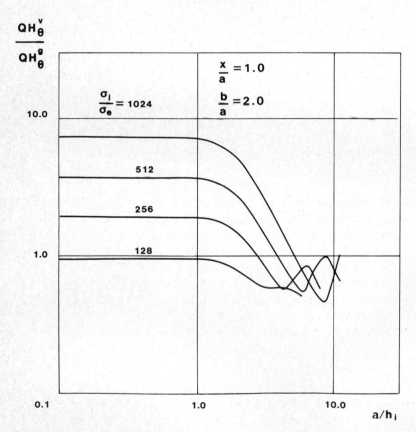

Fig. 6.5. The ratio of the vortex and galvanic parts of the quadrature component of the magnetic field, H_θ.

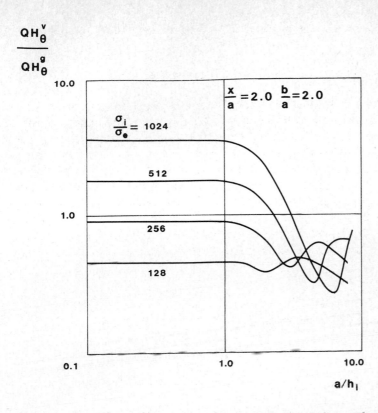

Fig. 6.6. The ratio of the vortex and galvanic parts of the quadrature component of the magnetic field, H_θ.

through the host medium becomes progressively greater. It is also apparent the higher the resistivity is in the host medium, the most significant will be the influence of currents induced in the sphere. The behavior of the curves shown in Figs. 6.5—6.8 can readily be explained. For example, at low frequencies, the density of induced currents as well as the surface density of electrical charges is directly proportional to frequency. Respectively, the ratio of the two magnetic field parts, H^v/H^g tends to a constant value which is in turn defined by geoelectric parameters.

For example, based on eq. 6.163 and taking $\theta = 0$ and $\phi = 0$, we have:

$$\frac{QH_\theta^v}{QH_\theta^g} = \frac{D_1}{2i\sigma_e\mu\omega\,br}, \quad \text{if} \quad \frac{\sigma_i - \sigma_e}{\sigma_i + 2\sigma_e} \tag{6.165}$$

Inasmuch as the function D_1 can be written as follows at low frequencies:

$$D_1 \approx -\frac{i\sigma_i\mu\omega a^2}{15} - \frac{2}{315}(\sigma_i\mu\omega a^2)^2 \tag{6.166}$$

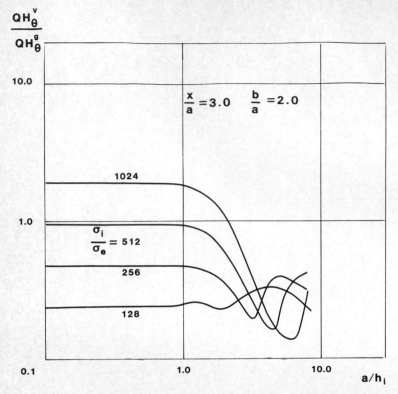

Fig. 6.7. The ratio of the vortex and galvanic parts of the quadrature component of the magnetic field, H_θ.

The ratio of the quadrature components tends to a constant value which in this case will be:

$$-\frac{1}{30}\frac{\sigma_i}{\sigma_e}\frac{a}{b}\frac{a}{r} \qquad (6.167)$$

With increase in frequency, and still with the condition that $a/h_i > 1$, is greater than 1, the quadrature component of the vortex part of the field will increase at a lesser rate than ω, causing the ratio QH^v/QH^g to decrease.

Because the magnetic field contributed by the galvanic part decreases more slowly with distance, r, than does the vortex part, the ratio QH^v/QH^g decreases as the distance r grows (see Fig. 6.8) except at very high frequencies.

The frequency spectrums for the vortex part of the magnetic field, normalized to the strength of the normal field:

$$(QH_0 + QH_s^v)/QH_0$$

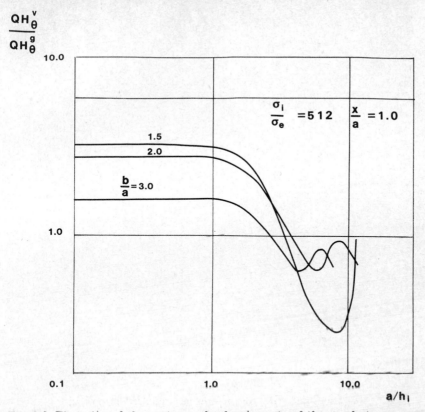

Fig. 6.8. The ratio of the vortex and galvanic parts of the quadrature component of the magnetic field, H_θ as a function of the distance b.

are shown in Fig. 6.9. The quantity QH_0 is the quadrature component of the magnetic field of a magnetic dipole source situated in a uniform conducting medium. These curves, as well as the developments given earlier in Chapter 5 demonstrate that at low frequency this function has maximal value which is independent of the frequency within this range. As may be seen from a comparison of the various curves in Fig. 6.5—6.9, over this frequency range, the effect of the galvanic part of the quadrature component is minimal. Moreover, one can say that low frequencies will be the most favorable range for classification of anomalies, when the quadrature component of the magnetic field is the quantity measured.

The ratio of the inphase components of the magnetic field

At this point, we will consider the ratio of the inphase components of the vortex and galvanic parts of the magnetic field, as demonstrated by the

528

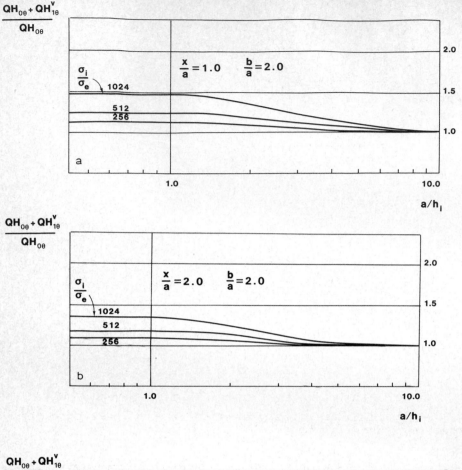

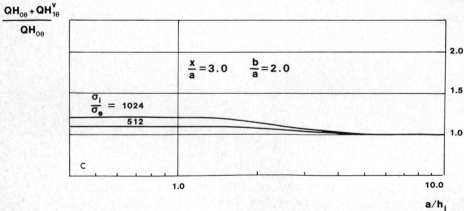

Fig. 6.9. Frequency response curves for the normalized tangential magnetic field. a. For the case $x/a = 1.0$. b. For the case $x/a = 2.0$. c. For the case $x/a = 3.0$.

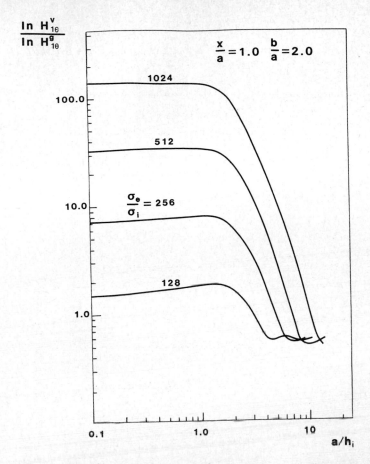

Fig. 6.10. Ratio of the vortex and galvanic parts of the inphase component of the magnetic field, H_θ.

various curves in Figs. 6.10—6.13. At low frequencies, and with the condition that a/h_i is less than unity, the function $\mathrm{In}H_\theta^v/\mathrm{In}H_\theta^g$ approaches a constant value in much the same manner as did the ratio of quadrature components. However, in the case of the inphase components, the ratio is significantly greater than that for the quadrature components, particularly for large values of conductivity. This behavior can be explained by comparing the expressions for the field given in eqs. 6.158—6.161. In fact, the inphase component of the vortex part of the magnetic field at low frequencies is proportional to the quantity $(\sigma_i\omega)^2$ while the galvanic part is proportional to the quantity $(\sigma_e\omega)^2$. Therefore, in the extreme, the ratio tends to a value which is greater than that for the quadrature component. This means that in measuring the inphase component for the magnetic field, one can significantly reduce the effect of the galvanic part of the

530

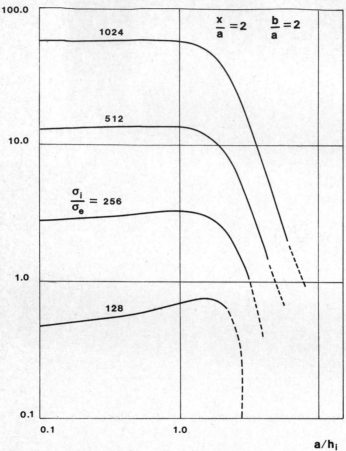

Fig. 6.11. Ratio of the vortex and galvanic parts of the inphase component of the magnetic field, H_θ.

field. As in the case for the quadrature component with an increase of frequency, $(a/h_i > 1)$, the influence of the magnetic field related with the presence of electric charges becomes more significant.

The various curves shown in Fig. 6.14 illustrate the behavior of the ratio of the vortex part of the inphase component of the secondary field to the strength of normal field:

$$(\ln H_{0\theta}^s + \ln H_{1\theta}^v)/\ln H_{0\theta}^s$$

$$-\frac{\ln H_{1\theta}^{v}}{\ln H_{1\theta}^{g}}$$

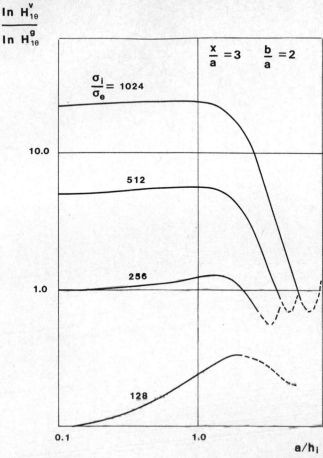

Fig. 6.12. The ratio of vortex and galvanic parts of the inphase component of the magnetic field, H_{θ}.

It can readily be shown that the maximum in this function is observed over an intermediate frequency range. At the same time, one can recognize that in this range, the influence of the vortex part is still dominant when $\sigma_i/\sigma_e \gg 1$.

Vortex and galvanic parts of the magnetic field in the dual-frequency method

In this section we will examine the ratio of the vortex and galvanic parts of the quadrature components of the magnetic field when the term proportional to ω is canceled instrumentally. Assume that the parameter:

$$-\frac{\ln H_{1\theta}^{v}}{\ln H_{1\theta}^{g}}$$

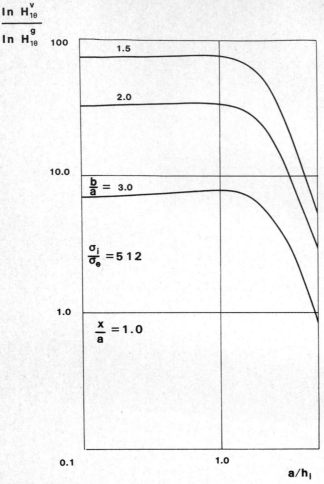

Fig. 6.13. The ratio of vortex and galvanic parts of the inphase component of the magnetic field as a function of the distance b.

$$\Delta QH = QH(\omega_1) - \frac{\omega_1}{\omega_2}\, QH(\omega_2)$$

is the quantity measured with a field system, and we will investigate the effect of the magnetic field of currents closed through the host medium (see the corresponding curves in Figs. 6.15—6.18). In order to understand the asymptotic behavior for the ratio $\Delta QH_s^v/\Delta QH_s^g$ at low frequencies, we will use eqs. 6.158—6.161. It is not difficult to see that the next term of the series, presenting both parts of the quadrature component is directly proportional to $\omega^{5/2}$. Correspondingly, the left asymptote of curves in Figs. 6.15—6.18 tends to a constant value other than zero. With increasing

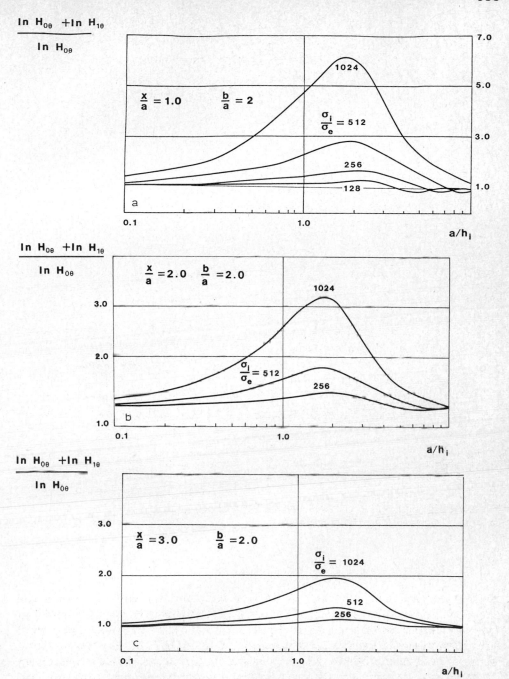

$$\frac{\ln H_{0\theta} + \ln H_{1\theta}}{\ln H_{0\theta}}$$

$$\frac{x}{a} = 1.0 \qquad \frac{b}{a} = 2$$

$$\frac{\sigma_i}{\sigma_e} = 512$$

1024

256

128

a

0.1   1.0

a/h_i

$$\frac{\ln H_{0\theta} + \ln H_{1\theta}}{\ln H_{0\theta}}$$

$$\frac{x}{a} = 2.0 \quad \frac{b}{a} = 2.0$$

1024

$$\frac{\sigma_i}{\sigma_e} = 512$$

256

b

0.1   1.0

a/h_i

$$\frac{\ln H_{0\theta} + \ln H_{1\theta}}{\ln H_{0\theta}}$$

$$\frac{x}{a} = 3.0 \qquad \frac{b}{a} = 2.0$$

$$\frac{\sigma_i}{\sigma_e} = 1024$$

512

256

c

0.1   1.0

a/h_i

Fig. 6.14. Frequency responses for the inphase component of the tangential magnetic field. a. For the case $x/a = 1.0$. b. For the case $x/a = 2.0$. c. For the case $x/a = 3.0$.

534

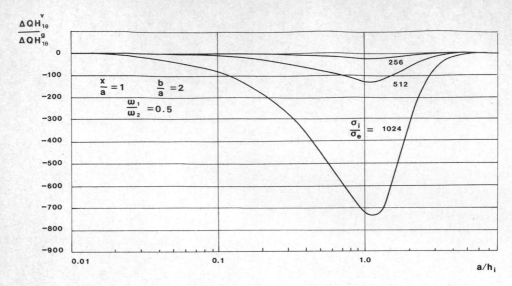

Fig. 6.15. Ratio of the vortex and galvanic parts of the function $\Delta Q H_{1\theta}$.

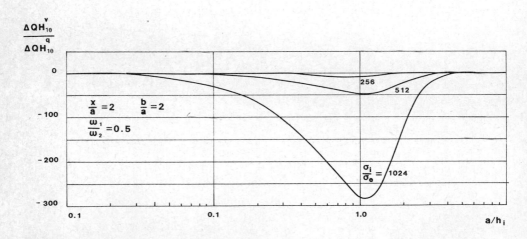

Fig. 6.16. Ratio of the vortex and galvanic parts of the function $\Delta Q H_{1\theta}$.

frequency, the relative contribution from the vortex part increases, reaches a maximum, and with further increase in frequency, becomes smaller, gradually approaching zero. It is important to note that the maximum reduction in the galvanic part of the magnetic field takes place over almost the same range of parameters, a/h_i, where the relative anomaly of the vortex part of the field expressed as a fraction of the normal field (see Fig. 6.19) approaches its maximum value.

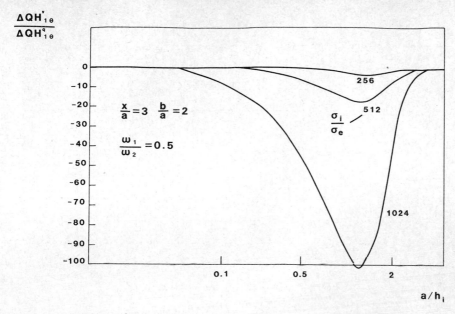

Fig. 6.17. Ratio of the vortex and galvanic parts of the function $\Delta QH_{1\theta}$.

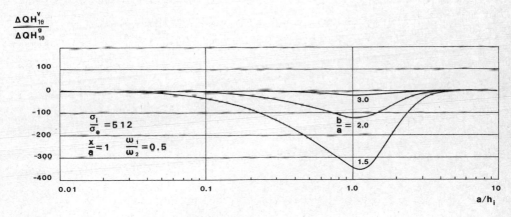

Fig. 6.18. Ratio of the vortex and galvanic parts of the function $\Delta QH_{1\theta}$ as a function of the distance b.

A comparison of the various curves in Figs. 6.15—6.19 shows that measuring the difference ΔQH permits one to reduce the effect of the galvanic part to a greater degree than when the quadrature or inphase components are measured.

536

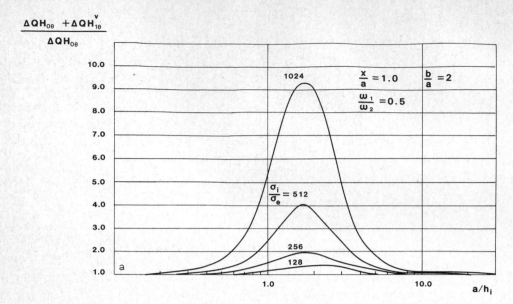

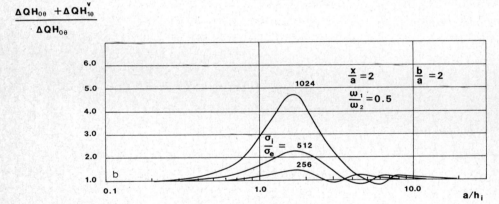

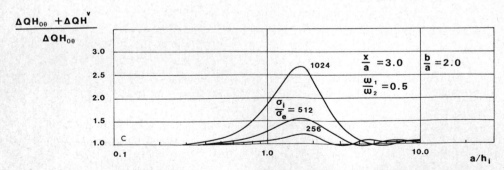

Fig. 6.19. Frequency responses for the normalized tangential magnetic field. a. For the case $x/a = 1.0$. b. For the case $x/a = 2.0$. c. For the case $x/a = 3.0$.

The ratio between the vortex and galvanic parts of a transient magnetic field

In this section, we will examine the transient response of the magnetic field. The transient responses for the ratio between the vortex and galvanic parts of the function $\partial B/\partial t$ have been calculated by Fourier transform, with the results being shown in Figs. 6.20—6.23 as functions of the parameter τ_i/a, where $\tau_i = (2\pi\rho_i t \times 10^7)^{1/2}$. During the very early stage of transient response, when the magnitude of the normal electrical field is much greater than that of the magnetic field, the influence of electric charges is dominant, and therefore, the galvanic part of the magnetic field will be dominant. With increase in time, this part of the field becomes equivalent to that of an electric dipole and can be written as follows (a transversal dipole and the θ-component)

$$\frac{\partial B_\theta^g}{\partial t} = M_{0t} \frac{\sigma_i - \sigma_e}{\sigma_i + 2\sigma_e} \frac{5(r^3 + b^3)a^3}{32\pi\sqrt{\pi}\,b^2 r^2} \frac{\sigma_e^{5/2}\mu^{7/2}}{t^{7/2}} \cos\phi, \quad \text{if } \tau_e/r \gg 1 \qquad (6.168)$$

As this last equation demonstrates, the galvanic part of the field decreases quite quickly because the rate of decay is controlled principally by the behavior of the current in a reasonably resistive host medium.

During the early stage of the transient response of induced currents in a conductor, the field decays relatively slowly, and therefore within this range with increase in time, the ratio $\dot{B}_\theta^v/\dot{B}_\theta^g$ gradually increases. At later times, as was demonstrated in the preceding chapter, the vortex part of the field is described by the equation:

$$\frac{\partial B_\theta^v}{\partial t} = \frac{M_{0t}\,\sigma_i(r^3 + b^3)\,a^5}{96\pi\sqrt{\pi}\,b^3 r^3} \frac{\sigma^{3/2}\mu^{7/2}}{t^{7/2}} \cos\theta \cos\phi \qquad (6.169)$$

In view of the exponential decay of vortex currents during the late stage behavior, the ratio $\dot{B}_\theta^v/\dot{B}_\theta^g$ approaches a maximum, and then begins to decrease as shown by the curves in Figs. 6.20—6.22.

As follows from eq. 6.169, with increase in time, the vortex part of the field is controlled mainly by secondary induced currents in the host medium, and it decays with time in the same manner as the galvanic part of the field. Therefore, at relatively late times, the ratio $\dot{B}_\theta^v/\dot{B}_\theta^g$ tends to a constant value defined by the geoelectric parameters.

Thus, considering all of these factors, we recognize that it is possible to distinguish the following ranges for the transient response of the function $\dot{B}_s^v/\dot{B}_s^g$.

(a) An early stage, during which the galvanic part of the field is dominant.

(b) A time range during which the influence of the galvanic part gradually decreases.

(c) A range of times over which this ratio takes on a maximum value.

(d) A time interval over which the influence of the galvanic part increases.

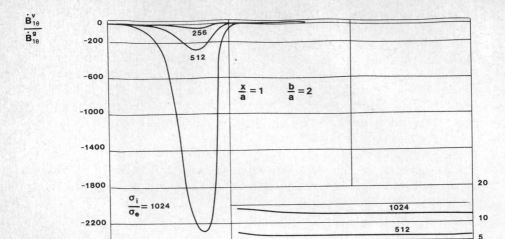

Fig. 6.20. The ratio of the vortex and galvanic parts of the transient response for the time rate of change of magnetic induction.

(e) A very late time range during which the ratio of the vortex and galvanic parts does not depend on time.

A comparison of curves for the transient response with the corresponding curves for the quadrature and inphase component as well as for function $\Delta QH(\omega)$ clearly demonstrates that in measuring the transient magnetic field within the range of the maximum, it is possible to reduce the effect of the galvanic part to a much greater extent. It is also fundamental to note that within this range, the anomaly in the vortex transient field $(\dot{B}_0 + \dot{B}_1^v)/\dot{B}_0$ is a maximum (see Fig. 6.24a, b, c).

The analysis of electromagnetic field behavior described in the last two paragraphs permit us to arrive at some useful generalizations.

(1) In the general case, when the primary electric field crosses a surface of an inhomogeneity, there are two sources for the electromagnetic field, namely currents and charges. Usually the charges arise at interfaces. Therefore, the secondary field consists of both vortex and galvanic parts.

(2) Over some range parameters, the vortex part of the secondary field is mainly controlled by the conductivity and by the radius of the sphere.

(3) In contrast to the vortex part, the galvanic part of the field depends on slightly on the conductivity in the sphere.

(4) The relationship between the vortex and galvanic parts of the secondary field depends primarily on a number of factors, including:

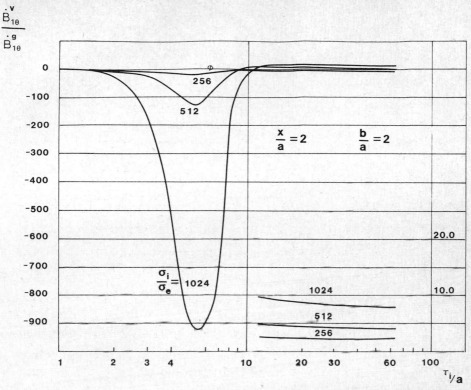

Fig. 6.21. The ratio of the vortex and galvanic parts of the transient response for the time rate of change of magnetic induction.

(a) The type of excitation for the field (whether it be plane wave, magnetic or electric dipole, or other type of a source).

(b) The zone in which measurements are carried out (that is whether it be the near, the intermediate, or the wave zone).

(c) The frequency or time.

(d) The particular field component which is measured (whether it is electric or magnetic).

(e) The geometric factors (the coordinance of the observation point, the position of the source of the primary field, and so on).

(5) The influence of the magnetic field of currents flowing through the conductor and through the host medium is minimum at low frequencies when the quadrature and inphase components are measured. At the same time, in measuring the inphase component of the magnetic field, the effect of the galvanic part can be reduced to a significant extent.

(6) A relatively stronger reduction of the galvanic part of the field can be achieved by using the dual frequency method over a specific range of frequencies where the relative anomaly of the vortex field is near its maximum.

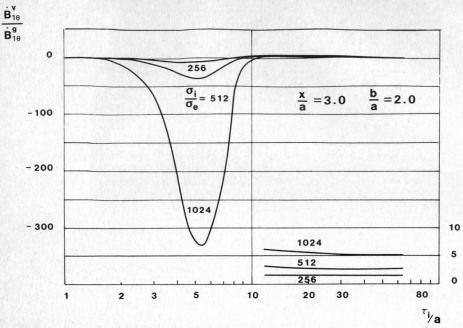

Fig. 6.22. The ratio of the vortex and galvanic parts of the transient response for the time rate of change of magnetic induction.

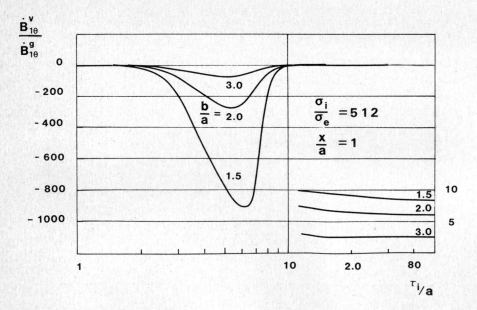

Fig. 6.23. The ratio $\dot{B}_{1\theta}^{v}/\dot{B}_{1\theta}^{g}$ has a function of the distance b.

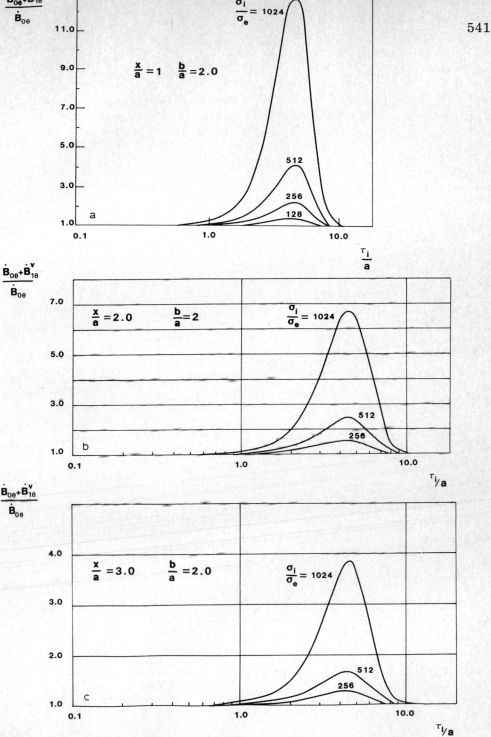

Fig. 6.24. Transient responses for the normalized time rate of change of magnetic induction. a. For the case $x/a = 1.0$. b. For the case $x/a = 2.0$. c. For the case $x/a = 3.0$.

(7) In the transient method, one can emphasize the influence of the vortex part of the secondary field to a much greater extent than is possible with conventional methods in the frequency domain. In order to achieve this, measurements must be made within a time interval over which the relative anomaly of the vortex part of the field approaches its maximum.

(8) In the wave zone, the possibility of classifying anomalies according to the conductivity is very poor. This conclusion is all the more valid when the electric field is measured.

6.4. A CONDUCTING CYLINDER IN THE FIELD OF AN INFINITELY LONG CURRENT FILAMENT PARALLEL TO THE AXIS OF THE CYLINDER

In the induction methods for mineral prospecting, usually non-grounded rectangular current-carrying loops are used for excitation of an electromagnetic field in the medium to be explored, and quite frequently, the long dimension of such a loop is laid parallel to the strike of the expected conductive ore bodies. In a first approximation in which we neglect the influence from the two short sides of the source loop, we can assume that the source of the normal field is two infinitely long current filaments with current flowing in opposite directions. In order to use such a field structure to explore the subsurface, it will be sufficient to derive expressions for an electromagnetic field from one infinitely long current filament located parallel to the strike of a cylinder, and then by applying the principle of superposition, the solution for two current filaments can be obtained.

Let us assume to begin with that the current flow is sinusoidal, that is:

$$I = I_0\, e^{i\omega t} \tag{6.170}$$

In this case, because of the change in the magnetic field with time, induced currents will arise in the host medium and in the target cylinder. In spite of the fact that the normal electric field does not cross the surface of the cylinder, there will be some effect of non-vortex currents observed. As a start, we can consider the normal field caused by current flowing in the line, as well as by currents induced in the host medium, when the cylinder is absent, and later, the secondary field will be considered.

Normal field of a current filament

We will start from Maxwell's equations, assuming that the field is quasi-stationary in character:

$$\begin{aligned}
\operatorname{curl} E &= -i\omega\mu\, H & \operatorname{div} E &= 0 \\
\operatorname{curl} H &= \sigma E & \operatorname{div} H &= 0
\end{aligned} \tag{6.171}$$

where σ is the conductivity of the medium. From this set of equations, we have:

$$\mathrm{curl\ curl}\,E \;=\; -i\omega\mu\,\mathrm{curl}\,H \;=\; -i\omega\mu\sigma E$$

or

$$\nabla^2 E - k^2 E \;=\; 0 \tag{6.172}$$

By analogy, we also have:

$$\nabla^2 H - k^2 H \;=\; 0 \tag{6.173}$$

where k is the wave number characterizing the uniform medium.

Inasmuch as the current filaments are directed along the z-axis as shown in Fig. 6.25 and in virtue of the symmetry characterizing the model, the resulting electric field has only a z-component. Therefore, eq. 6.172 can be written in cylindrical coordinates as follows:

$$\frac{\mathrm{d}^2 E_z}{\mathrm{d}R^2} + \frac{1}{R}\frac{\mathrm{d}E_z}{\mathrm{d}R} - k^2 E_z \;=\; 0$$

It can readily be shown that solutions to this equation are linear combinations of the modified Bessel functions $K_0(kR)$ and $I_0(kR)$. Considering that the electromagnetic field should decrease with increase in distance, R and $\lim I_0(kR) \to \infty$, as $R \to \infty$, we have the following expression for the electric field

$$E_z \;=\; M\,K_0(kR) \tag{6.174}$$

where M is unknown at this point. In order to define this constant, we must consider the strength of the magnetic field in the vicinity of the source.

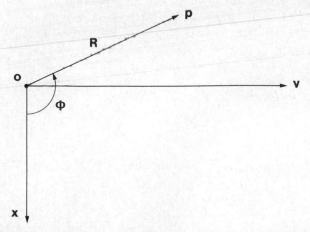

Fig. 6.25. Current filament in a uniform conducting medium.

Inasmuch as the magnetic field at points which are very close to the line is controlled mainly by the current in the source, we have:

$$\oint H\,dl = I_0 \quad \text{as } R \to 0$$

In view of the axial symmetry, the magnetic field has only a single component H_ϕ, so that:

$$H_\phi\, 2\pi R = I_0, \quad \text{as } R \to 0 \tag{6.175}$$

As follows from Maxwell's equations, the relationship between the components H_ϕ and E_z is:

$$H_\phi = \frac{1}{i\omega\mu}\frac{\partial E_z}{\partial R} \tag{6.176}$$

We obtain the following from eqs. 6.174—6.176 for points which are located near the source:

$$\frac{M}{i\omega\mu}\frac{d}{dR} K_0(kR) = \frac{I_0}{2\pi R}, \quad \text{as } R \to 0$$

Inasmuch as:

$$\frac{d}{dR} K_0(kR) = -k\,K_1(kR) \quad \text{and} \quad K_1(x) \approx \frac{1}{x}, \quad \text{as } x \to 0$$

we have:

$$-\frac{M}{i\omega\mu}\,k\,K_1(kR) = -\frac{M}{i\omega\mu}\frac{1}{R} = \frac{I_0}{2\pi R}$$

Thus:

$$M = -\frac{i\omega\mu\,I_0}{2\pi} \tag{6.177}$$

Correspondingly, for the two components of the electromagnetic field we have:

$$E_z = -\frac{i\omega\mu\,I_0}{2\pi} K_0(kR) \tag{6.178}$$

and

$$H_\phi = \frac{I}{2\pi}\,k\,K_1(kR) \tag{6.179}$$

We can express the normal field in the form:

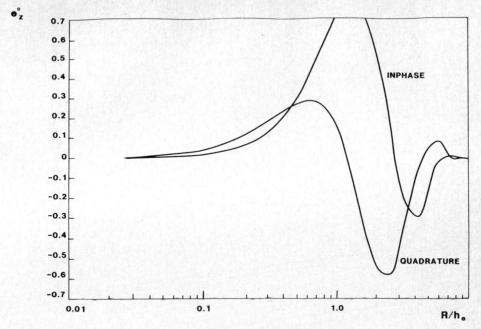

Fig. 6.26. Frequency response curves for the quadrature and inphase components of the function e_z^0.

$$E_z = -\frac{\rho I_0}{2\pi R^2} e_z \quad \text{and} \quad H_\phi = \frac{I}{2\pi R} h_\phi$$

where ρ is the resitivity in the uniform medium. Then, we have

$$e_z = k^2 R^2 K_0(kR), \quad h_\phi = kR \, K_1(kR) \tag{6.180}$$

The frequency response for the normal field, calculated using the formulas in 6.180 are shown in Figs. 6.26 and 6.27.

Before we examine the frequency response of the field, we will derive an expression for the current density. In accord with eq. 6.178, the current density, j_z, in the medium is

$$j_z = -\frac{I_0 k^2}{2\pi} K_0(kR) \quad \text{or} \quad j_z = -\frac{I_0}{2\pi R^2} e_z \tag{6.181}$$

It is a simple matter to calculate the total current in the medium. It is:

$$I = \int_S j_z \, dS = 2\pi \int_0^\infty j_z \, R \, dR = -I_0 k^2 \int_0^\infty R \, K_0(kR) \, dR$$

Substituting the variable $x = kR$, we obtain:

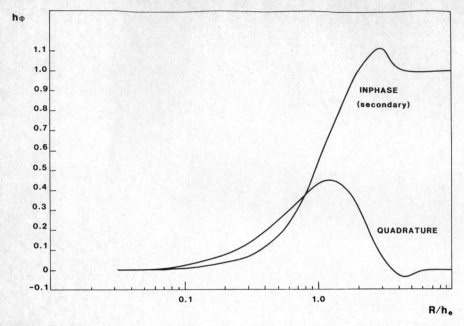

Fig. 6.27. Frequency response curves for the quadrature and inphase components of the function h_ϕ.

$$I = -I_0 \int\limits_0^\infty x \, K_0(x) \, \mathrm{d}x = -I_0 \tag{6.182}$$

This result can be interpreted as follows. The end points of the linear source are situated far from the observation point and the current I_0 passing through them is returned back through the host medium.

Now we will examine in detail the frequency response for the normal field. At low frequencies, and for $R/h_e < 1$ or $|kR| < 1$, the modified Bessel functions K_0 and K_1 can be replaced by the elementary functions:

$$K_0(kR) \approx -\ln kR - \frac{k^2 R^2}{4} \ln kR$$

and

$$K_1(kR) \approx \frac{1}{kR} + \frac{kR}{2} \ln kR \tag{6.183}$$

Substituting these expressions into eqs. 6.180, we have:

$$e_z \approx -k^2 R^2 \ln kR - \frac{k^4 R^4}{4} \ln kR =$$

$$= -i \left\{ \frac{\sigma\mu\omega R^2}{2} \ln \sigma\mu\omega R^2 - \frac{(\sigma\mu\omega R^2)^2}{16} \pi \right\}$$

$$+ \sigma\mu\omega \, R^2 \, \frac{\pi}{4} + \frac{1}{8} \, (\sigma\mu\omega R^2)^2 \, \ln \sigma\mu\omega R^2 \qquad (6.184)$$

Therefore, the inphase component of the electric field is:

$$\mathrm{In} E_z \approx -\frac{I_0 \mu\omega}{8} - \mu\omega \, \frac{I_0}{32\pi} \, (\sigma\mu\omega R^2) \ln \sigma\mu\omega R^2 \qquad (6.185)$$

The quadrature component can be written as follows if we neglect the second term:

$$Q E_z \approx \frac{I_0 \mu\omega}{4\pi} \ln \sigma\mu\omega R^2 \qquad (6.186)$$

It is important to note that at low frequencies the quadrature component of the electric field depends only slightly on the conductivity, but the inphase component to a first approximation is independent of both the conductivity and of the distance, R.

Expanding the function $K_1(kR)$, we arrive at the following expressions for the function h_ϕ:

$$\mathrm{In} h_\phi \approx \left(1 - \frac{\pi}{8} \, \sigma\mu\omega R^2 \right)$$

$$\qquad (6.187)$$

$$Q h_\phi \approx \frac{\sigma\mu\omega R^2}{4} \ln \sigma\mu\omega R^2$$

In the same way, we have the following expressions for the magnetic field:

$$\mathrm{In} H_\phi \approx \frac{I_0}{2\pi R} \left(1 - \frac{\pi}{8} \, \sigma\mu\omega R^2 \right)$$

$$\qquad (6.188)$$

$$Q H_\phi \approx \frac{I_0}{2\pi R} \frac{\sigma\mu\omega R^2}{4} \ln \sigma\mu\omega R^2$$

It is clear that the inphase component of the magnetic field is contributed by current flow in the source and by induced currents in the medium.

Comparing eqs. 6.185 and 6.186 with eq. 6.188, we can see that the secondary magnetic field is more sensitive to a change in the conductivity than is the electric field.

In order to derive asymptotic formulas for the high frequency part of the spectrum, we can use the expressions

$$K_0(kR) \approx \left(\frac{\pi}{2kR}\right)^{1/2} e^{-kR} \quad K_1(kR) \approx \left(\frac{\pi}{2kR}\right)^{1/2} e^{-kR}, \quad \text{when } |kR| \gg 1$$

Thus, we obtain the following for the functions e_z and h_ϕ:

$$e_z \approx k^2 R^2 \left(\frac{\pi}{2kR}\right)^{1/2} e^{-kR}, \quad h_\phi \approx kR \left(\frac{\pi}{2kR}\right)^{1/2} e^{-kR}$$

or

$$E_z \approx -\frac{i\omega\mu I_0}{2\pi} \left(\frac{\pi}{2kR}\right)^{1/2} e^{-kR}, \quad H_\phi \approx \frac{I_0 k}{2\pi} \left(\frac{\pi}{2kR}\right)^{1/2} e^{-kR} \tag{6.189}$$

The electromagnetic field decays very rapidly at high frequencies. In particular, for large values of the parameter $|kr|$, we have:

$$\frac{E_z}{H_\phi} \approx -\frac{i\omega\mu}{k} = -\rho k = Z$$

where Z is the impedance of a plane wave.

It is clear that with increasing frequency, the induced currents will concentrate near the source and the total field will decay exponentially.

At this point we will consider the transient response of the normal field in the case in which the current in the linear source filament varies with time as a step function:

$$I(t) = \begin{cases} 0 & t \leqslant 0 \\ I_0 & t \geqslant 0 \end{cases}$$

By using the Fourier transform, we have the following results for the electric field:

$$e_z = \frac{1}{2\pi} \int_{-\infty}^{\infty} \frac{e_z(\omega)}{i\omega} e^{i\omega t} \, d\omega = \frac{\mu\sigma}{2\pi} R^2 \int_{-\infty}^{\infty} K_0(kR) e^{i\omega t} \, d\omega \tag{6.190}$$

The integral in this last equation is tabulated, so that we have the following:

$$e_z(t) = \frac{1}{2\tau} e^{-1/4\tau} \tag{6.191}$$

or

$$E_z(t) = -\frac{\rho I_0}{2\pi R^2} \frac{1}{2\tau} e^{-1/4\tau} \tag{6.192}$$

where $\tau = t/\sigma\mu R^2$.

By analogy, we have the following for the magnetic field:

$$h_\phi(\tau) = e^{-1/4\tau}, \quad H_\phi(\tau) = \frac{I_0}{2\pi R} e^{-1/4\tau} \tag{6.193}$$

The curves in Fig. 6.28 show the behavior of the transient response for the components $e_z(\tau)$ and $h_\phi(\tau)$. During the early stage of the transient response when $\tau \to 0$, both components go quickly to zero as a consequence of skin effect. Just as was the case at high frequencies in the frequency domain, in this case, the current density is zero everywhere, except near the source.

In late-stage behavior (when $\tau \gg 1$) we have:

$$E_z \approx -\frac{\rho I_0}{4\pi R^2}\frac{1}{\tau}\left(1 - \frac{1}{4\tau}\right) = -\frac{I_0\mu}{4\pi t}\left(1 - \frac{1}{4\tau}\right), \quad \text{if } \tau > 1 \tag{6.194}$$

The electric field behaves during the late stage in very much the same manner as does the inphase component at the low-frequency part of the spectrum, in that it is practically independent of conductivity.

In accord with eq. 6.193 for the magnetic field, when the current is turned off, we have:

$$H_\phi(\tau) = \frac{I_0}{2\pi R}\,(1 - e^{-1/4\tau})$$

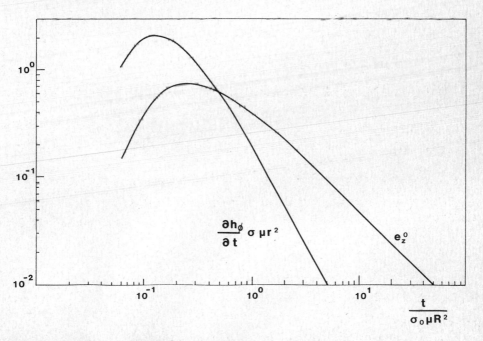

Fig. 6.28. Transient responses for the functions $e_z(\tau)$ and $h_\phi(\tau)$.

During the early stage when $\tau \to 0$, the field is the same as that for direct current $I_0/2\pi R$, but during the late stage, we have:

$$H_\phi \approx \frac{I_0}{2\pi R} \frac{1}{4\tilde{\tau}} = \frac{I_0 R \mu}{8\pi t} \sigma \tag{6.195}$$

Therefore, during the late stage, the magnetic field will be observed to be directly proportional to conductivity. The same dependence is inherent for the second term of the inphase component of the magnetic field at low frequencies in the frequency domain. It is important to note that during the late stage, both components of the field, H_ϕ and E_z have the same dependence on time, t. Therefore, the electromotive force varies as $1/t^2$ over this time range.

Frequency- and time-domain response of the secondary field caused by the presence of a cylinder

Let us now place a cylindrical conductor within the uniform space, as shown in Fig. 6.29. In order to find the secondary field, we must find functions satisfying the Helmholtz equations, the boundary conditions, and the condition at infinity. As is well known, the boundary conditions consists of a requirement for continuity of the tangential components of

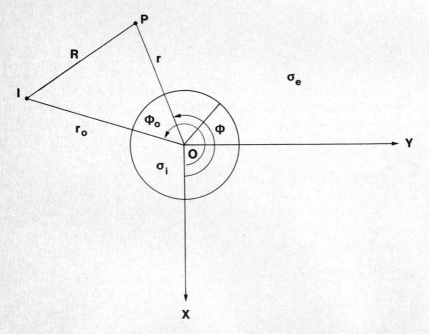

Fig. 6.29. A conducting cylinder in the electromagnetic field excited by a current filament.

the electromagnetic field, that is, of E_z and H_ϕ. To facilitate the application of these conditions and to find an appropriate expression for the secondary field, we will express the normal field in cylindrical coordinates with the z-axis lying along the axis of the cylinder as shown in Fig. 6.29. As follows from Fig. 6.29:

$$R = [r^2 + r_0^2 - 2r\, r_0 \cos(\phi - \phi_0)]^{1/2}$$

Applying the addition theorem, we have:

$$K_0(k_e R) = \sum_{n=0}^{\infty} \epsilon_n K_n(k_e r_0) I_n(k_e r) \cos n(\phi - \phi_0) \quad \text{if } r_0 > r \qquad (6.196)$$

where k_e is the wave number in the host medium:

$$\epsilon_n = \begin{cases} 1 & \text{as } n = 0 \\ 2 & \text{as } n > 0 \end{cases}$$

and $I_n(k_e r)$ and $K_n(k_e r)$ are the modified Bessel functions.

Therefore, the normal electric field can be written as:

$$E_z^0 = -\frac{i\omega\mu I_0}{2\pi} \sum_{n=0}^{\infty} \epsilon_n K_n(k_e r_0) I_n(k_e r) \cos n(\phi - \phi_0), \quad \text{if } r < r_0 \qquad (6.197)$$

This last equation suggests that we use the following form for the secondary field observed outside the cylinder:

$$E_z^s = -\frac{i\omega\mu I_0}{2\pi} \sum_{n=0}^{\infty} \epsilon_n A_n K_n(k_e r_0) K_n(k_e r) \cos n(\phi - \phi_0), \quad r > a \qquad (6.198)$$

where A_n are unknown coefficients at this stage. It should be clear that each term in eq. 6.198 will satisfy eq. 6.172, and that the function E_z^s tends to zero as the distance r goes to infinity. Inasmuch as the field must be finite everywhere inside the cylinder, we will seek a solution in the form:

$$E_z^i = -\frac{i\omega\mu I_0}{2\pi} \sum_{n=0}^{\infty} \epsilon_n B_n K_n(k_e r_0) I_n(k_i r) \cos n(\phi - \phi_0), \quad r < a \qquad (6.199)$$

where k_i is the wave number of the medium inside the cylinder, and B_n are unknown coefficients.

Using the boundary conditions at the surface of the cylinder:

$$E_z^e = E_z^i, \quad H_\phi^e = H_\phi^i, \quad \text{if } r = a$$

we have:

$$\sum_{n=0}^{\infty} \epsilon_n [K_n(k_e r_0) I_n(k_e a) + A_n K_n(k_e r_0) K_n(k_e a)] \cos n(\phi - \phi_0) =$$

$$= \sum_{n=0}^{\infty} \epsilon_n B_n K_n(k_e r_0) I_n(k_i a) \cos n (\phi - \phi_0) \tag{6.200}$$

and

$$\sum_{n=0}^{\infty} \epsilon_n k_e [K_n(k_e r_0) I_n'(k_e a) + A_n K_n(k_e r_0) K_n'(k_e a)] \cos n (\phi - \phi_0) =$$

$$= \sum_{n=0}^{\infty} k_i \epsilon_n B_n K_n(k_e r_0) I_n'(k_i a) \cos n (\phi - \phi_0) \tag{6.201}$$

By virtue of the orthogonality of trigonometric functions, eqs. 6.200 and 6.201 can be much simplified so that we obtain:

$$I_n(k_e a) + A_n K_n(k_e a) = B_n I_n(k_e a)$$

$$k_e [I_n'(k_e a) + A_n K_n'(k_e a)] = k_i B_n I_n'(k_e a)$$

Therefore, we have the following expressions for the coefficients A_n that describe the secondary field outside the cylinder:

$$A_n = - \frac{I_n'(\rho) I_n(m\rho) - m I_n(\rho) I_n'(m\rho)}{K_n'(\rho) I_n(m\rho) - m K_n(\rho) I_n'(m\rho)} \tag{6.202}$$

where

$$m = (\sigma_i/\sigma_e)^{1/2}, \quad \rho = k_e a$$

Inasmuch as:

$$H_r = - \frac{1}{i\omega\mu} \frac{1}{r} \frac{\partial E_z}{\partial \phi}, \quad H_\phi = \frac{1}{i\omega\mu} \frac{\partial E_z}{\partial r}$$

we have:

$$H_r^s = - \frac{I_0}{2\pi r} \sum_{n=1}^{\infty} n A_n K_n(k_e r_0) K_n(k_e r) \sin n (\phi - \phi_0)$$

$$\text{if } r > a \tag{6.203}$$

$$H_\phi^s = - \frac{I_0 k_e}{2\pi} \sum_{n=0}^{\infty} \epsilon_n A_n K_n(k_e r_0) K_n'(k_e r) \cos n (\phi - \phi_0)$$

In accord with eqs. 6.198 and 6.203, the secondary field is a sum of cylindrical harmonics. It will prove to be convenient to consider separately the fundamental and the rest of the harmonics. As follows from eqs. 6.198 and 6.203 the fundamental is independent of the angle ϕ, and therefore, it is contributed by linear current which has to be closed to the host medium. In contrast to the fundamental, the higher harmonics can be interpreted as being fields generated by currents closed within the cylinder. For this reason, it is only natural to expect that the two parts of the field will exhibit different dependencies on the parameters characterizing the medium, as well as on frequency and time.

Let us write both the secondary electric field E_z^s and the magnetic field H_ϕ^s in the forms:

$$E_z^s = E_{0z} + E_{1z}$$
$$H_\phi^s = H_{0\phi} + H_{1\phi}$$

(6.204)

where

$$E_{0z} = -\frac{i\omega\mu I_0}{2\pi} A_0 K_0(k_e r_0) K_0(k_e r)$$

(6.205)

$$E_{1z} = -\frac{i\omega\mu}{\pi} I_0 \sum_{n=1}^{\infty} A_n K_n(k_e r_0) K_n(k_e r) \cos n(\phi - \phi_0)$$

(6.206)

$$H_{0\phi} = \frac{I_0 k_e}{2\pi} A_0 K_0(k_e r_0) K_1(k_e r)$$

(6.207)

$$H_{1\phi} = -\frac{I_0 k_e}{\pi} \sum_{n=1}^{\infty} A_n K_n(k_e r_0) K_n'(k_e r) \cos n(\phi - \phi_0)$$

(6.208)

Comparing eqs. 6.205 and 6.207 with eqs. 6.178 and 6.179, respectively, it is clear that this part of the field E_{0z}, $H_{0\phi}$ is contributed by a current which is directed along the axis of the cylinder and closed through the host medium. At the same time, the rest of the secondary field described by eqs. 6.206 and 6.208, as will be shown later, is contributed by a system of induced currents which can be considered as being the sum of linear multi-poles.

Taking into account the completely different character of these two parts of the field, it is only reasonable that we examine them separately. First of all, let us consider the behavior of the secondary field corresponding to the fundamental. In accord with eq. 6.202 for the function A_0 which defines the main features of the frequency domain behavior, we have:

$$A_0 = \frac{I_1(\rho)I_0(m\rho) - m I_0(\rho) I_1(m\rho)}{K_1(\rho)I_0(m\rho) + m K_0(\rho) I_1(m\rho)}$$

(6.209)

since

$$I_0' = I_1 \quad \text{and} \quad K_0' = -K_1$$

where

$$\rho = k_e a, \quad m\rho = k_i a, \quad m = (\sigma_i/\sigma_e)^{1/2}$$

Expanding the Bessel functions in series in terms of small values for the parameters ρ and $m\rho$, we obtain an expression for the function A_0 which is appropriate at low frequencies:

$$A_0 \approx \frac{\rho^2 (1 - m^2)}{2} - \frac{\rho^4}{4} (1 - m^4) \ \text{In} \ (\alpha \rho) + \frac{\rho^4}{16} (1 - m^2) (3 - m^2) \quad (6.210)$$

here

$$\alpha = 0.8905362, \quad \text{if} \ |\rho| < 1 \ \text{and} \ |m\rho| < 1$$

Using the appropriate asymptotic expansions for the functions I_0, I_1, K_0, and K_1 for large arguments, we have:

$$A_0 \approx \frac{1 - m}{1 + m} \frac{I_0(\rho)}{K_0(\rho)}, \quad \text{if} \ |m\rho| \gg 1 \ \text{and} \ |\rho| \gg 1 \quad (6.211)$$

At this point, consider that the parameter ρ is small, while the skin effect in the cylinder is manifested strongly, that is, the parameter $|m\rho|$ is much greater than 1. Then, by using the appropriate expressions for the Bessel function, we obtain:

$$A_0 \approx \frac{1}{K_0(\rho)} \quad (6.212)$$

In accord with eqs. 6.207 and 6.210, we have the following result for the low-frequency behavior of the magnetic field component, $H_{0\phi}$:

$$QH_{0\phi} \approx \frac{I_0 \omega \mu}{4\pi r} a^2 (\sigma_i - \sigma_e) \tfrac{1}{2} \ln \sigma_e \mu \omega r_0^2$$

$$\text{In} H_{0\phi} \approx I_0 \frac{\omega \mu a^2}{16r} (\sigma_i - \sigma_e) \qquad (6.213)$$

Thus, the quadrature component is dominant over the inphase component, and its origin can be readily explained. In accord with eq. 6.186, it can be represented as

$$QH_{0\phi} \approx \frac{(\sigma_i - \sigma_e) QE_z^0 a^2}{2r} = \frac{I}{2\pi r} \quad (6.214)$$

where QE_z^0 is the primary vortex electric field directed along the axis cylinder, and:

$$I = (\sigma_i - \sigma_e) \pi a^2 QE_z^0$$

is the current flowing through a cross-section of the cylinder, assuming that the primary field QE_z^0 is uniform within the conductor.

Thus, under the action of the primary electric field which is directed along the axis of the cylinder, currents will arise and the current density for these anomalous currents will be:

$$j^a = (\sigma_i - \sigma_e) QE_z^0$$

In accord with the Biot-Savart law, the current I generates a magnetic field defined by eq. 6.214. It is clear that the same magnetic field will be present when a constant electric field is considered. In other words, this consideration has shown the quadrature component of the magnetic field, $QH_{0\phi}$ at low frequencies arises exactly in the same way as does the magnetic field contributed by direct current.

As follows from eq. 6.214, the quadrature component of the fundamental at low frequencies will be directly proportional to frequency and to the difference in conductivities between the cylinder and the host medium. If the cylinder is much more conductive than the host medium, this component of the magnetic field is practically independent of the conductivity of the host medium, and it will be directly proportional to the conductivity of the cylinder. It is appropriate at this point to note that such a dependence of conductivity also takes place under certain conditions in which a conductor with finite dimensions can be replaced by an infinitely long cylinder.

Later in this section, this question will be examined in more detail.

In order to compare the secondary field to the normal field, the normal field will be represented in a coordinate system based on the cylinder. As is seen from Fig. 6.29 and making use of eq. 6.188, we have:

$$\mathrm{In}H_\phi^0 = \frac{I_0\left[r - r_0\cos(\phi - \phi_0)\right]}{2\pi R^2}\left(1 - \frac{\pi}{8}\sigma_e\mu\omega R^2\right)$$

$$QH_\phi^0 = \frac{I_0\left[r - r_0\cos(\phi - \phi_0)\right]}{8\pi}\sigma_e\mu\omega\ \ln\left(\sigma_e\mu\omega\ R^2\right)$$

(6.215)

Comparing eq. 6.213 with eq. 6.215, we see that with a decrease of frequency, the ratio $QH_{0\phi}/QH_\phi^0$ tends to a constant value given by:

$$\left(\frac{\sigma_i}{\sigma_e} - 1\right)\frac{a^2}{r^2\left(1 - \frac{r_0}{r}\cos\phi\right)}$$

It should be obvious that for a very large ratio in conductivities, the effect of the host medium can be neglected.

In accord with eq. 6.213 and 6.215, the inphase component of the magnetic field, $H_{0\phi}$ and the corresponding component of the secondary normal field, $\mathrm{In}H_\phi^0$ are related to frequency in the same manner. For this reason, the ratio $\mathrm{In}H_{0\phi}/\mathrm{In}H_\phi^{0s}$ tends to a constant with decrease in frequency as well. Here, $\mathrm{In}H_\phi^{0s}$ is the inphase component of the normal field caused by induced currents in the host medium. This behavior of the relative anomaly in the inphase component can be related to this specific type of source and this specific model for the conductor.

As follows from eq. 6.207 and a physical point of view, at high frequencies, the normal field dominates, and therefore this part of the spectrum is of little practical interest.

At this point we can consider the low-frequency part of the spectrum for the magnetic field H_ϕ caused by induced currents which are closed within the cylinder. In accord with eq. 6.208 we have:

$$H_{1\phi} = -\frac{I_0 k_e}{\pi} \sum_{n=1}^{\infty} A_n K_n (k_e r_0) K'_n (k_e r) \cos n (\phi - \phi_0)$$

Initially, we will assume that the value for the parameter $|k_e a|$ is small, that is, that the skin depth in the host medium, h_e, is much greater than the radius of the cylinder, a. Replacing the Bessel functions with their appropriate asymptotic expansions, we obtain the following expression for A_n when n is greater than unity:

$$A_n \approx -\frac{\rho^{2n}}{2^{2n-1} n!(n-1)!} T_n(m\rho) \tag{6.216}$$

and

$$T_n = \frac{I_{n+1}(m\rho)}{I_{n-1}(m\rho)} \tag{6.217}$$

It is clear that T_n is a complex amplitude for the spectrum for each spatial harmonic, for the case in which the cylinder would be located in free space. In the same way, we have the following for the magnetic field $H_{1\phi}$:

$$H_{1\phi} = \frac{I_0 k_e}{\pi} \sum_{n=1}^{\infty} \frac{\rho^{2n}}{2^{2n-1} n!(n-1)!} T_n(m\rho) K_n(k_e r_0) K'_n(k_e r) \cos n (\phi - \phi_0) \tag{6.218}$$

Assuming that the skin depth, h_e, is much larger than the distances r_0 and r, we can make use of the following approximate expressions:

$$K_n(x) \approx \frac{(n-1)! \, 2^{n-1}}{x^n}, \qquad K'_n (x) \approx -\frac{n!}{x^{n+1}} 2^{n-1}$$

Then, we have:

$$H_{1\phi} \approx -\frac{I_0}{2\pi r} \sum_{n=1}^{\infty} T_n \left(\frac{a}{r_0}\right)^n \left(\frac{a}{r}\right)^n \cos n (\phi - \phi_0), \quad \text{if } |k_e r| < 1, |k_e r_0| < 1 \tag{6.219}$$

This last equation is precisely the same as that for a cylinder situated in free space. Therefore, one can expect that when the surrounding medium is relatively resistive, eq. 6.219 will describe the frequency-domain response with reasonable accuracy. In particular, for the first harmonic which corresponds to a uniform excitation of the field, we have:

$$H_{1\phi}^{(1)} \approx -\frac{I_0 T_1}{2\pi r_0} \frac{a^2}{r^2} \cos(\phi - \phi_0)$$

In contrast to the fundamental field component, the source of the field $H_{1\phi}$ as well as of H_{1r} are induced currents which arise as a consequence of the change in the normal magnetic field with time. This is the reason that this part of the field is called the vortex part. At low frequencies, by expanding the functions $I_{n+1}(m\rho)$ and $I_{n-1}(m\rho)$ in a series, we have:

$$H_{1\phi} \approx -\frac{I_0}{2\pi r} \sum_{n=1}^{\infty} \frac{(k_i a)^2}{4n(n+1)} \left[1 - \frac{k_i^2 a^2}{2n(n+2)}\right] \left(\frac{a}{r_0}\right)^n \left(\frac{a}{r_0}\right)^n \cos n(\phi - \phi_0)$$

$$(6.220)$$

For example, the quadrature and inphase components of the field corresponding to the first harmonics are:

$$QH_{1\phi}^{(1)} \approx -\frac{I_0}{2\pi r_0} \frac{\sigma_i \mu \omega a^2}{8} \frac{a^2}{r^2} \cos(\phi - \phi_0)$$

$$InH_{1\phi}^{(1)} \approx -\frac{I_0}{2\pi r_0} \frac{(\sigma_i \mu \omega a^2)^2}{48} \frac{a^2}{r^2} \cos(\phi - \phi_0)$$

$$(6.221)$$

It is clear that these expressions describe the secondary field when the primary field is uniform as was treated in Chapter 3 and with some approximation, the effect of the surrounding medium is absent. A more accurate expansion of the functions $K_n(k_e r_0)$ and $K'_n(k_e r_0)$, leads to the following expressions for $H_{1\phi}$:

$$H_{1\phi} = -\frac{I_0}{2\pi r} \left\{ \sum_{n=1}^{\infty} T_n \left(\frac{a}{r_0}\right)^n \left(\frac{a}{r}\right)^n \cos n(\phi - \phi_0) \right.$$

$$+ 2\frac{r}{a} \sum_{n=1}^{\infty} \left[\frac{\rho^{2n+1} m^2 \rho^2}{2^{2n-1} n!(n-1)! \, 4n(n+1)} \left(1 - \frac{m^2 \rho^2}{2n(n+2)}\right) \right]$$

$$\times (-1)^{n+1} \frac{1}{2\rho} \left(\frac{r^{n-1}}{r_0^n} \ln \alpha \rho r - \frac{r_0^n}{r^{n+1}} \ln \alpha \rho r_0\right) \right\}$$

$$(6.222)$$

The second sum in this last equation reflects the influence of the host medium at low frequencies, and in addition, it also defines the late-stage behavior in the transient response.

Now let us compare the vortex part of the field with that corresponding to the null-harmonic. For this purpose, we will make use only of the leading term in the expression for the vortex part of the field. Then in accord with eqs. 6.213 and 6.221, it is clear that the field from a linear current filament closed through the surrounding medium is dominant over the vortex part of the field at low frequencies. However, it is appropriate to emphasize again that the particular model used for the conductor and for the source of the

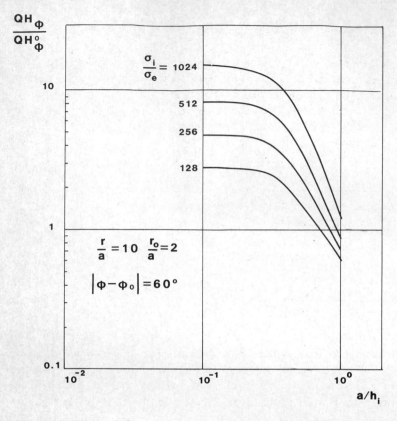

Fig. 6.30. Ratio of the quadrature component of the field, H_ϕ, to the quadrature component of the normal field.

normal field that is being considered in this paragraph can be applied for real conductors only under some circumstances.

So far, only the asymptotic behavior of the field has been examined. Now by making use of the exact solution (eq. 6.203) we can investigate the frequency-domain response of the quadrature and inphase components of the total field, H_ϕ, equal to $H_\phi^0 + H_\phi^s$, with respect to the corresponding components of the normal field H_ϕ^0. These various functions are shown by the curves in Figs. 6.30—6.34. Some obvious conclusions can be drawn by examination of these responses:

(1) The maximum in the relative anomaly of the quadrature component will be observed at low frequencies ($a/h_i < 0.3$).

(2) The frequency domain behavior for the relative anomaly in the inphase component has a maximum which is not significantly greater than the corresponding values for the quadrature component.

(3) When the observation point approaches the conductive cylinder, as well as with the source for the normal field being removed to greater distances,

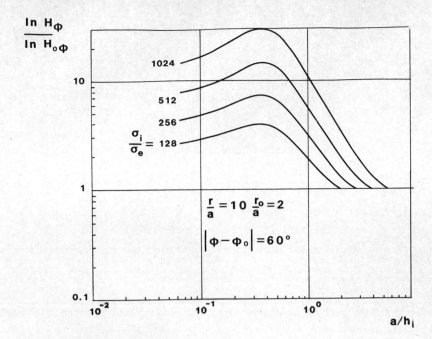

Fig. 6.31. Ratio of the inphase component of the total field, H_ϕ, to the inphase component of the normal field.

the relative anomaly in both the quadrature and inphase components will increase.

(4) With an increase in frequency, as a consequence of skin effect in the host medium, the relative anomalies decrease rapidly.

(5) Over all frequencies, the galvanic part of the field corresponding to the null-harmonic is essentially dominant and provides relatively large anomalies. This is the reason why at low frequencies, the relative anomalies in the quadrature and inphase components are almost directly proportional to the conductivity of the cylinder, provided that $\sigma_i/\sigma_e \gg 1$.

At this point it is appropriate to consider some features of the transient response of the electromotive source. In cylindrical coordinates, we have the following expression for the function $\partial H_\phi^0/\partial t$:

$$\frac{\partial H_\phi^0}{\partial t} = \frac{I_0\,\mu\sigma_e\,[r - r_0\cos(\phi - \phi_0]}{8\pi\,t^2}\,e^{-\sigma_e\mu R^2/t} \tag{6.223}$$

Therefore, during the late stage:

$$\frac{\partial H_\phi^0}{\partial t} \approx \frac{I_0\mu\sigma_e\,[r - r_0\cos(\phi - \phi_0)]}{8\pi t^2} \tag{6.224}$$

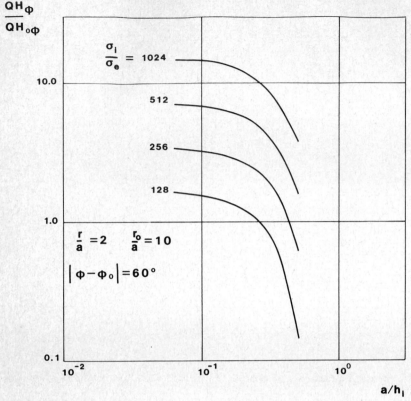

Fig. 6.32. Ratio of the quadrature component of the total field, H_ϕ, to the quadrature component of the normal field.

Using the low-frequency part of the spectrum and the Fourier transform, we obtain the following expression for the late-stage behavior of the function $\partial H_{0\phi}/\partial t$ which describes the galvanic part of the field:

$$\frac{\partial H_{0\phi}}{\partial t} \approx \frac{I_0(1-m^2)}{8\pi\,\sigma_e\,\mu a^2 r t^{*2}} \left[1 + \frac{1+m^2-r^2/a^2}{t^*}\ln t^*\right] \quad t^* > 1 \tag{6.225}$$

where

$$t^* = t/\sigma_e\mu a^2 \tag{6.226}$$

Respectively, we have the following for the first harmonic of the vortex part of the field, by applying the same approach:

$$\frac{\partial H_{1\phi}^{(1)}}{\partial t} \approx \frac{I_0}{32\pi\,\sigma_e\mu a^2 r_0 r^2}\left[\frac{m^2\,(r^2-r_0^2)}{t^{*3}} + \frac{64a^2}{m^2}\,e^{-t/\tau_0}\right]\cos(\phi-\phi_0) \tag{6.227}$$

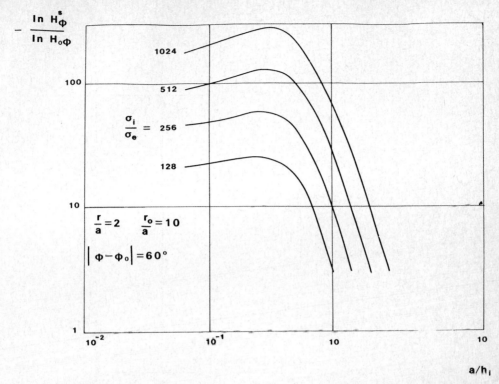

Fig. 6.33. Ratio of the inphase component of the total field, H_ϕ, to the component of the normal field.

where τ_0 is the time constant characterizing the cylinder.

As follows from eqs. 6.224—6.226, the derivatives of the normal field H_ϕ^0 and a null-harmonic field decay at the same rate during the late stage (as $1/t^2$), while the vortex part of the field decreases even more rapidly. Therefore, over a significant part of the transient response, including the late stage, the galvanic part of the field is dominant. For this part of the response, the electromotive force will be directly proportional to the conductivity of the cylinder and is practically independent of the conductivity in the host medium, provided that $\sigma_i/\sigma_e \gg 1$. Examples of the transient responses of the total electromotive force, normalized to the normal electromotive force, are shown by the various curves in Figs. 6.35 and 6.36.

In conclusion, we can consider the subject of the nature of the source and the character of the conductor which has been used here. It should be clear that with an infinitely long cylinder, and with the source of a normal field being an infinitely long current filament oriented parallel to the axis of the cylinder, the conditions are most favorable for the development of the galvanic part of the field which is characterized by the null-harmonic.

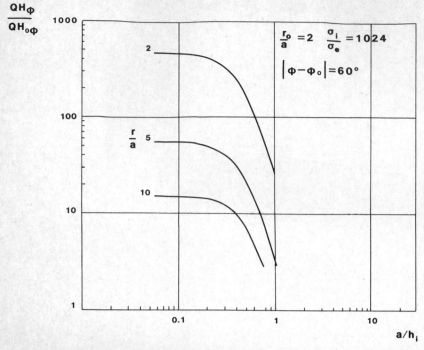

Fig. 6.34. The ratio of the quadrature component of the magnetic field H_ϕ to the quadrature component of the normal field.

Of course, it would be more interesting from the practical point of view to investigate this part of the field for cases in which the conductor would have finite dimensions. Unfortunately, up to this time, this more general problem has not been treated in enough detail to permit consideration of the contributions from the vortex and galvanic parts of the fields in either frequency or time domains. However, it is quite feasible to formulate the conditions under which the assumption about the existence of currents in the conductor being close to infinity does not lead to any significant error for both the late stage of the time-domain behavior and the low-frequency part of the frequency-domain behavior. We can assume that the normal field is uniform in the vicinity of a conductor, and results of calculations for direct current behavior can be used. As an example, we might consider the constant field in the case of an ellipsoid of revolution with semi-axes a and b, and a conductivity σ_i situated in a uniform host medium with a conductivity σ_e. The primary electric field is uniform and directed along the major axis, $2a$, is shown in Fig. 6.37. On excitation by the primary field, electric charges arise on the surface of the ellipsoid. These cause a secondary electric field. The charge density at point p on the surface of the spheroid is described by the equation:

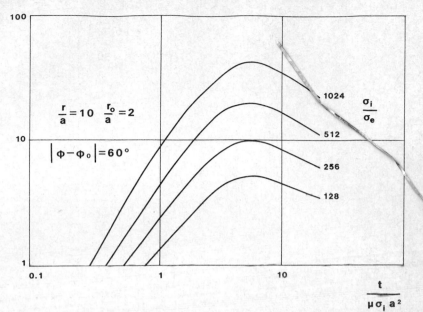

Fig. 6.35. Ratio of transient responses of the electromotive force of the total field to that of the normal field.

$$\Sigma(p) = \frac{1}{2\pi} \frac{\sigma_i - \sigma_e}{\sigma_1 - \sigma_e} E_n^{av}(p)$$

where E_n^{av} is the average of the values for the normal components of the electric field on the inside and outside of the surface at the point p.

With an increase in the dimension of the major semi-axis, a, the charge density on the lateral surface of the spheroid decreases and the principal part of the charge is concentrated near the ends of the axis, $2a$. It is clear that the field from these charges becomes smaller with increase in length, $2a$, and when measurements are made near the center of the spheroid, the total field is practically the same as the primary field, E_0. In other words, a marked elongation of an inhomogeneity in the direction of the primary field will not significantly distort the electric field, if measurements are made near the center of the ellipsoid at a distance from the center which is considerably smaller from the length of the semi-axis, a. However, the current density inside the conductor, and hence the magnetic field, will increase in direct proportion to the conductivity, σ_i. In this case only, the model for an infinitely long cylinder permits us to obtain the correct representation for the magnetic field contributed by currents in elongated conductors.

However, it must be stressed that the condition for applicability of this

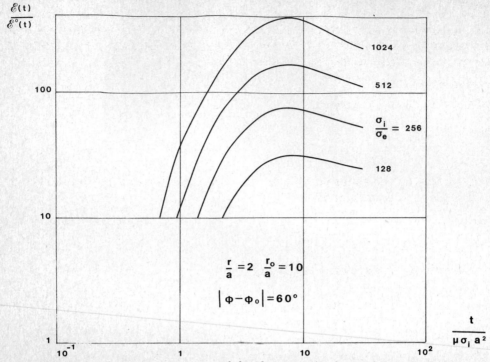

Fig. 6.36. Ratio of transient responses of the electromotive force of the total field to that of the normal field.

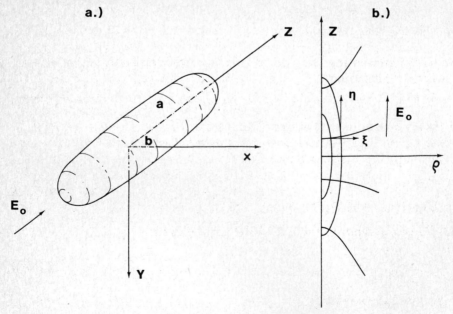

Fig. 6.37. Definition of a prolate spheroidal coordinate system.

particular model does not depend solely on the linear dimensions of the inhomogeneity, but will depend as well on the ratio of conductivities. Let us examine this question in some detail. We will introduce a prolate spheroidal coordinate system (ξ and η) that are related to the cylindrical coordinate systems as follows (see Fig. 6.37b):

$$z = c\eta\xi$$
$$\rho = c[(1-\eta^2)(\xi^2-1)]^{1/2} \tag{6.228}$$

and

$$c = (a^2 - b^2)^{1/2}$$

The surface of the spheroid characterized with semi-axes a and b is a coordinate surface ξ_0 equal a constant, and:

$$a = c\,\xi_0$$
$$b = c\,(\xi_0^2 - 1)^{1/2} \tag{6.229}$$

The metric for this system has the form:

$$h_\xi = c\left[\frac{\xi^2-\eta^2}{\xi^2-1}\right]^{1/2}, \quad h_\eta = c\left[\frac{\xi^2-\eta^2}{1-\eta^2}\right]^{1/2}, \quad h_\phi = \rho \tag{6.230}$$

The potential of the electric field inside and outside the spheroid will satisfy Laplace's equation as well as the boundary conditions:

$$U_i = U_e$$
$$\sigma_i \frac{\partial U_i}{\partial \xi} = \sigma_e \frac{\partial U_e}{\partial \xi} \quad \text{at } \xi = \xi_0 \tag{6.231}$$

The potential representing the primary electric field can be described using a single spheroidal harmonic:

$$U_0 = -E_0 z = -E_0 c\eta\xi = -E_0 c P_1(\eta)P_1(\xi) \tag{6.232}$$

where $P_1(x)$ is Legendre's function of the first order: $P_1(x) = x$.

Considering the expression for the potential inside and outside the spheroid given by eq. 6.232, we have:

$$U_i = -E_0 c B P_1(\xi) P_1(\eta)$$
$$U_e = -E_0 c [P_1(\xi) + A\,Q_1(\xi)]\, P_1(\eta) \tag{6.233}$$

where $Q_1(\xi)$ is a Legendre function of the second kind:

$$Q_1(\xi) = \frac{\xi}{2} \ln \frac{\xi+1}{\xi-1} - 1 \tag{6.234}$$

Applying eq. 6.231, we obtain:

$$A = \frac{[\sigma_i/\sigma_e - 1]\, \xi_0\, (\xi_0^2 - 1)}{1 + [\sigma_i/\sigma_e - 1]\, (\xi_0^2 - 1) \left\{\dfrac{\xi_0}{2}\ln\dfrac{\xi_0 + 1}{\xi_0 - 1} - 1\right\}} \tag{6.235}$$

$$B = \frac{1}{1 + [\sigma_i/\sigma_e - 1]\, (\xi_0^2 - 1) \left\{\dfrac{\xi_0}{2}\ln\dfrac{\xi_0 + 1}{\xi_0 - 1} - 1\right\}}$$

Thus, the electric field inside the spheroid, E_i, is uniform and directed along the z-axis:

$$E_i = BE_0 = \frac{E_0}{1 + (\sigma_i/\sigma_e - 1)\, L} \tag{6.236}$$

where L is a depolarization factor, which can be expressed as:

$$L = \frac{1 - e^2}{2e^3}\left(\ln\frac{1 + e}{1 - e} - 2e\right) \tag{6.237}$$

where

$$e = [1 - b^2/a^2]^{1/2} \tag{6.238}$$

The depolarization factor represents the effect of surface charges, and for markedly elongate bodies ($e \to 1$) we have:

$$L \approx \frac{b^2}{a^2}\left(\ln\frac{2a}{b} - 1\right) \ll 1 \tag{6.239}$$

In accord with eq. 6.236, with increase in major axis a ($L \to 0$) the electric field approaches the primary field E_0. With increase in values for the ratio σ_i/σ_e, the similarity of the fields for spheroids and for infinitely long cylinders is observed for progressively more elongate spheroids. In particular, if the host medium is insulating, the secondary field is equal to the primary field inside the spheroid and is directed in the opposite direction regardless of the linear dimensions (a phenomenon of electrostatic induction). A curve given the depolarization factor L is shown in Fig. 6.38. From eq. 6.236, the current density in the spheroid will be:

$$j = \frac{\sigma_i E_0}{1 + \left(\dfrac{\sigma_i}{\sigma_e} - 1\right) L} \tag{6.240}$$

For relatively small ratios of conductivity and with L much less than unity, as the conductivity of the spheroid is increased, the current density and therefore the magnetic field intensity will increase in direct proportion to σ_i, just as would be the case for an infinitely long cylindrical conductor.

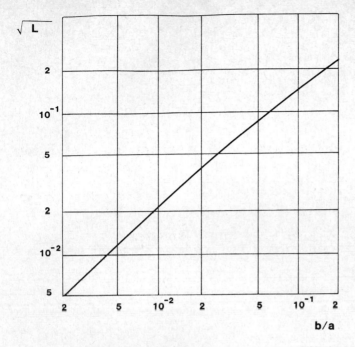

Fig. 6.38. The depolarization factor.

However, with a further increase in the conductivity σ_i, the secondary electric field increases and the total field E_i becomes smaller. When $(\sigma_i/\sigma_e)L$ is much greater than unity, the current density and the magnetic field intensity were practically independent of the conductivity in the spheroid, and we have

$$j = \frac{\sigma_e E_0}{L} \gg \sigma_e E_0 \qquad (6.241)$$

If the spheroid is more resistive, such that $\sigma_i/\sigma_e < 1$, for relatively small values of the ratio a/b, the current density will be proportional to the conductivity of the spheroid.

In determining the secondary magnetic field for which there is only the component $H_{1\phi}$, we find the secondary current passing through a circle of radius ρ for the case $\eta = 0$. Inside the spheroid, we have:

$$I_i = \sigma_i \pi b^2 B E_0 - \sigma_e \pi b^2 B E_0 = I_i' E_0 \qquad (6.242)$$

The secondary current through the elementary ring dS outside the spheroid (with $\eta = 0$) is:

$$dI_e = \sigma_e E_{1z}^e 2\pi\rho \, d\rho$$

where

$$E_{1z}^e = -\frac{1}{h_\eta}\frac{\partial}{\partial\eta}(U_e - U_0)\bigg|_{\eta=0} = A E_0 \frac{Q_1(\xi)}{\xi} \tag{6.243}$$

and

$$\xi = \frac{1}{c}(\rho^2 + a^2 - b^2)^{1/2}$$

Hence

$$I_e = 2\pi\sigma_e \int_b^\rho \rho E_{1z}^e \,d\rho = 2\pi\sigma_e E_0 A \int_b^\rho \rho\left(\tfrac{1}{2}\ln\frac{\xi+1}{\xi-1} - \frac{1}{\xi}\right)d\rho =$$

$$= 2\pi\,\sigma_e\,E_0\,A\,c^2 \int_{\xi_0}^\xi \left(\frac{\xi}{2}\ln\frac{\xi+1}{\xi-1} - 1\right)d\xi$$

This last equation has been written by taking into account that when $\eta = 0$, we have:

$$\rho\,d\rho = c^2\,\xi\,d\xi$$

Thus:

$$I_e = \pi\,\sigma_e E_0 A c^2 \left(\frac{\xi^2-1}{2}\ln\frac{\xi+1}{\xi-1} - \xi - \frac{\xi_0^2-1}{2}\ln\frac{\xi_0+1}{\xi_0-1} + \xi_0\right) = I_e' E_0 \tag{6.244}$$

From eqs. 6.242 and 6.244, at $\eta = 0$ we have:

$$\frac{H_{1\phi}^e}{H_0} = \frac{I_i' + I_e'}{2\pi\rho}\frac{E_0}{H_0} = \frac{\sigma_e(a^2-b^2)}{2\rho} A \left(\frac{\xi^2-1}{2}\ln\frac{\xi+1}{\xi-1} - \xi\right)\frac{E_0}{H_0} \tag{6.245}$$

The relationship between the magnetic field intensity and the parameters characterizing the medium is illustrated by the curves in Fig. 6.39. The manner of approach to the right-hand asymptotes for these curves defines the minimum value of a/b for which the magnetic field of currents in the spheroid is equivalent to that for currents in an infinitely long cylinder.

This consideration leads us to expect that at low frequencies as well as during the late stage of time-domain behavior, we may use the model of an infinitely long cylinder in determining the magnetic field if the following conditions are met:

$$\frac{\sigma_i}{\sigma_e}L \ll 1, \quad \lambda \gg l$$

or

$$\frac{\sigma_i}{\sigma_e}L \gg 1, \quad \tau \gg l \tag{6.246}$$

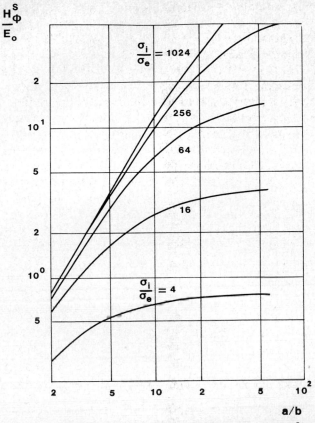

Fig, 6.39. Behavior of the secondary magnetic field H_ϕ^s normalized to the strength of the electric field E_0.

where $\lambda = 2\pi h$ and $\tau = (2\pi\rho t \times 10^7)^{1/2}$ are the minimum values for these parameters in the medium, l is a linear dimension of the conductive cylinder in the direction of the primary electric field, and L is the depolarization factor depending on the shape of the inhomogeneity. It is obvious that along with condition 6.246 it must be assumed that the normal electric field is uniform in the vicinity of the conductor. In the next section, we will examine the influence of the galvanic part of the field when the normal electric field varies along the axis of the cylinder.

6.5. FREQUENCY- AND TIME-DOMAIN BEHAVIOR FOR A MAGNETIC DIPOLE IN THE PRESENCE OF A CONDUCTING CYLINDER

We will next consider a more complicated problem, that in which a conducting cylinder is located within the field of a magnetic dipole source.

570

The surrounding medium has a finite resistivity. In this case, the normal electric field will intersect the surface of the cylinder and therefore, electrical charges will arise on the surface of the cylinder. Inasmuch as the surrounding medium is not insulating, the electric field from these charges as well as the vortex field will cause currents that flow through the cylinder and through the surrounding medium. In addition to these currents, we will observe induction currents that flow in closed paths within the cylinder and induction currents in the surrounding medium as well. In contrast to the previous cases, there is no analytical method that can be used to separate the galvanic and vortex parts of the field. For this reason, a complete solution for the field behavior will be numerically investigated. First,we will derive expressions for the field. Suppose that the magnetic dipole, which is oriented arbitrarily with respect to the axis of the cylinder, is located at point p_0 as shown in Fig. 6.40. The moment of the dipole will be:

$$M = M_0 \, e^{-i\omega t} n$$

where n is the unit vector indicating the direction of the dipole axis.

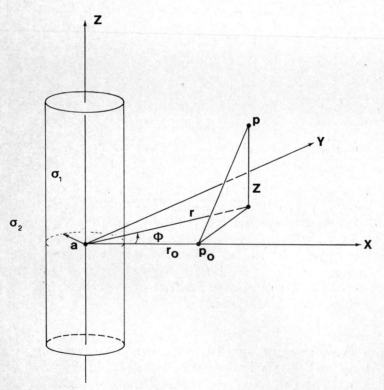

Fig. 6.40. A magnetic dipole source is arbitrarily oriented with respect to the axis of a conducting cylinder and is located at point p_0.

Let us select a coordinate system with an axis z that is coincident with the axis of the cylinder. We will assume that the coordinates of the source and of the observation point are $(r_0, 0, 0)$ and (r, ϕ, z) respectively. The conductivity of the cylinder is σ_1, the conductivity of the surrounding medium is σ_2, and all portions of the medium have the magnetic permeability of free space. Displacement currents will be neglected.

It is possible to write the total field in the host medium as the sum of two terms:

$$E_2 = E_0 + E_s, \quad H_2 = H_0 + H_s$$

where E_0 and H_0 are components of the normal field of the magnetic dipole source for a medium with conductivity σ_2, while E_s and H_s are components of the secondary field. The vectors E_0, H_0, E_s, and H_s, as well as the total fields E_1 and H_1 inside the cylinder satisfy the following equations:

$$\operatorname{curl} H_0 = \sigma_2 E_0 - j^{\text{ext, e}} \tag{6.247}$$

and

$$\operatorname{curl} E_0 = i\omega\mu H_0 - j^{\text{ext, m}} \tag{6.248}$$

where j is the external current density of electric and magnetic type.

The fields obviously can be separated into primary and secondary contributions as follows:

$$\begin{aligned} \operatorname{curl} H_1 &= \sigma_1 E_1 \\ \operatorname{curl} E_1 &= i\omega\mu H_1 \end{aligned} \quad \text{if } r < a \tag{6.249}$$

and

$$\begin{aligned} \operatorname{curl} H_2 &= \sigma_2 E_2 \\ \operatorname{curl} E_2 &= i\omega\mu H_2 \end{aligned} \quad \text{if } r > a \tag{6.250}$$

Boundary conditions at the surface of the cylinder are:

$$\begin{aligned} E_{z1} - E_{z2} &= E_{z0} \\ E_{\phi 1} - E_{\phi 2} &= E_{\phi 0} \end{aligned} \quad \text{if } r = a \tag{6.251}$$

and

$$\begin{aligned} H_{z1} - H_{z2} &= H_{z0} \\ H_{\phi 1} - H_{\phi 2} &= H_{\phi 0} \end{aligned} \quad \text{if } r = a \tag{6.252}$$

In accord with these last four equations, the problem of determining the behavior of the total electric magnetic field actually consists of two independent problems, namely: (1) determination of the normal field as described by eqs. 6.247 and 6.248; and (2) determination of the secondary field described by the two eqs. 6.249 and 6.250 along with the non-uniform boundary conditions expressed as eqs. 6.251 and 6.252.

At this point we will examine the solution of the second problem, inasmuch as the solution of the first problem is well known. Usually, the solution of the second problem is accomplished using two vector potentials, these being of the electric and magnetic types, which are related with the electric and magnetic fields as:

$$E = i\omega\mu \text{ curl } \mathbf{\Pi}^* \quad \text{and} \quad H = \text{curl } \mathbf{\Pi},$$

where in the cylindrical system of coordinates we have:

$$\mathbf{\Pi}^* = (0, 0, \Pi_z^*) \quad \text{and} \quad \mathbf{\Pi} = (0, 0, \Pi_z)$$

Following the method employed by Tabarovskiy (1975), we will examine further transform of the expressions for the vertical components of the electric and magnetic fields and use them as potentials for the electromagnetic field. For the z-component of and arbitrary function $A(r, \phi, z)$, the Fourier transform with respect to z will be:

$$A^*(r, \phi, m) = \int_{-\infty}^{\infty} e^{-imz} A(r, \phi, z) \, dz$$

$$A(r, \phi, z) = \frac{1}{2\pi} \int_{-\infty}^{\infty} e^{imz} A^*(r, \phi, m) \, dm$$

Proceeding from eqs. 2.249 and 2.250, it is a straight forward matter to demonstrate that the Fourier transforms for the horizontal components of the electric and magnetic fields can be expressed in terms of the transforms of the vertical components as follows:

$$E_{ri} = \frac{-1}{p_i^2} \left(\frac{i\omega\mu}{r} \frac{\partial H_{zi}^*}{\partial \phi} + im \frac{\partial E_{zi}^*}{\partial r} \right)$$

$$E_{\phi i}^* = \frac{1}{p_i^2} \left(i\omega\mu \frac{\partial H_{zi}^*}{\partial r} - \frac{im}{r} \frac{\partial E_{zi}^*}{\partial \phi} \right)$$

(6.253)

and

$$H_{ri}^* = -\frac{1}{p_i^2} \left(\frac{\sigma_1}{r} \frac{\partial E_{zi}^*}{\partial \phi} + im \frac{\partial H_{zi}^*}{\partial r} \right)$$

$$H_{\phi i}^* = \frac{1}{p_i^2} \left(\sigma_1 \frac{\partial E_{zi}^*}{\partial r} - \frac{im}{r} \frac{\partial H_{zi}^*}{\partial \phi} \right)$$

(6.254)

here

$$p_i^2 = m^2 - k_i^2, \quad k_i^2 = i\omega\mu\sigma_i; \quad i = 1, 2.$$

The Fourier transforms of the vertical component A_i^* satisfy the following differential equation:

$$\frac{\partial^2 A_i^*}{\partial r^2} + \frac{1}{r}\frac{\partial A_i^*}{\partial r} + \frac{1}{r^2}\frac{\partial^2 A_i^*}{\partial \phi^2} - p_i^2 A_i^* = 0 \qquad (6.255)$$

where

A_i^* is H_{zi}^* or E_{zi}^*, $\quad i = 1, 2.$

The boundary conditions for E_{zi}^* and H_{zi}^* satisfy eqs. 6.251 through 6.254 and have the form:

$$E_{z1}^* - E_{z2}^* = E_{z0}^*$$

$$H_{z1}^* - H_{z2}^* = H_{z0}^*$$

$$\frac{1}{p_1^2}\left(i\omega\mu\frac{\partial H_{z1}^*}{\partial r} - \frac{im}{r}\frac{\partial E_{z1}^*}{\partial \phi}\right) - \frac{1}{p_2^2}\left(i\omega\mu\frac{\partial H_{z2}^*}{\partial r} - \frac{im}{r}\frac{\partial E_{z2}^*}{\partial \phi}\right) =$$

$$= \frac{1}{p_2^2}\left(i\omega\mu\frac{\partial H_{z0}^*}{\partial r} - \frac{im}{r}\frac{\partial E_{z0}^*}{\partial \phi}\right) \qquad (6.256)$$

$$\frac{1}{p_1^2}\left(\sigma_1\frac{\partial E_{z1}^*}{\partial r} - \frac{im}{r}\frac{\partial H_{z1}^*}{\partial \phi}\right) - \frac{1}{p_2^2}\left(\sigma_2\frac{\partial E_{z2}^*}{\partial r} - \frac{im}{r}\frac{\partial H_{z2}^*}{\partial \phi}\right) =$$

$$= \frac{1}{p_2^2}\left(\sigma_2\frac{\partial E_{z0}^*}{\partial r} - \frac{im}{r}\frac{\partial H_{z0}^*}{\partial \phi}\right)$$

At this point, let us expand the functions E_{zi}^*, H_{zi}^*, E_{z0}^*, and H_{z0}^* in Fourier series in terms of the azimuthal coordinate, ϕ:

$$E_{zi}^* = \sum_{n=0}^{\infty} e_{ni}^c \cos n\,\phi + e_{ni}^s \sin n\,\phi$$

$$H_{zi}^* = \sum_{n=0}^{\infty} h_{ni}^c \cos n\,\phi + h_{ni}^s \sin n\,\phi$$

$$\qquad (6.257)$$

$$E_{z0}^* = \sum_{n=0}^{\infty} e_{n0}^c \cos n\,\phi + e_{n0}^s \sin n\,\phi$$

$$H_{z0}^* = \sum_{n=0}^{\infty} h_{n0}^c \cos n\,\phi + h_{n0}^s \sin n\,\phi$$

Substituting the expression 6.257 into eqs. 6.253—6.256, we obtain the following equations and boundary conditions for the functions $e_{ni}^{c,s}$ and $h_{ni}^{c,s}$:

$$\frac{\partial^2 e_{ni}^{c,s}}{\partial r^2} + \frac{1}{r}\frac{\partial e_{ni}^{c,s}}{\partial r} - \left(\frac{n^2}{r^2} + p_i^2\right) e_{ni}^{c,s} = 0$$

574

$$\frac{\partial^2 h_{ni}^{c,s}}{\partial r^2} + \frac{1}{r} \frac{\partial h_{ni}^{c,s}}{\partial r} - \left(\frac{n^2}{r^2} + p_i^2\right) h_{ni}^{c,s} = 0 \tag{6.258}$$

$$e_{n1}^{c,s} - e_{n2}^{c,s} = e_{n0}^{c,s}$$

$$h_{n1}^{c,s} - h_{n2}^{c,s} = h_{n0}^{c,s}$$

$$\frac{1}{p_1^2}\left(i\omega\mu\,\frac{\partial h_{n1}^{c,s}}{\partial r} \mp \frac{imn}{r}\,e_{n1}^{c,s}\right) - \frac{1}{p_2^2}\left(i\omega\mu\,\frac{\partial h_{n2}^{c,s}}{\partial r} \mp \frac{imn}{r}\,e_{n2}^{c,s}\right) =$$

$$= \frac{1}{p_2^2}\left(i\omega\mu\,\frac{\partial h_{n0}^{c,s}}{\partial r} \mp \frac{imn}{r}\,e_{n0}^{c,s}\right)$$

$$\frac{1}{p_1^2}\left(\sigma_1\,\frac{\partial e_{n1}^{c,s}}{\partial r} \mp \frac{imn}{r}\,h_{n1}^{s,c}\right) - \frac{1}{p_2^2}\left(\sigma_2\,\frac{\partial e_{n2}^{c,s}}{\partial r} \mp \frac{imn}{r}\,h_{n2}^{s,c}\right) =$$

$$= \frac{1}{p_2^2}\left(\sigma_2\,\frac{\partial e_{n0}^{c,s}}{\partial r} \mp \frac{imn}{r}\,h_{n0}^{s,c}\right) \tag{6.259}$$

The sign $(-)$ in a notation $(\mp)$ corresponds to the left upper index for e_{ni} and h_{ni} while the sign $(+)$ corresponds to the right upper index.

As is well known, the solution to eq. 6.258 must satisfy the field conditions as r becomes infinite and remains finite everywhere:

$$e_{n1}^{c,s} = C_{n1}^{c,s}\,I_n(p_1 r), \quad e_{n2}^{c,s} = C_{n2}^{c,s}\,K_n(p_2 r)$$

$$h_{n1}^{c,s} = D_{n1}^{c,s}\,I_n(p_1 r), \quad h_{n2}^{c,s} = D_{n2}^{c,s}\,K_n(p_2 r) \tag{6.260}$$

where I_n and K_n are modified Bessel functions.

Substituting these last expressions into the boundary condition of eq. 6.259 and transforming this system of linear equations, we can obtain two independent systems of the fourth order with respect to the unknown coefficients $C_{ni}^{c,s}$ and $D_{ni}^{c,s}$ ($i = 1, 2$). These two systems of equations have the property that on the right-hand side of one there will be only combinations of the functions e_{n0}^c and h_{n0}^s, while the right-hand side of the second system will contain only terms in e_{n0}^s and h_{n0}^c. For example, the first system is:

$$I_n(p_1 a)\,C_{n1}^c - K_n(p_2 a)\,C_{n2}^c = e_{n0}^c$$

$$I_n(p_1 a)\,D_{n1}^s - K_n(p_2 a)\,D_{n2}^s = h_{n0}^s$$

$$\frac{imn}{ap_1^2}\,I_n(p_1 a)\,C_{n1}^c + \frac{i\omega\mu}{p_1}\,I_n'(p_1 a)\,D_{n1}^s - \frac{imn}{ap_2^2}\,K_n(p_2 a)\,C_{n2}^c$$

$$- \frac{i\omega\mu}{p_2}\,K_n'(p_2 a)\,D_{n2}^s = \frac{1}{p_2^2}\left(i\omega\mu\,\frac{\partial h_{n0}^s}{\partial r} + \frac{imn}{a}\,e_{n0}^c\right) \tag{6.261}$$

$$\frac{\sigma_1}{p_1} I'_n(p_1 a) C^c_{n1} - \frac{imn}{ap_1^2} I_n(p_1 a) D^s_{n1} - \frac{\sigma_2}{p_2} K'_n(p_2 a) C^c_{n2} + \frac{imn}{ap_2^2} K_n(p_2 a) D^s_{n2} =$$

$$= \frac{1}{p_2^2} \left(\sigma_2 \frac{\partial e^c_{n0}}{\partial r} - \frac{imn}{a} h^s_{n0} \right)$$

The second system is identical to the first except that the indices "c" and "s" are interchanged, and the sign of the term imn/a reverses.

Solving this system for the quantities C_{n2} and D_{n2}, because we are concerned with the field outside the cylinder, we obtain:

$$C^{c,s}_{n2} = \frac{e^{c,s}_{n0} E^{(1)}_n \mp \dfrac{imn}{a} h^{s,c}_{n0} H^{(1)}_n}{\Sigma}$$

(6.262)

$$D^{s,c}_{n2} = \frac{\pm \dfrac{imn}{a} e^{c,s}_{n0} E^{(2)}_n + h^{s,c}_{n0} H^{(2)}_n}{\Sigma}$$

where

$$\Sigma = \frac{m^2 n^2}{a^2} I_n^2(p_1 a) K_n^2(p_1 a)(\sigma_2 - \sigma_1)(k_2^2 - k_1^2) + p_1^2 p_2^2 A_n B_n$$

$$E^{(1)}_n = \frac{p_1^2 p_2^2}{I_n(p_2 a)} B_n C_n + I_n^2(p_1 a) K_n(p_2 a) \frac{m^2 n^2}{a^2}(\sigma_2 - \sigma_1)(k_2^2 - k_1^2)$$

$$H^{(1)}_n = p_1 p_2 (k_2^2 - k_1^2) I_n(p_1 a) \left[\frac{K_n(p_2 a)}{I_n(p_2 a)} D_n + B_n \right]$$

$$E^{(2)}_n = (\sigma_2 - \sigma_1) p_1 p_2 I_n(p_1 a) \left[\frac{K_n(p_2 a)}{I_n(p_2 a)} C_n + A_n \right]$$

$$H^{(2)}_n = \frac{m^2 n^2}{a^2} I_n^{(2)}(p_1 a) K_n(p_2 a)(\sigma_2 - \sigma_1)(k_2^2 - k_1^2) + \frac{p_1^2 p_2^2}{I_n(p_2 a)} A_n D_n$$

(6.263)

$$A_n = \sigma_1 p_2 I'_n(p_1 a) K_n(p_2 a) - \sigma_2 p_1 K'_n(p_2 a) I_n(p_1 a)$$

$$B_n = p_2 I'_n(p_1 a) K_n(p_2 a) - p_1 K'_n(p_2 a) I_n(p_1 a)$$

$$C_n = \sigma_2 p_1 I'_n(p_2 a) I_n(p_1 a) - \sigma_1 p_2 I_n(p_2 a) I'_n(p_1 a)$$

$$D_n = p_1 I'_n(p_2 a) I_n(p_1 a) - p_2 I'_n(p_1 a) I_n(p_2 a)$$

Equations 6.261—6.263 along with eqs. 2.253, 2.254, 2.257 and 2.260 are a complete solution of a problem; that is, these equations completely define all the components of the electromagnetic field for an arbitrarily oriented source when the normal field is known.

In conclusion, let us write down the expressions for all the components of the electromagnetic field, for the three principal orientations of the magnetic dipole, M_z, M_r, and M_ϕ.

A. Dipole sources M_z and M_r

As has previously been shown with this form of excitation, the normal electric field can be expanded as a series in sinusoids only if the normal magnetic field can be expanded in a series of cosinusoids. For this reason, the field components are expressed in terms of the coefficients C_2^s and D_2^c as follows:

$$E_z^* = \sum_{n=0}^{\infty} C_2^s K_n(p_2 r) \sin n\,\phi$$

$$H_z^* = \sum_{n=0}^{\infty} D_2^c K_n(p_2 r) \cos n\,\phi$$

$$E_r^* = \frac{1}{p_2^2} \left\{ \sum_{n=0}^{\infty} \left(\frac{i\omega\mu}{r} D_2^c K_n(p_2 r)\, n - im\, p_2 C_2^s K_n'(p_2 r) \right) \sin n\,\phi \right\}$$

$$E_\phi^* = \frac{1}{p_2^2} \left\{ \sum_{n=0}^{\infty} \left(i\omega\mu\, p_2 D_2^c K_n'(p_2 r) - \frac{im}{r} C_2^s K_n(p_2 r)\, n \right) \cos n\,\phi \right\}$$ (6.264)

$$-H_r^* = -\frac{1}{p_2^2} \left\{ \sum_{n=0}^{\infty} \left(\frac{\sigma_2}{r} C_2^s K_n(p_2 r)\, n + im\, p_2 D_2^c K_n'(p_2 r) \right) \cos n\,\phi \right\}$$

$$H_\phi^* = \frac{1}{p_2^2} \left\{ \sum_{n=0}^{\infty} \left(\sigma_2 p_2 C_2^s K_n'(p_2 r) + \frac{im}{r} D_2^c K_n(p_2 r)\, n \right) \sin n\phi \right\}$$

B. Dipole source M_ϕ

In this case we have:

$$E_z^* = \sum_{n=0}^{\infty} C_2^c K_n(p_2 r) \cos n\,\phi$$

$$H_z^* = \sum_{n=0}^{\infty} D_2^s K_n(p_2 r) \sin n\,\phi$$

$$E_r^* = -\frac{1}{p_2^2} \left\{ \sum_{n=0}^{\infty} \left[\frac{i\omega\mu}{r} D_2^s K_n(p_2 r)\, n + im\, p_2 C_2^c K_n'(p_2 r) \right] \cos n\,\phi \right\}$$

$$E_\phi^* = \frac{1}{p_2^2} \left\{ \sum_{n=0}^{\infty} \left(i\omega\mu\, p_2 D_2^s K_n'(p_2 r) + \frac{im}{r} C_2^c K_n(p_2 r)\, n \right) \sin n\,\phi \right\}$$

$$H_r^* = \frac{1}{p_2^2} \left\{ \sum_{n=0}^{\infty} \left(\frac{\sigma_2}{r} C_2^c K_n(p_2 r) n - im\, p_2 D_2^s K_n'(p_2 r) \right) \sin n\,\phi \right\}$$

$$H_\phi^* = \frac{1}{p_2^2} \left\{ \sum_{n=0}^{\infty} \left(\sigma_2 p_2 C_2^c K_n'(p_2 r) - \frac{im}{r} D_2^s K_n(p_2 r) n \right) \cos n\,\phi \right\}$$

Now we have to find only the amplitudes of the harmonics of the vertical components of the normal field, $e_{n0}^{c,s}$ and $h_{n0}^{c,s}$. These amplitudes are well known for an arbitrarily situated point source, and given, for example, by Tabarovskiy (1975). For each of the three orientations of the magnetic dipole, the amplitudes will have the following forms:

a. *The dipole* M_z

$$e_{n0}^{c,s} = 0, \quad h_{n0}^s = 0$$

$$h_{n0}^c = \frac{M_z}{2\pi}\, (k_2^2 - m^2)\, \alpha_n K_n(p_2 r_0) I_n(p_2 a) \qquad (6.266)$$

b. *The dipole* M_r

$$e_{n0}^c = 0$$

$$e_{n0}^s = \frac{i\omega\mu\, M_r}{2\pi\, r_0}\, \alpha_n\, n\, K_n(p_2 r_0) I_n(p_2 a)$$

$$h_{n0}^s = 0 \qquad (6.267)$$

$$h_{n0}^c = \frac{M_r}{2\pi i}\, \alpha_n\, m\, p_2 I_n(p_2 a) K_n'(p_2 r_0)$$

c. *The dipole* M_ϕ

$$e_{n0}^s = 0, \quad e_{n0}^c = -\frac{i\omega\mu\, M_\phi}{2\pi}\, \alpha_n\, p_2 I_n(p_2 a) K_n'(p_2 r_0)$$

$$h_{n0}^c = 0, \quad h_{n0}^s = \frac{M_\phi}{2\pi\, i r_0}\, \alpha_n\, n\, m\, I_n(p_2 a) K_n(p_2 r_0) \qquad (6.268)$$

where $\alpha_0 = 1$, $\alpha_n = 2$, for $n = 1, 2, 3, \ldots$

Finally the transformation from the Fourier transforms to the components of the field is accomplished using the following formulas:

a. *Dipole* M_z

$$E_z = \frac{i}{\pi} \int_0^{\infty} E_z^* \sin mz\, dm \qquad H_z = \frac{1}{\pi} \int_0^{\infty} H_z^* \cos mz\, dm$$

$$E_r = \frac{1}{\pi} \int_0^\infty E_r^* \cos mz \, dm \qquad H_r = \frac{i}{\pi} \int_0^\infty H_r^* \sin mz \, dm \qquad (6.269)$$

$$E_\phi = \frac{1}{\pi} \int_0^\infty E_\phi^* \cos mz \, dm \qquad H_\phi = \frac{i}{\pi} \int_0^\infty H_\phi^* \sin mz \, dm$$

b. *Dipoles M_r and M_ϕ*

$$E_z = \frac{1}{\pi} \int_0^\infty E_z^* \cos mz \, dm \qquad H_z = \frac{i}{\pi} \int_0^\infty H_z^* \sin mz \, dm$$

$$E_r = \frac{i}{\pi} \int_0^\infty E_r^* \sin mz \, dm \qquad H_r = \frac{1}{\pi} \int_0^\infty H_r^* \cos mz \, dm \qquad (6.270)$$

$$E_\phi = \frac{i}{\pi} \int_0^\infty E_\phi^* \sin mz \, dm \qquad H_\phi = \frac{1}{\pi} \int_0^\infty H_\phi^* \cos mz \, dm$$

We have obtained expressions for all of the components of the electromagnetic field for an arbitrary orientation of the magnetic dipole source with respect to the axis of the cylinder. In order to illustrate some characteristic features of this field, we can assume that the magnetic dipole has but a single component lying along the axis ϕ, that is:

$$M = (0, M_\phi, 0)$$

Using eqs. 6.268–6.270, examples of the frequency responses of the quadrature and inphase components normalized to the primary field are presented (see Figs. 6.41–6.44), where $\sigma_i = \sigma_1$ and $\sigma_e = \sigma_2$. As may be seen from the curves in Figs. 6.41 and 6.42, we can recognize the following characteristic parts of the quadrature response:

(1) A low-frequency range over which the quadrature component increase gradually.

(2) A first maximum, the position of which is primarily determined by the conductivity in the host medium.

(3) A descending part of the response where the quadrature component rapidly decreases, becomes zero, and changes sign before reaching a minimum value.

(4) The minimum of the response, which is almost twice as deep as the maximum.

(5) A high-frequency part of the spectrum over which the magnitude of the quadrature component decreases with increase in frequency, oscillating as it approaches zero.

Generally speaking, the inphase component (see Fig. 6.43 and 6.44) behaves in much the same manner as the quadrature component, and one can also recognize five characteristic ranges. However, it should be noted

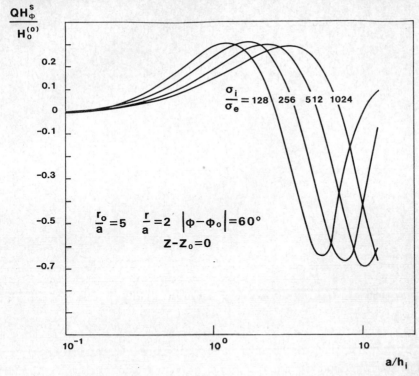

Fig. 6.41. Behavior of the normalized quadrature component of the magnetic field as a function of a/h_i for various conductivity contrasts.

that there are certain differences between the behaviors of the inphase and quadrature responses, as follows:

(1) At low frequencies, the quadrature component is dominant and increases in direct proportion to frequency, while the inphase component is much smaller and varies as the square of frequency.

(2) The magnitude of the maximum for the quadrature component is lower than the magnitude of the minimum for the inphase component.

(3) The inphase component of the secondary field changes sign at higher frequencies than does the quadrature component, and other differences might also be observed.

At this point, we can consider the ratio between the total field and the normal field for the quadrature and inphase components as a function of frequency, and of the conductivity of the host medium. It is clear that this ratio ultimately controls the depth of investigation and therefore an analysis of its behavior is of considerable practical interest. Examples of the frequency responses for the function $QH_\phi/QH_{0\phi}$ are shown in Figs. 6.45 and 6.46. The quantity QH_ϕ is the quadrature component of the total field,

580

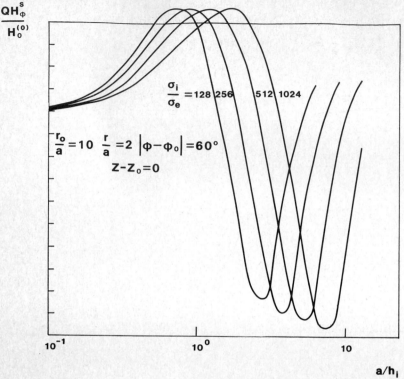

Fig. 6.42. Behavior of the normalized quadrature component of the magnetic field as a function of a/h_i for various conductivity contrasts for the case $r_0/a = 10$.

including the normal contribution, $QH_{0\phi}$, that is, the field when the cylinder is absent.

As follows from an examination of these responses, the maximum ratio is observed at low frequencies and does not depend significantly on the frequency. This asymptotic behavior holds approximately when the inequality:

$$a/h_i < 0.3 \tag{6.271}$$

is satisfied. This can be easily explained. At low frequencies, the quadrature components of the secondary and normal fields are both directly proportional to frequency, and as a consequence, the ratio $QH_\phi/QH_{0\phi}$ will be nearly constant.

With an increase in frequency, this ratio decreases because the secondary field decays more rapidly than the normal field.

A comparison with the case of axial symmetry which was examined in the preceding chapter clearly demonstrates several new significant features of the field caused by the presence of a conducting cylinder. Let us emphasize only three of these at this point:

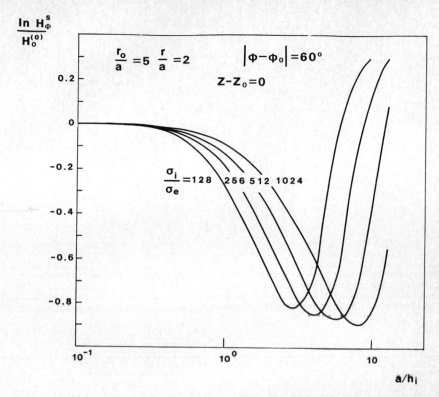

Fig. 6.43. Behavior of the normalized inphase component of the magnetic field as a function of a/h_i for the case $r_0/a = 5$.

(1) The relative anomalies are much greater than in the case of axial symmetry.

(2) The relative anomalies for the cases considered are not directly proportional to the ratio of the conductivities, σ_i/σ_e, even $QH_\phi/QH_{0\phi} \gg 1$.

(3) The low-frequency asymptote is observed at relatively lower values for the parameter a/h_i.

Next we will examine the behavior of the relative anomaly of the inphase component, $\mathrm{In}H_\phi/\mathrm{In}H_{0\phi}^s$ where $\mathrm{In}H_{0\phi}^s$ is the inphase component of the normal field with the primary field being canceled. Examples of the frequency response for this function are shown in Figs. 6.47 and 6.48. Frequency responses for the relative anomaly of the inphase component, $\mathrm{In}H_\phi/\mathrm{In}H_{0\phi}^s$ have the same features as those of the field caused only by vortex currents (see Chapter 5). With increase in frequency, the ratio $\mathrm{In}H_\phi/\mathrm{In}H_{0\phi}^s$ increases gradually, passes through a maximum, and with a further increase in frequency, decreases rapidly. At low frequencies, the secondary field $\mathrm{In}H_\phi^s$ is directly proportional to ω^2, while the normal field is proportional to $\omega^{3/2}$, and therefore, the relative anomaly

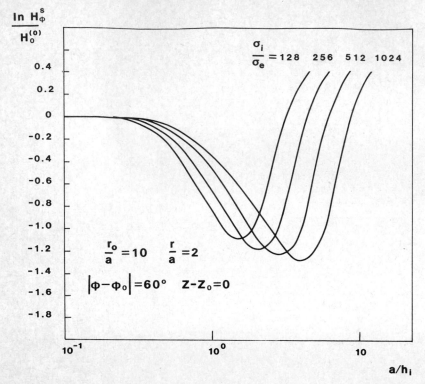

Fig. 6.44. Behavior of the normalized inphase component of the magnetic field as a function of a/h_i for the case $r_0/a = 10$.

$\mathrm{In}H_\phi/\mathrm{In}H^s_{0\phi}$ decreases with decrease in frequency. As follows from consideration of the inphase component of the secondary field (see Figs. 6.43 and 6.44), with increase in frequency, the inphase component increases progressively more slowly, which is the reason that the maximum is present in the ratio. The position of the maximum is almost independent of the conductivity of the host medium ($\sigma_i/\sigma_e \gg 1$) and it is observed that:

$$a/h_i \approx 0.3, \quad \text{if } \sigma_i/\sigma_e \gg 1 \tag{6.272}$$

It is clear that in this case the skin depth, h_e, in the host medium is much greater than the radius of the cylinder, a.

Considering the range of frequencies over which the relative anomalies in the quadrature and inphase components reach their maximum values, one can see that in measuring the inphase component, this ratio becomes greater. However, it should be noted that going to measurements of the inphase component in this case will not result in a strong increase in the relative anomaly as was observed when we investigated the fields for models characterized by axial symmetry. However, the most remarkable difference between these two cases is that in the latter, the relative anomaly in both

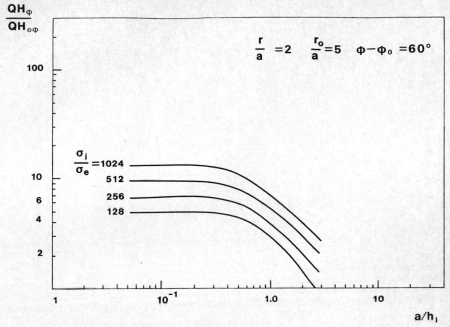

Fig. 6.45. Frequency response curves for the function $QH_\phi/QH_{0\phi}$ for the case $r_0/a = 5$, $z = z_0$.

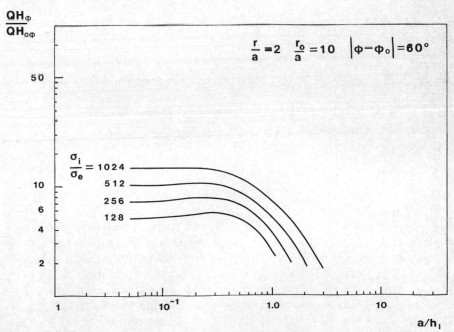

Fig. 6.46. Frequency response curves for the function, $QH_\phi/Qh_{0\phi}$ for the case $r_0/a = 10$, $z = z_0$.

584

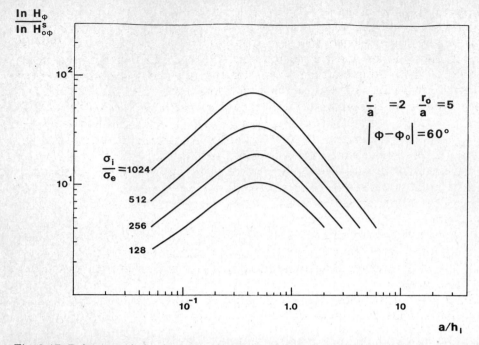

Fig. 6.47. Behavior of the inphase component of the normalized magnetic field for the case $r_0/a = 5$, $z = z_0$.

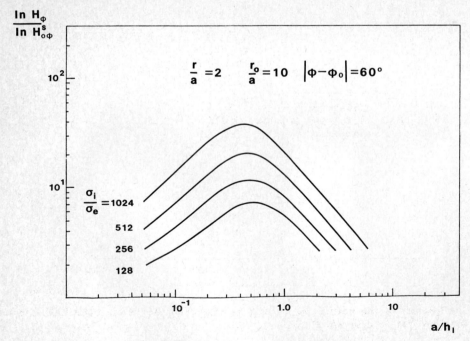

Fig. 6.48. Behavior of the inphase component of the normalized magnetic field for the case $r_0/a = 10$, $z = z_0$.

the quadrature and inphase components will be much stronger. This fundamental feature of the field behavior that is examined in this chapter is explained by the presence of surface charges as well as of currents flowing through the cylinder and the surrounding medium. In order to emphasize the role of these sources, let us consider the ratio of the quadrature and inphase components of the secondary field, QH_ϕ^s and InH_ϕ^s when the external medium is conductive in comparison with the components QH_ϕ $(\sigma_i, 0)$ and InH_ϕ^s $(\sigma_i, 0)$ for the case in which the surrounding medium is insulating. Curves showing these frequency responses are given in Figs. 6.49—6.52. It is clear that both components, QH_ϕ $(\sigma_i, 0)$ and InH_ϕ^s $(\sigma_i, 0)$ are caused only by the flow of eddy currents that are closed within the cylinder, because the surrounding medium is nonconductive.

From an examination of these curves we can arrive at the following conclusions:

(1) The galvanic part of the field related with the charges on the surface of the cylinder from the currents flowing through the cylinder on the host medium is much greater than the vortex part of the field for all of the cases considered here.

(2) At low frequencies, this ratio for both the quadrature and inphase components tends to a constant value which is independent of frequency.

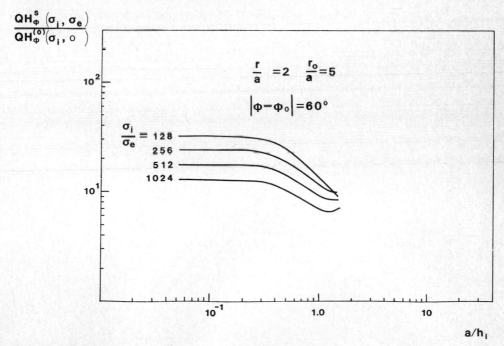

Fig. 6.49. Behavior of the normalized quadrature component of the magnetic field as a function of a/h_1 for the case $r_0/a = 5$, $z = z_0$.

586

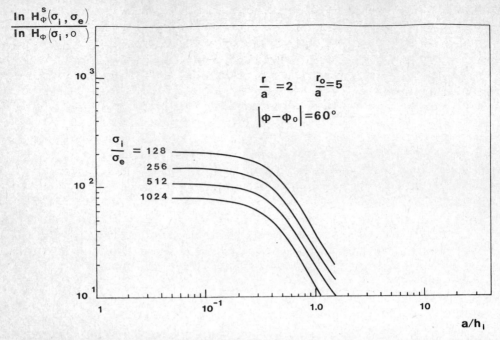

$$\frac{\ln H_\Phi^s(\sigma_i, \sigma_e)}{\ln H_\Phi(\sigma_i, 0)}$$

$$\frac{r}{a} = 2 \qquad \frac{r_o}{a} = 5$$

$$|\Phi - \Phi_o| = 60°$$

$$\frac{\sigma_i}{\sigma_e} = 128$$
256
512
1024

a/h_I

Fig. 6.50. The behavior of the inphase component of the normalized magnetic field as a function of a/h_1 for the case $r_0/a = 5$, $z = z_0$.

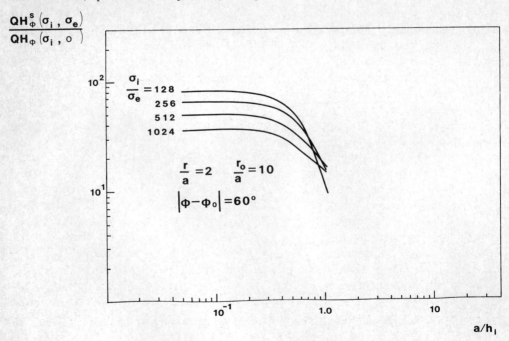

$$\frac{QH_\Phi^s(\sigma_i, \sigma_e)}{QH_\Phi(\sigma_i, 0)}$$

$$\frac{\sigma_i}{\sigma_e} = 128$$
256
512
1024

$$\frac{r}{a} = 2 \qquad \frac{r_o}{a} = 10$$

$$|\Phi - \Phi_o| = 60°$$

a/h_I

Fig. 6.51. Behavior of the quadrature component of the normalized magnetic field as a function of a/h_1 for the case $r_0/a = 10$, $z = z_0$.

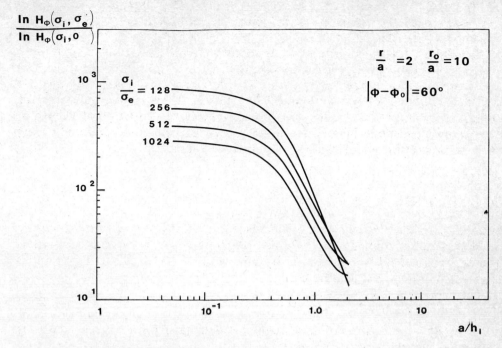

Fig. 6.52. Behavior of the inphase component of the normalized magnetic field for the case $r_0/a = 10$, $z = z_0$.

This means that the dependence on frequency for each component of the galvanic and vortex parts of the field is the same. For example, the quadrature component of the magnetic field caused by vortex currents flowing within the cylinder and currents closed through the host medium is directly proportional to frequency, while the inphase component is proportional to the square of frequency.

(3) In measuring the inphase component, the influence of the galvanic part of the field is significantly larger.

(4) The virtual independence of the field component ratio on frequency is observed practically within the same range of frequencies where the relative anomalies in the quadrature and inphase components are at a maximum.

(5) With increase in distance between the source dipole and the cylinder, r_0, the effect of the galvanic part of the field increases markedly. On the other hand, with a decrease in the conductivity of the host medium, the effect of currents intersecting the surface of the cylinder decreases, and the contribution of the galvanic part of the field becomes less.

Now we can obtain an asymptotic set of formulas for the galvanic part of the field, inasmuch as it is significantly greater than the vortex part. First, let us consider the electric field. In order to find this field, we will assume that:

(1) The secondary field is caused only by charges which appear on the surface of the cylinder.

(2) There is no interaction between current filaments.

(3) The vortex part of the field is negligible.

Inasmuch as we are dealing with the quasi-stationary field, the electric field from charges obeys Coulomb's law, that is, it has a potential character unlike the primary vortex electric field caused by the time rate of change of the primary magnetic field. Consequently, we can write expressions for the electric field inside and outside the cylinder as follows:

$$E^i = E_0 + E_s^i, \quad \text{if } r < a$$
$$E^e = E_0 + E_s^e, \quad \text{if } r > a \tag{6.273}$$

where E_0 is the primary vortex electric field and E_s^i and E_s^e are secondary fields inside and outside the cylinder respectively. In view of the fact that the secondary field is contributed by electric charges, it can be represented in terms of a potential as follows:

$$E_s^i = -\operatorname{grad} U^i \quad E_s^e = -\operatorname{grad} U^e \tag{6.274}$$

At the same time, the primary electric field, having the vortex origin, is expressed through a vector potential:

$$E_0 = -i\omega\mu \operatorname{curl} A_0 \tag{6.275}$$

Now we can formulate the boundary value problem for the secondary field as follows:

(1) The potentials U^i and U^e must satisfy Laplace's equation:

$$\nabla^2 U^i = 0, \quad \text{if } r < a$$
$$\nabla^2 U^e = 0, \quad \text{if } r > a \tag{6.276}$$

(2) At the surface of the cylinder, the potentials representing a secondary field and the normal components of the total current density are continuous functions:

$$U^e = U^i \tag{6.277}$$

and

$$J_r^e = J_r^i, \quad \text{if } r = a \tag{6.278}$$

Condition 6.277 provides for continuity of the tangential components of the electric field, while condition 6.278 can also be written as:

$$\sigma_i (E_{0r} + E_r^i) = \sigma_e (E_{0r} + E_r^e)$$

or

$$(\sigma_i - \sigma_e) E_{0r} = \sigma_i \frac{\partial U^i}{\partial r} - \sigma_e \frac{\partial U^e}{\partial r}$$

Thus, the boundary conditions become:

$$U^e = U^i$$

$$(\sigma_i - \sigma_e) E_{0r} = \sigma_i \frac{\partial U^i}{\partial r} - \sigma_e \frac{\partial U^e}{\partial r} \qquad (6.279)$$

where σ_i and σ_e are conductivities of the cylinder and surrounding medium, respectively.

(3) The secondary field is finite everywhere, and vanishes at infinity.

First let us consider the case in which the magnetic dipole moment is oriented along the z-axis. It should be clear that the primary vortex electric field will be located in planes lying perpendicular to the axis of the cylinder, and therefore, the positive and negative charges arise on opposite sides of the surface. As can readily be shown, the primary field can be described using only a single component of the vector potential A_{0z}, equal to $1/R$. In accord with results described previously in Chapter 3, we have:

$$A_{0z} = \frac{M}{\pi^2} \sum_{n=0}^{\infty} \int_0^\infty K_n(\lambda r_0) I_n(\lambda r) \cos \lambda z \, d\lambda \cos n \, \phi, \quad \text{if } r < r_0 \qquad (6.280)$$

As follows from eq. 6.275, we also have:

$$E_{0r} = -i\omega\mu \frac{1}{r} \frac{\partial A_z}{\partial \phi}$$

and therefore, eq. 6.281:

$$E_{0r} = \frac{i\omega\mu}{r} \frac{M}{\pi^2} \sum_{n=1}^{\infty} n \int_0^\infty K_n(\lambda r_0) I_n(\lambda r) \cos \lambda z \, d\lambda \sin n\phi, \quad \text{if } r < r_0 \qquad (6.281)$$

In order to satisfy the second boundary condition in eqs. 6.279 and considering the expression for E_{0z}, it is convenient to write the potentials in the form:

$$U^i = i\omega\mu \frac{M}{\pi^2} \sum_{n=1}^{\infty} \int_0^\infty C_n K_n(\lambda r_0) I_n(\lambda r) \cos \lambda z \, d\lambda \sin n\phi, \quad \text{if } r \leq a \qquad (6.282)$$

$$U^e = i\omega\mu \frac{M}{\pi^2} \sum_{n=1}^{\infty} \int_0^\infty D_n K_n(\lambda r_0) K_n(\lambda r) \cos \lambda z \, d\lambda \sin n\phi, \quad \text{if } r \geq a$$

where C_n and D_n are coefficients which are unknown at this time.

Using boundary condition 6.279, we obtain the following system of equations to be used in determining the values of C_n and D_n:

$$C_n I_n(\lambda a) = D_n K_n(\lambda a)$$

$$\frac{(\sigma_i - \sigma_e)nI_n(\lambda a)}{a} = \sigma_i \lambda I'_n(\lambda a)C_n - \sigma_e \lambda K'_n D_n$$

Whence

$$(\sigma_i - \sigma_e)nI_n(\lambda a) = \lambda a \left\{ \sigma_i I'_n(\lambda a) \frac{K_n(\lambda a)}{I_n(\lambda a)} - \sigma_e K'_n(\lambda a) \right\} D_n$$

and

$$D_n = \frac{1}{\lambda a} \frac{n(\sigma_i - \sigma_e)I_n^2}{\sigma_i I'_n K_n - \sigma_e I_n K'_n}$$

$$C_n = \frac{1}{\lambda a} \frac{n(\sigma_i - \sigma_e)I_n K_n}{\sigma_i I'_n K_n - \sigma_e I_n K'_n}$$

(6.283)

Making use of the identity:

$$I_n(x) K'_n(x) - I'_n(x) K_n(x) = -1/x$$

we have:

$$\sigma_i I'_n K_n - \sigma_e I_n K'_n - \sigma_e I'_n K_n + \sigma_e I'_n K_n =$$
$$= (\sigma_i - \sigma_e) I'_n K_n - \sigma_e (I_n K'_n - I'_n K_n) =$$
$$= (\sigma_i - \sigma_e) I'_n K_n + \frac{\sigma_e}{\lambda a}$$

Therefore:

$$D_n = n \frac{\left(\frac{\sigma_i}{\sigma_e} - 1\right) I_n^2(\lambda a)}{1 + \left(\frac{\sigma_i}{\sigma_e} - 1\right) \lambda a I'_n(\lambda a) K_n(\lambda a)}$$

(6.284)

and

$$U^e = i\omega\mu \frac{M}{\pi^2} \sum_{n=1}^{\infty} n \int_0^{\infty} \frac{\left(\frac{\sigma_i}{\sigma_e} - 1\right) I_n^2(\lambda a) K_0(\lambda r_0)}{1 + \left(\frac{\sigma_i}{\sigma_e} - 1\right) \lambda a I'_n(\lambda a) K_n(\lambda a)} K_0(\lambda r) \cos \lambda z \, d\lambda \sin n\phi$$

(6.285)

All of the components of the electric field can be calculated readily using eq. 6.274.

Next let us consider the special case in which the distance from the dipole to the cylinder, $r_0 \gg r$. In particular, when the ratio $r_0/a \gg 1$, is met, the primary field E_0 will be nearly uniform over a significant part of the cylinder.

If the condition $r_0/r \gg 1$ is met, the integrals in eq. 6.285 are defined mainly by the small values of the variables $\lambda(a/r_0)$ and $\lambda(r/r_0)$. In this case, the modified Bessel functions can be replaced by asymptotic representations such as:

$$I_n(x) \approx \frac{x^n}{2^n n!}, \qquad K_n(x) = \frac{2^{n-1}(n-1)!}{x^n}$$

$$I'_n(x) = \frac{x^{n-1}}{2^n(n-1)!}, \qquad K'_n(x) = -\frac{2^{n-1}n!}{x^{n+1}}$$

Therefore, we have the following expression for the function D_n:

$$D_n \approx \left(\frac{\sigma_i}{\sigma_e} - 1\right) \frac{x^{2n}}{2^{2n}(n!)^2 \left[1 + \left(\frac{\sigma_i}{\sigma_e} - 1\right) x \dfrac{x^{n-1}}{2^n(n-1)!} \dfrac{2^{n-1}(n-1)!}{x^n}\right]}$$

$$= \left(\frac{\sigma_i}{\sigma_e} - 1\right) \frac{x^{2n}}{2^{2n}(n!)^2 \left[1 + \left(\frac{\sigma_i}{\sigma_e} - 1\right) \dfrac{1}{2}\right]}$$

Whence

$$D_n K_n(\lambda r) \approx \frac{\sigma_i - \sigma_e}{\sigma_i + \sigma_e} \frac{\lambda^n}{n 2^n n!} \frac{a^{2n}}{r^n}$$

Consequently:

$$U^e \approx i\omega\mu \frac{\sigma_i - \sigma_e}{\sigma_i + \sigma_e} \frac{M}{\pi^2} \sum_{n=1}^{\infty} \frac{a^{2n}}{r^n n!} \int_0^{\infty} \lambda^n K_n(\lambda r_0) \cos \lambda z \, d\lambda \sin n\phi \qquad \text{if } \frac{r}{r_0} \ll 1$$

$$\hspace{12cm} (6.286)$$

In this approximation, the first harmonic term is dominant and therefore:

$$U^e \approx i\omega\mu \frac{\sigma_i - \sigma_e}{\sigma_i + \sigma_e} \frac{M}{\pi^2} \frac{a^2}{r} \sin \phi \int_0^{\infty} \lambda K_1(\lambda r_0) \cos \lambda z \, d\lambda$$

This integral can also be written as:

$$\int_0^{\infty} \lambda K_1(\lambda r_0) \cos \lambda z \, d\lambda = -\frac{\partial}{\partial r_0} \int_0^{\infty} K_0(\lambda r_0) \cos \lambda z \, d\lambda =$$

$$= -\frac{\pi}{2} \frac{\partial}{\partial r_0} \frac{1}{(r_0^2 + z^2)^{1/2}} = \frac{\pi}{2r_0^2}, \qquad \text{if } r_0 \gg z$$

Finally we obtain:

$$U^e \approx i\omega\mu \frac{\sigma_i - \sigma_e}{\sigma_i + \sigma_e} \frac{M}{2\pi r_0^2} \frac{a^2}{r} \sin \phi$$

or

$$U^e \approx \frac{\sigma_i - \sigma_e}{\sigma_i + \sigma_e} E_0 \frac{a^2}{r} \sin \phi, \quad \text{if } \frac{r_0}{r} \gg 1, \frac{r_0}{z} \gg 1 \tag{6.287}$$

where E_0 is the primary vortex electric field at the z-axis.

It is clear that this expression is exactly the same as that for the static case.

As follows from eq. 6.287, the electric field is essentially independent of the conductivity of the cylinder and of the host medium when the dipole moment is oriented along the z-axis. We can show that almost the same relationship with both conductivities will be observed for arbitrary values of the distances r_0 and r. This permits us to conclude that the magnetic field caused by currents flowing through the cylinder and the host medium depend mainly on the external conductivity, σ_e.

At this point, we should derive asymptotic expressions for the field in the case in which the source dipole has a moment with only the component M_ϕ. In this case, the primary vortex electric field has two components, E_{0r} and E_{0z}, and as a consequence, the distribution of charges on the surface of the cylinder is quite different from that which was observed with the orientation of the dipole treated in the preceding part of this paragraph.

As in the case considered above, in order to determine the electric field caused by charges, it is necessary to find an expression for the radial component of the primary electric field. Carrying out relatively complicated operations, we can obtain the following result:

$$E_{0r} = -\frac{i\omega\mu M}{2\pi^2} \sum_{n=0}^{\infty} \epsilon_n \int_0^{\infty} A_n \sin \lambda z \, d\lambda \cos n \phi, \quad \text{if } r < r_0 \tag{6.288}$$

where

$$A_n = \frac{n^2 I_n(\lambda r) K_n(\lambda r_0)}{\lambda r r_0} + \lambda I_n'(\lambda r) K_n'(\lambda r_0) \tag{6.289}$$

$$\epsilon_0 = 1, \quad \epsilon_1 = \epsilon_2 = \ldots = 2$$

Consequently, we will search for expressions for potentials that have the form:

$$U^i = -\frac{i\omega\mu M}{2\pi^2} \sum_{n=0}^{\infty} \epsilon_n \cos n \phi \int_0^{\infty} C_n I_n(\lambda r) \sin \lambda z \, d\lambda, \quad r \leqslant a$$

$$\tag{6.290}$$

$$U^e = -\frac{i\omega\mu M}{2\pi^2} \sum_{n=0}^{\infty} \epsilon_n \cos n \phi \int_0^{\infty} D_n K_n(\lambda r) \sin \lambda z \, d\lambda, \quad r \geqslant a$$

Using boundary condition 6.279, we have:

$$C_n I_n(\lambda a) = D_n K_n(\lambda a)$$

$$(\sigma_i - \sigma_e) A_n = \lambda \sigma_i C_n I'_n(\lambda a) - \lambda \sigma_e D_n K'_n(\lambda a)$$

Thus,

$$D_n = \frac{(\sigma_i - \sigma_e) A_n I_n(\lambda a)}{\lambda \{\sigma_i I'_n K_n - \sigma_e I_n K'_n\}}$$

Because

$$\sigma_i I'_n K_n - \sigma_e I_n K'_n = \left[\frac{\sigma_e}{\lambda a} + (\sigma_i - \sigma_e) I'_n K_n \right]$$

we will have:

$$D_n = \left(\frac{\sigma_i}{\sigma_e} - 1 \right) \frac{A_n a I_n(\lambda a)}{1 + \left(\frac{\sigma_i}{\sigma_e} - 1 \right) \lambda a I'_n(\lambda a) K_n(\lambda a)}$$

$$C_n = \left(\frac{\sigma_i}{\sigma_e} - 1 \right) \frac{A_n a K_n(\lambda a)}{1 + \left(\frac{\sigma_i}{\sigma_e} - 1 \right) \lambda a I'_n(\lambda a) K_n(\lambda a)} \tag{6.291}$$

where

$$A_n = \left[\frac{n_2 I_n(\lambda a) K_n(\lambda r_0)}{\lambda a r_0} + \lambda I'_n(\lambda a) K'_n(\lambda r_0) \right] \tag{6.292}$$

Equations 6.290 through 6.292 permit us to calculate the electric field inside and outside the cylinder. In particular, in the host medium we will have:

$$E^e_r = -\frac{\partial U^e}{\partial r} = \frac{i\omega\mu M}{2\pi^2} \sum_{n=0}^{\infty} \epsilon_n \cos n\phi \int_0^{\infty} \lambda D_n K'_n(\lambda r) \sin \lambda z \, d\lambda$$

$$E^e_\phi = -\frac{1}{r} \frac{\partial U^e}{\partial \phi} = -\frac{i\omega\mu M}{2\pi^2} \sum_{n=0}^{\infty} \epsilon_n n \sin n\phi \int_0^{\infty} D_n K_n(\lambda r) \sin \lambda z \, d\lambda \tag{6.293}$$

$$E^e_z = -\frac{\partial U^e}{\partial z} = -\frac{i\omega\mu M}{2\pi^2} \sum_{n=0}^{\infty} \epsilon_n \cos n\phi \int_0^{\infty} \lambda D_n K_n(\lambda r) \cos \lambda z \, d\lambda$$

Comparison of results of calculations based on the exact solution with that obtained using eq. 6.293 indicates that the eqs. 6.293 describes the quadrature component of the electric field at low frequencies with high accuracy provided that:

$$r_0/h_e < 1, \quad r/h_e < 1 \quad \text{and} \quad a/h_e < 0.3 \tag{6.294}$$

Before we start examining the asymptotic behavior of the magnetic field, it should be emphasized that the quadrature component of the electric field caused by charges is directly proportional to frequency while the quadrature component of the vortex part of the field varies as ω^3. In principle, this difference should be useful for detecting the effect of the galvanic part of the field.

According to Ohm's law, the electric field E^g causes a current that flows through the cylinder into the host medium. Applying the Biot-Savart law, we can calculate the magnetic field of these currents. Inasmuch as this approach involves tedious intergration over the conducting medium, we will use an alternate method based on the application of the vector potential Π.

From the basic theory governing electromagnetic fields, it is known that in the presence of a cylindrical conductor oriented along the z-axis, the field can be described using two vector potentials of the electric and magnetic types:

$$H = \text{curl } \Pi, \quad E = i\omega\mu \text{ curl } \Pi^*$$

with each of them having been a single component, the z-component. Thus, in cylindrical coordinates we have:

$$\Pi = (0, 0, \Pi_z), \quad \Pi^* = (0, 0, \Pi_z^*)$$

Considering that our objective is that of determining the magnetic field contributed by currents arising from the galvanic part of the electric field, E^g, we will make use of only the vector potential of the electric type:

$$\Pi = (0, 0, \Pi_z) \tag{6.295}$$

Substituting:

$$H = \text{curl } \Pi$$

in Maxwell's equation:

$$\text{curl } H = \sigma E$$

we will have:

$$\text{curl curl } \Pi = \sigma E$$

or

$$\text{grad div } \Pi - \nabla^2 \Pi = -\sigma \text{ grad } U$$

Letting:

$$\text{div } \Pi = -\sigma U \tag{6.296}$$

we have:

$$\nabla^2 \Pi_z = 0$$

Equation 6.296 permits us to find the proper expression for the vector potential Π_z. This is:

$$\text{div } \boldsymbol{\Pi} = \frac{\partial \Pi_z}{\partial z} = -\sigma U \tag{6.297}$$

Outide the cylinder, in accord with eq. 6.290:

$$U^e = -\frac{i\omega\mu M}{2\pi^2} \sum_{n=0}^{\infty} \epsilon_n \cos n\phi \int_0^\infty D_n K_n(\lambda r) \sin \lambda z \, d\lambda$$

Comparing this result with eq. 6.297 we obtain the following expression for the vector potential:

$$\Pi_z^e = -\frac{i\omega\mu M}{2\pi^2} \sigma_e \sum_{n=0}^{\infty} \epsilon_n \cos n\phi \int_0^\infty \frac{D_n}{\lambda} K_n(\lambda r) \cos \lambda z \, d\lambda \tag{6.298}$$

Inasmuch as:

$$H = \text{curl } \boldsymbol{\Pi}$$

we have the following expression for the azimuthal component of the magnetic field:

$$H_\phi^e = \frac{i\omega\mu M}{2\pi^2} \sigma_e \sum_{n=0}^{\infty} \epsilon_n \cos n\phi \int_0^\infty D_n K_n'(\lambda r) \cos \lambda z \, d\lambda \tag{6.299}$$

As comparisons with exact solutions will show, eq. 6.299 provides satisfactory results for the low-frequency asymptotic behavior of H_ϕ when the galvanic part of the field is dominant.

Let us again examine the case in which the ratio of distances r/r_0 is small. Then the integrals in eq. 6.299 are defined primarily by small values of the variables $\lambda(a/r_0)$ and $\lambda(z/r_0)$. Replacing the Bessel functions by their asymptotic expansions, we have:

$$A_0 = -\frac{\lambda^2 a}{2} K_1(\lambda r_0) \quad \text{if } n = 0$$

$$\tag{6.300}$$

$$A_n = \frac{(\lambda a)^{n-1}}{2^n (n-1)! r_0} [n K_n(\lambda r_0) + \lambda r_0 K_n'(\lambda r_0)] \quad \text{if } n > 0$$

and

$$D_0 = -\left(\frac{\sigma_i}{\sigma_e} - 1\right) \frac{\lambda^2 a^2}{2} K_1(\lambda r_0)$$

$$\tag{6.301}$$

$$D_n = \frac{\sigma_i - \sigma_e}{\sigma_i + \sigma_e} \frac{a}{r_0} \frac{(\lambda a)^{2n-1}}{2^{2n-1} n! (n-1)!} [n K_n(\lambda r_0) + \lambda r_0 K_n'(\lambda r_0)]$$

As can be seen from eq. 6.301, only the null-harmonic depends on the conductivities in the cylinder and the host medium when $\sigma_i/\sigma_e \gg 1$, while the higher harmonics are practically independent of both conductivities. It is also clear that with an increase in the ratio of r_0/r, the cylindrical null-harmonic becomes dominant and therefore it is the principal term that defines the field. Consequently, substituting D_0 from eq. 6.301 into eq. 6.299, and restricting ourselves to consideration of only the null-harmonic, we obtain:

$$H_\phi^e \approx \frac{i\omega\mu M}{2\pi^2} (\sigma_i - \sigma_e) \frac{a^2}{2} \int\limits_0^\infty \lambda^2 K_1(\lambda r_0) K_1(\lambda r) \cos \lambda z \, d\lambda$$

$$\approx \frac{i\omega\mu M}{4\pi^2} (\sigma_i - \sigma_e) \frac{a^2}{r} \int\limits_0^\infty \lambda K_1(\lambda r_0) \cos \lambda z \, d\lambda$$

$$= -\frac{i\omega\mu M}{8\pi} (\sigma_i - \sigma_e) \frac{a^2}{r} \frac{\partial}{\partial r_0} \frac{2}{\pi} \int\limits_0^\infty K_0(\lambda r_0) \cos \lambda z \, d\lambda$$

Since:

$$\frac{2}{\pi} \int\limits_0^\infty K_0(\lambda r_0) \cos \lambda z \, d\lambda = \frac{1}{(r_0^2 + z^2)^{1/2}}$$

we will finally have:

$$H_\phi^e = \frac{i\omega\mu(\sigma_i - \sigma_e) a^2 M}{8\pi r r_0^2}, \quad \text{if} \quad \frac{r_0}{r} \gg 1 \text{ and } \frac{r_0}{z} \gg 1 \tag{6.302}$$

and

$$H_\phi^e = \frac{(\sigma_i - \sigma_e) E_0 a^2}{2r} = \frac{I^s}{2\pi r} \tag{6.303}$$

where E_0 is the primary vortex electric field and:

$$I^s = (\sigma_i - \sigma_e) \pi a^2 E_0$$

Thus, the magnetic field is directly proportional to the algebraic difference in conductivities and coincides with that which would be caused by a linear current filament I^s.

Making use of eq. 6.299, we will examine the influence of the conductivities and distances on the quadrature component of the magnetic field, H_ϕ^e at low frequencies. As follows from eqs. 6.291 and 6.299, the magnetic field at low frequencies outside the cylinder is a function of the ratio of conductivities, σ_i/σ_e, and it is directly proportional to the conductivity of the host medium:

$$H_\phi = i\omega\mu \, \sigma_e \, H_0^{(0)} F \left(\frac{\sigma_i}{\sigma_e}, \frac{r}{a}, \frac{r_0}{a} \right)$$

In order to investigate the influence of the resistivity of the host medium, let us assume that the resistivity of the cylinder is unity. Curves illustrating

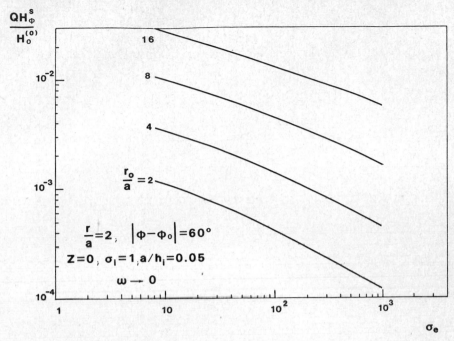

Fig. 6.53. Curves illustrating the behavior of the azimuthal component of the magnetic field as a function of the resistivity of the host medium for the case $r/a = 2$.

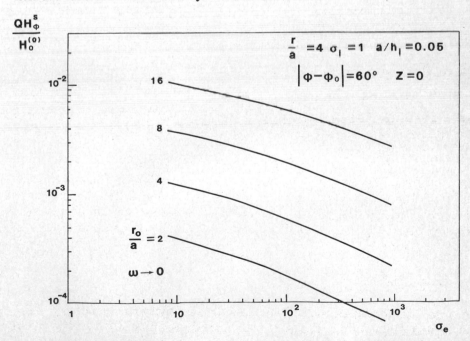

Fig. 6.54. Curves illustrating the behavior of the azimuthal component of the magnetic field as a function of the resitivity of the host medium for the case $r/a = 4$.

$$\frac{QH_\phi^s}{H_o^{(0)}}$$

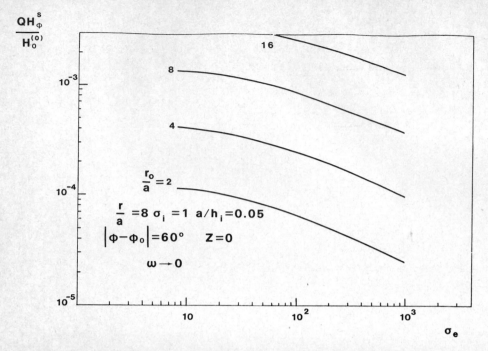

Fig. 6.55. Curves illustrating the behavior of the azimuthal component of the magnetic field as a function of the resistivity of the host medium for the case $r/a = 8$.

the behavior of the azimuthal component, H_ϕ, as a function of the resistivity of the host medium are shown in Figs. 6.53—6.56.

Above all, it is clear that with an increase in resistivity, ρ_e, the magnetic field decreases. This results from the fact that the current density flowing through the host medium decreases, and in the limit when the external medium is insulating, this part of the magnetic field vanishes. The relation between the field and the resistivity of the host medium depends primarily on geometrical factors. For example, the magnetic field depends more strongly on ρ_e when the magnetic dipole is located relatively close to the cylinder, but this dependence becomes weaker with an increase in the distance r_0. A similar behavior of the field as a function of the resitivity ρ_e is observed when the distance between the observation point and the cylinder, r, increases.

The curves shown in Figs. 6.57—6.60 show the effect of the conductivity of the cylinder on the magnitude of the field. As follows from these curves, the field H_ϕ increases monotonically with an increase of the conductivity of the cylinder. The behavior of the magnetic field is defined by the distribution of currents closed through the surrounding medium, and therefore by the electrical field caused by surface charges as well as the primary electrical field. From a qualitative point, some features of the field behavior,

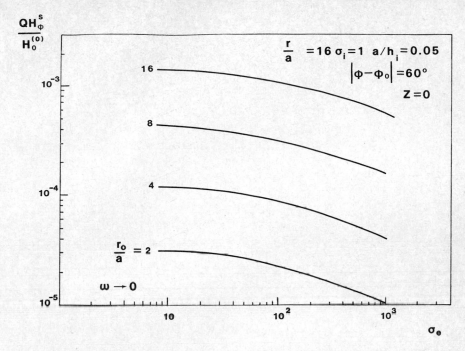

Fig. 6.56. Curves illustrating the behavior of the azimuthal component of the magnetic field as a function of the resistivity of the host medium for the case $r/a = 16$.

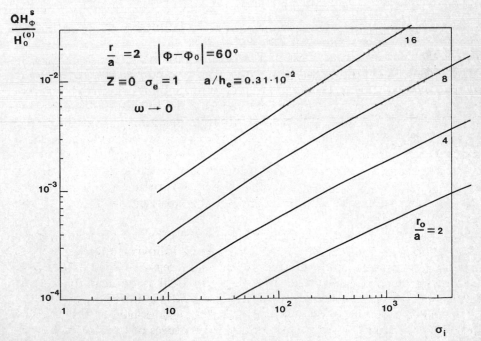

Fig. 6.57. Curves showing the effect of the conductivity of the cylinder on the magnitude of the field when $r/a = 2$.

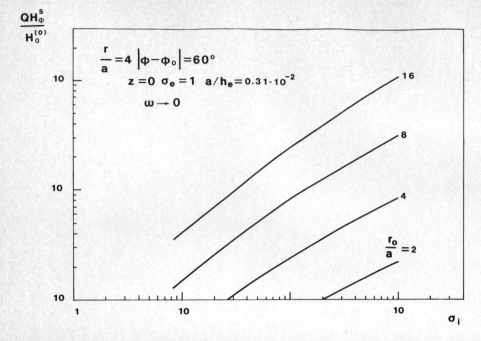

Fig. 6.58. Curves showing the effect of the conductivity of the cylinder on the magnitude of the field when $r/a = 4$.

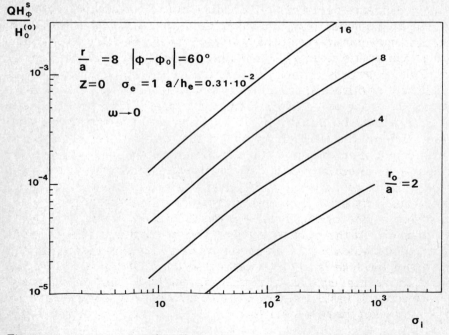

Fig. 6.59. Curves showing the effect of the conductivity of the cylinder on the magnitude of the field when $r/a = 8$.

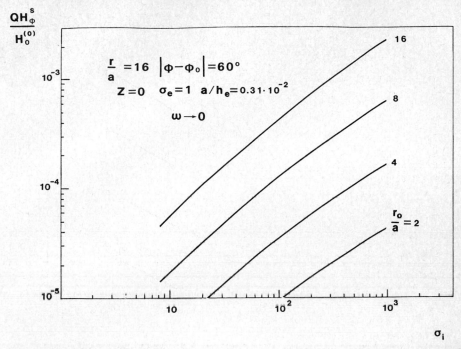

Fig. 6.60. Curves showing the effect of the conductivity of the cylinder on the magnitude of the field when $r/a = 16$.

such as the dependence on the ratio of conductivities and geometrical factors, r and r_0, can be explained using a model of the spheroid placed in a uniform electrical field and considered in the previous paragraph. However, unlike this case, the magnetic field, H_ϕ, increases without limit as the ratio σ_i/σ_e increases. This behavior is a consequence of the fact that on the two surfaces of the cylinder, positive and negative charges arise and their compensating effect is not as strong as that in the case of a spheroid in a uniform electrical field. It should be noted that at low frequencies, both the galvanic and vortex parts of the field, H, behave in the same manner as functions of the frequency. For example, the quadrature component of magnetic field increases in direct proportion to frequency, while the inphase component increases to ω^2. However, the quadrature component of the electrical field caused by surface charges, increases in direct proportion to ω, while the vortex part of this field, generated by the secondary magnetic field, varies as ω^3. At this point, it is appropriate to emphasize that development of methods which would allow one to recognize the presence of the galvanic part of the field is of great practical value. In fact, it is very important inasmuch as the current interpretation methods used in inductive mining prospecting is based mainly on the assumption that induced currents are the sole source of the secondary field.

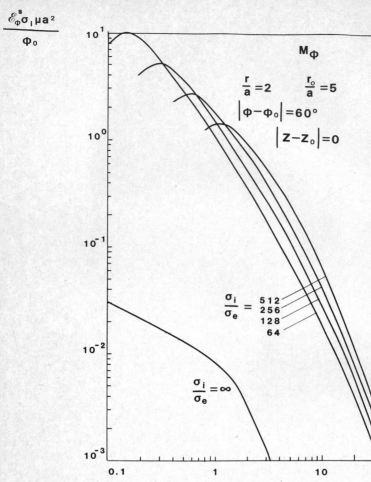

Fig. 6.61. Transient response of the scaled tangential electromotive force as a function of time for the case $r_0/a = 5$.

Now we will explore the behavior of the transient behavior of the magnetic field. As calculations show the function $\mathscr{E}_\phi^s$ initially is zero (skin effect), then decreases, reaches minimum and with an increasing time, change sign, approaches the maximum and then gradually decreases. The transient behavior of the function $(\sigma_i \mu a^2 / \phi_0)\mathscr{E}_\phi^s$, shown in Figs. 6.61–6.62, exhibits a maximum and a relatively slow decrease. For comparison we have also shown the transient response, for induced currents in a cylinder in an insulating medium ($\sigma_i/\sigma_e = \infty$).

From these curves one can draw the following conclusions:

(1) Except during the early stage, where the transient response of the

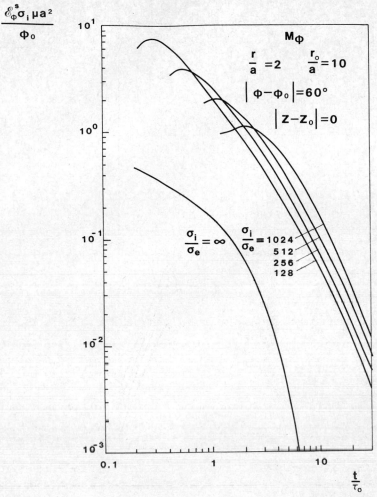

Fig. 6.62. Transient response of the scaled tangential electromotive force as a function of time for the case $r_0/a = 10$.

electromotive force changes sign, the magnitude of this function is much greater than in the case of the insulating host medium, $(\sigma_i/\sigma_e = \infty)$. Therefore, one can say the transient responses are defined mainly by current circults closed through the host medium.

(2) With a decrease in the conductivity of the host medium, the amplitude of the maximum increases and is observed at earlier times.

(3) There is a relatively wide range of times over which the transient electromotive force decays almost as $1/t^{3/2}$ and then at a very late stage it decreases as $1/t^{7/2}$.

(4) Unlike the transient responses caused by induced currents in the

604

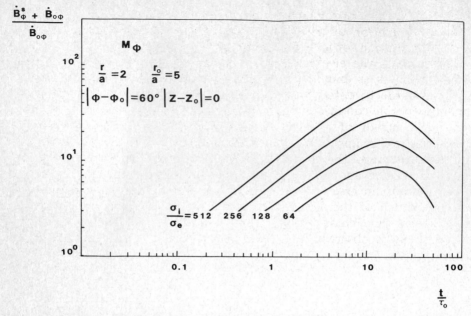

Fig. 6.63. The transient response of the relative anomaly in the time rate of change of magnetic induction for the case $r_0/a = 5$.

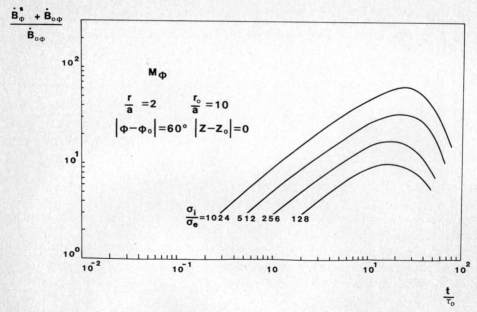

Fig. 6.64. The transient response of the relative anomaly in the time rate of change of magnetic induction for the case $r_0/a = 10$.

conductor only, the electromotive force due to currents intersecting the host medium, is a function of both conductivities, σ_i and σ_e, and consequently reduces the resolution for inductive methods.

(5) Inasmuch as the electromotive force caused by only the vortex magnetic field behaves as exponentially, one can expect that in many cases it would be possible to detect the presence of the galvanic part of the field and under favorable conditions, to eliminate its influence.

Transient behavior of the relative anomaly $(\dot{B}_\phi^s + \dot{B}_{0\phi})/\dot{B}_{0\phi}$ for various conductivities and geometrical parameters are shown in Figs. 6.63 and 6.64. During the early stage, the normal field $\dot{B}_{0\phi}$ dominates, but with an increase in time, the relative anomaly increases, reaches a maximum and begins to decrease in almost inverse proportion to time. Due to the fact that over the descending branch of the transient response of the secondary field, $\dot{B}_\phi^s$, its dependence on time changes from $1/t^{3/2}$ to $1/t^{7/2}$, the maximum of the relative anomaly is observed at relatively large times ($t/\tau_0 \approx 20$), where τ_0 is the time constant of the cylinder.

BIBLIOGRAPHY

Alpin, L. M., 1966, Sources of the field in the theory of electroprospecting by direct current. Appl. Geophys. 3.

Astrakhantzev, G. V., 1963. On the frequency domain and transient characteristics of the electromagnetic fields used in electrical prospecting. Tr., Inst. Geophys., Urals Affiliate Acad. Sci. USSR, no. 3.

Bathe, K. J. and Wilson, E. L., 1976. Numerical Methods in Finite Element Analysis. Prentice-Hall, Englewood Cliffs, N.J.

Best, M. E. and Shammas, B. R., 1979. A general solution for a spherical conductor in a magnetic dipole field. Geophysics, 44(4): 781—800.

Bhattacharya, B. K., 1955. Electromagnetic induction in a two-layer earth. J. Geophys. Res., 60: 279—288.

Bhattacharya, B. K., 1957. Propagation of an electric pulse through a homogeneous and isotropic medium. Geophysics, 22: 905—921.

Bhattacharya, B. K., 1959. Electromagnetic fields of a transient magnetic dipole on the earth's surface. Geophysics, 24: 89—108.

Bhattacharya, B. K., 1963. Electromagnetic fields of a vertical magnetic dipole placed above the earth's surface. Geophysics, 28: 408—425.

Brewitt-Taylor, C. R. and Weaver, J. T., 1976. On the finite difference solution of two-dimensional induction problems. Geophys. J. R. Astron. Soc., 47: 375—396.

Bursian, V. R., 1936. Theory of Electromagnetic Fields used in Electrical Prospecting. Nedra, Leningrad, 360 pp.

Chetaev, D. N., 1956a. On the reflection method of electrical prospecting. Izv. Akad. Nauk SSSR, Ser. Geofiz., no. 2.

Chetaev, D. N., 1956b. Theory of sounding by impulsive switching of a steady current into an ungrounded loop. Izv. Akad. Nauk SSSR, Ser. Geofiz., no. 5.

Chetaev, D. N., 1956c. Concerning transient electromagnetic fields in inhomogeneous media. Tr. Geofiz. Inst., Akad. Nauk SSSR, no. 32.

Chetaev, D. N., 1962. Concerning the field of a low frequency electric dipole lying on the surface of a uniform anisotropic conducting half space. Zh. Tekh. Fiz., 11: 754—757.

Coggon, J. G., 1971. Electromagnetic and electrical modelling by the finite element method. Geophysics, 36: 132—155.

Dey, A. and Morrison, H. F., 1973. Electromagnetic response of two-dimensional inhomogeneities in a dissipative half-space for Turam interpretation. Geophys. Prospect., 21: 340—365.

Dey, A. and Ward, S. H., 1970. Inductive sounding of a layered earth with a horizontal magnetic dipole. Geophysics, 35: 660—703.

Dmitriev, V. I., 1959. The influence of inhomogeneities of the earth on the field of a linear infinitely long cable. Izv. Akad. Nauk SSSR, Ser. Geofiz., 4.

Dmitriev, V. I., 1960a. Solution of a basic problem in the theory for induction methods of electrical prospecting. Izvestiya Akad. Nauk SSSR, Ser. Geofiz., 8.

Dmitriev, V. I., 1960b. A solution for a basic problem in the theory of the induction method of electromagnetic surveying. Bull. Akad. Nauk. SSSR, Ser. Geofiz., 8: 748—753.

608

Dmitriev, V. I., 1961. Screening influence of overburden on anomalous field. Izv. Akad. Nauk SSSR, Ser. Geofiz., 1: 46—53.

Dmitriev, V. I., 1962. Diffraction of an electromagnetic wave on a conducting layer embedded in a conducting medium. Izv. Akad. Nauk SSSR, Ser. Geofiz., no. 6.

Dmitriev, V. I., 1965. A method of computing the magnetotelluric field in an inhomogeneous layer with an arbitrary flexure in the lower surface. Prikl. Geofiz., no. 51, Nedra, Moscow.

Dmitriev, V. I., 1965. Diffraction of a plane electromagnetic field by a cylindrical body situated in a layered medium. In: Vichislitelniye Metodi i Programmirovaniye, 3: 307—316.

Dmitriev, V. I., 1966. Diffraction of an arbitrary electromagnetic field on a cylindrical body. In: Vichislitelniye Methodi i Programmirovaniye, 5: 253—259.

Dmitriev, V. I., 1968. On computing the magnetotelluric field in a layer with an arbitrary flexure for the case of H-polarization. Prikl. Geofiz., no. 41.

Dmitriev, V. I., 1969a. Electromagnetic Fields in Nonuniform Media. Moscow State University, Moscow, 131 pp.

Dmitriev, V. I., 1969b. Electromagnetic Fields in Inhomogeneous Media. Moscow State University, Moscow.

Dmitriev, V. I. and Kokotushkin, G. A., 1968. The integral equation method in the solution of the magnetotelluric field in a layer with varying thickness. In: Vichislitelniye Metodi i Programmirovaniye, 10: 25—31.

Dmitriev, V. I. and Kokotushkin, G. A., 1971. Collection of reference curves for magnetotelluric sounding in inhomogeneous layers. Moscow Stage University, Moscow.

Dmitriev, V. I. and Zakharov, E. V., 1965. Solution of some classes of integral equations on a Rolf-line. In: Vichislitelniye Metodi i Programmirovaniye, 3: 317—328.

Dmitriev, V. I. and Zakharov, Yu. V., 1966. Diffraction of a plane electromagnetic field on a perfectly conducting half-plane embedded in a homogeneous half-space. In: Vichislitelniye Metodi i Programmirovaniye. 5: 187—196.

Dmitriev, V I. and Zakharov, Yu. V., 1967. Diffraction of a plane electromagnetic field on a perfectly conducting band buried in a layered medium. Izv. Akad. Nauk SSSR, Fiz. Zemlya, 62—70.

Dmitriev, V. I. and Zakharov, Yu. V., 1968a. Concerning the resolution of inductive methods of electrical prospecting. Izvestiya Akad. Nauk SSSR, Ser. Fiz., 11: 88—93.

Dmitriev, V. I. and Zakharov, Yu. V., 1968b. On the numerical solution of some integral equation of Fredholm's first type. In: Vichislitelniye Metodi i Programmirovaniye, 10: 49—54.

Dmitriev, V. I. and Zakharov, E. V., 1969. Algorithm for the solution of the problem of diffraction of electromagnetic fields on a perfectly conducting band. In: Vichislitellniye Metodi i Programmirovaniye, 13: 158—165.

Dmitriev, V. I., Barishnikova, I. A. and Zakharov, E. V., 1971. Anomalous field of a perfectly conducting but vanishingly thin band in the induction methods of electrical prospecting. Metodi Razvedochnaya Geofiziki, no. 13.

Dmitriev, V. I., Barishnikova, I. A. and Zakarov, E. V., 1974. Conducting layer in a medium with finite resistivity in the field of plane electromagnetic waves. Izvestiya Akad. Nauk SSSR, Ser. Fiz. Zemlya, 2: 66—71.

Drobintza, V. V. and Tzetzokho, V. A., 1971. Method of computing a plane electromagnetic field with layers of variable thickness. Matematicheskiye Problemi Geofiz, 2: 251—284.

D'Yakonov, B. P., 1959a. Diffraction of electromagnetic waves on a circular cylinder in a homogeneous half-space. Izv. Akad. Nauk SSSR, Ser. Geofiz., no. 9.

D'Yakonov, B. P., 1959b. Diffraction of electromagnetic waves on a sphere embedded in a half-space. Izv. Akad. Nauk SSSR, Ser. Geofiz., no. 11.

609

D'Yakonov, B. P., 1959c. The diffraction of electromagnetic waves by a circular cylinder in a homogeneous half-space. Izv., Akad. Nauk. SSSR, Ser. Geofiz., 9: 950—955.

D'Yakonov, B. P., 1959d. Diffraction of electromagnetic fields on a sphere located in a uniform half-space. Izv. Akad. Nauk SSSR, Ser. Geofiz., no. 11.

Fok, V. A., 1946. A plane wave field close to the surface of a conducting body. Izv. Akad. Nauk SSSR, Ser. Geofiz., 10 (2).

Fok, V. A. and Bursian, V. P., 1926. The electromagnetic field of a varying current in a circuit with two grounds. Zh. Russkoyo Fiz-Khim Obshestva, Chast Fiz., vol. 58, no. 2.

Foster, R. M., 1931. Mutual impedance of grounded wires lying on the surface of the earth. Bell Sys. Tech. J., 10: 408—419.

Frischknecht, F. C., 1967. Fields about an oscillating magnetic dipole. Q., Colo. School of Mines, 62 (1).

Fuller, J. A. and Wait, J. R., 1976. A pulsed dipole in the earth. Topics Appl. Physics, 10: 237—269.

Gasanenko, L. B., 1958. The normal field of a vertical harmonic low frequency magnetic dipole. In: Voprosi Geofiziki, Ucheniye Zapiski Leningrad State University, no. 249, Ser. Geofiz. Geolog. Nauk, no. 10.

Gasanenko, L. B., 1959. The field of a vertical harmonic magnetic dipole on the surface of a multiple layered structure. In: Voprosi Geofiziki, Ucheniye Zapiski Leningrad State University, no. 286, Ser. Geofiz. Geolog. Nauk, no. 11.

Gasanenko, L. B., 1967. Normal field of an infinite linear cable (field in air). Ucheniye Zapiski Leningrad State University, no, 333, Probl. Geofiz., 17: 173—200.

Gasanenko, L. B. and Molochnov, G. V., 1958. The electromagnetic field of a horizontal magnetic dipole over a horizontally layered structure. In: Voprosi Geofiziki, Ucheniye Zapiski Leningrad State University, no. 249, Ser. Geofiz. Geolog. Nauk, no. 10.

Gasanenko, L. B. and Sholpo, G. P., 1959. On the question of computing the electromagnetic field of a vertical low frequency magnetic dipole on the surface of a two-layer structure. In: Voprosi Geofiziki, Ucheniye Zapiski Leningrad State University, no. 278, Ser. Geofiz. Geolog. Nauk, no. 11.

Gasanenko, L. B. and Sholpo, G. P., 1960. On the theory of electromagnetic sounding. In: Voprosi Geofiziki, Ucheniye Zapiski Leningrad State University, no. 286, Ser. Geofiz. Geolog. Nauk, no. 12.

Gasanenko, L. B., Molotkova, M. P. and Sapozhnikov, B. G., 1964. The normal field of an infinite linear cable (the field in air). In: Ucheniye Zapiski Leningrad State University, no. 324, Ser. Geofiz Geolog. Nauk, no. 15.

Geyer, R. G., 1972a. Electromagnetic response near a fault zone. Geophys. Prospect, 20: 828—846.

Geyer, R. G., 1972b. The effect of a dipping contact on the behavior of the electromagnetic field. Geophysics, 37: 337—350.

Goldman, M. M., 1977. Transient electromagnetic field of a disk in a horizontally layered medium. Geol. Geofiz., 2: 129—135.

Goldman, M. M. and Stoyer, C. H., 1983. Finite difference calculations of the transient field of an axially symmetric earth for vertical magnetic dipole excitation. Geophysics, 48 (7).

Gordon, A. N., 1951. The field induced by an oscillatory magnetic dipole outside a semi-infinite conductor. Q. J. Mech. Appl. Math., 4: 106—115.

Harrington, R. F., 1961. Time-Harmonic Electromagnetic Fields. McGraw-Hill, New York, N.Y., 480 pp.

Harrington, R. F., 1968. Field Computation by the Moment Method. MacMillan, New York, N.Y., 229 pp.

Hermance, J. F., 1982. Refined finite-difference simulations using local integral forms: Application to telluric fields in two dimensions. Geophysics, 47: 825—831.

610

Hill, D. A. and Wait, J. R., 1973. EM transient response of a spherical conducting shell over a conducting half-space. Int. J. Electron, 34: 795—805.

Hill, D. A. and Wait, J. R., 1974. EM pulses into the earth from a line source. IEEE Trans. Antennas Propagation, AP22 (1): 145—146.

Hjelt, S. E., 1971. The transient EM field on a two-layered sphere. Geoexploration, 9: 213—230.

Hohmann, G. W., 1971. Electromagnetic scattering by conductors in the earth near a line source of current. Geophysics, 36: 101—131.

Hohmann, G. W., 1975. Three-dimensional induced polarization and electromagnetic modelling. Geophysics, 40, 309—324.

Howard, A. Q., Jr., 1972. The electromagnetic fields of a subterranean cylindrical inhomogeneity excited by a line source. Geophysics, 36: 975—984.

Itskovitch, G. B. and Isaev, G. A., 1980. Transient magnetic field of a circular loop over a low resistivity disk beneath an S plane. Geol. Geofiz., no. 2.

Jones, F. W., 1973. Induction in laterally non-uniform conductors: theory and numerical models. Phys. Earth Planet. Inter., 7: 282—293.

Jones, F. W. and Thompson, D. J., 1974. A discussion of the finite difference method in computer modelling of electrical conductivity structures. Geophys. J. R. Astron. Soc., 37: 537—543.

Jones, F. W. and Vozoff, K., 1978. The calculation of magnetotelluric quantities for three-dimensional conductivity inhomogeneities. Geophysics, 43: 1167—1175.

Kan, T. K. and Clay, C. S., 1979. Hybrid-ray approximation in electromagnetic sounding. Geophysics, 44: 1846—1861.

Kamenetzkiy, F. M., 1969. Elements of the theory of inductive electrical prospecting by the transient method. Prikl. Geofiz., 57: 137—153.

Kamenetzkiy, F. M. and Kovalenko, V. F., 1962a. Transient surface currents in conductive covered deposits. Izv. Vuzov, Geol. Razvedka, no. 6.

Kamenetzkiy, F. M. and Kovalenko, V. F., 1962b. Transient field of an ungrounded coil in the presence of a conductive body, covered by overburden layers. Prikl. Geofiz., no. 35.

Kamenetzkiy, F. M., Kovalenko, V. F. and Yakubovskiy, Yu. V., 1963. The transient method. ONII, VIMS MGiON, SSSR, Moscow.

Kamenetzkiy, F. M., Maiarov, L. V. and Makagonov, P. P. 1966. Transient field of a magnetic dipole for three simple types of geoelectric sequence. Izv. Vuzov, Ser. Geol. Razvedka, 6: 101—111.

Kaufman, A. A. 1961. On the influence of the surrounding medium on results of inductive prospecting for ore deposits in the wave zone. Tr. Inst. Geol. Geofiz., Sib. Otdel. Akad. Nauk, SSSR.

Kaufman, A. A., 1961. Three methods of field excitation for low frequency exploration. Geol. Geofiz., no. 5.

Kaufman, A. A., 1964. Concerning the influence of the galvanic part of the field in inductive mining prospecting. Tr. MGRI.

Kaufman, A. A., 1972. Contribution to the theory of the inductive mining prospecting in a horizontally heterogeneous medium. Geol. Geofiz., no. 10.

Kaufman, A. A., 1973. Electromagnetic methods of prospecting in the near zone based on the use of harmonic fields. Fiz. Zemli , 11—12.

Kaufman, A. A., 1974. Basic Theory of Inductive Mineral Prospecting. Nauka, Novosibersk, 350 pp.

Kaufman, A. A., 1978. The frequency and transient responses of the electromagnetic fields created by currents in confined conductors. Geophysics, 43: 1002—1010.

Kaufman, A. A., 1978. The resolving capabilities of the inductive methods of electroprospecting. Geophysics, 43: 1392—1398.

Kaufman, A. A., 1979. Harmonic and transient fields on the surface of a two-layer medium. Geophysics, 44: 1208–1217.

Kaufman, A. A., 1981. The influence of currents induced in the host rock on electromagnetic response of a spheroid directly beneath a loop. Geophysics, 46: 1121–1136.

Kaufman, A. A., 1983. On the influence of the surrounding medium on results of inductive prospecting for ore deposits in the near zone. Tr. Inst. Geol. Geofiz. Sib. Otdel. Akad. Nauk SSSR.

Kaufman, A. A. and Goldman, M. M., 1973. An approximate theory of inductive prospecting for ores. Geol. Geofiz., no. 10.

Kaufman, A. A. and Goldman, M. M., 1973. A conductive spheroid in a homogeneous primary alternating magnetic field. Geol. Geofiz., no. 10.

Kaufman, A. A., Kamanetskiy, F. M. and Yakubovskiy, Yu. V., 1957a. On the inductive methods of exploration. Tr. MGRI, no. 36.

Kaufman, A. A., Kamenetskiy, F. M. and Yakubovskiy, Yu. V., 1957b. On the choice of the optimum frequency in inductive prospecting. Izv. Akad. Nauk SSSR, Ser. Geofiz, no. 2.

Kaufman, A. A., Kamenetskiy, F. M. and Yakubovskiy, Yu. V., 1960a. Concerning the capabilities of the transient method for prospecting for highly conductive bodies. Geol. Geofiz., no. 6.

Kaufman, A. A., Kovalenko, V. F. and Yakubovskiy, Yu. V., 1960b. Concerning the resolving capabilities of the transient method in minerals electrical prospecting. Geol. Razvedka, no. 3.

Kaufman, A. A., Tabarovskiy, L. A. and Goldman, M. M., 1971a. The electromagnetic field of a conductive spheroid in a two-layer medium (collection of reference curves). Trudi IGG, SOAN, SSSR.

Kaufman, A. A., Tabarovskiy, L. A. and Terentiev, S. M., 1971b. A conductive elliptical cylinder in a homogeneous primary harmonic magnetic field. Geol. Geofiz., no. 6.

Kaufman, A. A., Tabarovskiy, L. A. and Terentiev, S. M., 1971c. The electromagnetic field of a plane wave in a two-layer medium that contains an elliptical cylinder (E-polarization). Tr. Inst. Geol. Geofiz., Sib. Otdel, Akad. Nauk, SSSR.

Keller, G. V., 1971. Natural field and controlled source methods in electromagnetic exploration. Geoexploration, 9: 99–147.

Keller, G. V. and Frischknecht, F. C., 1966. Electrical Methods in Geophysical Prospecting. Pergamon, Oxford, 519 pp.

Kokotushkin, G. A. and Dmitriev, V. I., 1964. Magnetotelluric field in a layer with a steplike change in thickness. Prikl. Geofiz., 52: 104–110.

Kuo, J. T. and Cho, D. H., 1980. Transient time-domain electromagnetics. Geophysics, 45: 271–291.

Lajoie, J. J. and West, G. F., 1976. The electromagnetic response of a conductive inhomogeneity in a layered earth. Geophysics, 41: 1133–1156.

Lajoie, J., Afanso-Roche, J. and West, G. F., 1975. Electromagnetic response of an arbitrary source on a layered earth: A new computational approach. Geophysics, 40: 773–789.

Lee, K. H., Pridmore, D. F. and Morrison, H. F., 1981. A hybrid three-dimensional electromagnetic modeling scheme. Geophysics, 46: 796–805.

Lee, T., 1975a. Sign reversals in the transient method of electrical prospecting (one loop version). Geophys. Prospect, 23: 653–662.

Lee, T., 1975b. Transient electromagnetic response of a sphere in a layered medium. Geophys. Prospect., 23: 492–512.

Lee, T., 1979. Transient electromagnetic waves applied to prospecting. Proc. IEEE, 67: 1016–1021.

Lee, T., 1981a. Transient electromagnetic response of a sphere in a layered medium. Pure Appl. Geophys., 119: 307–338.

612

Lee, T., 1981b. Transient electromagnetic response of a polarizable ground. Geophysics, 46: 1037—1041.

Lee, T., 1982. Asymptotic expansions for transient electromagnetic fields. Geophysics, 47: 38—46.

Leontovich, M. A., 1948. On the approximate boundary conditions. In: V. A. Vvedensky (Editor), Akad. Nauk SSSR, Moscow. pp. 5—20.

Lewis, R. and Lee, T., 1981. The effect of host rock on transient electromagnetic fields. Bull. Aust. Soc. Explor. Geophys., 12, pp. 5—12.

Lines, L. R. and Jones, F. W., 1973. The perturbation of alternating geomagnetic field by three dimensional island structures. R. Astron. Soc., 32: 133—154.

Lodha, G. S. and West, G. F., 1976. Practical airborne Em (AEM) interpretation using a sphere model. Geophysics, 41: 1157—1169.

Mahmoud, S. F., Botros, A. Z. and Wait, J. R., 1979. Transient electromagnetic fields of a vertical magnetic dipole on a two-layer earth. Proc. IEEE, 67: 1022—1029.

Makagonov, P. P., 1977. Some questions in the analysis of two-dimensional fields by the method of naturally orthogonal components. Geol. Geofiz., 11: 47—51.

Mallick, K., 1971. Electromagnetic response of a layered transitional earth — infinite cable: Pure Appl. Geophys., 83: 102—110.

Mallick, K., 1972. Conducting sphere in electromagnetic input field. Geophys. Prospect., 20: 293—303.

Mallick, K., 1973. Conducting horizontal infinite cylinder in electromagnetic input field. Geophys. Prospect., 21: 102—108.

Mallick, K., 1978. A note on the decay pattern of magnetic field and voltage response of conducting bodies in an electromagnetic time-domain system. Geoexploration, 16: 303—307.

March, H. W., 1953. The field of a magnetic dipole in the presence of a conducting sphere. Geophysics, 18: 671—684.

Milne, W. E., 1970. Numerical Solutions of Differential Equations. Dover, New York, N.Y., 359 pp.

Morse, P. M. and Feshbach, H., 1953. Methods of Theoretical Physics. McGraw-Hill, New York, N.Y.

Nabighian, M. N., 1970. Quasi-static transient response of a conducting permeable sphere in a dipole field. Geophysics, 35: 303—309.

Nabighian, M. N., 1971. Quasi-static transient response of a conducting permeable two-layer sphere in a dipole field. Geophysics, 36: 25—37.

Nabighian, M. N., 1979. Quasi-static transient response of a conducting half-space — an approximate representation. Geophysics, 44: 1700—1705.

Negi, J. G., 1967. Electromagnetic screening due to a disseminated spherical zone over a conducting sphere. Geophysics, 32: 69—87.

Negi, J. G. and Verma, S. K., 1972. Time domain electromagnetic response of a shielded conductor. Geophys. Prospect., 20: 901—909.

Negi, J. G., Gupta, C. P. and Raval, V., 1973. Electromagnetic response of a permeable inhomogeneous conducting sphere. Geoexploration, 2: 1—20.

Nikitina, V. N., 1960. Concerning diffraction on a half-plane in a host medium. Izv. Akad. Nauk SSSR, Ser. Geofiz., no. 4.

Obukhov, G. G., 1965a. Magnetotelluric field over a buried structure (H-polarization). Prikl. Geofiz., no. 44.

Obukhov, G. G., 1965b. Magnetotelluric field over a buried elongate structure (E-polarization). Prikl. Geofiz., no. 46.

Ogujade, S. O., Ramaswamy, V. and Dosso, H. W., 1974. Electromagnetic response of a conducting sphere buried in a conducting earth. J. Geomagn. Geoelectr. 26: 417—427.

Ogujade, S. O. 1981. Electromagnetic response of an embedded cylinder for line current excitation. Geophysics, 46: 45—52.

Ott, R. H. and Wait, J. R., 1972. On calculating transient EM fields of a small current-carrying loop over a homogeneous earth. Pure Appl. Geophys. 95: 157—162.

Parry, J. R. and Ward, S. H., 1971. Electromagnetic scattering from cylinders of arbitrary cross-section in a conductive half-space. Geophysics, 36: 291—299.

Pascoe, L. J. and Jones, F. W., 1972. Boundary conditions and calculation of surface values for the general two-dimensional electromagnetic induction problem. Geophys. J. R. Astron. Soc., 27: 179—193.

Patra, H. P. and Mallick, K., 1980. Geosounding Principles, 2, Time-varying Geoelectric Soundings. Elsevier, New York, Amsterdam, 419 pp.

Poddar, M., 1982. A rectangular loop source of current on a two-layered earth. Geophys. Prospect., 30: 101—114.

Poddar, M., 1982. Very low-frequency electromagnetic response of a perfectly conducting half-plane in a layered half-space. Geophysics, 47: 1059—1067.

Poddar, M., 1983. Electromagnetic sounding near a large square loop source of current. Geophysics, 48: 107—109.

Praus, O., 1975. Numerical and analog modelling of induction effects in laterally non-uniform conductors. Phys. Earth Planet. Inter., 10: 262—270.

Pridmore, D. F., Hohmann, W. W., Ward, S. H. and Sill, W. R., 1981. An investigation of finite element modeling for electrical and electromagnetic data in three dimensions. Geophysics, 46: 1009—1024.

Rabinovitch, B. I. and Goldman, M. M., 1977. Apparent resistivity theoretical curves for the transient sounding method in the near zone with a disk embedded in a horizontally layered half space (album no. 10). SNIIGGIMS, Novosibersk, 84 pp.

Rabinovitch, B. I. and Goldman, M. M., 1981. Transient electromagnetic sounding over a disk whose axis coincides with that of the transmitter coil. Geol. Geofiz., no. 1.

Raiche, A. P., 1974. An integral equation approach to three-dimensional modeling: Geophys. J. R. Astron. Soc., 36: 363—376.

Raiche, A. P. and Spies, B. R., 1981. Coincident loop transient electromagnetic master curves for interpretation of two-layer earths: Geophysics, 46: 53—64.

Reddy, I. K., Rankin, D. and Phillips, R. J., 1977. Three-dimensional modeling in magnetotelluric and magnetic variational sounding. Geophys. J. Roy. Astron. Soc., 51: 818—826.

Roy, A., 1968. Continuation of electromagnetic fields, I. Geophysics, 33: 834—837.

Ryu, J., Morrison, H. F. and Ward, S. H., 1970. Electromagnetic fields about a loop source of current. Geophysics, 35: 862—896.

Shakhsuvarov, D. N., 1956. Method of interpreting the results of observing the electromagnetic field in dipole sounding. Izv. Akad. Nauk SSSR, Ser. Geofiz., no. 5.

Shatokhin, V. N. and Rabinovitch, B. I., 1974. Modelling curves for the transient electromagnetic method in the near zone in a medium with horizontal inhomogeneities (album no. 8). SNIIGGIMS, Novosibersk.

Shaub, Yu, B., 1971. Methods of aeroelectric prospecting based on application of artificial alternating electromagnetic fields. Nedra, Leningrad, 222 pp.

Sheinmann, S. M., 1947. About transient electromagnetic fields in the earth. Prikl. Geofiz., no. 3.

Sheinmann, S. M., 1970a. Method of calculating the pseudo-scalar electromagnetic field and its variation. Prikl. Geofiz., no. 60.

Sheinman,, S. M., 1970b. Electromagnetic field of a varying current, flowing in a disk. Prikl. Geofiz., no. 61.

Sheinmann, S. M., 1971. Calculation of electromagnetic fields by the method of linear superposition of excitation from small defomations of the bounding surfaces. Prikl. Geofiz., no. 62.

Sinha, A. K., 1968. Electromagnetic fields of an oscillating magnetic dipole over an anisotropic earth. Geophysics, 33: 346—353.

Sinha, A. K., 1969. Vertical electric dipole over an inhomogeneous and anisotropic earth. Pure Appl. Geophys., 72: 123—147.

Sinha, A. K. and Bhattacharya, P. K., 1967. Electric dipole over an anisotropic and inhomogeneous earth. Geophysics, 32: 652—667.

Singh, S. K., 1972. Transient electromagnetic response of a conducting infinite cylinder embedded in a conducting medium. Geofis. Int., 12: 7—21.

Singh, S. K., 1973. Electromagnetic transient response of a conducting sphere embedded in a conducting medium: Geophysics, 38: 864—893.

Slichter, L. B. and Knopoff, L., 1959. Field of an oscillating magnetic dipole on the surface of a layered earth. Geophysics, 24: 77—88.

Sommerfeld, A. N., 1949. Lectures on Theoretical Physics, Vol. 1: Partial Differential Equations in Physics. Academic Press, New York, N.Y.

Southwell, R. V., 1946. Relaxation Methods in Theoretical Physics. Clarendon, Oxford, 248 pp.

Smythe, W. R., 1950. Static and Dynamic Electricity. McGraw-Hill, New York, N.Y., 616 pp.

Stefanescu, S. S., 1935. On the basic theory of electromagnetic exploration with alternating current at very low frequency. Beitr. Angew. Geophys., 5: 182—192.

Stodt, J. A., Hohmann, G. W. and Ting, S. C., 1981. The telluric-magnetotelluric method in two- and three-dimensional environments. Geophysics, 46: 1137—1147.

Stoyer, C. H. and Greenfield, R. J., 1976. Numerical solutions of the response of a two-dimensional earth to an oscillating magnetic dipole source. Geophysics, 41: 519—530.

Stratton, J., 1941. Electromagnetic Theory. McGraw-Hill, New York, N.Y.

Sunde, E. D., 1949. Earth Conduction Effects in Transmission Systems. Van Nostrand, New York, N.Y., 373 pp.

Svetov, B. S., 1960. Some results of modelling studies of induction methods. Izv. Akad. Nauk SSSR, Ser. Geofiz., no. 1.

Svetov, B. S., 1961. On the role of the means for exciting the field in the low frequency induction method of electrical prospecting. Izv. Akad. Nauk SSSR, Ser. Geofiz., no. 12.

Svetov, B. S. and others, 1966. Electromagnetic Methods of Exploration in Mining Geophysics. Nedra, Moscow.

Svetov, B. S., 1973. Theory, Methodology and Interpretation of Surveys with Low-Frequency Inductive Electrical Prospecting. Nedra, Moscow.

Swift, C. M., Jr., 1971. Theoretical magnetotelluric and Turam response from two-dimensional inhomogeneities. Geophysics, 36: 38—52.

Tabarovskiy, L. A., 1968. Quasi-stationary electromagnetic fields in media with non-horizontal surfaces of separation: Geol. Geofiz., 9: 68—79.

Tabarovskiy, L. A., 1975a. Applying the Method of Integral Equations in the Problems of Geoelectrics. Nauka, Siberian Division, Acad. Sci. USSR.

Tabarovskiy, L. A., 1975b. Use of the Integral Equation Method in Geoelectric Problems. Nauka, Novosibersk.

Tabarovskiy, L. A., 1979. Collection of theoretical curves for sounding with transient fields over inhomogeneous layers. Inst. Geol. Geofiz., Sibersk Otdel. Akad. Nauk SSSR, Novosibersk.

Tabarovskiy, L. A. and Goldman, M. M., 1979. Use of various methods for solving certain problems in electrical prospecting and some questions of deep electrical sounding with transient fields. Inst. Geol. Geofiz., Sibersk. Otdel. Akad. Nauk SSSR, Novosibersk.

Tabarovskiy, L. A. and Itzkovitch, G. B., 1982. Transient methods for investigating kimberlite plugs. Inst. Geol. Geofiz., Sibersk. Otdel. Akad. Nauk SSSR, Novosibersk, no. 5.

Tabarovskiy, L. A. and Itzkovitch, G. B., 1983. Diffraction of a transient electromagnetic field on a thin disk (three-dimensional problem). Geol. Geofiz., 2: 99—109.

Tabarovskiy, L. A. and Krivoputzkiy, V. S., 1978. Solution of a problem of transient electromagnetic fields in axially symmetric models by the network method. Geol. Geofiz., no. 7.

Tikhonov, A. N., 1946. On transient electric currents in a homogeneous conducting half-space. Izv. Akad. Nauk SSSR, Ser. Geograf. Geofiz., 3: 243—247.

Tikhonov, A. N., 1950a. On transient electric currents in inhomogeneous layered media. Izv. Akad. Nauk SSSR, Ser. Geograf. Geofiz., 14 (3).

Tikhonov, A. N., 1950b. On determining the electrical characteristics of deep layers in the earth's crust. Dokl. Akad. Nauk SSSR, 73 (2).

Tikhonov, A. N., 1959a. On the asymptotic behavior of integrals containing Bessel's functions. Dokl. Akad. Nauk SSSR, vol. 125, no. 5.

Tikhonov, A. N., 1959b. On the propagation of a varying electromagnetic field in a layered anisotropic medium. Dokl. Akad. Nauk SSSR, vol. 126, no. 5.

Tikhonov, A. N. and Chetaev, D. N., 1956. On the possibility of using the radiation resistance of an antenna in electrical mapping. Izv. Akad. Nauk SSSR, Ser. Geofiz., no. 3.

Tikhonov, A. N. and Dmitriev, V. I., 1959. Concerning the question of the effect of noise in induction methods of aeroelectrical prospecting. Izv. Akad. Nauk SSSR, Ser. Geofiz., no. 6.

Tikhonov, A. N. and Mukhina, G. V., 1950. Determining the variable electric field in a layered medium. Izv. Akad. Nauk SSSR, Ser. Geograf. Geofiz., 2: 375—380.

Tikhonov, A. N. and Samarskiy, A. A., 1951. The Equations of Mathematical Physics. Gostekhteoretizdat, Moscow.

Tikhonov, A. N. and Shakhsuvarov, D. N., 1954. On the use of electromagnetic fields of radio stations for geophysical prospecting, Tr. Geofiz. Inst. Akad. Nauk, SSSR, no. 9.

Tikhonov, A. N. and Shakhsuvarov, D. N., 1956. A method of computing electromagnetic fields, excited by varying currents in layered media. Izv. Akad. Nauk SSSR, Ser. Geofiz., no. 3.

Tikhonov, A. N. and Shakhsuvarov, D. N., 1959. Electromagnetic field of a dipole in the far zone. Izv. Akad. Nauk SSSR, Ser. Geofiz., no. 7.

Tikhonov, A. N. and Shakhsuvarov, D. N., 1961. On the inequality of the asymptotic electromagnetic fields excited by a dipole with varying current in layered media. Izv. Akad. Nauk SSSR, Ser. Geofiz., no. 1.

Tikhonov, A. N. and Skugarevskaya, O. A., 1950. On transient electric currents in an inhomogeneous layered medium: Izv. Akad. Nauk SSSR, Ser. Geograf. Geofiz., 14 (4).

Tikhonov, A. N. and Skugarevskaya, O. A., 1951. On transient electric currents in inhomogeneous layered media. Izv. Akad. Nauk, Geofiz., no. 6.

Tikhonov, A. N. and Skugarevskaya, O. A., 1958. Concerning the interpretation of transient electromagnetic fields in layered media. Izv. Akad. Nauk SSSR, Ser. Geofiz., no. 3.

Tikhonov, A. N. and Skugarevskaya, O. A., 1959. Asymptotic behavior of transient electromagnetic fields. Izv. Akad. Nauk SSSR, Ser. Geofiz., no. 6.

Ting, S. C. and Hohmann, G. W., 1981. Integral equation modeling of three-dimensional magnetotelluric response. Geophysics, 46: 182—197.

Tsubota, K. and Wait, T. R., 1980. The frequency- and the time-domain response of a buried axial conductor. Geophysics, 45: 941—951.

Tzetzokho, V. A., 1964. Questions about the behavior of electromagnetic waves in layered media with axial symmetry. Vich. Metodi, 12: 52—78.

Velikin, A. B., 1971. The transient field from an induction coil in the presence of a conducting sphere. In: A. F. Fokin (Editor), The Transient Method in the Search for Sulfide Ore Deposits. Nedra, Leningrad.

616

Velikin, A. B. and Bulgakov, Yu. I., 1967. An Inductive Electrical Prospecting Method with Coincident Transmitter and Receiver. Nedra, Leningrad.

Velikin, A. B. and Frantov, G. S., 1962. The Electromagnetic Fields used in the Induction Methods of Electrical Prospecting. Gostoptekhizdat, Leningrad.

Verma, S. K., 1972. Quasi-static time-domain electromagnetic response of a homogeneous conducting infinite cylinder. Geophysics, 37: 92—97.

Verma, S. K., 1973. Time-dependent electromagnetic fields of an infinite conducting cylinder excited by a long current-carrying cable. Geophysics, 38: 369—379.

Verma, S. K., 1980. Master Tables for Electromagnetic Depth Sounding Interpretation. Plenum, New York, N.Y.

Verma, S. K. and Singh, R. N., 1970. Transient Electromagnetic response of an inhomogeneous conducting sphere. Geophysics, 35: 331—336.

Veshev, A. B., 1965. Electromagnetic Profiling with Constant or Alternating Currents. Nedra, Leningrad.

Veshev, A. B., and others. 1971. Electromagnetic Profiling; Nedra, Leningrad.

Veshev, A. B., Rediko, G. V. and Pertel, M. I., 1971. Normal field of a vertical electric dipole. Vcheniye Zapiski Leningrad State University, Problemi Geofiziki, no. 361, 21: 26—42.

Wait, J. R., 1951a. A conducting sphere in a time-varying magnetic field. Geophys, 16: 666—672.

Wait, J. R., 1951b. The magnetic dipole over a horizontally stratified earth. Can. J. Phys., 29: 577—592.

Wait, J. R., 1951c. Transient electromagnetic propagation in a conducting medium. Geophysics, 16: 213—221.

Wait, J. R., 1952a. Mutual inductance of circuits on a two-layer earth. Can. J. Phys., 30: 450—452.

Wait, J. R., 1952b. The cylindrical ore body in the presence of a cable carrying an oscillating current. Geophysics, 17: 378—386.

Wait, J. R., 1953a. Induction by a horizontal oscillating magnetic dipole over a conducting homogeneous earth. Trans. Am. Geophys. Union, 34: 185—189.

Wait, J. R., 1953b. A conducting permeable sphere in the presence of a coil carrying an oscillating current. Can. J. Phys., 31: 670—678.

Wait, J. R., 1954. Mutual coupling of loops lying on the ground. Geophysics, 13: 290—296.

Wait, J. R., 1955. Mutual electromagnetic coupling of loops over a homogeneous ground. Geophysics, 20: 630—637.

Wait, J. R., 1956. Mutual electromagnetic coupling of loops over a homogeneous ground. Geophysics, 21: 479—484.

Wait, J. R., 1957. Diffraction of a spherical wave pulse by a half-plane screen. Can. J. Phys., 35: 693—697.

Wait, J. R., 1958a. Induction by an oscillating magnetic dipole over a two-layer ground. Appl. Sci. Res., Sect. B, B-7: 73—80.

Wait, J. R., 1958b. Transmission and reflection of electromagnetic waves in the presence of stratified media. J. Res. Nat. Bur. Standards, 61: 205—232.

Wait, J. R., 1960. On the electromagnetic response of a conducting sphere to a dipole field. Geophysics, 25: 649—659.

Wait, J. R., 1962. A note on the electromagnetic response of a stratified earth. Geophysics, 27: 382—385.

Wait, J. R., 1966. Fields of a horizontal dipole over a stratified anistropic earth. IEEE Trans. Antennas Propagation, AP-14; 790—792.

Wait, J. R., 1969. Electromagnetic induction in a solid conducting sphere caused by a thin spherical conducting shell. Geophysics, 34: 753—759.

Wait, J. R., 1970. Electromagnetic Waves in Stratified Media. Pergamon Press, Oxford.

Wait, J. R., (Editor), 1971a. Electromagnetic Probing in Geophysics. Golem Press, Boulder, Colo.

Wait, J. R., 1971b. EM pulse propagation in a simple dispersive medium. Electron. Lett., 7: 285—286.

Wait, J. R., 1971c. On the theory of transient EM sounding over a stratified earth. Can. J. Phys., 50: 1055—1061.

Wait, J. R., 1972. Transient dipole radiation in a conducting medium. Rev. Bras. Technol., 3: 29—37.

Wait, J. R., 1977. Multiple scattering between a buried line conductor and the earth's surface. Geophysics, 42: 1470—1472.

Wait, J. R., 1980. Exact surface impedance for a spherical conductor. Proc. IEEE, 68: 279—280.

Wait, J. R., 1981. Wave propagation theory. Pergamon, Oxford.

Wait, J. R., 1982. Geo-Electromagnetism. Academic Press, New York, 268 pp.

Wait, J. R. and Fuller, J. A., 1972. Argand representations of the mutual electromagnetic coupling of loops on a two-layer earth. Geoexploration, 10: 221—227.

Wait, J. R. and Hill, D.A., 1972a. Transient EM fields of a finite circular loop in the presence of a conductive half space. J. Appl. Phys., 43: 4532—4534.

Wait, J. R. and Hill, D. A., 1972b. Transient magnetic fields produced by a step-function-excited loop buried in the earth. Electron. Lett., 8: 294—295.

Wait, J. R. and Spies, K. P., 1969. Quasi-static transient response of a conducting permeable sphere. Geophysics, 34: 789—792.

Ward, S. H., 1967. The electromagnetic method. Min. Geophys. 2: 224—272.

Ward, S. H., 1980. History of geophysical exploration, electrical, electromagnetic and magnetotelluric methods. Geophysics, 45: 1659—1666.

Ward, S. H., Ryu, J., Glenn, W. E., Hohmann, G. W. and Smith, B. D., 1974. Electromagnetic methods in conductive terraines. Geoexploration, 12: 121—183.

Watts, R. D., 1978. Electromagnetic scattering from buried wires. Geophysics, 43: 767—781.

Weaver, J. T., 1970. The general theory of EM induction in a conducting half-space. Geophys. J. R. Astron. Soc., 22: 83—100.

Weidelt, P., 1975. Electromagnetic induction in three-dimensional structures. J. Geophys., 41: 85—109.

Williamson, K., Hewlett, C., and Tammemagi, H. W., 1974. Computer modelling of electrical conductivity structures. Geophys. J. R. Astron. Soc., 37: 533—536.

Zaborovskiy, A. I., 1960. Varying Electromagnetic Fields in Electrical Prospecting. Moscow State University, Moscow.

Zakharov, E. V., 1968. Algorithm for computing the anomalous electromagnetic field of a perfectly conducting Rolf-plane buried in a conductive half-space. In Vichisl. Metodi Programm., 10: 32—48.

Zakharov, E. V., 1969. Concerning the diffraction of a plane electromagnetic field on a homogeneous cylindrical body embedded in a layered medium. Izv. Akad. Nauk SSSR, Fiz. Zemli, no. 1.

Zakharov, E. V. and Dmitriev, V. I., 1966. Diffraction of electromagnetic waves on a perfectly conducting half-plane buried in a layered medium. Izv. Akad. Nauk SSSR, Fiz. Zemli, 11: 83—99.

Zakharov, E. V. and Ilin, I. V., 1970. Integral representation of fields in inhomogeneous layered media. Izv. Akad. Nauk SSSR, Fiz. Zemli, 8: 62—71.

Zienkiewicz, O. C., 1977. The Finite Element Method in Engineering Sciences, 3rd Ed., McGraw-Hill, London.

SUBJECT INDEX